THE
ENVIRONMENT DICTIONARY

'Clearly written, comprehensive, balanced and up-to-date.'
Chris Park, Lancaster University

'The dictionary is excellent and will be a valuable reference work. The balance of entries is good and the quality of the definitions under each entry is extremely high.'
Martin Kent, University of Plymouth

'Well written and exciting. It is an impressive summary of key environmental terms and ideas which will assuredly serve as an essential source book for all subjects on the environmental theme.'
Richard Huggett, University of Manchester

'A comprehensive volume that will be a useful resource to undergraduate students.'
David Higgett, University of Durham

'Well written: the entries reflect the large and diversified field of environment and they have a standard that should be easy for students and others to adopt, understand and use.'
Tormod Klemsdal, University of Oslo

'*The Environment Dictionary* is an interesting and informative source book pitched at an accessible level for its intended audience.'
John McClatchey, Nene College of Higher Education

THE
ENVIRONMENT
DICTIONARY

DAVID D. KEMP

London and New York

First published 1998
by Routledge
11 New Fetter Lane, London EC4P 4EE

Simultaneously published in the USA and Canada
by Routledge
29 West 35th Street, New York, NY 10001

Typeset in Sabon by Solidus (Bristol) Ltd
Printed and bound in Great Britain by
T.J. International Ltd, Padstow, Cornwall

British Library Cataloguing in Publication Data
A catalogue record for this book is available from the British Library

Library of Congress Cataloguing in Publication Data
Kemp, David D.
The environment dictionary / David D. Kemp.
Includes bibliographical references and index.
1. Environmental sciences – Dictionaries.
2. Pollution – Dictionaries. I. Title.
GE10.K45 1998
363.7'003 – dc21 97–38997

ISBN 0–415–12752–1 (hbk)
ISBN 0–415–12753–X (pbk)

IN MEMORY OF DAVID KEMP
(1911–1996)

CONTENTS

PREFACE

This volume is designed to accommodate a broad interpretation of the term environmental studies, and to do so incorporates material from a wide range of disciplines. The interdisciplinary nature of environmental issues necessitates the inclusion of topics from physical sciences such as chemistry, physics, geology and biology alongside entries from atmospheric, engineering, earth and soil sciences. Since current environmental issues include a strong human element, entries from geography, demography, politics and economics also have their place. Within these broad groupings, the dictionary includes both technical and simple descriptive topics along with a number of boxed entries which provide an examination of selected current issues in greater depth. Along with a range of references from introductory texts through popular magazines to academic journals, this variety is designed to encourage the widest possible readership. As with any dictionary of this type, however, the nature, disciplinary distribution and depth of the entries will reflect, to a greater or lesser extent, the individual interests and interpretations of the author, which may or may not match exactly those of potential users. Nevertheless, the multifaceted approach, the abundance of cross-references and the reading lists provided will allow readers to develop particular interests and pursue specific topics, both within and outside the dictionary, to the level of complexity they require, or with which they are comfortable.

ACKNOWLEDGEMENTS

The completion of this dictionary would not have been possible without the direct and indirect efforts of a great number of individuals. At Routledge, Sarah Lloyd provided a nice mixture of editorial encouragement and advice, and faced the delays that bedevilled the production of the manuscript with remarkable patience and tolerance. Casey Mein, also at Routledge, managed the ebb and flow of drafts, reviews and enquiries with great efficiency, providing appropriate guidance at all stages of the production. In the Department of Geography at Lakehead University, Cathy Chapin provided her cartographic skills, often at short notice, to draw and redraw many of the maps and diagrams in the dictionary. The help of these three individuals is very much appreciated. The reviewers who commented on the choice and content of my topics, and my colleagues and students who suggested entries have my thanks also. The final choice was of course mine, and any errors, omissions or apparently superfluous entries are my responsibility.

I am grateful to my wife, Pat, for her patience and understanding during this volume's unexpectedly long gestation and to my daughters, Susan and Heather, for their contribution to the research process.

In the preparation of this volume, a number of other dictionaries were consulted. For technical and scientific terms, *The Penguin Dictionary of Science* was invaluable. *The Encyclopedic Dictionary of Physical Geography* provided direction on a variety of topics through the many references it contains. *The Dictionary of Global Climate Change*, compiled by J.W. Maunder as a contribution from the Stockholm Environmental Institute to the Second World Climate Conference was an excellent point of entry for information on the many government organizations involved in environmental issues. Most government agencies also have a presence on the World Wide Web, as do a wide range of environmental organizations. Since web addresses are subject to unannounced change, they are not included in the dictionary, but organizations can be easily accessed using one of the many available search engines.

My thanks are also due to the following for allowing me to reproduce copyright material:

Figure A-16 With permission from Kaufman, D.G. and Franz, C. (1993) *Biosphere 2000: Protecting our Global Environment*, New York: HarperCollins.
Figure A-17 Reprinted from Barrie, L.A. (1986) 'Arctic air pollution: an overview of current knowledge', *Atmospheric Environment* 20: 643–63, with kind permission from Elsevier Science Ltd, The Boulevard, Langford Lane, Kidlington OX5 1GB, UK.
Figures B-4, G-7 and R-3 With permission from Park, C.C. (1992) *Tropical Rainforests*, London/New York: Routledge.
Figures C-5, I-1, M-7 and O-3 With permission from Moore, P.D., Chaloner, B. and Stott, P. (1996) *Global Environmental Change*, Oxford: Blackwell Science.
Figures C-9, F-7 and S-10 With permission of the McGraw-Hill Companies from Enger, E.D. and Smith, B.F. (1995) *Environmental Science: A Study of Interrelationships* (5th edition), Dubuque, IA: Wm C. Brown.
Figures C-11 and H-3 With permission from Turco, R.P. (1997) *Earth under Siege*, Oxford/New York: Oxford University Press.

Figure D-5 With permission from Faughn, J.E., Turk, J. and Turk, A. (1991) *Physical Science*, Philadelphia: Saunders.

Figures E-1, O-11, P-4, P-10, V-2 and W-1 With permission from Goudie, A. (1989) *The Nature of the Environment* (2nd edition) Oxford: Blackwell.

Figure E-6 With permission from Nkenderim, L.C. and Budikova, D. (1996) 'The El Niño-Southern Oscillation has a truly global impact', *IGU Bulletin* 46: 30.

Figures E-10 and O-8 With permission from Miller, G.T. (1994) *Living in the Environment* (8th edition), Belmont, CA: Wadsworth.

Figure G-4 With permission from Jones, M.D.H. and Henderson-Sellers, A. (1990) 'History of the greenhouse effect', *Progress in Physical Geography* 14: 1–18. Copyright Edward Arnold (1990).

Figures H-2 and P-12 With permission from Oke, T. (1978) *Boundary Layer Climates*, London: Methuen.

Figure K-1 With permission of the Royal Meteorological Society from Jenkins, G.J., Johnson, D.W., McKenna, D.S. and Saunders, R.W. (1992) 'Aircraft measurements of the Gulf smoke plume', *Weather* 47 (6): 212–9.

Figure L-6 With permission from Meadows, D.H., Meadows, D.L., Randers, J. and Behrens, W.W. (1974) *The Limits to Growth*, London: Pan Books.

Figure L-7 With permission from IPCC (1990) *Climate Change: The IPCC Scientific Assessment*, Cambridge: Cambridge University Press.

Figure M-3 With permission from Mackie, R., Hunter, J.A.A., Aitchison, T.C., Hole, D., McLaren, K., Rankin, R., Blessing K., Evans, A.T., Hutcheon, A.W., Jones, D.H., Soutar, D.S., Watson, A.C.H., Cornbleet, M.A. and Smith J.F. (1992) 'Cutaneous malignant melanoma, Scotland 1979–89', *The Lancet* 339: 971–5. Copyright The Lancet Ltd (1992).

Figures M-8 and R-6 With permission from Park, C.C. (1997) *The Environment: Principles and Applications*, London: Routledge.

Figure N-3 with permission of the Minister of Public Works and Government Services Canada, 1998, from Environment Canada (1983) *Stress on Land in Canada*, Ottawa: Ministry of Supply and Services.

Figure O-5 With permission from Cutter, S.L., Renwick, H.L. and Renwick, W.H. (1991) *Exploitation, Conservation, Preservation: A Geographical Perspective on Natural Resource Use*, New York: Wiley.

Figures D-4, G-3, I-3, L-2, M-2 and P-1 Despite considerable effort, it has not been possible to contact the copyright holders of these figures prior to publication. The author and publishers apologize for the resulting omissions. If notified, the publisher will attempt to correct these at the earliest opportunity. The source of each of these figures is indicated where it appears in the text.

References

Goudie, A., Atkinson, B.W., Gregory, K.J., Simmons, I.G., Stoddart, D.R. and Sugden, D. (eds) (1994) *The Encyclopedic Dictionary of Physical Geography* (2nd edition), Oxford/Cambridge, MA: Blackwell.

Maunder, J.W. (1992) *Dictionary of Global Climate Change*, London/New York: Chapman and Hall.

Uvarov, E.B. and Isaacs, A. (1993) *The Penguin Dictionary of Science* (7th edition), London: Penguin.

ACRONYMS

ACMAD	African Centre for Meteorological Applications for Development
ADP	adenosine diphosphate
AEAB	Australian Environment Assessment Branch
AEC	Atomic Energy Commission
AGGG	Advisory Group on Greenhouse Gases
ANCA	Australian Nature Conservation Agency
AONB	Area of Outstanding Natural Beauty
AOSIS	Alliance of Small Island States
ATP	adenosine triphosphate
ATV	all-terrain vehicles
BAHC	Biological Aspects of the Hydrologic Cycle
BAPMoN	Background Air Pollution Monitoring Network
BNFL	British Nuclear Fuels Limited
BOD	biochemical oxygen demand
CANDU	Canadian Deuterium-Uranium
CCDP	Climate Change Detection Project
CEGB	Central Electricity Generating Board
CERCLA	Comprehensive Environment Response Compensation and Liability Act
CFCs	chlorofluorocarbons
CIAP	Climate Impact Assessment Program
CITES	Convention on International Trade in Endangered Species
CLIMAP	Climate Long-range Investigation Mapping and Predictions Project
COD	chemical oxygen demand
CRU	Climate Research Unit
CSCS	Comprehensive Soil Classification System
CSERGE	Centre for Social and Economic Research on the Global Environment
DALR	dry adiabatic lapse rate
DDT	dichlorodiphenyltrichloroethane
DMS	dimethyl sulphide
DNA	deoxyribonucleic acid
DOE	Department of Energy (United States)
DU	Dobson unit
DVI	dust veil index
EASOE	European Arctic Stratospheric Zone Experiment
ECRC	Environmental Change Research Centre
EIA	environmental impact assessment
ELR	environmental lapse rate
ENUWAR	Environmental Consequences of Nuclear War

ENSO	El Niño-Southern Oscillation
EPA	Environmental Protection Agency
EPA (UK)	Environmental Protection Act
ERTS	Earth Resources Technology Satellites

FAO	Food and Agriculture Organization
FBC	fluidized bed combustion
FCCC	Framework Convention on Climate Change
FEWS	Famine Early Warning System
FGD	flue gas desulphurization
FOE	Friends of the Earth

GAW	Global Atmosphere Watch
GCM	general circulation model
GCOS	Global Climate Observing System
GEMS	Global Environment Monitoring System
GEWEX	Global Energy and Water Cycle Experiment
GFDL	Geophysical Fluid Dynamics Laboratory
GIS	geographic information systems
GISS	Goddard Institute for Space Studies
GNP	gross national product
GO_3OS	Global Ozone Observing System
GVI	glaciological volcanic index

HCFCs	hydrochlorofluorocarbons
HDP	Human Dimensions of Global Environmental Change Program
HFCs	hydrofluorocarbons

IAEA	International Atomic Energy Agency
IBRD	International Bank for Reconstruction and Development
ICRP	International Commission on Radiological Protection
ICSU	International Council of Scientific Unions
IDA	International Development Association
IGACP	International Global Atmospheric Chemistry Project
IGBP	International Geosphere-Biosphere Program
IGCP	International Geological Correlation Program
IGU	International Geographical Union
IHP	International Hydrological Program
IIASA	International Institute for Applied Systems Analysis
IJC	International Joint Commission
IMF	International Monetary Fund
IMO	International Meteorological Organization
INCO	International Nickle Company
IOC	Intergovernmental Oceanographic Commission
IPCC	Intergovernmental Panel on Climate Change
IPM	integrated pest management
ITCZ	intertropical convergence zone
IUCN	World Conservation Union

| JGOFS | Joint Global Ocean Flux Study |

LIMB	lime injection multi-stage burning
LISA	low-input sustainable agriculture
LNG	liquefied natural gas

LPG	liquefied petroleum gas
LRTAP	long-range transportation of air pollution
LULU	locally unwanted land use
MAB	Man and Bisophere Program
MAC/MPC	maximum allowable/permissible concentration
MCS	multiple chemical sensitivity
MECCA	Model Evaluation Consortium for Climate Assessment
MHD	magnetohydrodynamic generator
MRCFE	Man's Role in Changing the Face of the Earth
MSA	methane sulphonic acid
MUSYA	Multiple Use Sustained Yield Act (United States Forest Service)
MVP	minimum viable population
NAAQS	National Ambient Air Quality Standards
NAPAP	National Acid Precipitation Assessment Program
NAS	National Academy of Sciences
NAWAPA	North American Water and Power Alliance
NCAR	National Center for Atmospheric Research
NCC	Nature Conservancy Council
NEF	noise exposure forecast
NEPA	National Environment Policy Act
NGOs	non-governmental organizations
NIMBY	not in my backyard
NNI	noise and number index
NPP	net primary productivity
NRDC	Natural Resources Defence Council
ODP	ozone depletion potential
OECD	Organization for Economic Co-operation and Development
OPEC	Organization of Petroleum Exporting Countries
PAN	peroxyacetyl nitrate
PCBs	polychlorinated biphenyls
PDSI	Palmer Drought Severity Index
PE	potential evapotranspiration
pH	potential hydrogen
PVC	polyvinyl chloride
QBO	quasi-biennial oscillation
RAN	Rainforest Action Network
RAINS	Regional Acidification Information and Simulation
RCRA	Resource Conservation and Recovery Act
rem	Roentgen equivalent man
SALR	saturated adiabatic lapse rate
SCEP	Study of Critical Environmental Problems
SCOPE	Scientific Committee on Problems of the Environment
SDWA	Safe Drinking-Water Act
SMD	soil moisture deficit
SMIC	Study of Man's Impact on Climate
SOE	State of the Environment Reporting
SSSI	sites of special scientific interest
SSTs	sea-surface temperatures

SSTs	supersonic transports
TNC	The Nature Conservancy
TOGA	Tropical Ocean and Global Atmosphere Project
TTAPS	Turco, Toon, Ackerman, Pollack, Sagan (nuclear winter scenario)
TVA	Tennessee Valley Authority
UKAEA	UK Atomic Energy Authority
UNCCD	United Nations Convention to Combat Desertification
UNCED	United Nations Conference on Environment and Development
UNCHE	United Nations Conference on the Human Environment
UNCOD	United Nations Conference on Desertification
UNDP	United Nations Development Program
UNEP	United Nations Environment Program
UNESCO	United Nations Educational, Scientific and Cultural Organization
USAID	United States Agency for International Development
VEI	volcanic exposivity index
WCASP	World Climate Applications and Services Program
WCP	World Climate Program
WCRP	World Climate Research Program
WHO	World Health Organization
WMO	World Meteorological Organization
WWF	Worldwide Fund for Nature
WWW	World Weather Watch
ZPG	zero population growth

A

ABIOTIC

Non-living. **Ecosystems** consist of abiotic or non-living components and **biotic** or living components. Abiotic components include physical factors such as **soil, water** and **climate**. The mix of these elements will influence the nature and number of living organisms in a particular **environment**. Human activities have altered abiotic components in many ecological systems; positively through soil improvement and irrigation in agriculture, for example, or negatively through **pollution** and other forms of environmental degradation.

Further reading
Chiras, D.D. (1994) *Environmental Science: Action for a Sustainable Future* (4th edition), Redwood City, CA: Benjamin/Cummings.

ABSOLUTE HUMIDITY

A measure of the amount of **water vapour** in the **atmosphere**. It is usually expressed as the mass of water vapour per unit volume of **air** – grams per cubic metre. Since it is based on volume, which responds to temperature and pressure changes, absolute humidity values (in a moving air mass, for example) will change, even when no water vapour is added or lost. To overcome this, meteorologists often use **mixing ratio** or **specific humidity** values as indicators of the amount of water vapour in the air. Since both involve mass rather than volume, they are not affected by changing temperature or pressure. Absolute humidity values are not easily determined through direct sampling, but can be calculated through **wet-bulb temperature** readings.

See also
Humidity, Relative humidity, Vapour pressure.

Further reading
Linacre, E. (1992) *Climate Data and Resources*, London: Routledge.

ABSOLUTE ZERO

The lowest value on the Kelvin or absolute scale of temperature. It is a theoretical value representing the lowest **temperature** attainable, and one at which all molecular motion is presumed to cease. Absolute zero is equivalent to $-273.16°C$.

See also
Kelvin.

ABSORPTION

The assimilation of one substance by another. Gases may be taken up by liquids or solids, for example, and liquids by solids. The process is essential to the working of the earth/atmosphere system. The absorption of atmospheric gases by plants and animals provides for the survival of life and the absorption of **precipitation** by **soil** gives plants easy access to the moisture they require. Absorption is also part of the energy flow in the system. The absorption of **solar energy** at the earth's surface, followed by the absorption of **terrestrial energy** by the **greenhouse gases** in the **atmosphere**, is basic to the working of the earth's **energy budget**. At the local scale, absorption has a major role in **pollution** control. The removal of noxious gases from **flue gas** emissions is commonly accomplished through absorption, and absorbent materials such as **peat** are frequently used to soak up liquid spills.

See also
Flue gas desulphurization.

Further reading
Turco, R.P. (1997) *Earth Under Siege: From Air Pollution to Global Change*, Oxford/New York: Oxford University Press.

ABSORPTION COEFFICIENT

A measure of the degree to which a substance is capable of absorbing **radiation**, usually expressed as the ratio of the energy absorbed by the substance to the amount falling on it. A perfect absorber, such as a **black body**, would have an absorption coefficient of 1.

ABYSSAL ZONE

The deepest part of the **oceans** lying more than 2000 m beneath the surface. The main physical feature of this zone is the abyssal plain, a region of low relief beyond the continental margins. Its surface is broken in places by volcanic hills or **seamounts** rising out of the plain, and, particularly in the Pacific Ocean, by linear trenches that reach depths in excess of 12,000 m beneath the sea surface. Sediments deposited in the upper levels of the oceans ultimately reach the abyssal zone, but little light penetrates, **energy** levels are low and plant or animal life is sparse. The **water** in the abyssal zone is considered to be in long-term storage, and its participation in the **hydrological cycle** has a time-scale of centuries.

See also
Oceans.

Further reading
Pickard, G.L. and Emery, W.J. (1990) *Descriptive Physical Oceanography*, Oxford: Pergamon Press. Strahler, A.H. and Strahler, A.N. (1992) *Modern Physical Geography* (4th edition), New York: Wiley.

ACETATE PLASTICS

Plastics made from cellulose acetate (etha-noate), produced through the action of acetic acid (ethanoate acid CH_3COOH) on wood pulp **cellulose**. Cellulose acetate is also the raw material for the production of rayon, the so-called artificial silk. Acetate plastics are thermoplastics – they can be heated and reformed and are therefore suitable for **recycling**.

Further reading
Elias, H.-G. (1993) *An Introduction to Plastics*, New York: Weinheim.

ACID

A compound containing **hydrogen** (H), which on solution in **water**, produces an excess of hydrogen **ions**. An acid reacts with a base or **alkali** to form a **salt** and water.

ACID LOADING

The addition of **acid** to waterbodies by way of deposition from the **atmosphere**. Waterbodies differ in their sensitivity to acid loading, but once a critical **pH** level has been surpassed, the cumulative effect is the gradual destruction of the aquatic ecosystem. Harmful effects will begin to be felt by most waterbodies when their pH falls to 5.3, although damage to the **ecosystem** will occur in some lakes before that level is reached, and some authorities consider pH 6.0 to be a more appropriate value.

Further reading
Park, C.C. (1987) *Acid Rain: Rhetoric and Reality*, London: Methuen.

ACID MINE DRAINAGE

The seepage of sulphuric acid (H_2SO_4) from mining operations into adjacent waterways. It is most common in coal-mining areas, but is also associated with nickel (Ni) and **copper** (Cu) mining where the **ores** contain sulphur (S) compounds. The most common source of sulphur is iron sulphide (pyrites), which on exposure to oxygen (O) and **water**, and with the help of bacteria such as *Thiobaccillus thioxidans*, is converted into sulphuric acid. In the process ferric hydroxide is precipitated, giving acid mine drainage its characteristic yellow-brown colour. The precipitate also builds up as a brown coating on rocks and sediments in stream beds. Acid mine drainage is highly acidic and corrosive, and contains toxic metals such as **aluminum** (Al), copper (Cu), **zinc** (Zn), **manganese** (Mn) and **beryllium** (Be) leached from the local bedrock by the acid. This combination of acid and minerals completely disrupts the aquatic ecosystem, and may render the water unsuitable for municipal, industrial or recreational use. In the United States, 2.7 million tonnes of acids are estimated to drain every year from existing and abandoned mine workings, polluting some 12,000 km of streams. Methods developed to reduce the acidity of mine drainage include the addition of an alkaline buffer such as **limestone** to the backfill when **strip-mining** areas are being rehabilitated and the storage of water from working mines in holding ponds, where it can be neutralized before release. Such methods can be incorporated relatively easily into current operations, but to apply them to

abandoned mine workings, which make a major contribution to acid mine drainage (in the Appalachian region of the United States, for example), would take many years and would be costly.

Further reading
Chiras, D.D. (1994) *Environmental Science: Action for a Sustainable Future*, Redwood City, CA: Benjamin/Cummings.
Griggs, G.B. and Gilchrist, J.A. (1983) *Geologic Hazards, Resources and Environmental Planning*, Belmont, CA: Wadsworth.

ACID PRECIPITATION

The **wet** or **dry deposition** of acidic substances of anthropogenic origin on the earth's surface. Although natural **precipitation** is commonly acidic, the term acid precipitation is normally retained for precipitation at least ten times more acid than normal. Commonly called **acid rain**, it also includes acid snow and acid fog.

ACID RAIN

Acid rain is normally considered to be a by-product of modern atmospheric **pollution**. Even in a pure, uncontaminated world, however, it is likely that the rainfall would be acidic. Carbonic acid (H_2CO_3) is formed when **carbon dioxide** (CO_2) is absorbed by atmospheric water. Nitric acid (HNO_3) is created during thunderstorms and sulphuric acid (H_2SO_4) is formed from the **sulphur dioxide** (SO_2) released during volcanic eruptions and forest fires or from the **sulphur** (S) emitted by phytoplankton during their seasonal bloom period. All of these are natural processes which contribute to make normal rain acid.

Current concern over acid rain is not with the naturally produced variety, but rather with that which results from modern industrial activity and the acid gases it produces. Considerable amounts of sulphur dioxide are released into the atmosphere as a by-product of metal smelting, particularly when non-ferrous **ores** are involved. The burning of **coal** and **oil** to provide **energy** for space heating or to fuel **thermal electric power stations** also produces sulphur dioxide. The

continuing growth of transportation systems using the **internal combustion engine** contributes to acid rain through the release of **oxides of nitrogen** (NO_x) into the atmosphere.

Acid precipitation produced by human activities differs from natural acid precipitation not only in its origins, but also in its quality. Anthropogenically produced acid rain tends to be many times more acidic than the natural variety. The difference is commonly of the order of 1.0 to 1.5 points on the **pH** scale used to measure acidity. In North America, for example, naturally acid rain has a pH of about 5.6, whereas measurements of rain falling in southern Ontario, Canada frequently provide values in the range of 4.5 to 4.0 (Ontario: Ministry of the Environment 1980). To put these values in perspective, vinegar has a pH of 2.7 and milk a pH of 6.6. Similar values for background levels of acidic rain are indicated by studies in Europe with the average annual pH of rain over Britain between 1978 and 1980 being within the range of 4.5 to 4.2 (Mason 1990). Remarkably high levels of

Figure A-1 Schematic representation of the formation, distribution and impact of acid rain

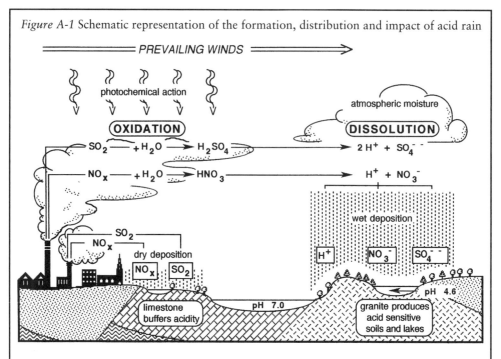

Source: After Kemp, D.D. (1994) *Global Environmental Issues: A Climatological Approach*, London/New York: Routledge

acidity have been recorded on a number of occasions on both sides of the Atlantic. In April 1974, for example, rain falling at Pitlochry, Scotland had a pH measured at 2.4 (Last and Nicholson 1982), and an extreme value of pH 1.5, some 11,000 times more acid than normal, was recorded for rain falling in West Virginia in 1979 (LaBastille 1981).

The quality of the rain is determined by a series of chemical processes set in motion when acidic materials are released into the **atmosphere**. Some of the sulphur dioxide and oxides of nitrogen emitted will return to the surface quite quickly, and close to their source, as **dry deposition**. The remainder will be carried up into the atmosphere, to be converted into sulphuric and nitric acid, which will eventually return to earth as acid rain. The processes involved are fundamentally simple. **Oxidation** converts the gases into acids, in either a **gas** or **liquid phase reaction**, at a rate which depends upon such variables as the concentration of **heavy metals** in airborne **particulate matter**, the

presence of ammonia and the intensity of sunlight. The metals and ammonia appear to act as **catalysts**, while the sunlight provides the energy required by the chemical reactions.

Acid rain remained mainly a local problem in the past. The introduction of the **tall stacks policy** in an attempt to reduce local pollution, however, increased its geographical extent. The release of pollutants at the greater altitudes possible with the taller smokestacks placed them outside the boundary layer circulation and into the larger scale atmospheric circulation system with its potential for much greater dispersal through the mechanisms involved in the **Long Range Transportation of Air Pollution** (LRTAP). Local problems were mitigated at the expense of the larger **environment**.

The main sources of acid rain are to be found in the industrialized areas of the northern hemisphere. North-eastern North America, Britain and Western Europe have received most attention, although their output of sulphur dioxide has been declining

Figure A-2 The geography of acid rain in
North America and Europe

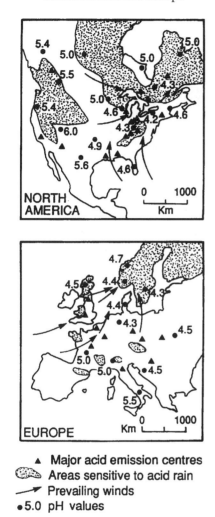

▲ **Major acid emission centres**
🝏 **Areas sensitive to acid rain**
↗ **Prevailing winds**
●**5.0 pH values**

Source: After Kemp (1994)

major industrial nations, but concern has been expressed over growing levels of air pollution, often associated with urban automobile exhaust emissions, which may already have provided a base for acid rain in some **Third World** countries (Park 1987).

Once the acid gases have been released into the atmosphere, they are at the mercy of the prevailing circulation patterns. With almost all of the areas currently producing large amounts of acidic pollution located within the mid-latitude westerly wind belt, emissions are normally carried eastwards, or perhaps north-eastwards, often for several hundred kilometres before being redeposited. In this way, pollutants originating in the US Midwest cause acid rain in Ontario, Quebec and the New England states. Emissions from the smelters and power stations of the English Midlands and the Ruhr contribute to the acidity of precipitation in Scandinavia, and acidity in the Arctic originates as far as 8000 km away to the south in North America and Eurasia (Park 1987). Thus the problem of acid rain transcends national boundaries, introducing political overtones to the problem and creating the need for international co-operation if a solution is to be found.

The impact of acid rain on the environment depends not only on the level of acidity in the rain, but also on the nature of the environment itself. Areas underlain by acidic rocks such as granitic or quartzitic bedrock, for example, are particularly susceptible to damage, lacking as they do the ability to 'buffer' or neutralize additional acidity from the precipitation. Acid levels therefore rise, the environmental balance is disturbed and serious ecological damage is the inevitable result. In contrast, areas which are geologically basic – underlain by **limestone** or chalk, for example – are much less sensitive, and may even benefit from the additional acidity. The highly alkaline **soils** and **water** of areas underlain by chalk or limestone ensure that the acid added to the environment by the rain is very effectively neutralized. The areas at greatest risk from acid rain in the northern

since the mid-1970s. Levels of oxides of nitrogen have not yet experienced significant decline and continue to rise in some areas. The emission of acid gases from Eastern Europe and the republics of the former USSR – Russia, Ukraine and Kazakhstan – remains high. In Asia, Japanese industries emit large quantities of sulphur dioxide (Park 1987), while the industrial areas of China are also major contributors. Acid emissions remain limited outside of the

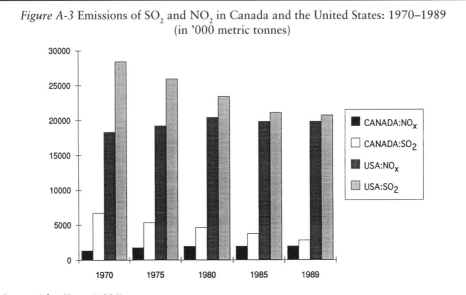

Figure A-3 Emissions of SO₂ and NO₂ in Canada and the United States: 1970–1989 (in '000 metric tonnes)

Source: After Kemp (1994)

hemisphere are the Precambrian Shield areas of Canada and Scandinavia, where the acidity of the rocks is reflected in highly acidic soils and water. The folded mountain structures of eastern Canada and the United States, Scotland, Germany and Norway are also vulnerable.

The impact of acid rain on the environment was first recognized in the lakes and rivers of these areas. Reduced pH values which indicated the rising acidity were accompanied by low levels of **calcium** (Ca) and **magnesium** (Mg), elevated sulphate concentrations and an increase in the amounts of potentially toxic metals such as **aluminum** (Al) (Brakke *et al.* 1988). When aquatic communities were examined in areas as far apart as New York State, Nova Scotia, Norway and Sweden there was clear evidence that increased surface water acidity had adverse effects on fish (Baker and Schofield 1985). In some cases the acidity was sufficiently high that mature fish died, but, more commonly, fish populations began to decline because of the effects of the increasing acidity on reproduction. Damage to the eggs during spawning and the inability of the young fry to survive the higher acidity, particularly during

the **spring flush**, ensured that the older fish were not replaced as they died. As a result, fish populations in many rivers and lakes in eastern North America, Britain and Scandinavia have declined noticeably in the last two to three decades and hundreds of lakes are now completely devoid of fish (Harvey 1989). Although the impact on the fish is most obvious, all aquatic organisms will decline in number and variety during progressive acidification. Waterbodies that have lost or are in the process of losing their flora and fauna are often described as 'dead' or 'dying'. This is not strictly correct, however, for some organisms are remarkably acid-tolerant, and even the most acid lakes have some life in them. **Species** of **protozoans** are found at pH levels as low as 2.0, for example (Hendrey 1985).

There is growing evidence that those areas in which the waterbodies have already succumbed to acidification must also face the effects of increasing acid stress on their forests and soils. The threat is not universally recognized, however, and there remains a great deal of controversy over the amount of damage directly attributable to acid rain. Reduction in forest growth in Sweden (LaBastille 1981), physical damage to trees in West Germany (Pearce 1982b), and the death of sugar maples in Quebec and Vermont (Norton 1985) have all been blamed on the increased acidity of the precipitation in these areas. Many of the impacts, such as the thinning of annual growth rings, reduction in **biomass** and damage to fine root systems, are only apparent after detailed examination, but others are more directly obvious and have been described as **dieback**. This involves the gradual wasting of the tree inwards from the outermost tips of its branches. The process is cumulative over several years until the tree dies. The symptoms of dieback have been recognized in the maple groves and red spruce forests of north-eastern North America, and across fifteen countries and some 70,000 km² of forest in Europe (Park 1991). In Germany, where it was first linked with acid rain, dieback, or **Wald-sterben**, was particularly extensive and in the 1980s was seen as a major threat to the survival of the German forests (Ulrich 1983). More recent research has indicated that forest decline in Europe is a multi-faceted process in which acid precipitation may be only one of a series of contributors along with tree harvesting practices, **drought** and fungal attacks, although its

Figure A-4 A statue damaged by acid rain – a common sight on the historic buildings of Europe

Photograph: The author

exact role remains controversial (Blank *et al.* 1988; Hauhs and Ulrich 1989).

Present concern over acid rain is concentrated mainly on its effect on the **natural environment**, but acid rain also contributes to deterioration in the built environment: it attacks limestone and marble used as building stone, for example. The faceless statues and crumbling cornices of the world's famous palaces, castles, abbeys and cathedrals, from the Parthenon in Greece to the Taj Mahal in India, attest to its power. By attacking the fabric of these buildings it not only causes physical and economic damage, but threatens the world's cultural heritage.

Acid rain in the form of **aerosols** or attached to **smoke** particles can cause respiratory problems in humans, as was the case with the infamous **London Smog** of 1952. There is no evidence that **wet deposition** is directly damaging to human health, although in its ability to mobilize metals, from **lead** (Pb) or **copper** (Cu) pipes, for example, it may have important indirect effects (Park 1987).

Solutions to the problem of acid rain are deceptively simple. In theory, a reduction in the emission rate of acid forming gases is all that is required to slow down and eventually stop the damage being caused by the acidification of the environment. Translating that concept into reality has proved difficult, however. One of the first approaches involved adaptation, such as the addition of lime to land or lakes to counteract the increased acidity, and to allow the recovery mechanisms to work more effectively. Experiments involving the liming of lakes in Canada and Sweden, for example, have provided encouraging results (Porcella *et al.* 1990). Artificial buffering of lakes in this way is only a temporary measure which may be likened to the use of antacid to reduce acid indigestion. The neutralizing effects of the lime may last longer than those of the antacid, but they do wear off in three to five years and re-liming is necessary as long as **acid loading** continues (Ontario: Ministry of the Environment 1980). Most of the current proposals for dealing with acid rain tackle the problem at its source. They attempt to prevent, or at least reduce, emissions of

acid gases into the atmosphere, and since sulphur dioxide makes the greatest contribution to acid rain in North America and Europe, it has received most attention in the development of abatement procedures. Sulphur dioxide is formed when coal and oil are burned to release energy, and the technology to control it may be applied before, during or after **combustion**. The exact timing will depend upon such factors as the amount of acid reduction required, the type and age of the system and the cost-effectiveness of the particular process (Ellis *et al.* 1990).

Flue gas desulphurization (FGD) is the most common approach to sulphur dioxide reduction. It uses **scrubbers** to neutralize the acidity of the flue gases before they are released into the atmosphere. With sulphur dioxide removal rates of between 80 and 95 per cent, FGD is more effective than methods such as **fuel switching** or **fuel desulphurization** applied prior to combustion and it is technically simpler than **fluidized bed combustion** (FBC) and **lime injection multi-stage burning** (LIMB) incorporated in the combustion process itself. Although FGD systems do little to reduce emissions of oxides of nitrogen they are the preferred method of dealing with acid rain, and future requirements to reduce sulphur dioxide levels will be met in large part by the installation of FGD equipment (Park 1991).

The study of acid rain has declined remarkably since the 1980s, when it was viewed by many as the major environmental problem facing the northern hemisphere. It is one environmental issue in which abatement programmes have met with some success. Sulphur dioxide levels continue to decline, the rain in many areas is considerably less acid than it was a decade ago, and the transboundary disputes which absorbed large amounts of time, energy and money in the 1980s have been resolved. There are indications that lakes and forests are showing some signs of recovery in the most vulnerable areas of North America and Europe, although this is still a matter of dispute.

Developments such as these have created the perception that the acid rain problem is being solved. It would be more accurate, however, to say that it has changed in nature and in geographical extent. In the developed nations, oxides of nitrogen are gradually making a larger contribution to acidity as levels of sulphur dioxide decline. In less developed areas, from Eastern Europe to China, the full impact of acid rain may still be in the future, and the research activities currently winding down in Western Europe and North America may have to be revived to deal with it.

References and further reading

Baker, J.P. and Schofield, C.L. (1985) 'Acidification impacts on fish populations: a review', in D.D. Adams and W.P. Page (eds) *Acid Deposition: Environmental, Economic and Political Issues*, New York: Plenum Press.

Blank, L.W., Roberts, T.M. and Skeffington, R.A. (1988) 'New perspectives on forest decline', *Nature* 336: 27–30.

Brakke, D.F., Landers, D.H. and Eilers, J.M. (1988) 'Chemical and physical characteristics of lakes in the northeastern United States', *Environmental Science and Technology* 22: 155–63.

Ellis, E.C., Erbes, R.E. and Grott, J.K. (1990) 'Abatement of atmospheric emissions in North America: progress to date and promise for the future', in S.E. Lindberg, A.L. Page and S.A. Norton (eds) *Acidic Precipitation, Volume 3, Sources, Deposition and Canopy Interactions*, New York: Springer-Verlag.

Harvey, H.H. (1989) 'Effects of acid precipitation on lake ecosystems', in D.C. Adriano and A.H. Johnson (eds) *Acidic Precipitation, Volume 2, Biological and Ecological Effects*, New York: Springer-Verlag.

Hauhs, M. and Ulrich, B. (1989) 'Decline of European forests', *Nature* 339: 265.

Hendrey, G.R. (1985) 'Acid deposition: a national problem', in D.D. Adams and W.P. Page (eds) *Acid Deposition: Environmental, Economic and Political Issues*, New York: Plenum Press.

Kemp, D.D. (1994) *Global Environmental Issues: A Climatological Approach* (2nd edition), London/New York: Routledge.

LaBastille, A. (1981) 'Acid rain – how great a menace?', *National Geographic* 160: 652–81.

Last, F.T. and Nicholson, I.A. (1982) 'Acid rain', *Biologist* 29: 250–2.

Mason, B.J. (1990) 'Acid rain – cause and consequence', *Weather* 45: 70–9.

Norton, P.W. (1985) 'Decline and fall', *Harrowsmith* 9: 24–43.

Ontario: Ministry of the Environment (1980) *The Case Against the Rain*, Toronto: Ministry of the Environment.

Park, C.C. (1987) *Acid Rain: Rhetoric and Reality*, London: Methuen.

Park, C.C. (1991) 'Trans-frontier air pollution: some geographical issues', *Geography* 76: 21–35.

Pearce, F. (1982) 'Science and politics don't mix at acid rain debate', *New Scientist* 95 (1313): 80.

Porcella, D.B., Schofield, C.L., Depinto, J.V., Driscoll, C.T., Bukaveckas, P.A., Gloss, S.P. and Young, T.C. (1990) 'Mitigation of acidic conditions in lakes and streams', in S.A. Norton, S.E. Lindberg and A.L. Page (eds) *Acidic Precipitation, Volume 4, Soils, Aquatic Processes and Lake Acidification*, New York: Springer-Verlag.

Ulrich, B. (1983) 'A concept of forest ecosystem stability and of acid deposition as driving force for destabilization', in B. Ulrich and J. Pankrath (eds) *Effects of Accumulation of Air Pollutants in Forest Ecosystems*, Dordrecht, Netherlands: Reidel.

ACTINIDES

A group of radioactive **elements** with **atomic numbers** from 89 to 103. A few, such as thorium (Th) and **uranium** (U), occur naturally, but most, such as **plutonium** (Pu) and neptunium (Np), are the products of **nuclear reactor** operations. Actinides are created when a uranium **atom** absorbs a **neutron**, but no **fission** takes place. The most common actinide is plutonium (^{239}Pu), a fissile element and therefore a potential fuel source. It is also highly **radioactive**, with a **half-life** of 24,000 years. Neptunium (^{237}Np), formed in the same way, has an even longer **half-life** – 2.1 million years. Thus **nuclear waste** containing actinides continues to release **radiation** for many years, creating storage and disposal problems for the nuclear industry.

See also
Periodic table.

Further reading
Friedman, A.M. (ed.) (1976) *Actinides in the Environment*, Washington, DC: American Chemical Society.

ACTIVATED CARBON

Carbon (C), usually in the form of **charcoal**, which has been treated to augment its capacity for **adsorption**. The process involves heat-treatment at high temperatures to remove the **hydrocarbons** and increase the **porosity** of the carbon. The consequent increase in surface area allows the carbon to adsorb large quantities of gases, vapours and colloidal solids. Activated carbon is used extensively in water and air filters for improving taste and controlling odours. It is also very effective in other forms of pollution control such as the reduction of **acid precipitation** through the removal of **sulphur dioxide** (SO_2) from **flue gas** emissions.

Further reading
Bansal, R.C., Donnet, J.-B. and Stoeckli, F. (1988) *Active Carbon*, New York: M. Dekker.

ACTIVATED SEWAGE SLUDGE

A by-product and active ingredient of secondary **sewage treatment** processes. It is solid organic material produced by the biological degradation of **sewage** through a combination of aeration and bacterial activity. The abundant **oxygen** (O) introduced during aeration is used by the micro-organisms as they feed on the organic matter in the sewage and convert it into sludge which settles out of the waste water. Since the sludge contains active **bacteria** and other micro-organisms, some of it is recycled to seed the raw sewage in the aeration tanks, and to speed up the treatment process.

ACTUARIAL (WEATHER) FORECAST

An estimation of the probability that a particular weather condition will occur, based upon past occurrences in the meteorological record. In the past, actuarial forecasts were often used in attempts at **drought prediction**. They have generally been superseded by forecasting methods involving some form of computer modelling.

ADENOSINE TRIPHOSPHATE (ATP)

An organic **phosphate** ($C_{10}H_{12}N_5O_3H_4P_3O_9$) present in all living organisms. It is an important source of chemical **energy** for **cell** growth, cell reproduction, muscle contraction and the synthesis of organic compounds in plants and animals. The energy is made available when one of the three phosphate groups held by ATP is released. Having lost one phosphate, the ATP becomes adenosine diphosphate (ADP), but the process is reversible, and when energy is available the ADP may be built up again into ATP.

$ATP \leftrightarrow ADP + P + energy$

ATP, formed from ADP and phosphate using light energy absorbed by the **chlorophyll** in green plants, has an integral role in **photosynthesis**. The energy stored in ATP is utilized in the reduction of **carbon dioxide** (CO_2) to **carbohydrates** and **oxygen** is released.

Further reading
Mauseth, J.D. (1995) *Botany: An Introduction to Plant Biology*, Philadelphia, PA: Saunders.

ADIABATIC PROCESS

A process that involves thermodynamic change in a **system** with no transfer of heat or **mass** between the system and its surroundings. It applies in the operation of such devices as air compressors, internal combustion engines and aerosol spray cans, but the impact of adiabatic change is particularly prominent as **air masses** travel through the **atmosphere**. In the absence of heat exchange, the **temperature** of an air mass is governed by changing **atmospheric pressure**. When an air mass rises it moves into an area of reduced pressure, and therefore expands. That expansion in turn leads to a fall in temperature which is proportional to the pressure change. A descending **air mass** will be compressed as the atmospheric pressure increases, and its temperature will rise. If the air mass is dry, the rate of change will be constant at the dry adiabatic lapse rate (DALR). If the air mass is saturated or becomes saturated, it will cool (or warm up) at the saturated adiabatic lapse rate (SALR), which is not constant. The relationship between the adiabatic lapse rates and the **environmental lapse rate** (ELR) determines the stability of the atmosphere, which has implications for the accumulation or dispersal of air pollutants.

See also
Atmospheric stability.

Figure A-5 Orographic uplift and the role of adiabatic processes in the formation of clouds and precipitation

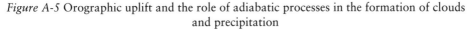

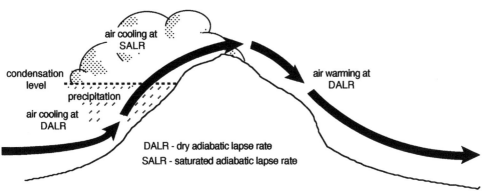

Further reading
Ahrens, C.D. (1993) *Essentials of Meteorology*, St Paul, MN: West Publishing.

ADSORPTION

The attachment of **solid, liquid** or gaseous **molecules** to the surface of a solid – the adsorbent – by physical or chemical bonds. Because of their ability to capture a wide range of impurities, adsorbents are frequently used in pollution control programmes. The adsorbing qualities of **activated carbon**, for example, are exceptional, and account for its extensive use in water purification and gaseous emission control systems. As the surface of the adsorbent is increasingly covered by the captured impurities, its ability to act as a filter

Figure A-6 Backwashing or reversal of flow in a filter to remove impurities adsorbed during the filtering process

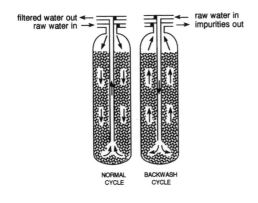

is reduced. However, since the impurities are not incorporated into the adsorbent but remain on the surface (compare with **absorption**), they can be removed relatively easily – by backwashing, for example – after which the filtering ability of the adsorbent is restored.

Further reading
Ruthven, D.M. (1984) *Principles of Adsorption and Adsorption Processes*, New York: Wiley.

ADVECTION

The horizontal transfer of **energy** or matter in a fluid. It refers particularly to the movement of **air** in the **atmosphere** which results in the redistribution of such elements as warm or cold air, moisture or pollutants. Although advection may also be used to describe vertical movement, the term **convection** is more common.

ADVISORY GROUP ON GREENHOUSE GASES (AGGG)

A group established as a result of the **Villach Conference** (1985) to ensure continued academic and public interest in the impact of rising levels of **greenhouse gases** on **climate**, the **environment** and human activities.

AEROBIC

Living or active only in the presence of free **oxygen** (O). Organisms which obtain oxygen from the **atmosphere** during **respiration** are aerobic organisms.

See also
Anaerobic.

AEROSOLS

Finely divided **solid** or **liquid** particles dispersed in the **atmosphere**. They include **dust**, **soot**, **salt** crystals, **pollen** grains, **spores**, **bacteria**, **viruses**, and a variety of other microscopic particles. Most aerosols are produced

Figure A-7 A comparison of the size range of common aerosols with radiation wavelength

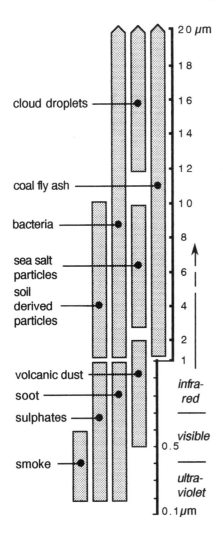

naturally by volcanic activity, forest and grass fires, **evaporation**, local atmospheric turbulence and biological processes, but human agricultural and industrial activities, plus increased **energy** consumption, have enhanced the anthropogenic contribution to the aerosol content of the atmosphere. Within the atmosphere, aerosols are redistributed by **wind** and pressure patterns, remaining in suspension for periods ranging from several hours to several years, depending upon particle size and altitude attained. Aerosols are responsible for **atmospheric turbidity**, and as a result disrupt the inward and outward flow of energy through the atmosphere. Some studies suggest that high aerosol levels contribute to global cooling through the **attenuation** of **solar radiation**, but there is also evidence that they may actually produce a slight warming through their ability to absorb and re-radiate outgoing **terrestrial radiation**. Resolution of this apparent contradiction will only be possible through systematic observation and monitoring of atmospheric aerosol levels and temperatures.

See also
Arctic Haze, Pollution.

Further reading
Shaw, R.W. (1987) 'Air pollution by particles', *Scientific American* 257: 96–103.
Thompson, R.D. (1995) 'The impact of atmospheric aerosols on global climate: a review', *Progress in Physical Geography* 19: 336–50.

AESTHETIC DEGRADATION

The deterioration of the visual quality of the **environment**. Since individual tastes determine what is or is not aesthetically pleasing, any definition of aesthetic degradation must remain broad. Pollutants such as **smoke** and **soot** particles emitted into the **atmosphere** and **sewage** released into rivers and lakes have an obvious impact on the visual quality of the environment, but degradation can also be brought about by planned changes to the built environment, such as the replacement of historic buildings by modern apartment blocks, or the introduction of industry into a previously undeveloped green site. It can include the deterioration associated with mining and

Figure A-8 Proliferation of street signs in Hong Kong – essential advertising for some, aesthetic degradation for others

Photograph: The author

heavy industry or the introduction of power lines or modern wind generators into a wilderness area. **Flora** and **fauna** must also be considered. In parts of North America, for example, the forest industry is required to leave a buffer zone of trees around lakes and other recreation areas, so that people using these areas are not directly exposed to the aesthetic degradation that **clear cutting** can produce. For many people, viewing wild animals in their natural **habitat** is a pleasing activity, and any action that compels the animals to leave an area might be seen as causing aesthetic degradation.

AFRICAN CENTRE OF METEOROLOGICAL APPLICATIONS FOR DEVELOPMENT (ACMAD)

An organization set up to promote applied **meteorology** and **climatology** in Africa. Centred in Niger, it began operations in 1990, but

had its roots in the response to the **droughts** that hit sub-Saharan Africa in the 1970s. Its activities include the development of **agrometeorology**, hydrometeorology, weather analysis and prediction with the aim of assisting African nations to improve food production, water-resource management and **energy** use.

AGENDA 21

A blueprint for **sustainable development** into the twenty-first century produced at the **United Nations Conference on Environment and Development** (UNCED) at Rio de Janeiro in 1992. It attempts to embrace the entire **environment** and development agenda, through four sections – social and economic dimensions, **conservation** and management of **resources** for development, strengthening the role of major groups, means of implementation – and forty chapters covering all aspects of the environment, including such issues as **climate change, ozone depletion,** transboundary air pollution, **drought** and **desertification.** It stresses the importance of improved international co-operation in observation and research, the need to develop an early warning system for change and the necessity to promote a stronger partnership between environmental and developmental agencies. Comprehensive as it is, Agenda 21 is far from ideal. Being the result of compromise among a large number of nations, the language tends to be weak and the recommmendations are not binding on the nations that adopted it. Problems remain in the areas of financing the transition to sustainable development, particularly in the developing nations, and in areas such as energy and forestry, vested interests prevented the agreements from being as comprehensive as they might have been.

Further reading
Parson, E.A., Hass, P.M. and Levy, M.A. (1992) 'A summary of the major documents signed at the earth summit and global forum', *Environment* 34: 12–15 and 34–6.
Spurgeon, D. (1993) *Agenda 21: Green Paths to the Future,* Ottawa: International Development Research Centre.
United Nations Conference on Environment and Development (1993) *Agenda 21: The United Nations Program of Action From Rio,* New York:

United Nations Department of Public Information.

AGENT ORANGE

A mixture of the **herbicides** 2, 4, 5-trichloro-phenoxyacetic acid (2,4,5-T) and 2, 4-dichlorophenoxyacetic acid (2,4-D) used between 1963 and 1971 by the US military during the Vietnam War and named after the orange-striped barrels in which it was delivered. Both 2,4,5-T and 2,4-D were developed in the 1940s to control broad-leaved weeds in crops and to remove unwanted plants from rangeland, forests and other non-croplands. In Vietnam, Agent Orange was sprayed on the jungles from low-flying aircraft to defoliate the trees and deny cover to the Viet Cong. It was also sprayed on cropland to reduce the food available to them. During the 1970s, concern grew over potential health hazards associated with these herbicides, and use of 2,4,5-T was prohibited in many countries from the early 1980s. The presence of the **dioxin** TCDD – the most toxic of the dioxin family – in 2,4,5-T raised the concern further since TCDD was a suspected **carcinogen**. Exposure to Agent Orange was identified as a possible reason for perceived increases in health problems such as chloracne (skin disease), **cancer** of the immune system and birth defects among rural Vietnamese, and US, Australian and New Zealand servicemen. Controlled medical studies produced conflicting evidence, but the US government now pays disability pensions to Vietnam War veterans suffering from chloracne, cancer of the immune system, certain soft tissue cancers and nerve disease.

See also
Organochlorides.

Further reading
Van Strum, C. (1983) *A Bitter Fog: Herbicides and Human Rights*, San Francisco: Sierra Club.

AGRARIAN CIVILIZATIONS

Generally refers to the civilizations that developed in Egypt, Mesopotamia, the Indus Valley and the Yellow River Basin (Hwang-He) in China between 7000 and 3000 BP (c.5000–1000 BC). (In Central America, the development of the Mayan civilization coincides with the later stages of that time period.) They are credited with the first serious domestication of plants and animals, which permitted sedentary agriculture and ultimately the development of permanent settlements and local urbanization. All of these civilizations were located on riverine plains in areas that experienced dry conditions for part of the year. Natural **irrigation** provided by seasonal overbank flooding and artificial irrigation using small dams, cisterns and ditches to redistribute the **water**, allowed year-round cropping and the accumulation of a food surplus. This in turn permitted a greater division of labour and the development of social, cultural and economic activities not possible in a migratory community or one dependent upon **subsistence agriculture**. Accompanying this was an increase in the level of human intervention in the **environment** associated with accelerated population growth. Natural vegetation was replaced by cultivated crops, the **aquatic environment** was altered, and the beginnings of **soil** degradation in the form of siltation and **salinization** became apparent in some areas.

See also
Fertile Crescent, Harappan civilization.

Further reading
Lamberg-Karlovsky, C.C. and Sabloff, J.A. (1979) *Ancient Civilizations: The Near East and Mesoamerica*, Menlo Park, CA: Benjamin/Cummings.

Figure A-9 Location of the original agrarian civilizations

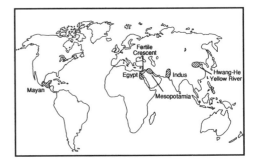

AGRICULTURAL REVOLUTION

A period of rapid change in agricultural activities in Britain between about 1750 and 1850. New land management and cultivation techniques had been introduced as early as the sixteenth century in some areas, but the pace of change increased sharply after the middle of the eighteenth century. Improvements were introduced in all aspects of farming, leading to greater efficiency and allowing a substantial increase in food production. Greater attention was paid to maintaining and increasing the quality of the **soil**, by adding lime and manure. Land that was previously too wet to be used was brought into production by improving drainage, and soils in areas too dry or light were treated with marl, **clay** rich in **calcium carbonate** (CaCO$_3$), to improve their texture. New crops such as turnips, potatoes and clover were grown more frequently and crop rotation was introduced. Experiments with livestock breeding increased the quality and quantity of meat and wool. New mechanized or semi-mechanized implements were developed to deal with all aspects of cultivation, from ploughing – better designed steel ploughs – and planting – horse-drawn seed drill – to harvesting – horse-drawn reaper. These innovations paralleled similarly innovative changes taking place in industry and contributed to them by providing a food surplus for the growing workforce in the new industrial towns. By the mid-nineteenth century, even the increased production from domestic agriculture could not keep up with the demand from a rapidly growing **population**, and Britain had to import foodstuffs. The Agricultural Revolution changed the landscape of Britain, replacing natural vegetation with crops, and open fields with enclosures surrounded by hedges. In places, it also contributed to environmental degradation in the form of **soil erosion**, where the enthusiasm for improvement brought land unsuitable for arable agriculture into production.

Further reading
Hoskins, W.G. (1955) *The Making of the English Landscape*, London: Hodder & Stoughton.
Mannion, A.M. (1991) *Global Environmental Change: A Natural and Cultural Environmental History*, London: Longman.

Simmons, I.G. (1996) *Changing the Face of the Earth* (2nd edition), Oxford: Blackwell.

AGROCLIMATOLOGY

The study of the role of **climate** in all forms of agriculture, commonly including hydrological as well as climatological factors. Agricultural climatology is very much an applied discipline in which the general aim is to use the knowledge obtained in the studies to improve the quality and quantity of agricultural output.

See also
Agrometeorology.

Further reading
Monteith, J.L. (ed.) (1975) *Vegetation and the Atmosphere*, London/New York: Academic Press.

AGROFORESTRY

The treatment of trees as an agricultural crop to provide for the planned production of fuelwood, timber, animal fodder and food. Agroforestry may also involve the integration of tree cultivation into existing or planned agricultural activities. When interplanted with farm crops, trees help to regulate **temperature** extremes and moisture availability. Leaf litter adds organic matter and **nutrients** to the soil, and leguminous trees help to improve soil fertility through their ability to fix **nitrogen** (N) from the **air**. In semi-arid regions and areas subject to **desertification**, agroforestry techniques have been used to prevent **soil erosion**, reduce water loss and restore soil fertility.

Further reading
Ong, C.A. and Huxley, P.A. (eds) (1996) *Tree–Crop Interactions: A Physiological Approach*, Wallingford: CAB International/ International Centre for Research in Agroforestry.
Pegorie, J. (1990) 'On-farm agroforestry research: case study from Kenya's semi-arid zone', *Agroforestry Today* 2: 4–7.

AGROMETEOROLOGY

Similar to **agroclimatology**, in that it involves the study of the relationships between atmospheric processes and agricultural activity. The main difference is in scale. Agrometeorologists tend to emphasize the impact of shorter term weather phenomena (e.g. daily conditions) in their studies, whereas agroclimatologists are more involved with longer term planning using mean data sets as their sources. Traditionally, agrometeorologists have also been more concerned with micro-scale processes, including conditions in the **soil** environment. However, the distinction between the two disciplines is not always clear, and the terms can often be used interchangeably without confusion.

AGRONOMY

The study of land management and rural economy.

AIR

The common name for the combination of gases that make up the earth's **atmosphere**. Air itself is not a **gas,** but rather a mixture of individual gases, each of which retains its own particular properties.

AIR MASS

A large body of **air** in which there is little or no horizontal variation in **temperature, humidity** or **atmospheric stability**. Since an air mass can cover an area of more than 1000 km², conditions within it cannot be completely uniform, but internal variations are much less than the differences across the air mass boundaries. The air mass approach to **climatology,** which links the **climate** of an area with the nature, timing and frequency of the air masses it experiences, was first introduced in 1928 by the Norwegian meteorologist, Tor Bergeron.

The classification of air masses is based on the characteristics they display in their source regions. Temperatures are linked to **latitude** (i.e. tropical, polar, Arctic/Antarctic), **humidity** is a product of surface conditions (i.e. land, **water, ice**) and stability is a reflection of the way in which the other two elements combine. Seasonal variations alter the characteristics of some air masses, particularly those originating in higher latitudes in the northern hemisphere.

As an air mass moves away from its source region, it carries the characteristics of that region with it. For example, cold air masses originating in the Canadian north move south every winter to bring low temperatures and **snow** as far south as Florida. Ultimately, thermal and dynamic factors, such as heating and cooling or uplift and **subsidence,** will modify the original properties of the air mass. Cold continental polar (cP) air moving westwards from Scandinavia and Russia, for example, will be heated as it passes over the relatively warm North Sea. When it reaches

Table A-1 Air mass characteristics and sources

AIR MASS	CHARACTERISTICS	SOURCE REGION
Maritime tropical (mT)	Warm, very moist, potentially unstable	Tropical and sub-tropical oceans
Continental tropical (cT)	Hot, dry, variable stability	Sub-tropical deserts
Maritime polar (mP)	Cool, moist, variable stability	Mid- to high-latitude oceans
Continental polar (cP)	Cold, dry, very stable in winter Cool, less dry, potentially unstable in summer	Central North America and Siberia in winter. Sub-arctic regions in summer
Maritime Arctic or Antarctic (mA or mAA)	Cold, moist, potentially unstable	Arctic and Antarctic Oceans
Continental Arctic or Antarctic (cA or cAA)	Cold, dry, very stable	Antarctica, frozen Arctic Ocean, Greenland

the United Kingdom, it is forced to rise as it crosses the coast, and the moisture picked up as it crosses the sea is precipitated, usually in the form of snow. Thus a cold, dry, stable air mass has been rendered warmer, moister and less stable in its movement away from the source region.

See also
Polar front.

Further reading
Barry, R.G. and Chorley, R.J. (1992) *Atmosphere, Weather and Climate* (6th edition), London/New York: Routledge.

AIR POLLUTION

See environmental pollution.

AIR QUALITY

A measure of the degree to which air is polluted – air quality is low when pollution levels are high and vice versa. Quality may be judged subjectively through the colour, clarity or smell of the air, but it can also be related to invisible pollutants such as **carbon monoxide** (CO) or **sulphur dioxide** (SO$_2$). Low or declining air quality levels have

Figure A-10 An air mass climatology of Canada

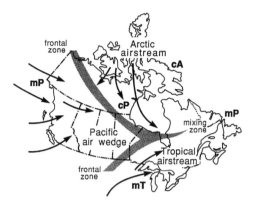

Source: After Kemp, D.D. (1994) 'Global Warming and the Provincial Norths' in M.E. Johnston (ed.) *Geographic Perspectives on the Provincial Norths*, Thunder Bay, On: Lakehead University Centre for Northern Studies/Copp Clark Longman

Table A-2 Changing ambient air quality in selected cities

CITY, COUNTRY*	SULPHUR DIOXIDE (SO$_2$) AVERAGE ANNUAL GROWTH RATE 1979–1990 (%)	SUSPENDED PARTICULATE MATTER AVERAGE ANNUAL GROWTH RATE 1979–1990 (%)
Beijing, China	3.5 (8)	–2.7 (8)
Tokyo, Japan	–8.9 (17)	–4.9 (13)
Hong Kong	47.3 (4)	14.9 (4)
New Delhi, India	12.0 (6)	–0.3 (7)
Sydney, Australia	–10.9 (11)	2.2 (11)
Sao Paulo, Brazil	–7.5 (12)	–9.1 (6)
New York City, United States	–5.8 (9)	–2.2 (11)
Frankfurt, Germany	–7.2 (17)	0.5 (17)
Lisbon, Portugal	–3.0 (10)	0.4 (9)

* All values refer to city centre locations
The figure in parentheses is the total of years in average.

Source: World Bank (1992) *Development and Environment*, New York: Oxford University Press

adverse effects on people, animals, vegetation and materials, and standards have been established to help maintain air quality at a level which is not detrimental to public health or welfare.

See also
Air quality standards.

Further reading
Godish, T. (1991) *Air Quality* (2nd edition), Chelsea, MI: Lewis.

AIR QUALITY STANDARDS

Maximum allowable concentrations of air pollutants in ambient outdoor **air**. Primary standards apply to human beings and are designed to protect public health, whereas secondary standards are established to limit damage to plants and materials. The main pollutants involved are **sulphur dioxide** (SO_2), **oxides of nitrogen** (NO_x), **carbon monoxide** (CO), ground-level **ozone** (O_3) and airborne particles. In addition, there are as many as 150 to 160 other toxic chemicals, such as volatile **hydrocarbons**, or **metals** such as **lead** (Pb), which may be present from time to time. The concentration values are derived empirically. They vary from pollutant to pollutant and specific values may differ from jurisdiction to jurisdiction, but they are usually expressed as an average over a specific time period. The **WHO** guidelines for NO_x, for example, are a mean of 400 $\mu g/m^3$ for a one-hour period and a mean of 150 $\mu g/m^3$ over a 24-hour period. The European Community (EC) standards for NO_x are expressed somewhat differently, requiring that 98 per cent of the mean hourly values over one year do not exceed 200 $\mu g/m^3$. Standards are also expressed in parts per million (ppm) or parts per billion (ppb). In Ontario, Canada, for example, air quality is considered poor if ground-level ozone values exceed 80 ppb. Since the standards are driven by the effects of the pollutants on people and other elements in the environment, they undergo regular revision as knowledge of their impact increases. Once the standards are established, it is important to take the necessary steps to monitor and reduce pollution levels so that they can be maintained.

Further reading
Smith, Z.A. (1995) *The Environmental Policy Paradox*, Englewood Cliffs, NJ: Prentice-Hall.

AIRSHED

A concept developed by analogy with 'watershed' to represent the catchment area for pollutants around known emission sources. By comparing the output of pollutants with the expected air movement and rates of chemical change, it is possible to arrive at a crude estimation of the likely pollution concentrations within the boundaries of the air shed.

ALBEDO

A measure of the reflectivity of a body or surface. It is the total **radiation** reflected by a body divided by the total incident radiation, expressed as a fraction or a percentage. The average albedo of the earth measured at the outer edge of the **atmosphere** – the planetary albedo – is about 30 per cent, but the presence or absence of **clouds** (some of which have an albedo of 90 per cent) will cause significant differences in that value. Surface values also vary considerably. Newly fallen **snow** reflects most of the **solar radiation** falling on it and has a high albedo – between 75 and 95 per cent – whereas a black asphalt road surface, which readily absorbs radiation, has a low albedo – less than 10 per cent. The albedo of vegetation varies between 10 and 25 per cent depending upon the nature of the vegetation and the season of the year. Changes in the earth's reflectivity, brought about by variations in such elements as cloud, snow cover and the distribution of vegetation, may contribute to climate variability through their disruption of the earth's **energy budget**.

See also
Atmospheric turbidity.

ALCOHOLS

A group of organic compounds derived from **hydrocarbons** through the replacement of a **hydrogen atom** in the **hydrocarbon molecule** by a hydroxyl group. For example, **methane** (CH$_4$) becomes **methanol** or methyl alcohol (CH$_3$OH).

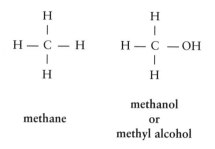

methane   methanol or methyl alcohol

Methanol is the simplest of the alcohols and is sometimes called 'wood alcohol' since it can be produced through the heating and **anaerobic** decomposition of wood. One of the most common alcohols is **ethanol** or ethyl alcohol (C$_2$H$_5$OH), produced by the **fermentation** of natural **sugars** in grain and fruit, and is the base for many alcoholic beverages. Both methanol and ethanol are chemically relatively simple alcohols. Greater complexity is provided in other examples through the extension and branching of the basic **carbon** (C) chain and the addition of hydroxyl groups.

Alcohols are widely used as **solvents** in the food and beverage, pharmaceutical, printing and chemical industries. Both methanol and ethanol also have some potential as **fuels**. Combined with **gasoline** to form **gasohol** (10 to 20 per cent alcohol by volume), they help to improve fuel economy and reduce automobile exhaust emissions. In the 1970s, ambitious schemes to reduce **petroleum** imports by producing alcohol from grain were set up in the United States. A similar system using sugar cane met with some success in Brazil, but as long as gasoline remains relatively cheap and plentiful, gasohol is unlikely to become a major fuel.

Since alcohol can be obtained from a wide range of products, including urban and agricultural waste, its large-scale production also has the potential to contribute to the solution of **waste disposal** problems.

See also
Hydrogen oxides.

Further reading
Zumdahl, S.S. (1993) *Chemistry* (3rd edition), Lexington, MA: D.C. Heath.

ALDEHYDES

Organic compounds formed through the **oxidation** of primary **alcohols** such as **methanol** and **ethanol**, which are converted into formaldehyde and acetaldehyde respectively. Further oxidation converts them into **acids**.

$$\text{methanol} \xrightarrow{\text{oxidation}} \text{formaldehyde} \xrightarrow{\text{oxidation}} \text{formic acid}$$
$$\text{ethanol} \xrightarrow{\text{oxidation}} \text{acetaldehyde} \xrightarrow{\text{oxidation}} \text{acetic acid}$$

Aldehydes are used in the production of resins, dyes and pharmaceuticals. As by-products of **combustion**, they can also be air pollutants, giving off unpleasant **odours** and irritating eyes and noses.

ALGAE

A group of simple plants which inhabit moist **environments**. Most are aquatic, but they are also found in damp locations on land – they give damp walls their characteristic green coloration, for example. Since they contain **chlorophyll** they are capable of **photosynthesis**.

See also
Algal bloom, Eutrophication.

ALGAL BLOOM

The rapid growth in the number of **algae** in an environment in which nutrient enrichment or **eutrophication** has taken place. Eutrophication supplies the abundant chemical energy necessary for rapid reproduction. Algal blooms occur naturally in spring and early summer, when the rate of reproduction of the algae outstrips that of their consumers. However, the most serious algal blooms are associated with human activities. **Phosphates** and **nitrates**, carried into waterways in **sewage**, agricultural **fertilizers** and **detergents**, provide the nutrients that cause explosive growth in the algae population. In 1993/94, for example, more than 1000 km of the River Darling in Australia were affected. The algae discolour the water in which they are living, but the biggest problems arise when they die. Masses of dead and putrifying algae – more than a million tonnes a year are dredged from the lagoon at Venice in Italy – pollute beaches, release obnoxious odours, poison the water and create such a **biochemical oxygen demand** (BOD) that other aquatic organisms die.

See also
Red tides.

Further reading
Pearce, F. (1995) 'Dead in the water', *New Scientist* 145 (1963): 26–31.

ALKALI

A compound, usually a soluble **hydroxide** of a **metal**, such as **sodium hydroxide** (NaOH), which on **solution** with **water** produces an excess of hydroxyl **ions** (OH⁻). Alkalis neutralize **acids** to form **salts** and release water formed when **hydrogen** (H⁺) and hydroxyl ions (OH⁻) combine. Acidified lakes may be neutralized, for example, by the addition of calcium hydroxide (Ca(OH)₂).

$$H_2SO_4 \; + \; Ca(OH)_2 \; \rightarrow \; CaSO_4 \; + \; 2H_2O$$

sulphuric	calcium	calcium	water
acid	hydroxide	sulphate	

See also
Acid rain, Base.

ALLIANCE OF SMALL ISLAND STATES (AOSIS)

A group of states, occupying low-lying islands mainly in the South Pacific and Indian Oceans and in the Caribbean, that consider themselves most likely to suffer the consequences of the rising sea-level and increased storminess expected to accompany **global warming**. Islands such as the Maldives in the Indian Ocean, with a maximum elevation of 6 m and the Marshall Islands in the west central Pacific, with an average altitude of only 3 m above sea-level, would face serious problems of flooding and coastal erosion, and might well become uninhabitable.

ALLOTROPY

The existence of elements in several different physical forms, but with the same chemical properties. Graphite and diamond, for example, are allotropic forms of **carbon** (C), and **ozone** (O₃) is an allotrope of **oxygen**.

ALLOY

A combination of two or more **metals**, usually to obtain properties not present in the individual constituents. For example, bronze, an alloy of **copper** (Cu) and tin (Sn), and one of the first alloys to be produced, was used for tools and weapons because it was hard enough to carry an edge, a property not available from tin or copper. Alloys are commonly grouped into those containing **iron** (Fe) as the elemental metal (ferrous alloys) or those containing a metal other than iron (nonferrous alloys). The term alloy is also used for combinations of metals and non-metals. Steel, for example, may be considered an alloy of iron and **carbon** (C). Modern industry is highly dependent upon alloys, particularly those which include steel. The addition of such elements as chromium (Cr), tungsten (W) and **nickel** (Ni) to carbon steel has improved a variety of its qualities including **corrosion** resistance, hardness and flexibility.

ALLUVIUM

Sediment deposited by flowing **water** in stream and river beds, or on adjacent land when rivers overflow their banks. Alluvial sediments are commonly considered to include fine-grained **clay**, **silt** and **sand**, but grain size can vary. For example, alluvial fans often include coarse gravel and cobbles, as a result of the sudden change in **velocity** of the stream and the consequent rapid deposition when the stream emerges from a mountain valley on to the adjacent plain. Fine-grained alluvium is common in the lower reaches of rivers, which, together with the **nutrients** incorporated from the entire drainage basin, produces easily worked, fertile and productive **soils**. Although alluvium is a natural product of the erosion/transporatation/deposition cycle in the environment, human activity may provide additional material or increase the rate at which it becomes available. **Deforestation** or the mismanagement of agricultural

Figure A-11 An alluvial fan formed where a flowing stream deposits sediments into the standing waters of a lake

Photograph: The author

land, for example, can lead to accelerated **soil erosion**, which increases the availability of alluvium and disrupts the environment through such processes as siltation, flooding and the alteration of aquatic ecosystems.

Further reading
Miall, A.D. (1996) *The Geology of Fluvial Deposits: Sedimentary Facies, Basin Analysis and Petroleum Geology*, Berlin/New York: Springer-Verlag.
Ritter, D.F., Kochel, R.C. and Miller, J.R. (1995) *Process Geomorphology* (3rd edition), Dubuque, IA: Wm C. Brown.

ALUMINUM/ALUMINIUM (AL):

A light grey or white **metal** which is the third most abundant **element** in the earth's crust. Its main **ore** is bauxite, from which it is extracted by **electrolysis**. The process is extremely energy intensive and requires tight pollution controls. Being light, strong and

corrosion resistant, aluminum and its **alloys** are used in aircraft construction and for other uses where a high strength/weight ratio is desirable. It is also a good conductor of electricity, replacing **copper** (Cu) for some purposes in the electrical industry. In areas suffering from the effects of **acid rain**, high aluminum levels in lakes contribute to fish kills.

Further reading
Cronan, C.S. and Schofield, C.L. (1979) 'Aluminum leaching response to acid precipitation: effects on high elevation watersheds in the north-east', *Science* 204: 304–6.
Hatch, J.A. (ed.) (1984) *Aluminum: Properties and Physical Metallurgy*, Metals Park, OH: American Society for Metals.

AMINO ACIDS

Organic compounds which combine together in chains to form **proteins**, the basic building blocks of living matter. Structurally, amino acids include a carboxyl group (–COOH) and an amino group (–NH$_2$). The linking of the carboxyl group of one **acid** with the amino group of another allows the buildup of the long chains characteristic of proteins. There are some twenty common amino acids which are considered essential for human life. Twelve of these can be synthesized by the human body, but the remaining eight – the so-called essential amino acids – must be obtained through the consumption of plants and animals in the **environment**. There are a few complete proteins, such as human milk, which contain all the essential amino acids, but a varied diet is usually necessary to supply the amino acids required for normal bodily health and development. During **digestion**, the proteins are broken down into their constituent amino acids, of which some are used to synthesize new proteins, some are used to supply **energy** and some are excreted unused.

Further reading
Expert Advisory Committee on Amino Acids (Canada) (1990) *Report of the Expert Advisory Committee on Amino Acids*, Ottawa: Health and Welfare Canada.

ANABOLISM

That part of **metabolism** that involves the buildup of complex substances from simpler materials, often produced during **catabolism**. Through anabolism, animals are able to grow, reproduce, repair damaged tissue and store **energy**.

See also
Amino acids, Proteins.

ANAEROBIC

Living or active only in the absence of free **oxygen** (O). Organisms which do not require access to oxygen to produce **energy** during **respiration** are anaerobic organisms.

See also
Aerobic, Anaerobic decay.

ANAEROBIC DECAY

The breakdown of organic material in the absence of **oxygen** (O). Brought about by anaerobic **bacteria**, the process commonly causes the production of **methane** (CH$_4$), a greenhouse gas. **Oxides of nitrogen** (NO$_x$) and ammonium may be released during the anaerobic decay of material containing **nitrates,** and **hydrogen sulphide** (H$_2$S) may be released from materials containing **sulphates**. Anaerobic processes are used in the processing of **sewage** effluent, with the methane produced being used as a **fuel**.

See also
Synfuels.

ANALOGUE

A **system** or situation with properties equivalent to or closely resembling the properties of some other system or situation. In environmental studies, analogue models are often developed to simulate the workings of earth/atmosphere systems, allowing predictions of environmental change to be made more rapidly and without the limitations imposed by the **natural environment**. The impact of **pollution** on a water body, for example, can be studied using an analogue model without subjecting the water body to real pollution.

ANALOGUE CLIMATE MODELS

Models developed to predict future changes in an existing situation using analogues from the historical record. The documented development of the historical situation is taken as an indication of possible changes in the existing situation. Prior to the development of computerized prediction, the analogue method was commonly used in weather forecasting. The accuracy of such forecasting is limited by the impossibility of finding an exact comparison of the present with the past.

Further reading
Critchfield, H.J. (1983) *General Climatology* (4th edition), Englewood Cliffs, NJ: Prentice-Hall.

ANHYDRIDE

A **compound** formed throught the **dehydration** of an **acid** or a **base**. A basic anhydride is the oxide of a **metal** – calcium oxide (CaO) is the anhydride of calcium hydroxide ($Ca(OH)_2$) – whereas an acid anhydride is the oxide of a non-metal – sulphur trioxide (SO_3) is the anhydride of sulphuric acid (H_2SO_4). The addition of **water** to the anhydride reverses the process as follows:

basic anhydride

CaO	+	H_2O	$\leftrightarrow$	$Ca(OH)_2$
calcium oxide		water		calcium hydroxide

acid anhydride

SO_3	+	H_2O	$\leftrightarrow$	H_2SO_4
sodium trioxide		water		sulphuric acid

ANIMAL COMMUNITY

A group of animals occupying a common environment. They vary in number and type according to the particular elements in the environment – a tropical animal community will differ from one in the high Arctic, for example – but the various members of the community interact with each other, as well as with the plants and micro-organisms that are present in the environment, and as such the animal community is an integral part of any **ecosystem**.

ANION

A negatively charged **ion**. During electrolysis, anions are attracted to the positively charged **anode**.

ANODE

A positive **electrode** or pole. Being positively charged, it attracts negative **ions** or **anions** during **electrolysis**. In a primary cell or battery, it is the electrode that carries the positive charge.

See also
Cathode.

ANTARCTIC OZONE HOLE

Despite the name, not a true 'hole', but rather

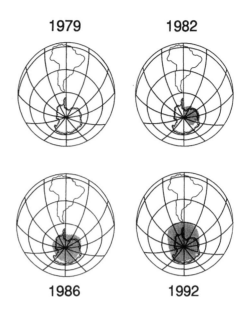

Figure A-12 The enlargement of the ozone hole over the Antarctic between 1979 and 1992

Source: Environment Canada (1993) *A Primer on Ozone Depletion,* Ottawa: Environment Canada

an intense thinning of the stratospheric **ozone** (O_3) layer above Antarctica. Seasonal thinning of the ozone above the Antarctic during the southern spring was long considered part of the normal variability of the **atmosphere** in high southern latitudes. The marked increase in the intensity of the thinning was first reported in the early 1980s by scientists of the British Antarctic Survey at Halley Bay. The hole commonly became evident in late August and intensified into mid- and late October, before beginning to fill again in November. By the mid-1980s, however, thinning was persisting into December. Record levels of ozone destruction above the Antarctic were recorded in the first half of the 1990s and the geographical extent of the hole increased, allowing the thinning to reach the more southerly parts of Australia, New Zealand and South America. Bans on the production and use of **CFCs** and other ozone-destroying chemicals should help the Antarctic **stratosphere** to recover, but since their rate of removal from the atmosphere is low, the

Figure A-13 A diagrammatic representation of the sources and types of anthropogenic aerosols

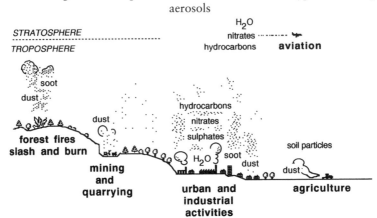

annual enlargement of the Antarctic ozone hole is likely to persist, at least in the short term.

See also
Arctic ozone hole, Heterogeneous chemical reactions, Montreal Protocol, Ozone depletion, Ozone Protection Act.

Further reading
Farman, J.C., Gardiner, B.G. and Shanklin, J.D. (1985) 'Large losses of total ozone in Antarctica reveal seasonal ClO_x/NO_x interaction', *Nature* 315: 207–10.
Gribbin, J. (1993) *The Hole in the Sky*, New York: Bantam.
Johnson, B.J., Deshler, T. and Zhao, R. (1995) 'Ozone profiles at McMurdo Station, Antarctica, during the spring of 1993: record low ozone season', *Geophysical Research Letters* 22: 183–6.

ANTARCTIC TREATY

A treaty that came into effect in 1961, signed by the sixteen nations involved in Antarctic research at that time. It confirmed the use of Antarctica for peaceful purposes only, and contained provision for the protection of the **flora** and **fauna** in the Antarctic **environment**. The Treaty was subsequently reinforced by agreements on the **conservation** of seals and other marine organisms in the area.

Further reading
Herr, R.A., Hall, H.R. and Haward, M.E. (eds) (1990) *Antarctica's Future: Continuity or Change*, Hobart, Tasmania: Tasmanian Government Printing Office/Australian Institute of International Affairs.

ANTHROPOGENIC AEROSOL SOURCES

Sources of **aerosols** released into the **atmosphere** by human activities, such as mining and quarrying, agriculture and industrial processes.

ANTIBIOTICS

Chemicals produced by living micro-organisms such as **bacteria** and moulds, which are able to destroy or inhibit the growth of other micro-organisms. Their development allowed the first effective control of diseases caused by **bacteria** and **fungi**. Although mouldy substances have been used for centuries in traditional medicine to treat infection, the modern development of antibiotics began with the discovery of penicillin in 1928. Since then, numerous antibiotics have been developed. Many continue to be derived from moulds and bacteria, but some are now produced synthetically, and others combine both forms of production. Antibiotics are used extensively to combat infection in humans, animals and plants, but such large-scale use has also allowed the development of resistant bacteria, no longer sensitive to the antibiotics. New or improved antibiotic strains are regularly required to overcome such resistance.

Further reading
Coghlan, A. (1996) 'Animal antibiotics threaten hospital epidemics', *New Scientist* 151 (2040): 4.

ANTIBODY

A protein produced by an animal in automatic response to the presence of a foreign substance (antigen) in the body. The presence of **bacteria** or a **virus** in the blood, for example, will promote the production of specific antibodies which kill or neutralize such antigens, by combining with them to form less harmful substances which are eventually expelled from the body. The process provides long-term immunity to specific antigens since the **cells** that produce the antibodies remain active in the bloodstream. This is the basis of disease control by inoculation, in which limited numbers of antigens are injected into the body to stimulate the production of antibodies and provide protection against future infection. Although antibodies are primarily beneficial, some may lead to secondary allergic reactions such as those associated with hay fever or asthma. There is evidence that the incidence of such allergic reactions is increasing along with the rising level of **environmental pollution**.

ANTICYCLONE

A zone of high **atmospheric pressure** created by the cooling of air close to the earth's surface (cold anticyclone) or the sinking of air from higher levels in the **atmosphere** (warm anticyclone). Cold anticyclones are shallow domes of cold, dense air caused by radiative cooling. They are a common feature of the atmospheric circulation over North America and Asia during the winter when radiative cooling over the snow-covered landscape is strong. Warm anticyclones are semi-permanent features of sub-tropical regions where the air from the tropical **Hadley Cells** sinks towards the surface. Adiabatic warming causes the **temperature** of the subsiding air to rise and its **relative humidity** to decrease. The combination of a steady flow of air from the upper atmosphere, warming and drying as it descends, inhibits **precipitation**, and the world's major sub-tropical **deserts** are located beneath warm anticyclones. The surface circulation of air in an anticyclone is clockwise in the northern hemisphere and counter-clockwise in the southern hemisphere, but winds are generally light and variable towards the centre. As a result, mixing in the lower atmosphere tends to be poor and pollutants readily accumulate. Warm anticyclonic conditions contribute to **smog** in Los Angeles, for example, whereas a persistent cold anticyclone had a major role in the **London Smog** of 1952.

See also
Adiabatic process.

Figure A-14 The formation of warm and cold anticyclones

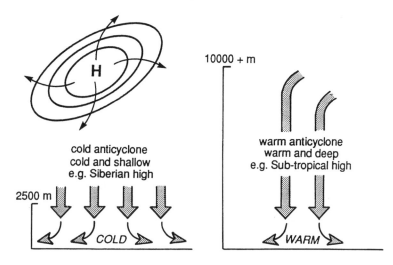

Plan view: northern hemisphere circulation

H

cold anticyclone
cold and shallow
e.g. Siberian high

2500 m

COLD

10000 + m

warm anticyclone
warm and deep
e.g. Sub-tropical high

WARM

Further reading
Ahrens, C.D. (1993) *Essentials of Meteorology*, St Paul, MN: West Publishing.

APPROPRIATE TECHNOLOGY

Technology designed to meet the needs of a group of people at a specific time in their development without imposing unmanageable stress on **resources** and the **environment**. The concept applies particularly to developing nations, where environmentally compatible devices – hand tools, **methane** (CH_4) digesters, **wind** and water-powered machinery – using locally available resources are encouraged. The term has also been used in developed nations to describe technology that is relatively simple, locally adaptable, resource-efficient and environmentally friendly. Modern **recycling** technology would fit that category, for example.

Further reading
Barbour, I.G. (1993) *Ethics in an Age of Technology*, San Francisco, CA: Harper.
Getis. J. (1991) *You Can Make a Difference: Help Protect the Earth*, Dubuque, IA: Wm C. Brown.

AQUATIC BIOTA

Organisms living in the **hydrosphere** – the water-based component of the **biosphere**.

AQUATIC ENVIRONMENT

That part of the **environment** that includes the **water** resting on or flowing over the earth's surface and the plants and animals that live in it. The quality of the aquatic environment is being threatened by pollutants originating in human and industrial **waste**. The **aquatic biota** are harmed by such **pollution**, and their numbers are further threatened by **overfishing** and excessive hunting.

See also
Atmospheric environment, Terrestrial environment.

AQUIFER

A layer of rock beneath the earth's surface sufficiently porous and **permeable** to store significant quantities of **water**. Aquifers provide water for human settlement in those areas where surface water is absent or inadequate. Any water loss from an aquifer is made up through **precipitation**, but in most of the world's major aquifers, such as those in the Ogallala formation in the United States and the Great Artesian Basin of Australia, the **groundwater** supply is declining because society is able to remove the water faster than it can be replenished. In parts of Kansas, Oklahoma, New Mexico and Texas, served by the Ogallala aquifer, water is being withdrawn at a rate twenty times faster than it is being replenished, and water levels have declined by as much as 30–60 m since the 1940s. The removal of groundwater without **recharge** can also lead to land subsidence, as in Mexico City, where depletion of the underlying aquifers has allowed the land to sink, and caused structural damage to buildings and roads.

See also
Artesian well, Permeability, Porosity, Recharge.

Further reading
Owen, O.S. and Chiras, D.D. (1995) *Natural Resource Conservation: Management for a Sustainable Future*, Englewood Cliffs, NJ: Prentice-Hall.

Figure A-15 The characteristics of a simple aquifer

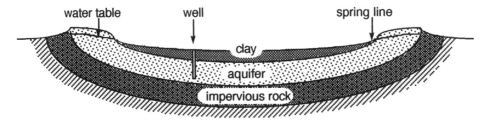

ARAL SEA

Once the fourth largest inland lake in the world, covering 64,000 sq km, but now much depleted, the Aral Sea occupies a drainage basin in arid Central Asia, on the border of Kazakhstan and Uzbekistan. In the 1960s, when the area was still part of the Soviet Union, major **irrigation** schemes were developed to increase cotton production. This involved diverting most of the **water** entering the sea through the Amu Dar'ya and Syr Dar'ya river systems. Over the thirty years since then, the declining amount of water entering the Aral Sea has caused it to shrink by 30 per cent and created what is considered by many environmentalists to be one of the world's greatest environmental disasters. A fishery that once provided large catches of carp, perch and sturgeon has ended because the waters are now too salty and shallow. **Salts**, along with toxic **herbicides** and **pesticides**, blow off the exposed sea bed to contaminate the surrounding land, and cause respiratory and other health problems for the local inhabitants. Proposed rehabilitation schemes include reduction of the amount of water being diverted from the rivers flowing into the basin or diversion of water from rivers such as the Irtysh that flow north to the Arctic. Given the economic climate in the states of the former Soviet Union, neither approach seems likely to be implemented in the near future.

Further reading
Brown, L.R. (1991) 'The Aral Sea: going, going . . .', *World-Watch* 4 (1): 20–7.

ARCTIC HAZE

The **pollution** of the Arctic atmosphere, mainly in winter, by **aerosols** such as **dust**, **soot** and **sulphate particles** originating in Eurasia. The haze is most pronounced between December and March for several reasons, including the increased emission of pollutants at that season, the more rapid and efficient poleward transport in winter and the longer residence time of haze particles in the highly stable Arctic air at that time of year. Arctic air

Figure A-16 The shrinking Aral Sea

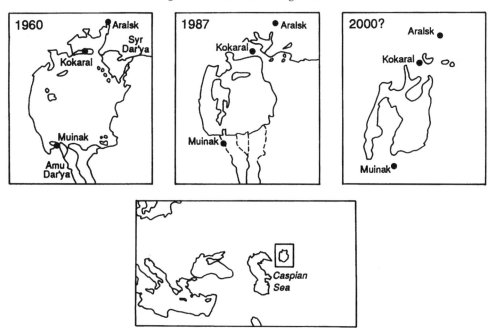

Source: Kaufman, D.G. and Franz, C. (1993) *Biosphere 2000: Protecting our Global Environment*, New York: HarperCollins

28

Figure A-17 An estimate of the mean vertical profile of the concentration of anthropogenic aerosol mass in the high Arctic during March and April. (C(H)/C(O) is the concentration at a specific altitude divided by the concentration at the surface.)

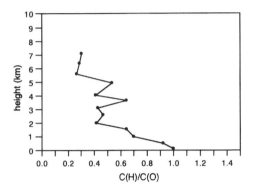

Source: After Barrie, L.A. (1986) 'Arctic air pollution: an overview of current knowledge', *Atmospheric Environment* 20: 643–63

pollution has increased since the mid-1950s in parallel with increased aerosol emissions in Europe, and the net result has been a measurable reduction in visibility and perturbation of the regional radiation budget.

Further reading
Barrie, L.A. (1986) 'Arctic air pollution: an overview of current knowledge', *Atmospheric Environment* 20: 643–63.
Shaw, G.E. (1980) '*Arctic Haze*', Weatherwise 33: 219–21.

ARCTIC OZONE HOLE

Following the discovery of the major thinning of the **ozone layer** over the Antarctic, scientists began to examine the possibilty of a similar development over the Arctic. The European Arctic Stratospheric Ozone Experiment (EASOE) was established in the northern winter of 1991–1992 to establish the nature and extent of ozone depletion over the Arctic. Preliminary results indicated the absence of a distinct hole over the Arctic, but decreases of 10–20 per cent in Arctic ozone were detected. Because the Arctic stratosphere is generally warmer than its Antarctic counterpart, polar stratospheric clouds are less ready to form, and ozone destruction is less efficient. The less developed Arctic circumpolar vortex also allows the loss of ozone at the pole to be offset to some extent by the influx of ozone from more southerly latitudes. Thus, it seemed unlikely that a distinct hole would develop over the Arctic. Into the mid-1990s, however, the Arctic circumpolar vortex has become stronger and more persistent, and stratospheric ozone concentrations over the Arctic and adjacent high latitudes in Europe and North America have been as much as 20–40% below the 1979–1986 spring averages.

Further reading
Pyle, J. (1991) 'Closing in on Arctic ozone', *New Scientist* 132 (1794): 49–52.

AREA OF OUTSTANDING NATURAL BEAUTY (AONB)

An area in England and Wales which has been recognized by the Countryside Commission as worthy of preservation because of its natural beauty. Such areas are outside national parks, and the AONB designation allows local authorities to pass specific legislation to protect them.

ARIDITY

Figure A-18 Distribution of arid lands by continent

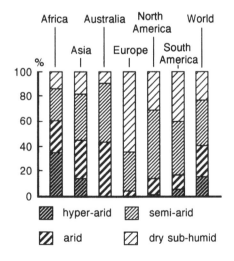
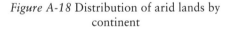

Source: UN Web Page

Permanent dryness caused by low average rainfall, often (but not always) in combination with high **temperatures**. The **deserts** of the world are permanently arid, with rainfall amounts of less than 100 mm per year, and **evapotranspiration** rates well in excess of that amount.

See also
Drought.

Further reading
Beaumont, P. (1993) *Drylands: Environmental Management and Development*, London: Routledge.

ARRHENIUS, S.

A Swedish chemist usually credited with being the first to recognize that an increase in atmospheric carbon dioxide (CO_2) would lead to **global warming**. He published his findings in the late nineteenth and early twentieth centuries at a time when the environmental implications of the **Industrial Revolution** were just beginning to be appreciated. Little attention was paid to the potential impact of increased levels of carbon dioxide on **climate** for some time after that, and the carbon dioxide-induced **temperature** increases estimated by Arrhenius in 1903 were not bettered until the early 1960s.

Further reading
Bolin, B. (1972) 'Atmospheric chemistry and environmental pollution', in D.P. McIntyre (ed.) *Meteorological Challenges: A History*, Ottawa: Information Canada.

ARSENIC

A highly toxic element which exists in three allotropic forms – grey, black and yellow arsenic. It occurs naturally in the **environment,** being released from arsenic-bearing rocks through **weathering**. Human activities such as **coal** burning and the refining of sulphide-rich minerals add arsenic to the atmosphere. Arsenic is used in the production of **alloys** and pigments and in semi-conductors. Its toxicity has allowed it to be used against infection in medicine and as a **pesticide** against a variety of organisms from insects to rats. Modern synthetic chemicals have replaced arsenic for many pesticide uses, but it remains

a component of some **herbicides**. Arsenic accumulates in the environment so that small doses, relatively harmless individually, may eventually kill organisms. This has been the basis of fictional and non-fictional murder plots, but consumption of **water** containing as little as 50 μg per litre of arsenic will cause sickness and perhaps death for some organisms. Accumulation of arsenic in **soil** can also poison plants and soil organisms.

See also
Allotropy.

ASBESTOS

The name given to a group of fibrous silicate minerals which are resistant to heat, fire and chemicals, and which provide electrical and thermal **insulation**. Chrysolite, a fibrous form of serpentine mined mainly in Quebec, Canada is the most common type of asbestos. Processed into textiles or combined with other materials, asbestos has been used in a great variety of products, including automobile brake pads, roofing and flooring tiles, protective clothing for fire-fighters, pipe and boiler **insulation**. Despite its many advantages, the use of asbestos is now banned or strictly regulated in many countries, because it is considered to be a major health hazard. The inhalation of asbestos fibres can lead to asbestosis – chronic lung damage characterized by breathlessness, coughing and chest pains – as well as **cancer** of the lung and chest cavity. Health hazards associated with exposure to asbestos have been recognized since at least the 1930s, but it is only since the 1960s and 1970s that its production and use have been strictly controlled. Thousands of products containing asbestos remain in use, and many buildings retain sound-proofing and fire-proofing asbestos in their walls and ceilings. Uncontrolled disposal of asbestos products and the demolition of buildings continues to release asbestos fibres into the atmosphere, sometimes in large quantities. An exceptional situation followed the Kobe **earthquake** in Japan in 1995, when demolition of unsafe buildings and the clean-up of debris led to short-term atmospheric asbestos levels some twenty-five times the national average.

Further reading
Baarschers, W.H. (1996) *Eco-facts and Eco-fiction*, London/New York: Routledge.

Selikoff, I. J. and Lee, D.H.K. (1978) *Asbestos and Disease*, New York: Academic Press.

ATMOSPHERE

A thick blanket of gases, containing suspended **liquid** and **solid** particles, that completely envelops the earth, and together with the earth forms an integrated environmental **system**. As part of the system, it performs several functions that have allowed humankind to survive and develop almost anywhere on the earth's surface. First, the atmosphere provides and maintains the supply of **oxygen** (O) required for life itself. Second, it controls the earth's **energy budget** through such features as the **ozone layer** and the **greenhouse effect**, and – by means of its internal circulation – distributes heat and moisture across the earth's surface. Third, it has the capacity to dispose of waste material or pollutants generated by natural or human activity. Society has interfered with all of these elements, and, through ignorance of the mechanisms involved or lack of concern for the consequences of its actions, has created or intensified problems that are now causing concern on a global scale.

The constituents of the atmosphere are collectively referred to as **air**, a mixture of individual gases, **water** and **aerosols**.

Accounting for more than 99 per cent of the volume of the gaseous atmosphere, **nitrogen** (N) and **oxygen** (O) are the major gases. Some of the remaining so-called minor gases have an importance far beyond their volume. **Carbon dioxide** (CO_2) and **methane** (CH_4) are important **greenhouse gases**, for example. From time to time, other gases such as **sulphur dioxide** (SO_2), oxides of nitrogen (NO_x), hydrogen sulphide (H_2S) and carbon monoxide (CO), along with a variety of more exotic **hydrocarbons**, may become constituents of the atmosphere. Even in small quantities they can be harmful to the **environment**, and although they may be produced naturally, their presence is increasingly associated with **pollution** from industrial or vehicular sources.

Lists of atmospheric gases normally refer to dry air, but the atmosphere is never completely dry. The proportion of **water vapour** in the atmosphere in the humid tropics may be as much as 4 per cent and even above the world's driest **deserts** there is **water** present, if only in fractional amounts. Water is unique among the constituents of the atmosphere in that it is capable of existing as solid, liquid or **gas**, and of changing readily from one state to another.

In addition to the gaseous components of the atmosphere and the water in its various forms, there are also solid or liquid particles dispersed in the **air**. These are called aerosols, and include **dust**, **soot**, **salt** crystals, **spores**, **bacteria**, **viruses** and a variety of other microscopic particles. Collectively, they are often regarded as equivalent to **air pollution**, although many of the materials are produced

Table A-3 The average gaseous composition of ambient air

GAS	% BY VOLUME	PARTS PER MILLION
Nitrogen	78.08	780,840.00
Oxygen	20.95	209,500.00
Argon	0.93	9,300.00
Carbon dioxide	0.0345	345.00
Neon	0.0018	18.00
Helium	0.00052	5.20
Methane	0.00014	1.40
Krypton	0.00010	1.00
Hydrogen	0.00005	0.50
Xenon	0.000009	0.09
Ozone	Variable	Variable

naturally by volcanic activity, forest and grass fires, **evaporation**, local atmospheric turbulence and biological processes.

See also
Atmospheric circulation, Atmospheric layers, Atmospheric turbidity.

Further reading
Ahrens, C.D. (1994) *Meteorology Today*, St Paul, MN: West Publishing.
Moran, J.M. and Morgan, M.D. (1997) *Meteorology: The Atmosphere and Science of Weather*, Englewood Cliffs, NJ: Prentice-Hall.

ATMOSPHERIC CIRCULATION

The large-scale movement of **air** around and above the earth, associated with complex but distinct patterns of pressure systems and wind belts. It is driven by latitudinal variations in the earth's **energy budget**, and through the redistribution of **energy** helps to reduce these variations. George **Hadley** first proposed the classic model of the general circulation of the atmosphere in 1735 as a simple convective system, based on the concept of a non-rotating earth with a uniform surface, that was warm at the equator and cold at the poles. Warm buoyant air rising at the equator spread north and south in the upper **atmosphere**, eventually returning to the surface in high or polar latitudes. From there, it flowed back across the surface towards the equator to close the circulation.

Hadley's original model, with its single **convection** cell in each hemisphere, was eventually replaced by a three-cell model as technology advanced and additional information became available, but his contribution was recognized in the naming of the tropical cell. The three-cell model continued to assume a uniform surface, but the rotation of the earth was introduced, and with it the **Coriolis effect**, which causes moving objects to swing to the right in the northern hemisphere and to the left in the southern. Thus, the winds became westerly or easterly in this new model, rather than blowing north or south as in the one-cell version. The three cells and the Coriolis effect, in combination, produced alternating bands of high and low pressure, separated by wind belts which were easterly in equatorial and polar regions and westerly in mid-latitudes. Although only theoretical, elements of this model can be recognized in existing global wind and pressure patterns, particularly in the

southern hemisphere, where the greater expanse of **ocean** more closely resembles the uniform surface of the model.

In the late 1940s and 1950s, as knowledge of the atmosphere improved, it became increasingly evident that the three-cell model oversimplified the general circulation. The main problem arose with the mid-latitude cell. Observations indicated that most energy transfer in mid-latitudes was accomplished by horizontal cells – such as travelling high and low pressure systems – rather than the vertical cell indicated by the model. Modern interpretations of the general circulation of the **atmosphere** retain the tropical **Hadley cell**, but horizontal eddies have come to dominate mid-latitudes, and have even replaced the simple thermal cell of polar latitudes.

Conditions in the upper atmosphere are an integral part of modern studies of the atmospheric circulation. The upper atmospheric circulation is quite complex in detail, but in

Figure A-19 Simple convective circulation on a uniform, non-rotating earth, heated at the equator and cooled at the poles

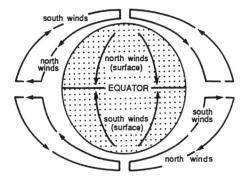

Source: After Kemp, D.D. (1994) *Global Environmental Issues: A Climatological Approach*, London/New York: Routledge

Figure A-20 Global atmospheric circulation patterns for January and July

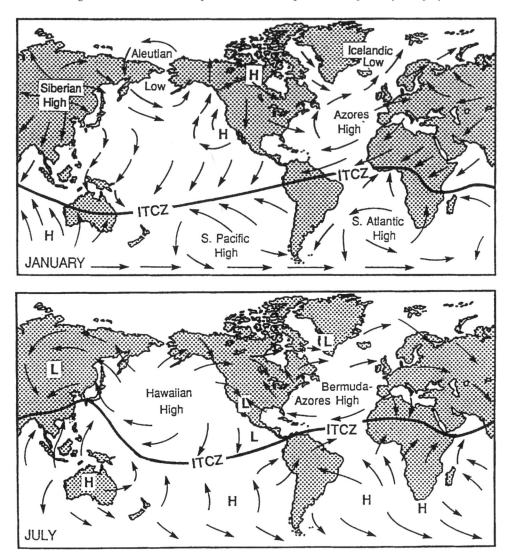

Source: After Kemp (1994)

general terms it is characterized by an easterly flow in the tropics and a westerly flow in mid- to high latitudes. The upper westerlies include a pattern of waves, called **Rossby waves**, which vary in amplitude in a quasi-regular sequence represented by the **index cycle**. During the run of any one cycle there is significant latitudinal energy transfer. Within these broad airflows, at the **tropopause**, there are relatively narrow bands of rapidly moving air called **jet streams**, in which wind

speeds may average 125–130 km per hour, although much higher speeds may occur.

Modern representations of the general circulation of the atmosphere take into account the non-uniform nature of the earth's surface, with its mixture of land and **water**, and include consideration of seasonal variations in energy flow. Differences in their physical properties ensure that land and sea warm up and cool down at different rates. This creates significant **temperature** differences between land and water, and produces a series of pressure cells rather than the simple belts of the original models. By altering the regional airflow, such pressure differences cause disruption of the theoretical wind patterns. The changing location of the zone of maximum **insolation** with the seasons also causes variations in the location, extent and intensity of the pressure cells and wind belts. Although such changes are repeated year after year they are not completely reliable, and this adds an additional element of variability to the representation of the atmospheric circulation.

See also
Atmospheric models, Geostrophic wind.

Further reading
Barry, R.G. and Chorley, R.J. (1992) *Atmosphere, Weather and Climate* (6th edition), London/New York: Routledge.
Lorenz, E.N. (1967) *The Nature and Theory of the General Circulation of the Atmosphere*, Geneva: WHO.

ATMOSPHERIC ENVIRONMENT

The gaseous envelope that surrounds the earth, which includes not only gases, but also a variety of **liquids** and **solids**. It also incorporates the various processes that create weather conditions. The atmospheric environment is being threatened at all scales, from local to global, by the addition of visible pollutants such as **dust, soot** and **smoke** as well as by invisible gases such as **carbon dioxide** (CO_2) and **sulphur dioxide** (SO_2). Many current environmental issues – **acid rain, ozone depletion, global warming** – are associated with the disruption of the atmospheric environment.

See also
Aquatic environment, Atmosphere, Terrestrial environment.

ATMOSPHERIC LAYERS

Although its gaseous components are quite evenly mixed, the **atmosphere** is not physically uniform throughout. Differences in **temperature** and **pressure** provide it with form and structure, for example, and the most

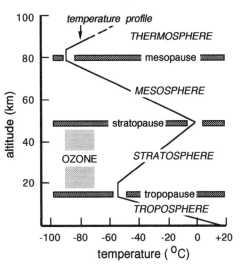

Figure A-21 A vertical profile of the atmosphere

common differentiation of the atmosphere into a series of layers is based on temperature. The lowest layer is the **troposphere**, which ranges in thickness from about 8 km at the poles to about 16 km at the equator. Within the troposphere, temperatures characteristically decrease with altitude at a rate of 6.5°C per km. This tropospheric **lapse rate** is quite variable, particularly close to the surface, which helps to produce instability in the system and makes the troposphere the most turbulent of the atmospheric layers.

The **tropopause** marks the upper limit of the troposphere. Beyond it, in the **stratosphere**, isothermal conditions prevail; temperatures remain constant, at or about the level reached at the tropopause, up to an altitude of about 20 km. Above that level the temperature begins to rise again, reaching a maximum some 50 km above the surface, at the **stratopause**, where temperatures close to or slightly above 0°C are common. This is caused by the presence of **ozone** (O_3), which absorbs **ultraviolet radiation** from the sun, and warms the middle and upper levels of the stratosphere, creating a **temperature inversion**. The combination of that inversion with the **isothermal layer** in the lower stratosphere creates very stable conditions so that the stratosphere has none of the turbulence associated with the troposphere.

Temperatures again decrease with height above the stratopause and into the **mesosphere**, falling as low as –100°C at the **mesopause**, some 80 km above the surface. The **thermosphere** stretches above this altitude with no obvious outer limit. In this layer, temperatures may exceed 1000°C, but such values are not directly comparable to temperatures in the stratosphere and troposphere, because of the rarified nature of the atmosphere at very high altitudes.

For the climatologist and the environmentalist, the troposphere and the stratosphere are the most important structural elements in the atmosphere. The main conversion and transfer of energy in the earth/atmosphere system takes place within these two layers of the lower atmosphere, and interference with the mechanisms involved has contributed to the creation and intensification of current problems in the atmospheric environment.

See also
Environmental lapse rate.

Further reading
Kemp, D.D. (1994) *Global Environmental Issues: A Climatological Approach* (2nd edition), London/New York: Routledge.

ATMOSPHERIC MODELS

Physical or mathematical representations of the workings of the **atmosphere**, ranging from regional models such as the mid-latitude cyclonic models used in weather forecasting to **general circulation models** which attempt to represent global circulation patterns.

ATMOSPHERIC PRESSURE

The **pressure** exerted by the weight of the constituents of the **atmosphere** on the earth's surface and anything on that surface. Atmospheric pressure decreases with height, but not at a constant rate. The rate is greatest within the first 5 km above the surface, where most of the atmosphere is concentrated, and much less above 20 km. Average sea-level pressure is 101.32 kp, but varies considerably in time and place, with pressure being particularly responsive to **temperature** changes. Various

units have been used to measure atmospheric pressure including inches of **mercury** (29.92 in Hg) and millibars (1013.2 mb), but currently the kilopascal (101.32 kp) is the standard **SI unit**. Pressure differences lead to air movement at all scales, from local to continental, and drive the **atmospheric circulation**.

See also
Vapour pressure.

ATMOSPHERIC STABILITY/ INSTABILITY

The stability or instability of the **atmosphere** is determined by its response to change. A stable atmosphere will resist change, whereas an unstable one will allow the change to develop. The terms are most commonly applied to vertical motion in the atmosphere. If a parcel of **air** is forced to rise in a stable atmosphere, for example, it will return to its original position once the force which caused it to rise has been removed. In an unstable atmosphere, the parcel would continue to rise away from its original position. The stability or instability of the atmosphere can be determined by comparing the **environmental lapse rate** (ELR) with the **dry** or **saturated adiabatic lapse rates** (DALR or SALR). If the DALR is greater than the ELR, then a dry parcel of air rising in the atmosphere will always be cooler and therefore denser than the surrounding air and will therefore tend to fall once the force causing it to rise ceases. The atmosphere is therefore stable. Conversely, if the DALR is less than the ELR the atmosphere will be unstable. A similar relationship between the ELR and the SALR will produce stability or instability also. As air rises and cools, however, its **relative humidity** increases and it may ultimately become saturated. In that case it will cool at the SALR, which is always less than the DALR. It is possible therefore that a parcel of air will begin to ascend cooling at a DALR greater than the ELR, but following saturation will cool at a rate less than the ELR, which leads to instability. Such instability which depends upon saturation is called conditional instability. The stability or instability of the atmosphere is important in

Figure A-22 Atmospheric lapse rates and atmospheric stability

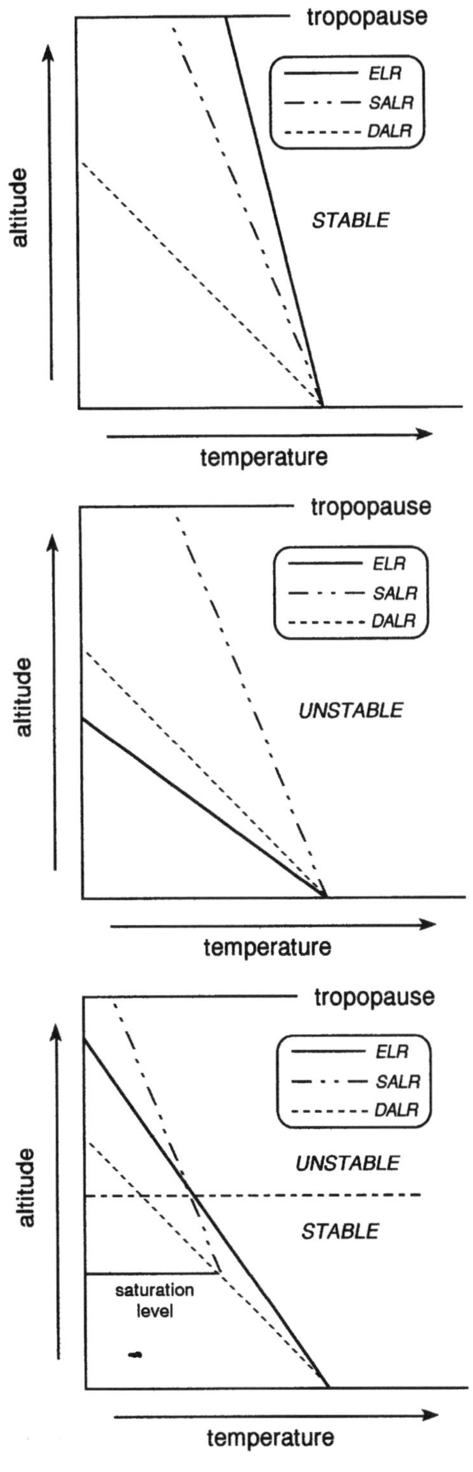

air **pollution** climatology. Pollutants tend to accumulate under stable conditions and are more easily dispersed when the atmosphere is unstable.

Further reading
Oke, T.R. (1987) *Boundary Layer Climates* (2nd edition), London: Methuen.

ATMOSPHERIC TURBIDITY

A measure of the atmosphere's aerosol content as indicated by the **attenuation** of **solar radiation** passing through it. For most purposes, it can be considered as an indication of the dustiness or dirtiness of the atmosphere. Volcanic activity, dust storms and a variety of physical and organic processes provide **aerosols** which are incorporated into the atmosphere. Once there, they are redistributed by way of the **wind** and **pressure** patterns, remaining in suspension for periods ranging from several hours to several years depending upon particle size and the altitude attained. Human industrial and agricultural activities also help to increase turbidity levels, but the anthropogenic production of aerosols cannot match the volume of material produced naturally. The presence of aerosols in the atmosphere disrupts the inward and outward flow of **energy** through the earth/atmosphere **system**. Studies of periods of intense volcanic activity suggest that the net effect of increased atmospheric turbidity is cooling, and some of the coldest years of the **Little Ice Age** have been correlated with major volcanic eruptions. The eruption of **Mount Pinatubo** in 1991 was blamed for the cool summer of 1992 in North America and was ultimately linked with a global **temperature** reduction of 0.5°C. Present opinion sees increased atmospheric turbidity actually producing a slight warming.

See also
Aerosols, Dust Veil Index, Little Ice Age, Nuclear winter, Pollution, Volcano.

Further reading
Fennelly, P.F. (1981) 'The origin and influence of airborne particles', in B.J. Skinner (ed.) *Climates, Past and Present*, Los Altos, CA: Kauffmann.

Groisman, P.Y. (1992) 'Possible regional climate consequences of the Pinatubo eruption: an empirical approach', *Geophysical Research Letters* 19: 1603–6.
Lamb, H.H. (1970) 'Volcanic dust in the atmosphere; with a chronology and assessment of its meteorological significance', *Philosophical Transactions of the Royal Society,* A 266: 435–533.

ATOM

The smallest unit of an **element** that retains the characteristics of that element and can take part in a chemical reaction. An atom consists of a central core or **nucleus** which consists of two types of stable particles – **protons** and **neutrons**. Protons are positively charged, neutrons are electrically neutral. The positive charges of the protons are balanced by an equal number of **electrons**, which are negatively charged particles distributed in orbits or shells around the nucleus. Since the number of electrons equals the number of protons, the atom is electrically neutral.

Figure A-23 A diagrammatic representation of the structure of a hydrogen atom and an oxygen atom

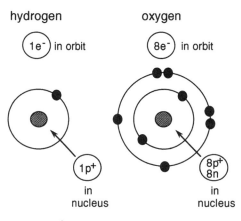

See also
Atomic number, Isotope, Nuclear fission.

Further reading
Das, A. and Ferbel, T. (1994) *Introduction to Nuclear and Particle Physics*, New York: Wiley.

ATOMIC BOMB

An explosive device in which the explosive power is provided by the fission of radioactive elements such as [235]**uranium** (U) or [239]**plutonium** (Pu). When two subcritical masses of the appropriate elements are brought together to create a **critical mass**, an uncontrolled **chain reaction** takes place, and the **energy** released is equivalent to the explosion of thousands of tonnes of **TNT**. In addition to producing instantaneous destruction, the impact of atomic bombs can continue for many years through the **radioactivity** that they release into the **environment**. The first – and, as yet, only – atomic bombs to be used as weapons were dropped on Hiroshima and Nagasaki, Japan in 1945.

See also
Nuclear fission, Thermonuclear device.

Further reading
Wilson, J. (ed.) (1975) *All in our Time: The Reminiscences of Twelve Nuclear Pioneers*, Chicago: Bulletin of Atomic Scientists.

ATOMIC ENERGY COMMISSION (AEC)

A US federal agency established in 1946 to control the development of **nuclear energy**. The AEC was responsible for overseeing major growth in the US nuclear power industry from the mid-1950s to the mid-1970s. In 1977, some of its development responsibilities were taken over by the newly created **Department of Energy**, and its regulatory function was assumed by the Nuclear Regulatory Commission.

ATOMIC NUMBER

The number of **protons** in the **nucleus** of an **atom** and the number of **electrons** in orbit around the nucleus. The atomic number allows a distinction to be made between the atoms of different elements.

ATTENUATION

As applied to **solar radiation**, it is the diminution in the intensity of the radiation as a result of its **scattering**, **reflection** and **absorption** by **aerosols** in the **atmosphere**. The attenuation of solar radiation provides a measure of **atmospheric turbidity**.

AUSTRALIAN ENVIRONMENT ASSESSMENT BRANCH (AEAB)

The Environment Assessment Branch is involved in the implementation of the provisions of the Environment Protection (Impact of Proposals) Act passed originally in 1974 and updated in 1995. It protects the **environment** by assessing proposals for development that

1 involve the Commonwealth (federal) government;
2 are funded through grants to state governments;
3 involve the export of primary products which require Commonwealth approval;
4 include foreign investment.

The AEAB also undertakes assessments on development proposals referred to it by other agencies.

AUSTRALIAN NATURE CONSERVATION AGENCY (ANCA)

Formerly the Australian National Parks and Wildlife Service, the ANCA was formed in 1975 as the principal nature conservation agency of the government of Australia. Its responsibilities include the **conservation** of the natural and cultural heritage of marine, terrestrial and freshwater areas of continental Australia and its external territories.

AUTECOLOGY

The study of the relationship between an individual organism or single **species** and its **environment**.

AUTOMBILE EMISSION CONTROLS

Controls on the nature and volume of pollutants emitted from automobile exhausts.

The term can be applied to the physical devices, such as **catalytic converters**, which limit the emissions or to the legislative devices that set the levels for acceptable emissions. Controls are needed because burning **gasoline** at high temperatures in an **internal combustion engine** leads to the production of a wide variety of pollutants including **oxides of nitrogen** (NO_x), **carbon monoxide** (CO) and volatile organic compounds (VOCs). These pollutants may become the constituents of **photochemical smog**. In the past, **lead** (Pb) was also common in automobile exhaust emissions, but it has been banned in North America since 1989 because of the health risks it posed and also because it impaired the efficiency of catalytic converters designed to reduce emissions of the other pollutants. Most developed nations have legislation designed to impose controls on automobile emissions. The situation in the United States is particularly important. Decisions made there have an effect beyond North America, since Asian and European manufacturers must meet US emission controls to enter the large North American market. Legislated EC emission levels are now comparable to those in the United States, in part because of this. In the introduction of emission controls, government agencies decide what levels are desirable, and it becomes the responsibility of automobile manufacturers and fuel producers to develop the engineering needed to meet the requirements. Despite opposition from the automobile industry, this approach has led to the introduction of catalytic converters, improved fuel efficiency, cleaner burning engines and cleaner gasoline. In some jurisdictions, controls exceed those imposed by the national government. California has much more stringent regulations than the rest of the United States, for example, and British Columbia has an inspection programme for exhaust emissions from cars and light trucks not required in other parts of Canada. The benefits of such controls are generally less than might be expected because the number of motor vehicles continues to rise and many which were built before tighter regulations were introduced continue to pollute at higher levels than new vehicles. Many environmentalists see the problem continuing without radical changes such as the introduction of electric cars (the first mass-production electric car was introduced by General Motors in December 1996) or the use of **hydrogen** (H) as a **fuel**, plus lifestyle changes which involve less use of automobiles.

Further reading
Baarschers, W.H. (1996) *Eco-facts and Eco-fiction*, London/New York: Routledge.
Marcus, A.A. and Jankus, M.C. (1992) 'The auto emissions debate: the role of scientific knowledge', in R.E. Buchholz, A.A. Marcus and J.E. Post (eds) *Managing Environmental Issues: A Casebook*, Englewood Cliffs, NJ: Prentice-Hall.

AUTOMOBILE TYRE RECYCLING

Automobile tyres are a major component of **solid waste** in most developed nations. About 280 million tyres are discarded every year in the United States. Not being biodegradable, at least in the short term, they tend to accumulate in large dumps, creating **aesthetic pollution,** providing breeding grounds for mosquitoes in the water that accumulates in them and representing a fire hazard. Fires which begin in tyre dumps are very difficult to extinguish. They produce toxic **air pollution** and the heat of the fire releases liquid **hydrocarbons** that flow into and contaminate the local surface and **groundwater** systems. Retreading of old tyres was once common, but modern radial tyres are costly to retread, and the process is now rarely undertaken. Some are recycled into garbage cans, mats and other rubber products; others are shredded and incorporated in asphalt road surfacing. The **energy** available from burning tyres is used directly in thermal electric power stations and in cement works. Using **pyrolysis,** the hydrocarbons in the tyres can be liquefied

to produce heating **oil** and **gasoline** additives. However, the total number of tyres recycled or used as a source of energy is less than 20 per cent of those discarded. Even the introduction of tyre taxes in some jurisdictions has had little impact, and the disposal of waste tyres remains a growing problem.

Further reading
Miller, G.T. (1994) *Living in the Environment: Principles, Connections and Solutions*, Belmont, CA: Wadsworth.
National Research Council (US) (1992) *Recycled Tire Rubber in Asphalt Pavements*, Washington, DC: Transportation Research Board: National Research Council.

AUTOVARIATION

Environmental change produced when one component of the **environment** responds automatically to change in another. This is made possible by the integrated nature of the elements that make up the environment. The response of vegetation to changes in moisture availability is an autovariation, for example.

Further reading
Trewartha, G.T. and Horn, L.H. (1980) *An Introduction to Climate*, New York: McGraw-Hill.

AZOTOBACTER

The most important group of nitrogen-fixing **bacteria.** They are **aerobic** and obtain their **energy** by breaking down **carbohydrates** in the **soil.** Being an inert gas, **nitrogen** (N) does not react readily with other **elements** and such bacteria make a major contribution to the **nitrogen cycle** by moving the **gas** from the **atmosphere** into the soil, where it is made available to plants.

B

BACILLUS

A rod-shaped bacterium, characteristic of the genus *Bacillus*. Bacilli are responsible for such diseases as anthrax (*B. anthracis*) and tetanus (*B. tetani*), but the genus also includes beneficial **bacteria** such as *B. subtilis*, which synthesizes vitamin B_{12} in many organisms, and *B. thuringiensis* (BT), used as a non-chemical **pesticide** in the control of leaf-eating caterpillars and other insects.

Further reading
Barjac, H. de and Sutherland, D.J. (eds) (1990) *Bacterial Control of Mosquitoes and Black Flies: Biochemistry, Genetics and Applications of Bacillus thuringiensis israeliensis and Bacillus sphaericus*, New Brunswick, NJ: Rutgers University Press.

BACKGROUND AIR POLLUTION MONITORING NETWORK (BAPMON)

A global monitoring system established by the **WMO** in 1968 to collect baseline and regional data on background levels of **pollution** in the **troposphere**. The network monitors suspended particulate matter, **CFCs**, **carbon dioxide** (CO_2), **methane** (CH_4) and **atmospheric turbidity** levels at more than 200 stations worldwide. The data are being used to evaluate trends in the concentrations of these pollutants.

BACKSCATTER

The redirection of **solar radiation** back into space as a result of its interception by atmospheric **aerosols**. Being reflected, with no change in wavelength, backscattered **radiation** does not contribute to the heating of the earth/**atmosphere** system.

BACTERIA

Unicellular, microscopic organisms that multiply by simple division. They take various forms, but are generally classified into four different types based on their shape; some are spherical (cocci), some are rod-shaped (bacilli), some are spiral-shaped (spirelli) and some are filamentous (actinomycetes). Bacteria can also be subdivided according to their requirement for free **oxygen** (O), into **aerobic** and **anaerobic** groups or, depending upon their **energy** source, into autotrophic and heterotrophic groups. Autotrophic bacteria oxidize inorganic material to obtain energy, whereas heterotrophic bacteria depend upon the decomposition of organic material for their energy. Some bacteria are sensitive to acidity in the environment, and the destruction of **soil** bacteria is common in areas where the **terrestrial environment** is subject to **acid rain**. Soil bacteria help to maintain soil fertility by breaking down organic matter and converting **nitrogen** into a form usable by plants. Bacteria also have an important role in **pollution** control, being able to decompose a great variety of organic wastes – from **sewage** to crude **oil** – into simpler **compounds** that can be reincorporated into the **natural environment**. Pollution problems arise, in part, because of society's ability to produce waste in amounts that exceed the capacity of natural decomposers such as bacteria. Despite performing such essential functions, bacteria are also responsible for a great variety of diseases including anthrax, tetanus, plague and tuberculosis. Most of these respond to treatment with **antibiotics**, but strains of bacteria that are resistant to antibiotics have evolved, raising fears of a resurgence of diseases such as tuberculosis, once thought to have been conquered. The apparent increased frequency of uncommon diseases such as the so-called 'flesh-eating disease' caused by a streptococcal bacterium may also be indicative of the ability of new strains of bacteria to resist antibiotics.

See also
Activated sewage sludge, Bacillus, Biochemical oxygen demand, Nitrogen cycle.

Further reading
Bisset, K.A. (1970) *The Cytology and Life History of Bacteria*, Edinburgh: Livingstone.
Wheatley, A. (ed) (1990) *Anaerobic Digestion: A Waste Treatment Technology*, London/New York: Elsevier Applied Science.

BARIUM (Ba)

A silvery-white soft **metal**, which tarnishes readily when exposed to **air**. It is used in the manufacture of paints, glass and some electronics products. Barium products are poisonous. Ingested through drinking **water**, they have toxic effects on the nervous and circulatory system. Barium also exists as a

hazardous, radioactive **fission** product, which may be released into the **environment** following a nuclear test or reactor accident.

BAROGRAPH

A recording **barometer**. A pen attached to the indicating arm of an aneroid barometer traces a continuous record of **atmospheric pressure** on chart paper attached to a revolving drum.

BAROMETER

An instrument which measures **atmospheric pressure**. There are two types in common use: the **mercury** (Hg) barometer and the aneroid barometer. The former is based on the principle – first evaluated by Torricelli, a student of Galileo, in 1643 – that the **atmosphere** is capable of supporting a column of mercury in a glass tube. The standard atmosphere can support a column of mercury 29.92 in. (76 cm) in height, and changes in that value indicate changes in atmospheric pressure. An aneroid barometer consists of a partially evacuated, thin-walled, metal box (or series of boxes) which expand(s) as pressure falls and contract(s) as pressure rises. Such changes are relatively small, and are amplified through a lever system attached

Figure B-1 The principles and structure of a mercury and an aneroid barometer

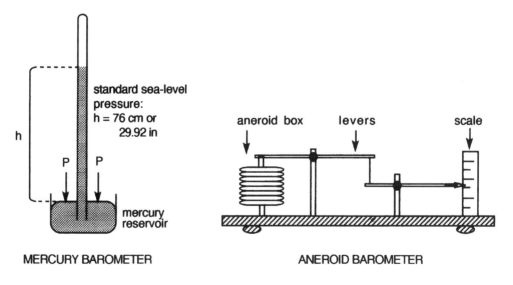

MERCURY BAROMETER   ANEROID BAROMETER

to a pointer that indicates the pressure on a scale. The main advantage of the aneroid over the mercury barometer is its portability. The aneroid barometer may be used as a crude weather forecasting instrument if descriptive **weather** terms such as 'stormy', 'rain', 'fair' and 'dry' are indicated on the pressure scale. Higher pressure is generally considered to be associated with fair, dry weather, and lower pressure with inclement weather. Since pressure changes with height in the atmosphere, aneroid barometers, calibrated to indicate altitude rather than pressure, can be used as altimeters in aircraft.

Further reading
Lutgens, F.K. and Tarbuck, E.J. (1989) *The Atmosphere: An Introduction to Meteorology,* Englewood Cliffs, NJ: Prentice-Hall.

BASE

A **compound** that releases hydroxyl ions (OH⁻) when dissolved in **water**. Bases include the oxides and **hydroxides** of **metals** and other compounds such as ammonium hydroxide (NH_4OH) that release hydroxyl ions on **solution**. Bases react with **acids** to form **salt** and water, neutralizing the acid in the process.

See also
Alkali.

BASE EXCHANGE

The exchange or transfer of **cations** in **solution**. Base exchange (or cation exchange) is a natural process, but the principle has been developed for agricultural and industrial purposes. **Nutrients** such as **calcium** (Ca), **magnesium** (Mg), **sodium** (Na) and **potassium** (K) made available through base exchange are essential for the growth of plants. Cations of these **elements** are normally loosely bonded to the surface of **clay** and **humus** particles in the **soil**, and when they are replaced by hydrogen **ions** or other cations they are free to be absorbed by plants. By adding **fertilizer** to the soil, farmers attempt to maintain soil fertility by replacing the **nutrient** ions which have been displaced from the clay and humus

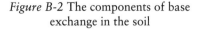

Figure B-2 The components of base exchange in the soil

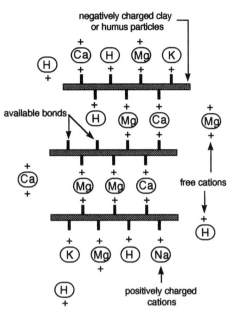

particles. However, every soil has a unique cation exchange capacity – the total exchangeable cations that a soil can absorb. Once that capacity has been filled, additional nutrient ions remaining in the soil are unbonded to particles and therefore susceptible to **leaching**. Where soils are subject to **acid rain**, the abundant hydrogen ions introduced into the soil may replace the other cations so rapidly that nutrients are leached out of the soil before they can be absorbed by the plants, thus reducing soil fertility. Cation exchange is used in **hard water** areas to soften water for residential and industrial use. When hard water is passed through a water softener, the calcium ions in the water are replaced by sodium ions, and the water is softened. When the exchange capacity of the softener is full, it can be recharged by reversing the process. For example, brine pumped through the system will cause the captured calcium ions to be replaced by sodium ions and the softener will be ready for use again.

Further reading
Trudgill, S.T. (1988) *Soil and Vegetation Systems* (2nd edition), Oxford: Oxford University Press.

BATHING WATER QUALITY

See **water quality.**

BATHOLITH

A large igneous geological feature, formed by the intrusion of molten material into existing country rock where it subsequently cooled and solidified. Individual batholiths cover thousands of square kilometres and extend deep into the earth's crust. Commonly composed of granite, that cooled and solidified relatively slowly beneath the earth's surface, they have a very uniform structure and composition. Batholiths are geologically stable and as a result they have been identified as potential locations for **nuclear waste disposal** facilities.

BECQUEREL

The unit used in the **SI** system to indicate the radioactivity of a substance. Named after Henri Becquerel (1852–1908), it has replaced the **curie** as a measure of **radioactivity.** One becquerel (Bq) is equivalent to the radioactivity of an **element** undergoing radioactive decay at a rate of 1 disintegration per second.

BENEFICIATION

The concentration of low-quality **ores** prior to smelting. The process involves the separation of the ore from the country rock by crushing, magnetic separation and flotation, to produce concentrated ore and large amounts of rock waste. Beneficiation is usually carried out at the mine site to save transportation costs. The large amounts of **waste** produced can create local environmental problems such as **air pollution** in the form of **dust** from the finely crushed **tailings** produced in the process and **water pollution** from the tailings ponds in which the fine sediments are allowed to settle.

Further reading
Wills, B.A. (1992) *Mineral Processing Technology: An Introduction to the Practical Aspects of Ore Treatment and Mineral Recovery* (5th edition), Oxford/New York: Pergamon.

BENTHOS

The bottom of a body of **water,** from the edge of a river or lake to its deepest point, or from the high water mark to the deepest part of the **ocean.** Benthic organisms are those that occupy this zone.

BENZENE (C_6H_6)

A clear, colourless, flammable **liquid hydrocarbon** present in **coal** tar and **petroleum.** The simplest of the aromatic hydrocarbons, all of which have a structure and properties similar to benzene, it is widely used as an industrial and laboratory **solvent,** and in the manufacture of styrene, varnishes and paints. Benzene is highly toxic and carcinogenic.

See also
Carcinogen.

BERYLLIUM (Be)

A hard, white **metal** used to produce light, strong, corrosion-resistant **alloys** for the aerospace and **nuclear** industries. The inhalation of airborne beryllium causes severe respiratory disease (berylliosis), and may lead to malignant growths in the lungs. Exposure to beryllium dust can also cause temporary dermititis. As a result, the use of beryllium is now strictly controlled and in the United States it is classified as a hazardous pollutant.

BEST PRACTICE/BEST PRACTICABLE MEANS

An approach to **air pollution** control under the Alkali and Clean Air Inspectorate in Britain, which takes into account the economic and technological realities of the operation of specific industrial plants. It attempts to attain the best possible level of control, given the constraints imposed by existing economics and technology.

See also
Environmental Protection Act (UK).

BHOPAL

An industrial city of more than 800,000 people in central India which was the site of a major industrial accident in 1984. Some 40 tonnes of methyl isocyanate, a highly toxic chemical used in the manufacture of carbamate **pesticides**, leaked from a tank in a Union Carbide chemical plant. The noxious **gas** that formed drifted over residential areas close to the plant, killing an estimated 2500 people by direct exposure, and perhaps as many again as a result of the after-effects. A further 200,000 suffered respiratory problems, temporary blindness and severe vomiting, and more than a year after the event thousands were still suffering side-effects from exposure to the gas. Five years of inquiries and litigation produced the conclusion that the disaster was the result of poor plant design combined with questionable operating and safety procedures, although the company claimed that sabotage might have been involved. Union Carbide was ordered to pay $470 million in compensation to the victims. Studies by an international medical commission suggest that poor medical practices in the decade after the accident contributed to the victims' suffering.

See also
Carbamates.

Further reading
Day, M. (1996) 'Bad medicine deepens Bhopal's misery', *New Scientist* 152 (2060): 6.
Shrivastava, P. (1987) *Bhopal: Anatomy of a Crisis*, San Francisco: Harper & Row.
Walker, G. (1990) 'Bhopal: five years on', *Geography* 75 (2): 158–60.

BIENNIAL OSCILLATION

See quasi-biennial oscillation.

BIOACCUMULATION

The retention of non-biodegradable chemicals, or those that decay only slowly, in the bodies of organisms. Toxic metals such as **lead** (Pb) and **mercury** (Hg) and **pesticides** such as DDT are absorbed more rapidly than they are excreted and therefore accumulate in the body. Accumulation often takes place in specific locations, including bones, fatty tissues and organs, such as liver and kidneys. The impact of bioaccumulation varies with the chemical and the organism involved, but it can cause death, prevent reproduction or increase susceptibility to disease. The dynamics of **food chains**, in which organisms at a specific level in a chain consume large numbers of the organisms in the preceding level, ensure that the products of bioaccumulation are amplified (biological amplification). Thus the chemical levels in the tissues of predators at the head of a food chain may be as many as 10 million times that in the tissues of the producer organisms at the base of the chain.

Further reading
Miller, G.T. (1994) *Living in the Environment* (8th edition), Belmont, CA: Wadsworth.

BIOCHEMICAL OXYGEN DEMAND (BOD)

A measure of **pollution** in a body of **water** based on the organic material it contains. The organic material provides food for **aerobic bacteria** which require **oxygen** (O) to be able to bring about the **biodegradation** of such pollutants. The greater the volume of organic material, the greater will be the numbers of bacteria and therefore the greater the demand for oxygen. Thus, the BOD value gives an indication of organic pollution levels in the water. It provides no information on other pollutants such as suspended mineral sediments or **heavy metals**. If the BOD exceeds the available dissolved oxygen in the water, oxygen depletion occurs, and aquatic organisms will suffer. Fish kills are not uncommon under such circumstances.

See also
Oxygen sag curve, Pulp and paper industry, Sewage.

Further reading
Nebel, B.J. and Wright, R.T. (1993) *Environmental Science: The Way the World Works*, Englewood Cliffs, NJ: Prentice-Hall.

BIODEGRADATION

The breakdown of materials in the environment by natural decay. Accomplished mainly by **aerobic bacteria**, biodegradation is essential for the **recycling** and replenishment of raw materials in a closed material **system** such as the earth/**atmosphere** system. Most natural materials and products manufactured from organic substances are biodegradable. Problems arise when biodegradable substances are produced at a rate which exceeds the ability of the system to cope, or when non-biodegradable materials are introduced into the **environment**. Crude **oil** is biodegradable, for example, but when millions of litres are spilled in an **oil tanker** accident, the microorganisms which would normally break down the oil are overwhelmed. For similar reasons, **sewage** remains a major **pollution** problem in many densely populated areas. Such materials are degraded eventually, but some products of modern technology such as **plastics** remain unaltered as **solid wastes**, while others such as **insecticides** and **heavy metals** can cause serious problems by accumulating in the environment.

See also
Biochemical oxygen demand, Carbon cycle, Nitrogen cycle.

Further reading
Allsopp, D. and Seal, K.J. (1986) *Introduction to Biodeterioration*, London/Baltimore: Edward Arnold.

BIODIVERSITY

The variety of life forms that inhabit the earth. Biodiversity involves **habitat** diversity, plant and animal **species** diversity within the various habitats and the genetic diversity of individual species. The large-scale slaughter of wild animals, the overharvesting of trees and other plants and the destruction of habitat worldwide are threatening biodiversity at a time when its importance is becoming increasingly apparent. Particular concerns are expressed over the habitat destruction in areas such as the tropical **rainforest** or the near extinction of such marine species as the northern cod and the blue whale, but these are perhaps only the more extreme examples of a ubiquitous problem.

See also
Biodiversity Convention.

Further reading
Barbier, E.B., Burgess, J.C. and Folke, C. (1994) *Paradise Lost: The Ecological Economics of Biodiversity*, London: Earthscan.
Gaston, K.J. (ed.) (1996) *Biodiversity: A Biology of Numbers and Differences*, Cambridge, MA: Blackwell Science.
Reid, W.V. and Miller, K.R. (1989) *Keeping Options Alive: The Scientific Basis for Conserving Biodiversity*, Washington, DC: World Resources Institute.

BIODIVERSITY CONVENTION

A product of the **United Nations Conference on Environment and Development** (UNCED), outlining policies aimed at combining the preservation of natural biological diversity with **sustainable development** of biological resources. It was signed by all participants except the United States. The document includes no deadlines, and the language is generally considered to be weak, but some of the signatories to the Convention have begun to put its policies into action. The Australian government, for example, has developed a National Strategy for the Conservation of Australia's Biological Diversity, which is intended to ensure that by the year 2000 the decline of remnant native vegetation in that country will have been arrested and reversed.

Further reading
Hawksworth, D.L. (1995) *Biodiversity: Measurement and Estimation*, New York: Chapman and Hall.
Parson, E.A., Hass, P.M. and Levy, M.A. (1992) 'A summary of the major documents signed at the earth summit and the global forum', *Environment* 34: 12–15 and 34–6.

BIOGEOCHEMICAL CYCLE

The combination of mechanisms by which an element is circulated through the earth/**atmosphere system**. Within the cycle, an element moves between sources and **sinks**, along well-established pathways. The process involves living organic and inorganic phases, and the net result is that the element may be used and reused, which is essential in a closed material flow system such as the earth/atmosphere system. Although any given cycle is ultimately balanced, flow through is normally uneven. Quantities of an element may be shunted out

Figure B-3 The sulphur cycle in the environment

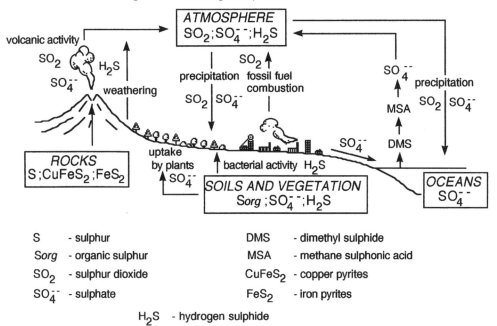

S	- sulphur	DMS	- dimethyl sulphide
Sorg	- organic sulphur	MSA	- methane sulphonic acid
SO_2	- sulphur dioxide	$CuFeS_2$	- copper pyrites
SO_4^{--}	- sulphate	FeS_2	- iron pyrites
H_2S	- hydrogen sulphide		

of the cycle for periods of time. **Carbon** (C) and **phosphorous** (P), for example, are regularly stored in the **oceans,** perhaps for years at a time, before becoming active in their respective cycles again. Similarly, **fossil fuel** deposits represent carbon that has not been directly involved in the **carbon cycle** for millions of years. Current global issues involve disruption of biogeochemical cycles. Enhancement of the **greenhouse effect** has come about as a result of inadvertent human interference in the carbon cycle, and modern intensive agricultural production, with its concentration on **nitrogen**-rich **fertilizers,** has contributed to the disruption of the nitrogen cycle in many areas.

Further reading
Bolin, B. and Cook, R. (eds) (1983) *The Major Biogeochemical Cycles and their Interactions*, New York: Wiley.
Kronberg, B. (1993) 'Response of major North American ecosystems to global change: a biogeochemical perspective', in E.F. Mooney, E.F. Fuentes and B. Kronberg (eds) *Earth System Responses to Global Change*, San Diego: Academic Press.
Kwon, O.-Y. and Schnoor, J.L. (1994) 'Simple global carbon model: the atmosphere–terrestrial biosphere–ocean interaction', *Global Biogeochemical Cycles* 8: 295–305

BIOGEOGRAPHY

The geography of living organisms, past and present. It includes the study of the distribution patterns of plants and animals and the processes responsible for these patterns. Biogeographers also study the impact of human activities on the **biosphere.**

Further reading
Brown, J.H. and Gibson, A.C. (1983) *Biogeography*, St Louis: C.V. Mosby.
Cox, C.B. and Moore, P.D. (1993) *Biogeography: An Ecological and Evolutionary Approach* (5th edition), Oxford: Blackwell Science.

BIOGEOPHYSICAL FEEDBACK MECHANISM

The hypothesis developed to explain **degradation-induced drought** in the **Sahel.** Overgrazing and woodcutting increased the surface **albedo,** which in turn disrupted the regional **radiation** balance. Reduced surface heating retarded convective activity and limited **precipitation.** With less precipitation, vegetation cover decreased and the albedo of the surface was further enhanced. This is an example of positive **feedback.**

Further reading
Charney, J. (1975) 'Dynamics of deserts and drought in the Sahel', *Quarterly Journal of the Royal Meteorological Society* 101: 193–202.

BIOLOGICAL AMPLIFICATION

See bioaccumulation.

BIOLOGICAL DIVERSITY

See biodiversity.

BIOMASS

The total weight of living organic matter in a given area. The contribution of the terrestrial environment to the earth's overall biomass is much greater than that of the aquatic environment, but values vary in time and space. Plant biomass, for example, increases rapidly during the growing season, when the energy for biological production is available, and decreases when the plants die and decompose. Spatial variations in biomass are large, the biomass in a tropical rainforest being as much as five to six times that of a similar area in a desert. Animal biomass in the terrestrial environment reaches a peak in the tropical rainforest and in the savanna grasslands of Africa. In the aquatic environment, the greatest concentration of animal biomass is in the shallow margins of the tropical seas.

BIOMASS ENERGY

Energy available from organic material in the environment, that originated as solar energy absorbed by plants and was converted into chemical energy by photosynthesis. Biomass energy, mainly in the form of wood, was the main source of energy prior to the development of fossil fuels. It includes the energy available in wood, crops, crop residues, industrial and municipal organic waste, food-processing waste and animal wastes. Some of these, such as wood and various waste products, can be used to provide direct energy – in wood stoves or incinerators, for example – but more often they are converted before being used. Grain crops or crop residues such as the waste from sugar cane harvesting (bagasse) can be used to produce alcohol; human and animal wastes can be digested anaerobically to produce methane (CH_4) (biogas). Some 40,000 small-scale, local biogas plants are currently active in India. Although biomass energy is renewable, relatively inexpensive to produce, and requires no advanced technology, it does have certain drawbacks. The conversion of solar energy by plants, for example, has an efficiency rate that rarely exceeds 2 per cent. Thus, to produce large quantities of biomass energy, large areas need to be planted or harvested. Although biomass energy is often considered environmentally appropriate, in common with fossil fuels it produces smoke and soot particles and carbon dioxide (CO_2). Biomass energy is renewable, but can be consumed faster than it can be replaced. In Nepal and adjacent parts of northern India, for example, the use of wood as the primary fuel has denuded hillslopes, leaving them open to erosion. Similarly, in the Sahel, the removal of scrubby vegetation for use as fuel has contributed to desertification in some areas. The future of biomass energy is unclear. Its use is likely to remain highest in the developing nations, where it supplies about 40 per cent of the energy used. Among the industrialized nations biomass energy accounts for about 5 per cent of the energy used.

Further reading
Aubrecht, G. (1989) *Energy*, Columbus, OH: Merrill.
World Bank (1992) *Development and the Environment*, New York: Oxford University Press.

BIOME

A community of plants and animals in equilibrium with the environmental characteristics – climate, soils, hydrology – of a major geographical area. A biome is usually considered to represent a climax community, that is, the most likely combination of flora and fauna for a given set of environmental conditions. Although animals are an integral part of any biome, major biome types are normally described and designated in terms of their plant assemblages – for example, **equatorial rainforests, grasslands, taiga.** Climate is the major environmental element responsible for the development of the char-

acteristic flora in a specific biome. As a result, there is a strong latitudinal element in their distribution. Similarly, the changing climatic conditions associated with increasing altitude

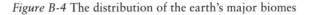

Figure B-4 The distribution of the earth's major biomes

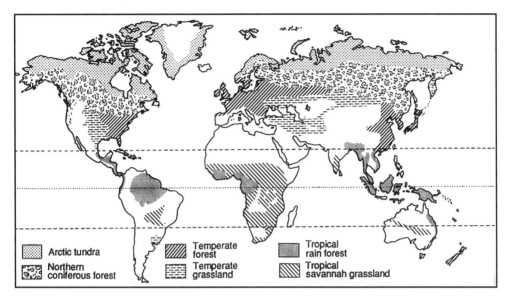

Source: Park, C.C. (1992) *Tropical Rainforests*, London/New York: Routledge

Figure B-5 The boreal forest in northern Canada

Photograph: The author

in mountainous areas have an impact in the distribution of biomes. Although biomes are commonly considered as terrestrial communities, some ecologists also denote the communities of intertidal zones, estuaries, coral reefs and freshwater environments as aquatic biomes. Most biomes are capable of absorbing and recovering from a considerable amount of natural disturbance – fires or insect infestation, for example – but increasingly society is permanently transforming large areas of many biomes through such activities as agriculture and forestry.

Further reading
Furley, P. and Newey, W.W. (1983) *Geography of the Biosphere*, London: Butterworth.
Stirling, P.D. (1992) *Introductory Ecology*, Englewood Cliffs, NJ: Prentice-Hall.
Walter, H. (1979) *Vegetation of the Earth* (2nd edition), Berlin: Springer-Verlag.

BIOMETEOROLOGY

The study of the interactions of living organisms with **weather** and **climate**. An interdisciplinary concept, requiring consideration of such elements as plant and animal physiology and biochemistry, as well as **climatology** and **meteorology** at all scales. Many human activities such as agriculture include a biometeorological element, and human biometeorology involves the study of the impact of climate on clothing, shelter and health.

Further reading
Tromp, S.W. (1980) *Biometeorology*, London: Heyden.

BIOREMEDIATION

The use of micro-organisms, usually **bacteria**, to break down organic chemical **waste** into less hazardous substances. They may be used under controlled conditions such as on a **sewage** farm or in waste stabilization ponds, but they also have the potential to clean up accidental spills of waste material hazardous to the **environment**. Following the **Exxon Valdez** oil spill in Alaska in 1989, for example, bacteria were sprayed on some of the contaminated beaches. Encouraged to grow by the application of water-soluble **fertilizer**,

the bacteria fed on the oil and within a month had made a measurable contribution to the clean-up. Bioremediation provides an attractive approach to environmental clean-up, since the materials used are relatively inexpensive, there is little disruption of the contaminated site and the by-products of the processes involved are usually harmless – often only **carbon dioxide** (CO_2) and **water**. The effectiveness of bioremediation may be restricted by limitations on the availability of **nutrients** or **oxygen** (O) in the contaminated zone.

Further reading
Kaufman, D.G. and Franz, C.M. (1993) *Biosphere 2000*, New York: HarperCollins.

BIOSPHERE

The zone of terrestrial life, sometimes called the ecosphere, including the earth's land and **water** surfaces plus the lowest part of the **atmosphere** and the upper part of the **soil** and water layers. Life is present throughout the **ocean** basins, but it is concentrated in the upper 100–200 m, and it is that layer which is commonly considered to be part of the biosphere. The biosphere is very much an interactive layer incorporating elements of the **atmosphere**, **lithosphere** and **hydrosphere** integrated through the activities of the various life forms in the **environment**.

Further reading
Furley, P. and Newey, W.W. (1983) *Geography of the Biosphere*, London: Butterworth.
Huggett, R.J. (1995) *Geoecology: An Evolutionary Approach*, London: Routledge.

BIOTIC

The living components of an **ecosystem**. They are usually classified into producers and consumers. Producers – green plants, for example – can synthesize the organic materials they need from inorganic **compounds** in the **environment**. Consumers are unable to manufacture their own food directly from inorganic compounds, and depend upon the producers to provide what they need.

See also
Abiotic, Food chain, Photosynthesis.

BIOTIC INDEX

A rating used to assess the quality of the **environment** by examining the diversity of the **species** it contains. It can provide an indication of the level of **pollution** in an **ecosystem**. A water body with a high biotic index, for example, would be relatively pollution-free, with a wide variety of species. In contrast, a badly polluted river or lake would support only a few pollution-tolerant species, and therefore would have a low biotic index.

BIOTIC POTENTIAL

The theoretical reproductive capacity of a **species** under ideal environmental conditions. The biotic potential of many organisms such as **bacteria**, insects and small mammals is so great that they could easily exceed the **carrying capacity** of their **environment**. Such a situation is seldom achieved, however, because of the presence of natural checks such as limited food supply, predators and **parasites**, that combine to check reproduction or limit the survival of the offspring. Despite concern for the rapidly growing human population of the earth, even in those parts of the world where populations are growing most rapidly the rate of increase is well below the biotic potential of the human **species**.

Further reading
Enger, E.D. and Smith, B.F. (1995) *Environmental Science: A Study of Interrelationships* (5th edition), Dubuque, IA: Wm C. Brown.

BLACK BODY

A body capable of absorbing all of the **radiation** falling upon it and reflecting none. Such a perfect absorber exists only in theory, but a dull, black surface may absorb as much as 90 per cent of the incident/incoming radiation. Black bodies are also good emitters of radiation, and a black body at a given **temperature** will emit the maximum intensity of radiation possible for that temperature.

Further reading
Faughn, J.S., Turk, J. and Turk, A. (1991) *Physical Science*, Philadelphia, PA: Saunders.

BLACK LIST

A European Commission list of dangerous pollutants commonly discharged into the **aquatic environment**. It includes more than 120 pollutants considered sufficiently hazardous that priority must be given to their elimination from effluent discharged into **water** bodies.

See also
Red List.

BLACK LUNG DISEASE

See **pneumonoconiosis**.

BLOCKING

A situation in which tropospheric circulation patterns become static for periods of days or weeks. Common in mid- to high latitudes in the northern hemisphere, blocking is associated with large amplitude stationary waves in the upper **atmosphere**. Its surface expression often takes the form of a blocking **anticyclone** which prevents the movement of other **weather** systems or causes them to deviate from their normal paths. Blocking anticyclones regularly develop in the eastern and central parts of the Canadian prairies in the spring, causing moisture-bearing low pressure systems to take a more northerly track than normal. The resulting low **precipitation** disrupts spring planting on the prairies and creates ideal conditions for forest fires in the adjacent **boreal forest**. Weather conditions in Britain are often influenced by the development of blocking anticyclones over Scandinavia.

Further reading
Chandler, T.J. and Gregory, S. (eds) (1976) *The Climate of the British Isles*, London: Longman.
Knox, J.L. and Hay, J.E. (1985) 'Blocking signatures in the northern hemisphere: frequency, distribution and interpretation', *Journal of Climatology* 5: 1–16.
Trewartha, G.T. and Horn, L.H. (1980) *An Introduction to Climate*, New York: McGraw-Hill.

BLUE BOX

An approach to curbside **recycling**, common in North America, in which recyclable material such as **plastic** bottles, **aluminum** and tin cans are placed in containers (blue boxes) to be collected by local authorities or private companies. Pre-sorting by the householder reduces the time, effort and cost of separating recyclable material from **domestic waste** or **garbage**.

BLUE-GREEN ALGAE

See **cyanobacteria**.

BLUE LIST

A list of bird **species** experiencing non-cyclic **population** decline and therefore having the potential to become endangered species. Published periodically by the **National Audubon Society**, a private **conservation** organization in the United States, the list is seen as providing early warning of potential problems for the species listed. In theory, recognition of the situation should allow appropriate steps to be taken before the crisis stage is reached.

See also
Red Data Books.

Further reading
Cox, G.W. (1993) *Conservation Ecology: Biosphere and Biosurvival*, Dubuque, IA: Wm C. Brown

BLUEPRINT FOR SURVIVAL

A statement on global environmental problems prepared by the editors of *The Ecologist* and published originally in that journal in 1972. It was also published in book form and received worldwide attention. Concerned with the rate at which **resources** were being exploited and the ways in which their use threatened the **environment**, the authors of the document called for a reappraisal of a system based on constant expansion and consumption. In its place, they advocated a stable society in which there would be minimum disruption of ecological processes, maximum **conservation** of materials and **energy**, a **zero-population growth** and a social system in which an individual could enjoy these conditions rather than be restricted by

them. The ideas set out in *Blueprint for Survival* continued to be developed and can be recognized in part in the concept of **sustainable development**.

See also
World Commission on Environment and Development.

Further reading
The Ecologist (1972) *Blueprint for Survival*, London: Penguin.

BOG

See **peat/peatlands**; **wetlands**.

BOREAL FOREST

See **taiga** and Figure B-5.

BOREHOLE

A hole drilled or dug into the earth's surface through which **groundwater** can be withdrawn. A well.

BOSCH PROCESS

A process used for the industrial production of **hydrogen** (H). A mixture of hydrogen and **carbon monoxide** (CO), called water gas, is treated with steam in the presence of a **catalyst**. This oxidizes the carbon monoxide to **carbon dioxide** (CO_2), which is removed in **solution** to release the hydrogen. The Bosch Process is no longer widely used since **natural gas** has replaced water gas as the main source of hydrogen.

BOTTLE LAWS

Laws aimed at encouraging the reuse or **recycling** of beverage containers, usually through the addition of a returnable deposit to the price of the product. Originally applied to glass bottles, they now include aluminum cans and a variety of plastic containers. The glass bottles can be refilled whereas the others are recycled. Bottle laws help to reduce **waste** that would otherwise be added to sanitary **landfill** sites. They also reduce the demand for the raw materials and the **energy** needed to produce the containers and, by encouraging consumers to return used bottles, provide a steady supply of material for the recycling industry.

BRITISH NUCLEAR FUELS LIMITED (BNFL)

Created in 1971 from the Production Group of the **UK Atomic Energy Authority** (UKAEA), BNFL is involved in all aspects of nuclear **fuel** supply. It has expertise in fuel manufacture and enrichment, fuel reprocessing, transportation and waste management plus the decommissioning of nuclear facilities, and is also very active in research and development. Based in Britain, BNFL operates internationally through offices in the US, Japan, South Korea and China. Used fuel from a number of countries, but particularly Japan, is reprocessed at the company's plant at Sellafield in north-west England.

See also
Nuclear reactor, Nuclear waste, Plutonium, Uranium.

BRITISH THERMAL UNIT (Btu)

The amount of **energy** required to raise the **temperature** of one pound (1 lb) of **water** through 1°F – technically from 59.5°F to 60.5°F. For many purposes it has been superseded by the **joule** (J), but it remains in use in the US and UK as a unit of energy associated with the production or transfer of heat. One Btu is equivalent to 1055.06 joules or 252 **calories**.

BROMINE (Br)

A dark red liquid with a pungent odour. Along with **fluorine** (F), **chlorine** (Cl) and **iodine** (I), it is a member of the **halogen** group of elements. Since bromine is a highly reactive chemical it is never found free in nature, but can be extracted from brine, from highly saline waterbodies such as the Dead Sea and from underground **salt** deposits. Marine animals and shellfish also concentrate bromine in their bodies, and can be used as sources of the element. Bromine is used as a disinfectant and in the manufacture of **compounds** used in photography and in medicine. Being a powerful **oxidizing agent** it may cause ignition of combustible materials on contact. Thus in its free form it is a moderate to high fire hazard, yet in combination with other chemicals such as **fluorocarbons** it is a very effective fire suppressant. Unfortunately, as a constituent of the **halons** used in fire extinguishers, bromine has been implicated in **ozone depletion**.

See also
Bromofluorocarbons, Methyl bromide.

BROMOFLUOROCARBONS

Synthetic chemical **compounds**, commonly referred to as **halons**, containing **bromine** (Br), **fluorine** (F) and **carbon** (C). Similar in properties to **chlorofluorocarbons**, and used mainly in fire extinguishers, they decompose to release bromine which contributes to the destruction of the **ozone layer**. So effective are halons as ozone destroyers that their production was banned in 1994. However, the potential for future damage remains, in the form of hundreds of tonnes of bromofluorocarbons which remain unused in fire extinguishers.

Further reading
MacKenzie, D. (1992) 'Agreement reduces damage to ozone layer', *New Scientist* 136 (1850): 10.
National Academy of Sciences (1976) *Halocarbons: Effects on Stratospheric Ozone*, Washington, DC: National Academy Press.

BROWNIAN MOTION

The random motion of small, inanimate particles suspended in a fluid medium, often observed in microscopic **pollen** grains in suspension in an aqueous solution, for example, or in **smoke** particles suspended in **air**. Named after Robert Brown, an English botanist, who first recognized the process in 1827, the erratic motion is the result of continuous irregular collisions between the particles and the **molecules** that make up the surrounding fluid medium.

BRUNDTLAND COMMISSION

See **World Commission on Environment and Development.**

BUFFER SOLUTION

A **solution** to which large amounts of acidic or alkaline materials may be added without markedly altering its **hydrogen ion concentration** (pH). An acid buffer consists of a weak acid and a **salt** of that acid: a basic buffer consists of a weak **base** and a salt of that base. The absence or inadequacy of buffer solutions has contributed to the impact of **acid rain** on the environment in areas such as Scandinavia. With little or no buffering, for example, the lakes in such areas become increasingly more acidic.

BUFFERING AGENTS

Alkaline or basic materials capable of reducing or neutralizing acidity. Natural buffering agents such as **limestone** help to reduce the environmental impact of **acid rain**. Lime added to the **fossil fuel** combustion process or employed in flue gas **scrubbers** helps to reduce the acidity of emissions. The addition of lime is also a well-established means of improving the quality and increasing the productivity of acid soils.

See also
LIMB.

BUSINESS-AS-USUAL SCENARIO

One of the scenarios used to predict the future

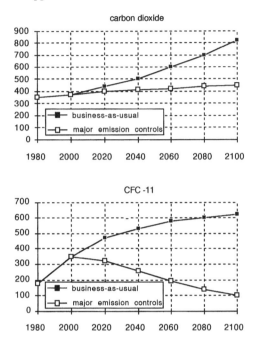

Figure B-6 The business-as-usual scenario applied to carbon dioxide and CFC-11

Source: After IPCC (1990) *Climate Change: The IPCC Scientific Assessment,* Cambridge: Cambridge University Press

status of environmental issues such as **global warming**. It is based on the maintenance of the status quo with no attempt made to reduce the output of the agents of change or mitigate their impact.

Further reading
Houghton, J.T., Jenkins, J.H. and Ephraums, J.J. (eds) (1990) *Climate Change: The IPCC Scientific Assessment,* Cambridge: Cambridge University Press.

C

C₃, C₄ AND CAM PLANTS

Three general groups into which the earth's vegetation can be classified. They differ mainly in the biochemistry of their photosynthetic processes. The C_3 group makes up about 95 per cent of the earth's **biomass** and includes important grain crops such as wheat and rice as well as most trees. C_4 plants include maize, sorghum and millet. The CAM or crassulacean acid metabolism plants are mainly succulents and few are important agriculturally. In the investigation of **global warming** the groups have received particular attention because of the potential impact of rising levels of atmospheric **carbon dioxide** (CO_2) and higher temperatures on their productivity. Although all will react positively to these new conditions, C_3 plants will be most responsive.

See also
Photosynthesis.

Further reading
Graves, J. and Reavy, D. (1996) *Global Environment Change: Plants, Animals and Communities*, Harlow, Essex: Longman.

CADMIUM (Cd)

A soft, silvery-white **heavy metal** which usually occurs in association with **zinc** (Zn) ore. It is used in the production of batteries, paints and fusible **alloys**. The control rods in **nuclear reactors** are commonly made of cadmium since it is a good absorber of **neutrons**. Because of its association with zinc, it appears in the waste products from electroplating plants. Cadmium is a toxic substance that causes

kidney damage, hypertension (high blood pressure), increased bone **porosity** and the inhibition of bone repair mechanisms. Cadmium pollution has been blamed for the occurrence of **itai-itai** disease in Japan.

CALCIUM (Ca)

A soft, white **metal** which tarnishes easily when exposed to **air**. It is essential to life, being a major constituent of bones and teeth. In the human body it is obtained from dairy products and a variety of green vegetables with the help of **vitamin D**. Calcium is most commonly found in compound form, with **calcium carbonate** ($CaCO_3$) being the most abundant. Along with a variety of other compounds it is important in agriculture and industry. Calcium carbide (CaC_2), for example, is a source of the welding gas, acetylene; calcium cyanamide ($CaCN_2$) is a **fertilizer** and used in the manufacture of **plastics**; calcium phosphate ($Ca_3(PO_4)_2$) is an ingredient of the artificial fertilizer superphosphate; calcium silicate ($CaSiO_3$) is an important constituent of glass and cement; calcium sulphate or gypsum ($CaSO_4.2H_2O$) is widely used in the ceramics, paper and paint industries.

CALCIUM CARBONATE (CaCO₃)

The most common compound of **calcium** (Ca): a white, insoluble solid that occurs naturally as **chalk, limestone**, marble and calcite. Most chalk and limestone deposits are of biological origin, being the accumulated skeletal remains of marine organisms. Calcium carbonate is a major component of the bedrock geology of many parts of the earth, contributing to

Figure C-1 Chalk exposed in the White Horse at Westbury, Wiltshire, England

Photograph: The author

landforms such as the relatively subdued chalk landscapes of Salisbury Plain in southern England, but also being incorporated in the rugged peaks of the Alps and Himalayas. Although calcium carbonate is insoluble in pure **water**, it can be dissolved by rainwater, which is commonly slightly acidic. This has helped to produce the peculiar landscapes of limestone areas, with their steep-sided valleys, caves and underground streams, referred to as karst topography. Calcium carbonate is used in the manufacture of paints, rubber goods, **plastics** and medicines, and in the form of limestone and marble it is a common building stone. Cement produced from limestone is also an essential raw material in the construction industry. Calcium carbonate has a role in a number of environmental issues. Although susceptible to damage by **acid rain**, it can be used as a **buffering agent** to reduce the acidity of **soils** and waterbodies. Used in **scrubbers**, it can lower the acidity of **flue gas** emissions. Calcium carbonate also acts as a **sink** for **carbon dioxide** (CO_2), and through this has an influence on the level of that **gas** in

the earth/atmosphere system. Any reduction in the number of marine organisms that sequester carbon dioxide in the calcium carbonate of their skeletons, or any change in their ability to do so, could help to raise the level of atmospheric carbon dioxide and therefore contribute to **global warming**. Any increase in the processes which break down calcium carbonate and release carbon dioxide would have a similar effect.

Further reading

Sweeting, M.M. (1972) *Karst Landforms*, London: Macmillan.
Taylor, H.F.W. (1990) *Cement Chemistry*, London/ Toronto: Academic Press.

CALORIE

The amount of **energy** required to raise the **temperature** of 1 gm of **water** through 1°C – technically from 15.5 to 16.5°C. This gram calorie is a very small quantity and for most practical purposes – for example, in dealing with food energy – the kilocalorie (1000 gram calories) is used. Kilocalories are sometimes referred to as 'large' calories or written with a capital C as Calorie, but the latter tends to cause confusion – the capital is often ignored or left out inadvertently by dieticians and food producers. For many purposes the calorie is being replaced by the **joule** (J), one calorie being equivalent to 4.1855 joules

CANCER

A rapid, uncontrolled growth of **cells** that can invade adjacent tissues, creating tumours or destroying body organs, causing incapacity or death. Often initially site specific, tumours may release malignant cells that travel in the body fluids to infect other parts of the body. Although some individuals may be predisposed to certain cancers because of their inherited genetic make-up, there is increasing evidence that the introduction of **carcinogens** into the **environment** by human activities can be a major contributing factor to the development of cancers.

Further reading
Oppenheimer, S.B. (1985) *Cancer, a Biological and Clinical Introduction*, Boston: Jones and Bartlett.

CANADIAN DEUTERIUM-URANIUM (CANDU)

A Canadian designed commercial **nuclear reactor** that uses heavy water (**deuterium**) as a moderator and **coolant**, each occupying separate circuits for efficiency and safety. The effectiveness of heavy water as a moderator allows natural (unenriched) **uranium** (^{235}U) to be used as a **fuel**, which contributes to the economy of the **system**. The fuel is contained in pressure tubes through which the coolant circulates to carry the heat produced to a separate steam generator for the production of electricity. The design of the CANDU reactor allows on-power fuelling that limits the time it is out of service and gives CANDU reactors a very high net capacity factor (i.e. **energy** actually delivered over a period of time compared to the maximum that could be delivered). Most operating CANDU reactors are in the province of Ontario, Canada, but systems have also been constructed in India,

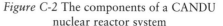

Figure C-2 The components of a CANDU nuclear reactor system

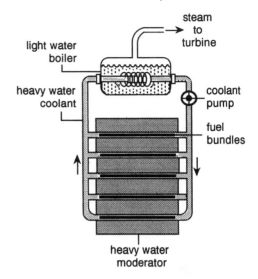

Pakistan, Korea and Argentina. In 1996, Atomic Energy of Canada signed a contract to build four CANDU reactors in China. The economy and the relatively simple technology of the CANDU system make it an **appropriate technology** for nations of intermediate economic and industrial capacity, but with the demand for new sources of nuclear energy being negligible, CANDU seems unlikely to reach its full potential.

See also
Nuclear fission.

Further reading
Robertson, J.A.L. (1978) 'The CANDU reactor system: an appropriate technology', *Science* 199: 657–64.

CANOPY

The uppermost layer of vegetation in a forest or woodland, usually consisting of the foliage of the tallest trees. The nature of the canopy will vary according to such factors as the type and age of the trees and the time of year. It has an important microclimatological role because of its ability to disrupt the flow of **energy** and moisture. Where the canopy is dense, for example, it may intercept as much as 80 per cent of the incoming **radiation** and more than 40 per cent of the **precipitation**. As a result, the canopy not only has its own microclimate, but also exerts a strong influence on the **climate** of the forest floor. Removal of the canopy through forest clearing has significant climatological consequences for the area. In studies of urban climatology, the canopy layer is the layer of the urban fabric below roof level, in which the microclimatology is governed by the nature and distribution of the individual buildings and the activities taking place at street level.

See also
Deforestation, Equatorial rainforest.

Further reading
Oke, T.R. (1987) *Boundary Layer Climates* (2nd edition), London: Methuen.

CAPILLARY FLOW

The movement of **water** through narrow spaces or tubes (capillaries) against the force of **gravity**. It is made possible by a combination of surface tension and the adsorptive forces between the **water molecules** and the capillary walls. Water in the **soil** can be carried upwards through pore spaces above the **water table** as a result of capillary action. The smaller the pore spaces, the greater the height to which the water will rise. This may provide plants with at least some of the moisture they need during dry conditions and return leached **nutrients** to the root zone. However, it can also lead to soil deterioration in the form of **salinization** when water brought from some depth in the **soil profile** and evaporated near the surface, deposits the **salts** it contains.

See also
Adsorption, Leaching, Salinization.

CARBAMATES

Chemical compounds derived from carbamic acid (NH_2COOH), used as **insecticides, fungicides** and **herbicides**. Their development was encouraged because of the growing resistance of some organisms to **chlorinated hydrocarbons** such as **DDT** and concern for the toxicity of **organophosphorous compounds**

used as **pesticides**. The dangers posed by the persistence of chlorinated hydrocarbons in the **environment** led to their being banned in many countries and carbamates were seen as a less persistent alternative. They have the advantage that they are not stored in the bodies of animals (i.e. no **bioaccumulation** takes place) and after use they do not remain long in the **soil**. Carbamates are not perfect, however. Their toxicity for humans and other animals is variable, but they are highly toxic to bees and fish. The commercial insecticide Sevin is one of the most frequently used carbamates.

Further reading
Kuhr, R.J. and Dorough, H.W. (1976) *Carbamate Insecticides: Chemistry, Biochemistry and Toxicology*, Cleveland: CRC Press.
Smith, G.J. (1993) *Toxicology and Pesticide Use in Relation to Wildlife*, Boca Raton, FL: C.K. Smoley.

CARBOHYDRATES

Organic compounds containing **carbon** (C), **hydrogen** (H) and **oxygen** (O), with the general formula $C_x(H_2O)_y$. Glucose, a simple carbohydrate, has the formula $C_6H_{12}O_6$. Carbohydrates include monosaccharides (e.g. glucose), disaccharides (e.g. sucrose and lactose) and polysaccharides (e.g. **starch**, glycogen and **cellulose**). The polysaccharides are composed of simpler carbohydrates combined together in branched or unbranched chains. Cellulose, for example, consists of long unbranched chains of glucose. Carbohydrates are an essential part of the lives of all living organisms. Glucose is produced during **photosynthesis**, and is the basic **energy** source for **metabolic processes** in plants and animals; glycogen (animal starch) is the main form in which energy is stored in animals; cellulose is the principal structural material of plants.

Further reading
Ketchmer, N. and Hollenbeck, C. (eds) (1991) *Sugars and Sweeteners*, Boca Raton, FL: CRC Press.
McIlroy, R.J. (1967) *Introduction to Carbohydrate Chemistry*, London: Butterworth.

CARBON (C)

A non-metallic **element** that exists in several allotropic forms (e.g. diamond and graphite) and as amorphous carbon (e.g. lampblack). Carbon, one of the most common **elements** in the **environment**, is present in all organic substances and is a constituent of a great variety of **compounds**, ranging from relatively simple gases such as **methane** (CH_4) to very complex derivatives of **petroleum hydrocarbons** such as peroxybenzoyl nitrate (PBzN). The complexity of carbon-based **molecules** is made possible by the carbon atom's valency of 4. Each **atom** has four available bonds that allow it to unite easily with other atoms to create the complex chains of **molecules** characteristic of the **carbohydrates** and hydrocarbons. The carbon in the environment is mobile, readily changing its affiliation with other elements in response to biological, chemical and physical processes, the mobility being controlled through a natural **biogeochemical cycle**.

See also
Allotropy, Carbon cycle.

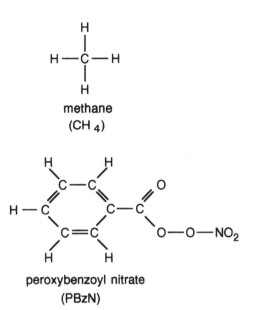

Figure C-3 Examples of the chemical structure of simple and complex carbon compounds

methane
(CH_4)

peroxybenzoyl nitrate
(PBzN)

Figure C-4 The carbon cycle in the environment

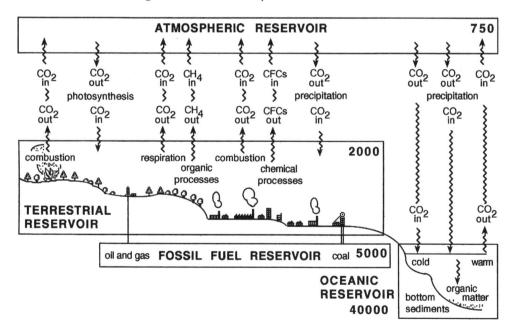

Source: After Kemp, D.D. (1994) *Global Environmental Issues: A Climatological Approach*, London/New York: Routledge

Further reading
Nimmo, W.S. (1997) *The World of Carbon: Organic Chemistry and Biochemistry*, Toronto: Wiley Canada.

CARBON CYCLE

A natural **biogeochemical cycle** through which the flow of **carbon** (C) in the earth/atmosphere system is regulated. Carbon moves through the **system** mainly as **carbon dioxide** (CO_2) – although other gases such as **methane** (CH_4) are also involved – driven by such processes as **photosynthesis, respiration** and **combustion**, that control the release of carbon **compounds** from their sources and their absorption in **sinks**. The cycle is normally considered to be self-regulating, but with a time-scale of the order of thousands of years. Over shorter periods it appears to be unbalanced. The **carbon** in the system moves between the major reservoirs indicated in Table C-1.

Table C-1 Carbon reservoirs in the earth/atmosphere system

RESERVOIR	CARBON RESERVES (1×10^9 TONNES)
Atmospheric	750
Terrestrial	2,000
Oceanic	40,000
Fossil fuel	5,000

Storage of carbon in the atmospheric and terrestrial reservoirs is usually short-term, but the carbon sequestered in the **oceans** may be effectively removed from the cycle for periods of several centuries, while the carbon in the **fossil fuel** reservoir has not been active for millions of years. The growing demand for **energy** in modern society has led to the reintroduction of fossil fuel carbon into the cycle mainly in the form of carbon dioxide, at a rate of some 5 billion tonnes per year. The system cannot cope immediately with the reactivation of this volume of carbon and becomes unbalanced. The natural sinks are unable to absorb the new carbon dioxide as rapidly as it is being produced. The excess remains in the

Figure C-5 Forests as carbon sinks. Only forests which are increasing their biomass act as sinks. Stable, equilibrium forests are carbon neutral

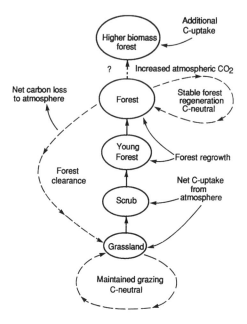

Source: Moore, P.D., Chaloner, B. and Stott, P. (1996) *Global Environmental Change*, Oxford: Blackwell Science

atmosphere to intensify the **greenhouse effect** and thus contributes to **global warming**.

Further reading
Bolin, B. (1970) 'The carbon cycle', *Scientific American* 223 (3): 124–32.
Houghton, R.A. (1995) 'Land-use change and the carbon cycle', *Global Change Biology* 1: 275–87.
Schimel, D.S. (1995) 'Terrestrial ecosystems and the carbon cycle', *Global Change Biology* 1: 77–91.

CARBON CYCLE MODELS

Computer-based **models** which simulate the workings of the **carbon cycle**. They are used in conjunction with climate models to investigate the impact of atmospheric carbon imbalance on global **climate**.

CARBON DIOXIDE (CO_2)

One of the variable atmospheric gases, and a

greenhouse gas. CO_2 is produced by the complete **combustion** of carbonaceous (carbon-rich) substances, by the **aerobic** decay of organic material, by **fermentation** and by the action of **acid** on **limestone** ($CaCO_3$). Its current level in the **atmosphere** is 353 parts per million by volume (ppmv), but growing. Despite its low volume, CO_2 is important to life on earth because of its participation in **photosynthesis** and its contribution to the **greenhouse effect**. It becomes solid at −78.5°C to form 'dry ice', once commonly used as a refrigerant when no other forms of refrigeration were available. Carbon dioxide is also used in fire extinguishers since it is heavier than air and does not support **combustion**. Nor does it lead to the **ozone** (O_3) destruction associated with **halon**-based extinguishers.

See also
Calcium carbonate, Carbon cycle, Global warming.

Further reading
Bach, W., Crane, A.J., Berger, A.L. and Longhetto, A. (eds) (1983) *Carbon Dioxide*, Dordrecht: Reidel.
Hansen, L.D. and Eatough, D.J. (1991) *Organic Chemistry of the Atmosphere*, Boca Raton, FL: CRC Press.

CARBON MONOXIDE (CO)

A colourless, odourless, tasteless **gas** produced by the incomplete **combustion** of carbonaceous material. Although naturally occurring, in many areas the level of the gas in the **atmosphere** is determined by human activities. **Internal combustion engines** are a common source of carbon monoxide, for example, and as a result it is a common component of street level urban air. Cigarette smokers inhale the **gas** directly and may have levels of carbon monoxide in their bloodstreams five to ten times greater than that of non-smokers. Carbon monoxide readily forms a stable **compound** (carboxyhaemoglobin) with the haemoglobin in blood. This effectively reduces the ability of the haemoglobin to carry **oxygen** (O) around the body and the functions of the brain, lungs and heart are impaired. High doses of carbon monoxide are lethal, but the effects of long-term exposure to low levels of the gas are less well understood.

Further reading
Baarschers, W.H. (1996) *Eco-facts and Eco-fiction*, London/New York: Routledge.
Newell, R.E., Reichle, H.G. and Seiler, W. (1989) 'Carbon monoxide and the burning earth', *Scientific American* 261 (4): 58–64.

CARBON TAX

A policy that would tax **fossil fuels** according to the amount of **carbon** (C) they contained. The resultant increases in costs would reduce the demand for **fossil fuels** in general and cause a realignment away from **coal** to **natural gas**. The latter, having a higher energy content, could be burned in smaller quantities than the coal to produce the same amount of **energy**, and the net result would be to reduce **carbon dioxide** (CO_2) emissions and slow the enhancement of the **greenhouse effect**. As with any new tax, however, the carbon tax faces political and economic opposition.

Further reading
Green, C. (1992) 'Economics and the greenhouse effect', *Climatic Change* 22: 265–91.

CARBON TETRACHLORIDE (CCl_4)

An industrial **solvent**, once widely used as a dry-cleaning agent, but largely replaced by other compounds because of its toxicity. The **vapour** is an irritant causing headaches and dizziness, and prolonged exposure can cause kidney and liver damage. It is also a **carcinogen**. Carbon tetrachloride has been identified as contributing directly to the decay of the **ozone layer** through the release of **chlorine** (Cl), and indirectly as a feedstock in the production of **chlorofluorocarbons** (CFCs). Carbon tetrachloride has an **ozone depletion potential** (ODP) of 1.1, and contributes about 8 per cent to global ozone depletion. Production and use of carbon tetrachloride was phased out in 1996.

Further reading
MacKenzie, D. (1992) 'Agreement reduces damage to ozone layer', *New Scientist* 136 (1850): 10.

CARCINOGEN

A chemical or physical agent capable of causing **cancer**. The general consensus is that a large proportion of human cancers are directly or indirectly linked to environmental conditions. Human activities can introduce chemical carcinogens such as **chlorinated hydrocarbons** or physical carcinogens such as **asbestos** or **ionizing radiation** into the environment, but natural carcinogens such as the toxic metals **beryllium** (Be), **cadmium** (Cd) and selenium (Se) are also present. There are two types of carcinogens: **DNA**-reactive carcinogens that alter the DNA of cells, and epigenetic carcinogens that do not react with DNA but may alter the immune system, cause hormonal imbalances or cause chronic tissue injury, all of which could lead to cancer. Carcinogens may be inhaled, ingested or absorbed through the skin, and their impact may be direct or indirect – working through the production of an intermediate chemical, for example. There is no simple link between exposure to a carcinogen and the development of cancer. Some individuals may be physiologically more able to cope with exposure to carcinogens than others.

See also
Radioactivity.

Further reading
World Health Organization (1972) *Health Hazards of the Human Environment*, Geneva: WHO.
Chiras, D.D. (1994) *Environmental Science: Action for a Sustainable Future*, Redwood City, CA: Benjamin/Cummings.

CARRYING CAPACITY

The maximum number of organisms that can be supported by a particular **environment**. If that number is exceeded, some form of environmental disruption will follow. Although originally an ecological concept, it can also be applied in a socioeconomic context. In agriculture, for example, the carrying capacity of an area will determine the number of grazing animals it will support, or when applied to recreational land use it represents the number of people and types of activity that can be

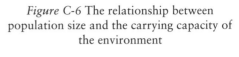
Figure C-6 The relationship between population size and the carrying capacity of the environment

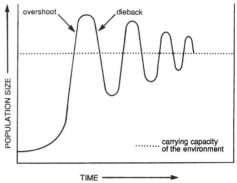

accommodated without ecological deterioration. Under natural conditions, it represents a theoretical equilibrium state. If the carrying capacity of an **ecosystem** is exceeded because of the rapid growth of the number of organisms in the system, for example, there will be insufficient **resources** to support the excess **population**, and it will decline until equilibrium between the resource base and the population is re-established. If the **species** in the area are fewer than could be supported by the carrying capacity, populations will tend to increase until they reach some form of balance with the resources available. Human interference frequently causes the carrying capacity of an area to be exceeded. Introducing extra grazing animals on rangeland or allowing too many people to use a recreational area will eventually lead to a deterioration of the environment. Although the concept of carrying capacity is usually applied at the ecosystem level, there is also a theoretical carrying capacity for the earth as a whole, applied to the maximum human population that it might support. A realistic value is probably impossible since factors other than resource availability – acceptable quality of life, for example – will help to decide the final carrying capacity.

See also
Sustainable development.

Further reading
Mungall, C. and McLaren, D.J. (eds) (1990) *Planet Under Stress: The Challenge of Global Change*,

Toronto: Oxford University Press.
Botkin, D.B. and Keller, E.A. (1995) *Environmental Science: Earth as a Living Planet*, New York: Wiley.

CARSON, R.

See **Silent Spring**.

CASH CROPPING

Growing crops for monetary return rather than direct food supply. It may contribute to environmental deterioration in marginal lands where the perceived financial incentive leads to the cropping of land unsuitable for development.

See also
Subsistence farming.

CATABOLISM

That part of **metabolism** in which complex **organic compounds** are broken down into simpler ones that can be more easily used by the organisms involved. In the process, **energy** is released.

See also
Anabolism, Digestion.

CATALYST

A substance that facilitates a chemical reaction, yet remains unchanged when the reaction is over. Being unchanged, it can continue to promote the same reaction again and again, as long as the reagents are available, or until the catalyst itself is removed. This is a catalytic chain reaction. **Enzymes** are organic catalysts.

See also
Catalytic converter, Ozone depletion.

Further reading
Butt, J.B. and Peterson, E.E. (1988) *Activation, Deactivation and Poisoning of Catalysts*, San Diego: Academic Press.
Dotto, L. and Schiff, H. (1978) *The Ozone War*, Garden City, NY: Doubleday.

CATALYTIC CONVERTER

A pollution control device attached to automobile exhaust systems to control the output of pollutants. The presence of platinum (Pt) as a **catalyst** causes unburned **hydrocarbons** and **carbon monoxide** (CO) to be converted into **carbon dioxide** (CO_2). More complex systems including platinum, palladium (Pd) and rhodium (Rh) also convert **oxides of nitrogen** (NO_x) to **nitrogen** (N). Since the presence of **lead** (Pb) reduces the catalytic action, catalytic converters can only be used with unleaded **gasoline**. Most jurisdictions in developed nations now require catalytic converters to be fitted to new vehicles.

See also
Selective catalytic reduction.

CATCHMENT

A drainage basin, or the area drained by a particular river system. Adjacent drainage basins are separated by watersheds. In North America the term **watershed** refers to the entire drainage basin, and the height of land between basins referred to as a divide.

CATENA

A group of **soils** formed from similar parent materials, but displaying different characteristics as a result of local topography, soil toposequence and/or internal drainage conditions. The changing succession of **soil profiles** from the top of a slope to the base provides the classic example of a catena.

CATHODE

A negative **electrode** or pole. In a primary cell or battery it is the pole from which the **electrons** are discharged before travelling to the **anode**.

CATION

A positively charged **ion**. During **electrolysis**, cations are attracted to the negatively charged **cathode**.

See also
Base exchange, Ion exchange.

CATION EXCHANGE CAPACITY

The total exchangeable cations that a **soil** can absorb.

See also
Base exchange, Cation.

CAUSTIC CHEMICALS

A term applied to corrosive alkaline solutions such as sodium hydroxide (NaOH) – caustic soda – or potassium hydroxide (KOH). Flesh exposed to caustic substances is severely burned, and other organic materials can be completely or partially destroyed. Caustic chemicals are used in a wide variety of industrial processes. Caustic soda, for example, is used in the **pulp and paper industry** to separate the **cellulose** fibres in wood by destroying the chemicals that bind them together.

CELL

A small unit of living matter or **protoplasm**. Each cell includes a **nucleus** distinct from the other protoplasm and is surrounded by a thin membrane. In plants, each cell is surrounded by a cell wall usually made of **cellulose**. Many micro-organisms – **bacteria**, for example – are unicellular, with the one cell providing for all of the needs of the organism. In multicellular organisms, however, the cells tend to be specialized to allow them to undertake specific functions. If a cell or group of cells is destroyed, their functions will be lost and the entire organism may suffer.

Further reading
Darnell, J.E., Lodish, H.F. and Baltimore, D. (1990) *Molecular Cell Biology* (2nd edition), New York: Scientific American Books.
Swanson, C.P. and Webster, P.L. (1985) *The Cell* (5th edition), Englewood Cliffs, NJ: Prentice-Hall.

CELLULOSE

A **carbohydrate** that occurs widely in nature. It has a fibrous form and is the major constituent of the **cell** walls of trees and other higher plants. Structurally, it is a polysaccharide, consisting of long unbranched chains of **glucose**, built up during the process of **photosynthesis**. An estimated 1×10^{11} tonnes of cellulose is produced yearly, by natural and cultivated plants. The **carbon** (C) sequestered in the cellulose is removed from direct participation in the **carbon cycle** until the cellulose is decomposed and **carbon dioxide** (CO_2) is released back into the **atmosphere**. Cellulose obtained from wood pulp, cotton and a variety of other plant sources is used in the manufacture of paper, rayon, **plastics** and explosives.

Further reading
Hon, D.N. and Shiraishi, N. (eds) (1991) *Wood and Cellulose Chemistry*, New York: M. Dekker.

CELSIUS SCALE

A **temperature** scale devised by the Swedish astronomer Anders Celsius (1701–1744) in which the melting point of ice is 0° and the boiling point of water is 100°. Formerly called the 'centigrade scale', it is the most commonly used scale in international weather reporting and forecasting, and in the scientific community.

See also
Fahrenheit scale, Kelvin.

CENTRAL ELECTRICITY GENERATING BOARD (CEGB)

The English public utility identified in the 1970s and 1980s as the main source of **acid**

Figure C-7 Diagrammatic representation of the components of a cell

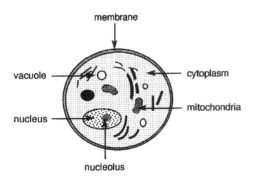

rain falling in Scandinavia. Acid gases released from the Board's **thermal electric power stations** were carried towards Norway and Sweden on the prevailing westerlies, being converted into acid rain on the way. Initially the CEGB protested against the allegations, but ultimately had to accept its contribution to acid rain and take steps – such as the introduction of **scrubbers** – to deal with the problem. The CEGB has since been privatized and split into two separate companies.

Further reading
Central Electricity Generating Board (1979) *Effects of Sulphur Dioxide and its Derivatives on Health and Ecology*, Leatherhead: CEGB.
Kyte, W.S. (1988) 'A programme for reducing SO2 emissions from U.K. power stations – present and future', *CEGB: Corporate Environment Unit*, Paper 1–2.

CENTRE FOR SOCIAL AND ECONOMIC RESEARCH ON THE GLOBAL ENVIRONMENT (CSERGE)

A research institute based at University College London and the University of East Anglia in Britain. Funded by the UK Economic and Social Research Council, its purpose is to study the social and economic aspects of global environmental change. This is a very broad mandate and has allowed the centre to research a wide range of topics including global **climate change**, **biodiversity**, **waste disposal**, **recycling** and **sustainable development**.

CENTRIFUGAL FORCE

The force directed away from the centre of curvature of an object moving in a circular path. For many purposes it can be considered as a reaction to the centripetal force that causes an object to move in a curved path and is directed inward towards the centre of curvature. In the **environment**, these forces apply to **air** in the **atmosphere** following circular paths as in a **cyclone** or **tornado**.

CHAIN REACTION

A reaction which, once begun, is self-sustaining as long as the reactants remain available. It

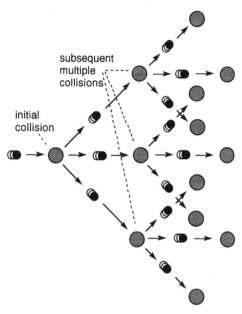

Figure C-8 A branching chain reaction similar to that initiated by nuclear fission

may be straight or branching. The destruction of the **ozone layer** by **chlorine** (Cl) atoms, for example, is a straight line chain reaction in which the same **atom** is used again and again, whereas **nuclear fission** involves branching chain reactions. When a **neutron** splits the **nucleus** of an atom, it releases several neutrons, which in turn proceed to split additional atoms. Thus, at each step, the amount of fission increases. Each fission is accompanied by the release of **energy** and in a branching chain reaction the release of energy may be so rapid that a nuclear explosion will occur. A controlled chain reaction is the basis of the commercial production of **nuclear energy**.

See also
Catalyst, Nuclear fission, Ozone depletion.

CHALK

Soft, white, pure **limestone** formed mainly from the calcareous skeletons of the microscopic marine organism foraminifera.

See also
Calcium carbonate.

CHANGING ATMOSPHERE CONFERENCE (TORONTO 1988)

One of a series of international conferences held between the **World Commission on Environment and Development** (1983) and the **UN Conference on Environment and Development** (1992) dealing with the causes, impacts and mitigation of global issues associated with the **atmospheric environment**. One of the aims of the Toronto Conference was to raise the level of awareness of these issues among policy makers. It focused on **ozone depletion** and the changing concentrations of **greenhouse gases**, proposing an Action Plan for the Protection of the Environment, which included support for the **Montreal Protocol** and called for energy policies aimed at reducing levels of **carbon dioxide** (CO_2) emissions.

Further reading
McKay, G.A. and Hengeveld, H. (1990) 'The changing atmosphere', in C. Mungall and D.J. McLaren (eds) *Planet Under Stress: The Challenge of Global Change*, Toronto: Oxford University Press.

CHAOS THEORY

Perhaps more appropriately termed 'complex systems theory', chaos theory is at odds with most other theories of environmental **systems** which include a high degree of order and balance. It involves the study of non-linear systems characterized by variability and unpredictable behaviour. Specific inputs into such systems do not produce the same results every time. In his pioneering work on chaos theory in the 1960s, Edward Lorenz hypothesized that in non-linear, chaotic systems, relatively minor initial differences between similar systems are magnified with time to create increasingly significant differences as the systems run. Such characteristics make chaotic systems difficult to model, and limits their use in prediction. Lorenz saw **weather** systems as chaotic, and that may explain why the accuracy of weather forecasting is limited to a few days. Chaos theory has also been applied to other physical and biological systems in the environment and has economic and industrial applications.

See also
Steady state.

Further reading
Coveney, P. and Highfield, R. (1995) *Frontiers of Complexity: The Search for Order in a Chaotic World*, New York: Fawcett Columbine.
Gleick, J. (1987) *Chaos: Making a New Science*, New York: Viking.
Lorenz, E.N. (1995) *The Essence of Chaos*, Seattle: University of Washington Press.

CHARCOAL

Impure **carbon** (C) produced by the **destructive distillation** of organic matter, particularly wood. Charcoal was formerly widely used as a **fuel** in the **iron** (Fe) and steel industry before being superseded by **coke** during the **Industrial Revolution**. It has an important role in many environmental issues because of its efficiency as a filtering medium. Being porous, a limited volume of charcoal provides a large surface area for the **adsorption** of impurities, and, as a result, charcoal filters are common in **water** purification and gaseous emission control systems. **Activated carbon** is charcoal treated to increase its adsorptive capacity.

Further reading
Speight, J.G. (ed.) (1990) *Fuel Science and Technology Handbook*, New York: M. Dekker.

CHEMICAL CHANGE

A change involving an interaction between chemicals which results in the alteration of the chemical composition of the participating **elements** or **compounds**. Such changes involve the creation or breakup of chemical bonds, and may or may not be reversible.

CHEMICAL MODELS

Models which simulate chemical processes. In studies of the **atmospheric environment**, they are being developed to investigate the role of trace gases in atmospheric processes.

CHEMICAL OXYGEN DEMAND (COD)

An indication of the pollution level in a **water** sample determined by boiling the sample in an oxidizing agent, usually potassium dichromate ($K_2Cr_2O_7$), and measuring the amount of **oxygen** (O) consumed by the sample over a given period of time.

CHEMICAL PRECIPITATION

The formation of an insoluble substance – a precipitate – as a result of a chemical reaction in a **solution**. Precipitation can be used in **waste** treatment to remove hazardous chemicals, or in refining to allow a commercially important substance to be separated out of a solution.

See also
Coagulation.

CHERNOBYL

The site of the world's most serious **nuclear reactor** accident. In 1986, one of four nuclear reactors in a nuclear power plant at Chernobyl, Ukraine (part of the USSR at that time) exploded as a result of human error, lax operating practices and a poorly designed system. The force of the explosion was so great that the top of the containment vessel

Figure C-9 The spread of radioactive pollutants in the atmosphere following the nuclear accident at Chernobyl

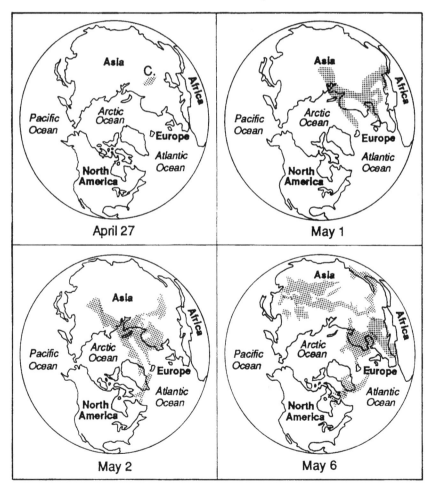

Source: After Enger, E.D. and Smith, B.F. (1995) *Environmental Science: A Study of Interrelationships* (5th edition), Dubuque, IA: Wm C. Brown

was destroyed, releasing large quantities of radioactive material into the **environment**. In addition, the reactor's graphite moderator caught fire, contributing to the continued emission of **radioactivity**. Over a ten-day period hazardous fission products escaped into the **atmosphere**, forming a radioactive cloud which gradually spread over adjacent parts of Europe. The fire was eventually extinguished and the leak plugged by dropping several thousand tonnes of **limestone**, **sand** and boron (B) into the damaged reactor from helicopters. Thirty-one people were killed by the explosion, by burns and by the immediate effects of exposure to high levels of radioactivity, but it is estimated that several thousand will die prematurely over the next 50 years as a result of their exposure to the **radiation**. The greatest effects were and will continue to be in the immediate vicinity of the plant. Some 135,000 people were evacuated from the area, but an estimated 130,000 of them received significant radiation doses, and the incidence of radiation-induced diseases such as thyroid malfunction, anaemia, **cancer** and birth defects rose sharply in the years after the accident. Contamination spread to adjacent areas of Russia, Poland, Belarus and beyond to Scandinavia and Western Europe. To reduce health risks, many countries banned the importation and consumption of food from contaminated **soil**. Even in Britain, which was at the outer edge of the main cloud, the direct contamination of pasture, plus the recycling of **radionuclides** from the soil into vegetation, was sufficient to cause the banning of meat and dairy products. With no past experience of any such event, it is impossible to know what the ultimate effect of the Chernobyl explosion will be in environmental terms. The area around the reactor may never again be safely workable, and even in areas reasonably far removed from the source of radioactivity, the land may remain contaminated for decades or even centuries. The wrecked Chernobyl reactor is now sealed in a concrete sarcophagus, but that has already been weakened by the radioactivity remaining in the site and will need remedial work. Reactors similar to the one which exploded are still in operation in other parts of the former Soviet Union.

See also
Nuclear energy, Nuclear fission, Radiation sickness, Three Mile Island.

Further reading
Anspaugh, L.R., Catlin, R.J. and Goldman, M. (1988) 'The global impact of the Chernobyl reactor accident', *Science* 242 (4885): 1513–19.
Park, C.C. (1989) *Chernobyl; The Long Shadow*, London: Routledge.

CHINA SYNDROME

See **meltdown**.

CHITIN

A tough, flexible polysaccharide, structurally similar to **cellulose**. It is an essential component of the exoskeletons of crustaceans and insects and is also present in the **cell** walls of some **fungi**.

See also
Carbohydrates.

CHLORINATED HYDROCARBONS

See **organochlorides**.

CHLORINE (Cl)

A greenish-yellow **gas** with an irritating smell. Along with **bromine** (Br), **fluorine** (F) and **iodine** (I) it is a member of the **halogen group** of elements. When inhaled it causes choking, and for that reason was used as a weapon during the First World War. Chlorine is a very reactive **element** and is not commonly found free in the **environment**. It occurs mainly in the form of **sodium chloride** (NaCl) or as the chlorides of metals such as **potassium** (K) and **magnesium** (Mg). Most free chlorine used in industry is produced by the **electrolysis** of brine (NaCl). It is used in combination with many organic chemicals to produce **pesticides** (e.g. **organochlorides**), solvents (e.g. **carbon tetrachloride**) and **plastics** (e.g. **PVC**). Chlorine is a strong bleaching agent and is commonly used as a disinfectant in municipal water supply systems. As a constituent of **chlorofluorocarbons** (CFCs), chlorine has been implicated in the destruction of the **ozone layer**.

Further reading
Solarski, R.S. and Cicerone, R.J. (1974) 'Stratospheric chlorine. Possible sink for ozone', *Canadian Journal of Chemistry* 52: 1610–15.

CHLORINE MONOXIDE (CIO)

A compound containing **chlorine** (Cl) and **oxygen** (O), that has been implicated in the destruction of stratospheric **ozone** (O_3). Most chlorine monoxide in the atmosphere is a byproduct of the breakdown of **chlorofluorocarbons** (CFCs). ClO is an integral part of the chlorine catalytic chain, a remarkably efficient destroyer of ozone.

CHLOROFLUOROCARBONS (CFCs)

A group of chemicals containing **chlorine** (Cl), **fluorine** (F) and **carbon** (C), sometimes referred to by their trade name **Freon**. Their stability and low toxicity made them ideal for use as propellants in aerosol spray cans and foaming agents in the production of **polymer** foams. They are widely used in refrigeration and air-conditioning systems. Inert at surface **temperature** and **pressure**, they become unstable in the **stratosphere**, where they break down to release chlorine. The chlorine in turn initiates a catalytic **chain reaction** which leads to the destruction of **ozone** (O_3). The signing of the **Montreal Protocol** in 1987 was the first of a series of agreements aimed at eliminating the production and use of CFCs by the year 2000. CFCs are also powerful **greenhouse gases**.

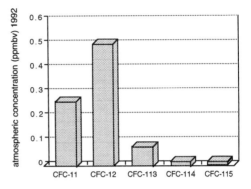

Figure C-10 Atmospheric concentrations of selected CFCs Press

Source: Based on data in IPCC (1996) *Climate Change 1995: The Science of Climate Change*, Cambridge: Cambridge University

See also
Ozone depletion.

Further reading
Downing, R.C. (1988) *Fluorocarbon Refrigerants Handbook*, Englewood Cliffs, NJ: Prentice-Hall.
Molina, M.J. and Rowland, F.S. (1974) 'Stratospheric sink for chlorofluoromethanes: chlorine atom-catalysed destruction of ozone', *Nature* 249: 810–12.
Wofsy, S.C., McElroy, M.B. and Sze, N.D. (1975) 'Freon consumption: implications for atmospheric ozone', *Science* 187: 535–7.

Table C-2 Characteristics and use of some common CFCs

CFC	OZONE DEPLETING POTENTIAL (ODP)	CURRENT OR FORMER USE	PHASE-OUT DATE
CFC-11	1.0	Foaming agent for rigid and flexible foams	1996
CFC-12	1.0	Refrigeration: air conditioning	1996
CFC-113	0.8	Solvent	1996
CFC-115	0.6	Refrigeration: air conditioning	1996

Source: Kemp, D.D. (1994) *Global Environmental Issues: A Climatological Approach* (2nd edition), London/NewYork: Routledge

CHLOROFORM (TRICHLOROMETHANE CHCl₃)

A volatile, sweet-smelling **liquid** once used extensively as an anaesthetic. However, the dosage was critical and it is now considered too dangerous for general use. Exposure to chloroform can lead to liver damage and there is some evidence that it is a **carcinogen**. It remains in use as a **solvent** in the rubber and **plastics** industries, but it is to be phased out there also, since it contributes to **ozone depletion** through the release of **chlorine** (Cl).

CHLOROPHYLL

A green and yellow pigment found mainly in the leaves of plants which makes **photosynthesis** possible through its ability to absorb **solar energy**. Chlorophyll is chemically complex and exists in two forms: chlorophyll-a $(C_{55}H_{72}O_5N_4Mg)$ and chlorophyll-b $(C_{55}H_{70}O_6N_4Mg)$. Plants, such as red and brown **algae** or copper beech trees, which are not green owe their colour to the presence of pigments which mask the greenness of the chlorophyll.

See also
Chloroplast.

CHLOROPLAST

Chloroplasts are complex structures located within plant **cells**. They contain the **chlorophyll** pigments that give green plants their colour, and are therefore the main sites of **photosynthesis**.

CHOLERA

A bacterial disease caused by drinking **water** contaminated by **sewage** or eating food that has not been washed or is inadequately cooked. It is common in areas where the growth in **population** has outstripped the development of facilities for providing clean water or disposing of sewage. The problem is most severe in warm climates, where high temperatures encourage the growth of the cholera **bacteria**. Cholera can erupt following **floods** or **earthquakes** when drinking-water becomes con-taminated by sewage. It can be controlled through the provision of piped and chlorinated drinking-water supplies and the construction of sewage disposal facilities.

See also
Water quality standards.

Further reading
Banarjee, B. and Hazra, J. (1974) *Geoecology of Cholera in West Bengal: A Study in Medical Geography*, Calcutta: K.P. Bagchi & Co.

CHROMOSOME

A strand of genetic material consisting of **DNA** and **protein** found in the **cell** nuclei of plants and animals. Each chromosome consists of a series of **genes** which carry the inherited characteristics of an organism, passed on from one generation to the next via the reproductive cells. Chromosomes occur in pairs, and each **species** has a specific number of chromosomes – twenty-three pairs in humans – carried in every body cell.

CIRCUMPOLAR VORTEX

A band of strong **winds** circling the poles in the upper **atmosphere**. The vortex is mainly a winter phenomenon and best developed around the South Pole. The vortex contributes to the development of the **Antarctic Ozone Hole** by restricting airflow into and out of the polar region.

Figure C-11 The conditions responsible for the creation of the southern hemisphere polar vortex

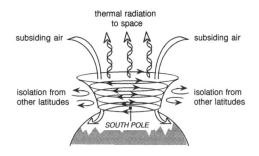

Source: After Turco, R.P. (1997) *Earth Under Siege*, Oxford/New York: Oxford University Press

CLAY

A fine-grained natural sediment with grain sizes less than 2 μm. Such sediments may contain small grains (< 2 μm) of other minerals such as quartz in addition to the more common clay particles.

Further reading
Velde, B. (1995) *Origin and Mineralogy of Clays*, Berlin/New York: Springer-Verlag.

CLAY-HUMUS COMPLEX

A mixture of the finest mineral (**clay**) and organic (**humus**) components of the **soil**. Particles of humus and clay tend to carry negative electrical charges on their surfaces which allows them to attract positively charged nutrient **ions** such as **potassium** (K^+), **calcium** (Ca^{++}) and ammonium (NH_4^+). This attachment prevents such **cations** from being leached out of the **soil profile** by percolating **water**, and makes them available as **nutrients** for growing plants. As well as contributing to soil fertility, the presence of the clay-humus complex also helps to improve the water capacity of the soil, through the ability of both clay and humus to absorb moisture.

CLAY MINERALS

Secondary minerals formed through the **weathering** of bedrock. They can be classified into two main groups – the silicate clays containing mainly **aluminum** (Al) and **magnesium** (Mg) formed in mid-latitudes and the hydrous oxide clays containing **iron** (Fe) and aluminum formed as a result of tropical weathering. Clay minerals have a platy structure that allows them to pack closely together, creating dense impermeable soils with high plasticity. Their finely divided nature also allows them to hold and exchange nutrient **ions**. Thus the presence of clay particles in a soil adds to soil fertility. Clay is also an important industrial mineral. Different forms are used for different purposes. Kaolin or 'china clay' is an important raw material in the ceramics and paper-making industries. Bentonite is used as a drilling mud and as a liner for ditches and settling ponds to seal against **water** loss. Other clays are employed in brick and tile making and the so-called fireclays capable of withstanding very high temperatures are used in the production of refractory materials.

CLEAN AIR LEGISLATION

Various laws, acts and ordinances designed to bring about the reduction of atmospheric **pollution**. When the first modern clean air laws were introduced in the 1950s, most were aimed at controlling visible pollutants such as **smoke** and **soot,** but more recent legislation has included gaseous pollutants such as the **sulphur dioxide** (SO_2) and **oxides of nitrogen** (NO_x) responsible for **acid rain.** In Britain, for example, the original Clean Air Act was passed in 1956, in part as a result of the impact of the **London Smog** of 1952. It included provision for the establishment of smoke control zones in which the type of **fuel** used would be restricted. A later act in 1968, which introduced the **tall stacks policy,** was also mainly concerned with visible pollutants, and it was not until the mid-1980s that legislation was introduced by the UK government to control emissions of sulphur dioxide. In the United States, the initial Clean Air Act was passed in 1963, but was amended in 1970, 1977 and 1990. In federal jurisdictions, such as the US, Australia and Canada, it is common for the federal government to pass blanket legislation that establishes **air quality standards** or timelines for the reduction of **air pollution,** and to expect the states or provinces to work towards meeting them. The states, provinces and (occasionally) municipalities then pass legislation to meet the requirements. In some cases the state and provincial regulations are more stringent than those produced at the federal level. In California, for example, tough air quality standards have been in place since at least the 1940s. As a result of Clean Air legislation, average ambient air pollution levels have fallen in most developed nations. Pollution episodes still occur, however, when emissions are higher than normal or weather conditions prevent the dispersal of the pollutants. Among the developing nations, the maintenance or improvement of air quality lags behind that in the developed world, even where legislation exists. As a result, cities such as Bangkok, New Delhi and Beijing experience serious air pollution.

See also
Photochemical smog, Smog.

Further reading
Edmonds, R.L. (1994) *Patterns of China's Lost Harmony*, London/New York: Routledge.
Smith, Z. A. (1995) *The Environmental Policy Paradox*, Englewood Cliffs, NJ: Prentice-Hall.
Thackery, T.O. (1971) 'Pittsburgh: how one city did it', in R. Revelle, A. Khosla and M. Vinovskis (eds) *The Survival Equation*, Boston, MA: Houghton Mifflin.

CLEAR CUTTING

The complete removal of trees from an area during commercial forestry. Although it is economically the most efficient way to harvest trees, it is no longer considered environmentally sound. Local moisture and **energy** regimes are changed completely and the **soil** is exposed to deterioration and possible **erosion**. Wildlife **habitat** is destroyed and the aesthetic quality of the landscape is reduced. Clear cutting makes **sustainable development** difficult, because the changes it produces in the local **hydrology** and microclimatology mean that it is not always easy or even possible to re-establish the forest. With block or strip cutting, in which sections are left uncut, the environmental changes are much less drastic and natural regeneration is encouraged. Environmental groups such as **Earthfirst** and the **Sierra Club** have fought against clear cutting in North America, and in some areas of old growth forest have worked to prevent cutting of any kind.

Further reading
Maini, J.S. (1990) 'Forests: barometers of environment and economy', in C. Mungall and D.J. McLaren (eds) *Planet under Stress: The Challenge of Global Change*, Toronto: Oxford University Press.

Figure C-12 The edge of a clear cut in the boreal forest. The surface debris or 'slash' is typical.

Photograph: The author

CLIMATE

The combination or aggregate of **weather** conditions experienced in a particular area. It includes averages, extremes and frequencies of such meteorological elements as **temperature, atmospheric pressure, precipitation, wind, humidity** and **sunshine,** measured over an extended period of time – usually a minimum of thirty years.

Further reading
Barry, R.G. and Chorley, R.J. (1992) *Atmosphere, Weather and Climate* (6th edition), London: Routledge.

CLIMATE CHANGE

Because of the integrated nature of the earth/atmosphere system, climatic conditions are inherently variable. Variations in such meteorological conditions as **temperature, precipitation** and **atmospheric pressure** from season to season and from year to year are common, for example. If the variations continue in one direction for a number of years, they can be seen as an expression of climate change. Evidence of past change is available in many forms: for example, biological, stratigraphical, archaeological, agricultural, glaciological or historical. All these sources provide **proxy data,** but change is also indicated in the instrumental record that began to take a more reliable form in the

nineteenth century. Modern research methods also make use of computerized climate models to look forward as well as back. The earth's climate appears to have fluctuated over a wide range in the past, with most areas being, at some time or other, warmer or drier or wetter or colder than they are now. During the **Pleistocene,** for example, higher latitudes experienced a series of **ice ages,** separated by periods of amelioration. Over the past 10,000 years, since the end of the ice ages, variations have continued, with periods of improvement such as the **Climatic Optimum** and periods of deterioration as represented by the **Little Ice Age.** All these changes were generated by natural processes, but in recent years, there are growing indications that human activities have also contributed to change. For many scientists, human-induced climate change ranging from the development of **urban heat islands** to **global warming** is now of more concern than the natural variety. The importance of this attitude can be seen in the definition of climate change adopted by the UN **Framework Convention on Climate Change** (FCCC), which sees climate change as 'attributed directly or indirectly to human activity that alters the composition of the global atmosphere and which is in addition to natural climate variability observed over comparable time periods'. Much of the activity in the study of climate change reflects this interest in the human element. As a result, attention is increasingly being focused

Figure C-13 Sources of evidence of climate change

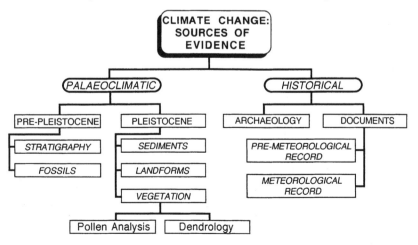

on present and future changes, and their potential impact on society.

See also
Dendroclimatology, General circulation models, Interglacial, Palynology.

Further reading
Bradley, R.S. and Jones, P.D. (1992) *Climate since A.D. 1500*, London: Routledge.
IPCC (1996) *Climate Change 1995: The Science of Climate Change*, Cambridge: Cambridge University Press.
Lamb, H.H. (1977) *Climate: Present, Past and Future, Volume 2, Climate History and the Future*, London: Methuen.
Parry, M. (1990) *Climate Change and World Agriculture*, London: Earthscan.
SMIC (1971) *Inadvertent Climate Modification: Report of the Study of Man's Impact on Climate*, Cambridge, MA: MIT Press.

CLIMATE CHANGE CONVENTION

See **Framework Convention on Climate Change** (FCCC).

CLIMATE CHANGE DETECTION PROJECT (CCDP)

A **WMO** project initiated in 1989 to provide reliable analyses of climate trends, suitable for use in decision making. National meteorological organizations provide data from which global climate baseline data sets can be constructed. These data sets are then analysed to establish the status of global climates, identify any anomalies that may be present and detect possible indications of **climate change**.

CLIMATE IMPACT ASSESSMENT PROGRAM (CIAP)

A program commissioned by the US Department of Transportation in the mid-1970s to study the effects of **supersonic transports** (SSTs) on the **ozone layer**.

CLIMATE LONG-RANGE INVESTIGATION MAPPING AND PREDICTIONS PROJECT (CLIMAP)

A multi-university research project set up in the United States in 1971 to study global climate over the past million years. Most of the data for the project were obtained from ocean sediments and ice cores, and were fed into numerical **models** to produce some of the earliest computer-generated maps of global climate.

Further reading
Calder, N. (1974) *The Weather Machine and the Threat of Ice*, London: BBC.
CLIMAP Project Members (1976) 'The surface of the ice-age earth', *Science* 191: 1131–7.

CLIMATIC OPTIMUM

Period of major warming during the immediate post-glacial period between 5000 and 7000 years ago, also referred to as the altithermal or hypsithermal. **Temperatures** were perhaps 1–3°C higher than at present and changes in **precipitation** amounts and distribution also occurred. As a result of recent **global warming**, current world temperatures are approaching those of the Optimum.

Further reading
Lamb, H.H. (1977) *Climate Present, Past and Future, Volume 2, Climate History and the Future*, London: Methuen.
Lamb, H.H. (1982) *Climate, History and the Modern World*, London: Methuen

CLIMATE RESEARCH UNIT (CRU)

A research institute, based at the University of East Anglia in England, involved in the study of **climate change**, past, present and future. The unit is active in climate modelling studies with the Hadley Centre of the UK Meteorological Office and with the major European modelling centres. Scientists from CRU are participating in a study to reconstruct climatic conditions in Eurasia over the past 10,000 years, using dendroclimatological techniques, and, using historical instrumental data, are assessing ways of establishing the nature and extent of climate variability in Europe over the past 215 years. Research into modern climatological conditions includes participation in the Mediterranean Desertification and Land Use research project. The Climate Research Unit publishes a quarterly journal – *Climate*

Monitor – which provides current summaries of global climatic conditions and maintains an information site – *Tiempo: Global Warming and the Third World* – on the World Wide Web.

See also
Dendroclimatology, Models.

CLIMAX COMMUNITY

A mature community which represents the final stage in a natural **succession,** and which reflects prevailing environmental conditions such as **soil** type and **climate.** Climax communities are characterized by a diverse array of **species** and an ability to use **energy** and to recycle chemicals more efficiently than immature communities. In theory, they are capable of indefinite self-perpetuation under given climatic and edaphic conditions, but if these conditions change, the community will change also. Climatic variations, **fire** and disease are common instigators of natural change, but existing climax communities are being damaged increasingly by human agricultural and industrial activities.

See also
Biome, Edaphic factors, Sere, Succession.

Further reading
Stiling, P.D. (1992) *Introductory Ecology*, Englewood Cliffs, NJ: Prentice-Hall.
Woodward, F.I. (1987) *Climate and Plant Distribution*, Cambridge/New York: Cambridge University Press.

CLONE

A group of genetically identical **cells** or organisms descended from a single individual through asexual reproduction. Among simpler organisms such as **bacteria, algae** and **yeasts,** cloning is brought about by the growth and division of cells, and in some higher plants and animals, such as dandelions and aphids, cloning may result from parthenogenesis – the development of an unfertilized egg into an adult organism. The identical siblings produced during multiple births in humans and other animals are also examples of clones. Artifical cloning has been successful in plant and animal breeding, and important medicines such as insulin and human growth hormone have been produced using the natural cloning properties of bacteria. Cloning has also some potential in the treatment of certain **cancers,** but the possibility of cloning complete human beings through genetic engineering raises many ethical questions.

See also
Chromosome, Gene.

Further reading
Griffin, H. and Wilmut, I. (1997) 'Seven days that shook the world', *New Scientist* 153 (2074): 49.
Wilmut, I, Schnieke, A.E., McWhir, J., Kind, A.J. and Campbell, K.H.S. (1997) 'Viable offspring derived from fetal and adult mammalian cells', *Nature* 385: 810–13.

CLOSED SYSTEM

See **systems.**

CLOUDS

Dense, visible concentrations of suspended **water** droplets or **ice** crystals ranging in diameter from 20–50 μm. Nearly all clouds occur in the **troposphere,** where they are formed when **air** is cooled below its saturation point, usually as a result of the adiabatic expansion which accompanies uplift. In the presence of **condensation nuclei,** the **water vapour** condenses to form droplets of water. The visible cloud base represents the level at which **condensation** begins. Clouds may also be formed by radiative cooling which lowers the **temperature** of the **air** close to the ground sufficiently to cause condensation and produce **fog.** The ability of clouds to reflect or absorb **radiation** gives them an important role in the earth's **energy budget.** Clouds take an infinite variety of forms, but there are sufficient similarities among these forms to allow them to be grouped into a workable classification. The first attempt at classification was made in 1803 by Luke Howard, an English scientist, who grouped clouds according to their form (cumulus – heap; stratus – layer; nimbus – rain; cirrus – wispy) and these descriptors have been included in all subsequent classifications. Current cloud

Table C-3 Ten basic cloud types

	DESCRIPTION
Cirrus (Ci)	High detached fibrous clouds forming filaments, patches or bands
Cirrocumulus (Cc)	Thin, white patches or complete sheets of cloud with regularly arranged ripples or bumps.
Cirrostratus (Cs)	Transparent, white veil cloud, fibrous or smooth in appearance and often producing a halo around the sun or moon.
Altocumulus (Ac)	White and/or grey cloud composed of regularly arranged lumpy, round masses or rolls.
Altostratus (As)	Grey or bluish sheet or layer cloud largely or totally covering the sky, and thin enough in places to reveal the sun.
Nimbostratus (Ns)	Grey, dark layer cloud thick enough to blot out the sun. Lower edge usually diffuse because of falling rain and snow.
Stratocumulus (Sc)	Grey and/or white patches or complete layers of cloud composed of tessellations, rounded masses or rolls.
Stratus (St)	Grey layer cloud with a generally uniform base, producing drizzle, ice crystals or snow grains.
Cumulus (Cu)	Dense, detached clouds with sharp outlines. They develop vertically as rising mounds or domes from a nearly horizontal base. Sharp contrast between the brilliant white sunlit parts of the cloud and the dark base.
Cumulonimbus (Cb)	Dark, dense cloud well developed vertically into high towers. The upper part of the cloud is usually smooth or fibrous and flattened out into the shape of an anvil. The base of the cloud is dark with ragged clouds and heavy precipitation.

classification is based on that derived by the **World Meteorological Organization** (WMO) and published in its *International Cloud Atlas* (1956). It identifies ten basic cloud types that incorporate descriptive and altitudinal references (strato – low level; alto – mid level; cirro – high level). Thus, a high level layer cloud would be classifed as cirrostratus.

Further reading
Meteorological Office (UK) (1986) *Cloud Types for Observers*, Edinburgh: HMSO.
WMO (1956) *International Cloud Atlas*, Geneva: WMO.

CLUB OF ROME

A group of economists, politicians, scientists, educators, industrialists, humanists and civil servants who came together in 1968 to bring their collective intelligence and experience to bear on the present and future predicament of mankind. They aimed to foster better understanding of the various interdependent components – natural and human – that make up the global system, and through that understanding encourage policy makers and the public to initiate policies that would benefit society, and in some cases might be necessary for its survival. The Club initiated a 'Project on the Predicament of Mankind', which produced a computer analysis of the status of the world in terms of its **population, resources,** industrial production and **pollution – Limits to Growth** – and a follow-up report that examined alternative patterns of world development – **Mankind at the Turning Point**.

Further reading
Meadows, D.H., Meadows, D.L., Randers, J. and Behrens, W.W. (1972) *Limits to Growth*, New

York: Universe Books.
Mesarovic, M. and Pestel, E. (1975) *Mankind at the Turning Point*, London: Hutchinson.

COAGULATION

The clustering of finely divided particles in suspension to form larger particles which may then be sufficiently heavy to fall out of suspension. Coagulation is one of the environment's cleansing mechanisms, contributing to the removal of **soot** particles from the **atmosphere**, for example.

COAL

A black or brown combustible material composed of **carbon** (C), various carbon **compounds** and other minerals such as sulphur (S). The most abundant of the **fossil fuels**, it was formed through the accumulation of vegetable matter over millions of years in **environments** (e.g. swamps, deltas) which reduced the rate of decay of the organic material and allowed the preservation of the **solar energy** absorbed by it when it was growing. When coal is burned it is that energy which is released. Most coal was formed from plant material – the humic coals – but some resulted from the accumulation of finely divided vegetable matter such as **spore** cases, **algae** and fungal material – the sapropelic coals. The compression and **temperature** increase brought about by the deposition of many metres of sediment on top of such organic matter caused the progressive elimination of moisture and **volatile** constituents and a subsequent increase in the proportion of carbon in the deposit. The classification of coal is based on its carbon content, with **peat** having least carbon and the proportion increasing through lignite and bituminous coal to anthracite. The increased use of coal as a **fuel** and an industrial raw material during the **Industrial Revolution** laid the foundations for modern society. Coal use peaked initially at the time of the First World War, after which it suffered in competition with **petroleum** and electricity. In the mid-1970s, its position improved again as a result of the 1973 oil crisis, and its use in developing nations such as China and India has continued to grow. At the present rate of use it is estimated that there is enough coal to last for another 300 to 500 years. The environmental costs of coal use have been high, both in extraction and use. Underground mining is costly in human terms and surface mining has despoiled the landscape in many areas. **Acid mine drainage** is common to both. The use of coal has contributed to such major environmental issues as **acid rain**, **atmospheric turbidity** and **global warming**. Although some improvement has been made in dealing with the environmental problems of coal, near-term solutions are unlikely to be found as long as modern industrial societies continue to depend on coal as an energy source and a raw material.

Table C-4 Carbon content of main types of coal

CARBON (%)	
Peat	<57
Lignite	*c.*70
Bituminous coal	*c.*85
Anthracite	*c.*94

Table C-5 World coal reserves (percentage by region)

REGION	RESERVES (%)
Russia, Ukraine, Eastern Europe	31
Western Europe	8
North America	23
Latin America	1
Africa	6
China	23
Australia, Asia	8

Further reading
Berkowitz, N. (1993) *An Introduction to Coal Technology* (2nd edition), San Diego: Academic Press.

Kleinbach, M.H. and Salvagin, C.E. (1986) *Energy Technologies and Conversion Systems*, Englewood Cliffs, NJ: Prentice-Hall.
Nisbet, E.G. (1991) *Leaving Eden: To Protect and Manage the Earth*, Cambridge: Cambridge University Press.

COAL GASIFICATION

The heating or cooking of **coal** – sometimes in place – to release **volatile** gases such as **methane** (CH_4), a cleaner, more efficient **fuel** than the coal itself.

See also
Synfuels.

Further reading
Supp, E. (1990) *How to Produce Methanol from Coal*, Berlin/New York: Springer-Verlag.

COAL LIQUEFACTION

The conversion of **coal** into a **liquid, petroleum**-type **fuel** by a combination of heating and the use of **solvents** and **catalysts**. The resulting synthetic fuel is generally cleaner burning than the original coal.

See also
Synfuels.

Further reading
Schlosberg, R.H. (1985) *Chemistry of Coal Conversion*, New York: Plenum Press.

COEVOLUTION

The **evolution** of traits in two or more **species** to allow mutually beneficial interactions among them. For example, the relationship between certain insects and the types of flowers that they are able to pollinate may result from coevolution.

COGENERATION

The combined production of heat and **power** from one plant. The power is usually in the form of electricity and the heat in the form of steam produced from a single **fuel** source. In a standard thermal electricity generating plant, the heat in the high pressure steam used to power the generators is allowed to escape into the **environment** unused. In a cogeneration plant, that heat would be used for industrial processing, space heating or to provide hot **water** for domestic use. Cogeneration can also be applied to diesel or gas turbine electricity generators in which the hot exhaust gases are used to heat water. In the case of the gas turbine, the exhaust gases are hot enough to produce high pressure steam that can be used to generate additional electricity. Cogeneration increases the efficiency of the fuel conversion process. Coal-fired generators, for example, convert as little as 30-40 per cent of the **energy** available in the **coal** into electricity. In a cogeneration plant, as much as 80 per cent of the energy in the coal is made available as heat or electricity.

Further reading
Hay, N.E. (ed.) (1992) *Guide to Natural Gas Cogeneration*, Lilburn, GA: Fairmont Press.
Kleinbach, M.H. and Salvagin, C.E. (1986) *Energy Technologies and Conversion Systems*, Englewood Cliffs, NJ: Prentice-Hall.

COKE

The porous, solid residue that remains when tar and **gas** have been driven off during the **destructive distillation** of **coal**. In the process, one tonne of coal will produce about 0.7 tonnes of coke which is between 80 and 90 per cent **carbon** (C). Originally introduced as a replacement for **charcoal** in the **iron** (Fe) and steel industry during the **Industrial Revolution**, it is still used in that and other metallurgical industries. It is also used as a smokeless **fuel**.

COLD WAR

The period between the end of the Second World War and the late 1980s, when the nations of the North Atlantic Treaty Organization and the Warsaw Pact maintained an antagonistic relationship without becoming involved in a major conflict, or 'hot' war.

See also
Nuclear winter.

COLIFORM BACTERIA

Bacteria found in the intestinal tract of human

beings and other mammals, and used as an indicator of **water quality**. The presence of coliforms in a water system indicates that the water is polluted, possibly by human **sewage**, and that there may be other pathogenic bacteria present. Cases of food poisoning caused by the consumption of meat and vegetables contaminated by Escherichia coli are not uncommon, and larger scale outbreaks, such as those in Japan and Scotland in 1996 and 1997, can lead to widespread sickness and even death.

Further reading
Neidhardt, F.C. (ed.) (1987) *Escherichia coli and Salmonella typhimurium Cellular and Molecular Biology*, Washington, DC: American Society of Microbiology.

COLLAGEN

A fibrous **protein** which is a major constituent of bone, cartilage, skin and other connective tissue in animals. Tendons are almost pure collagen. Collagen molecules are arranged in helical chains which provide both strength and flexibility. Damage to the collagen fibres in connective tissue can lead to physiological problems such as rheumatoid arthritis. When boiled down with **water**, collagen is converted into a gelatin, which is used in the food, photographic and textile industries.

COLLOID

Very small organic or inorganic particles – usually finer than 0.01 µm – capable of remaining indefinitely in suspension in a **liquid**. Such a colloidal suspension acts in many ways like a **solution**, although the colloids are much larger than the particles in a true solution. **Clay** and humic colloids are important in maintaining soil fertility. Being negatively charged and with a large surface area per unit mass, they are capable of attracting and holding nutrient **cations**, making them available to growing plants.

See also
Humus.

COMBINED SEWER SYSTEM

A sewer system in which the pipes carry domestic and industrial **sewage** as well as **rain** and storm water. Heavy rain, sudden storms or rapid **snow** melting can overwhelm the system, leading to flooding, damage to sewage plants and **pollution**. Most modern sewage systems carry **waste** water and rainwater in separate pipes.

See also
Separate sewage systems.

COMBUSTION

A chemical reaction in which a substance combines with **oxygen** (O) (**oxidation**) to release **energy** in the form of heat and light. The oxygen may be free or in the form of a **compound** such as nitric acid (HNO_3) or hydrogen peroxide (H_2O_2). At the molecular level, the process begins with the formation of **free radicals** and the initiation of the **chain reactions** that release the energy. If the reactions are very rapid, the energy may be released with explosive force. Most of the energy currently used by society is provided by the combustion of **fossil fuels**, in which the **carbon** (C) and **hydrogen** (H) in the **fuels** are converted into **carbon dioxide** (CO_2) and **water** respectively, as follows:.

$$\text{energy+}$$
$$C + O_2 \rightarrow CO_2$$
$$\text{energy+}$$
$$2H_2 + O_2 \rightarrow 2H_2O \text{ (steam)}$$

These are the products of complete combustion. If combustion is incomplete – for example, because of insufficient oxygen – **environmental pollution** may occur in the form of unburned carbon, excess **smoke** or **carbon monoxide** (CO) gas.

See also
Metabolism.

Further reading
Bartok, W. and Sarofim, A.E. (eds) (1991) *Fossil Fuel Combustion: A Source Book*, New York: Wiley.

COMMENSALISM

A close association between organisms of different **species** in which one species benefits whereas the other is apparently neither helped nor harmed. Clown fish, for example, live unharmed and protected among the poisonous tentacles of the sea anemone, feeding on the food remnants unconsumed by the sea anemone. The sea anemone seems neither to suffer nor to benefit from the relationship.

Further reading
Gotto, R.V. (1969) *Marine Animals: Partnerships and other Associations*, London: English University Press.

COMMONWEALTH EXPERT GROUP ON CLIMATE CHANGE

A group of experts brought together by the Secretary-General of the British Commonwealth in 1988 to examine the implications of global **climate change**, with particular emphasis on its impacts on Commonwealth countries. The report of the group highlighted the problems that would be faced by developing nations, and provided practical

Figure C-14 A backyard composter. Provided free of charge or at minimal cost in many parts of North America to encourage waste minimization through recycling

Photograph: The author

suggestions on how the impacts could be mitigated. As with many similar groups, its report emphasized the importance of high-quality data collection in the monitoring and analysis of the changes.

COMMUNITY

A general term for a group of organisms living and interacting in a common **environment** at a particular time. The size of the group and the degree of interaction are variable.

See also
Animal community, Plant community.

Further reading
Brewer, R. (1993) *The Science of Ecology* (2nd edition), Philadelphia, PA: W.B. Saunders.

COMPOST

A **soil** conditioner and **fertilizer** produced by the controlled decay of organic matter. It depends upon the ability of **aerobic bacteria** and other micro-organisms to decompose leaves, grass, domestic vegetable refuse and other organic **waste**. The process may take place in unconfined piles or in manufactured wooden or plastic composters. The end-product, after six months to several years of decomposition, is **humus**. Compost is used mainly by gardeners, but large-scale composting has been used with some success in tropical countries such as India, where the need to maintain soil fertility is crucial, and where high **temperatures** promote rapid decomposition. National governments in Europe – for example, the Netherlands, Germany, Sweden and Italy – have built large-scale composting plants, while many North American municipalities have provided composters free or at minimal cost, to encourage environmentally appropriate behaviour and reduce the amount of **garbage** sent to **landfill** sites.

See also
Recycling.

Further reading
Polprasert, C. (1989) *Organic Waste Recycling*, New York: John Wiley.

COMPOUND

A substance containing two or more **elements** united by chemical bonds. Compounds are created by chemical reactions and cannot be separated by mechanical or physical techniques. The elements involved lose their individual properties when they form a compound. For example, common **salt** (NaCl), a white crystalline **solid**, is formed by the combination of a highly reactive **metal** (**sodium** (Na)) and a poisonous **gas** (chlorine (Cl)), yet displays none of the characteristics of its constituents. Specific compounds contain a fixed proportion of elements and exhibit a constant set of properties. **Water**, for example, always consists of two **hydrogen** (H) **atoms** and one **oxygen** (O) atom and will always freeze at 0°C.

See also
Mixture.

COMPREHENSIVE ENVIRONMENT RESPONSE COMPENSATION AND LIABILITY ACT (CERCLA)

See **Superfund**.

CONCORDE

The more successful of the two types of **supersonic transport** built in the 1970s. It was the centre of much controversy at the time of its introduction because of its perceived threat to the **environment** in general and the **ozone layer** in particular.

See also
Sonic boom, Tu-144.

CONDENSATE

Light **petroleum hydrocarbons** (pentanes) present in an **oil** reservoir as **gas**, but separating out as a **liquid** when pumped to the surface.

CONDENSATION

The conversion of a **vapour** into a **liquid**. A common process in the chemical and petroleum industries, condensation in the environment is most frequently associated with the change of water vapour into liquid water in the atmospheric environment. It is generally brought about by cooling, which reduces the ability of the atmosphere to absorb and retain water vapour. If a parcel of air containing a specific volume of water vapour is progressively cooled, it will reach a temperature at which it is completely saturated – the dewpoint temperature of that parcel of air. Any additional cooling beyond the dewpoint will cause the condensation of some of the vapour. This process frequently accompanies adiabatic cooling as air rises or is associated with the passage of relatively warm air across a cold surface. Condensation is accompanied by the release of latent heat.

See also
Clouds, Condensation nuclei, Humidity.

Further reading
Moran, J.M. and Morgan, M.D. (1997) Meteorology: The Atmosphere and Science of Weather, Englewood Cliffs, NJ: Prentice-Hall.

CONDENSATION NUCLEI

Small particles in the atmosphere around which water vapour condenses to form liquid droplets. In the absence of condensation nuclei, droplets will not form even when the temperature falls below the dewpoint. The air then becomes supersaturated – holding more moisture than theoretically possible at that temperature. Although natural condensation nuclei such as dust or salt particles have always been present in the environment, human activities now make a significant contribution to the production of nuclei.

CONDUCTION

The transmission of thermal energy directly through matter from places or objects of higher temperature to those of lower temperature. For conduction to take place between objects, they must be in direct contact. Transfer of energy takes place at the atomic or molecular level and continues until both objects reach the same temperature. Some substances are better conductors than others – air or materials containing a high proportion of air are poor conductors, for example, whereas most metals are good conductors. Electrical conduction involves the free flow of an electric current through a conductor. As with thermal conduction, transmission takes place at the atomic level, the current flow being associated with the movement of electrons along the conductor.

See also
Insulation.

Further reading
Kakah, S. and Yener, Y. (1993) Heat Conduction (3rd edition), Washington, DC: Taylor & Francis.

CONSERVATION

Originally, conservation involved the maintenance of the status quo through the preservation and protection of natural resources such as flora, fauna and physiological features. That concept has come to be seen as having negative connotations, and the modern approach to conservation includes not only preservation and protection, but also the planning and management of resources to allow both use and continuity of supply. It may also involve the enhancement of the quality of some resources (for example, soils) and attempts to return mismanaged resources to their former state. Although still very much associated with natural materials, the conservation ethic is also applied to human resources such as historic buildings and artefacts.

See also
Recycling, Resources, Soil conservation.

Further reading
Adams, W.M. (1996) Future Nature: A Vision for Conservation, London: Earthscan.
Owen, O.S. and Chiras, D.D. (1995) Natural Resource Conservation (6th edition), Englewood Cliffs, NJ: Prentice-Hall.

CONSUMPTIVE USE

The use of a resource in such a way that its form or content is changed, and it is no longer available for its original use. The consumptive use of **water**, for example, results in the water being evaporated, rather than returned to the system as surface water or **groundwater**. The use of **fossil fuel** energy resources is also entirely consumptive.

Further reading
Skinner, B. (1986) *Earth Resources* (3rd edition), Englewood Cliffs, NJ: Prentice-Hall.

CONTINENTAL TROPICAL AIR MASS (CT)

An **air mass** originating over tropical to subtropical continental areas, and therefore hot and dry. It is commonly associated with the subsiding **air** beneath the poleward side of tropical **Hadley Cells**, which also contributes to the warmth and dryness. Continental tropical air is well developed over the Sahara Desert.

CONTINGENT DROUGHT

One of the four types of **drought** identified by **C.W. Thornthwaite**. It is characterized by irregular and variable **precipitation** in areas that normally have an adequate supply of moisture to meet crop needs. Serious problems arise because the agricultural system is not set up to cope with unpredictable and sometimes lengthy periods of inadequate precipitation. Droughts such as those in Britain in 1975–1976 and 1988–1992 fit the contingent category.

Further reading
Gregory, K.J. and Doornkamp, J.C. (1980) *Atlas of Drought*, London: Institute of British Geographers.

CONTOUR PLOUGHING

A method of cultivation employed to reduce **soil erosion**. Ploughing to prepare the **soil** for planting is carried out across the slope rather than up and down it. This slows the flow of **water** down the slope and reduces the potential for soil erosion.

See also
Soil conservation.

CONTROLLED TIPPING

See **sanitary landfill**.

CONVECTION

The vertical transfer of heat through a **liquid** or **gas** by the movement of the liquid or gas. The fluid in contact with the heat source warms up, expands, becomes less dense and rises through the surrounding gas or liquid. The process is common in the **atmosphere**, where **air** heated by **conduction** from the warm land surface rises. As it rises, it also cools adiabatically and ultimately returns to the surface to complete the convective circulation. If the air rises above the condensation level, **clouds** will form and **rain** may follow. Convective activity is a common cause of **precipitation**, particularly in tropical areas, where heating is intense and uplift therefore rapid. Tropical **Hadley Cells** are created by convective circulation.

See also
Advection.

Further reading
Bejan, A. (1984) *Convection Heat Transfer*, New York: Wiley.

Figure C-15 The cause and characteristics of convection

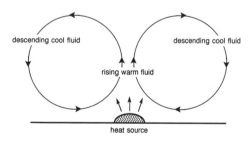

CONVENTION ON INTERNATIONAL TRADE IN ENDANGERED SPECIES (CITES)

An agreement drawn up in 1973 to protect a wide variety of animals and plants thought to be at risk of **extinction**. It has been signed by 125 nations, and protects nearly 90 **species** of plants and 400 species of animals, by prohibiting trade in the species or products made from them. Protection can never be complete, but CITES has been instrumental in restricting international trade in ivory, rhinoceros horn, exotic skins and animal parts used in medicine. The publicity received when the restrictions are enforced has helped to increase public support for the convention.

CONVERGENCE

Horizontal motion in the **atmosphere** that leads to the confluence of air streams and induces uplift. For example, two **air masses** of similar strength and **temperature** approaching from different directions will be forced to rise when they meet. If one of the air masses is warmer, it will rise above the other when the two masses converge. Convergence can also take place if one part of a moving air mass is forced to slow down – for example, as a result of **friction**. The more rapidly moving following air will push into the rear of the slower moving air forcing it to rise. The circulation of air around a low pressure system has a convergent motion to it at the surface, which causes a net inflow of air into the centre of the low and contributes to vertical motion in the system. As a result of the associated uplift, convergence in the atmosphere is often associated with cooling, **condensation** and **precipitation**.

COOLANT

A fluid used to extract heat from a heat source and thus cause cooling. For example, the water/antifreeze coolant mixture circulating in an **internal combustion engine** removes excess heat from the engine block and transfers it to the **atmosphere** via a **radiator**. The **gas**, **water** or **liquid sodium** (Na) coolants used in **nuclear reactors** not only maintain the reactors at a suitable operating temperature, but also carry fission-generated heat from the reactor core to the steam plant to produce electricity. Failure of the coolant supply will allow the system to overheat with consequences that might range from a seized engine to a reactor **meltdown**.

See also
Chernobyl, Three Mile Island.

COPPER (Cu)

A red-brown **metal** that is readily malleable and ductile. A very efficient conductor of electricity, it occurs as free **metal** – native copper – or in a number of **ores**, including cuprite (Cu_2O), chalcocite (Cu_2S) and chalcopyrite ($CuFeS_2$). Most copper is extracted from the sulphide ores, by a combination of roasting, smelting and electrolytic refining. The **sulphur dioxide** (SO_2) produced during roasting and smelting contributes to **acid rain**. Copper is widely used in the electrical industry, and is a constituent of a number of **alloys** such as bronze (copper plus tin (Sn)) and brass (copper plus zinc (Zn)). Occurring free in the environment and easily worked, it was one of the first metals used by many primitive societies. In the form of malachite (copper carbonate – $CuCO_3.Cu(OH)_2$), a bright green striped mineral, it is also used as a gemstone.

Further reading
Bowen, R. and Gunatilaka, A. (1977) *Copper, its Geology and Economics*, London: Applied Science Publishers.

CORIOLIS EFFECT

The effect, caused by the rotation of the earth, that brings about the apparent deflection of objects moving across the earth's surface or in the **atmosphere**. Deflection is to the right in the northern hemisphere and to the left in the southern hemisphere. The effect is greater at higher latitudes. Although sometimes referred to as the 'Coriolis force', it is not a force in the true sense of the word, since no outside agency is involved in the apparent deflection. It is named after the French engineer Gaspard de Coriolis (1792–1843).

CORROSION

The deterioration of a substance by chemical or electrochemical action. It applies to rocks in the **natural environment**, but it is most commonly used to describe the deterioration of **metals**. The rusting of **iron** (Fe) is an ubiquitous form of corrosion caused by the **oxidation** of the iron into hydrated iron oxide ($Fe_2O_3.H_2O$). On some metals, the process is self-limiting. The corrosion of **aluminum** (Al), for example, produces a surface layer of aluminum oxide (Al_2O_3) which prevents additional **oxygen** (O) from reaching the surface and therefore inhibits further corrosion. Since the corrosion products are generally weaker than the original **metal**, corrosion destroys the integrity of metal structures. Protection against corrosion is possible using paint, **plastic** and rubber as sealants to exclude moisture and oxygen, but the repair and replacement of corrosion-damaged structures worldwide costs billions of dollars annually.

Further reading
Trethewey, K.R. and Chamberlain, J. (1988) *Corrosion for Students of Science and Engineering*, Harlow, Essex/New York: Longman/Wiley.

COSMIC RADIATION

High-energy **radiation** reaching the earth from outside the solar system – galactic cosmic rays – or from the sun – solar cosmic rays. The galactic cosmic rays are high speed sub-atomic particles, mainly **protons** that may represent the remnants of **supernova** explosions. The action of cosmic rays on gases in the upper atmosphere may produce sufficient **oxides of nitrogen** (NO_x) to contribute to **ozone depletion**.

Further reading
Ruderman, M.A. (1974) 'Possible consequences of nearby supernova explosions for atmospheric ozone and terrestrial life', *Science* 184: 1079–81.

COUPLED MODELS

Models which combine two or more simulations of **elements** in the earth/atmosphere system. **Carbon cycle models**, for example, have been combined with oceanic and atmospheric circulation models in the search for a better understanding of **global warming**. In theory, such combinations should provide a more accurate representation of the earth's **climate**, but the coupled models retain any errors included in the individual simulations.

This remains an important constraint for coupled models.

See also
Coupled ocean/atmosphere climate models.

COUPLED OCEAN/ATMOSPHERE CLIMATE MODELS

The most common combination in **coupled models**. The main problem with these models is the difference in time scales over which atmospheric and oceanic phenomena develop and respond to change. Because of the relatively slow response time of the oceanic circulation, the oceanic element in most coupled models is much less comprehensive than the atmospheric element.

See also
Slab models.

CRITICAL MASS

The minimum mass of fissionable material required to sustain a **chain reaction** in a **nuclear bomb** or **nuclear reactor**. At less than the critical mass, too many **neutrons** escape without splitting additional **atoms** for the reaction to be self-sustaining.

See also
Nuclear fission.

CROP ROTATION

An **arable agriculture** system in which a field is planted with different crops in a regular sequence over a set period of years. Since different crops have different demands on the **nutrients** in the **soil**, this ensures that the soil is not depleted of specific nutrients, as would happen if no rotation was followed. A corn crop, which has high **nitrogen** (N) demands, for example, would be followed by a crop of **leguminous plants** such as clover, peas or beans to help restore nitrogen levels in the soil. The inclusion of grass in the rotation helps the soil to rest and improves its **crumb structure**. Crop rotation also helps to prevent the growth of pests in the soil.

Further reading
Follet, R.F. and Stewart, B.A. (eds) (1985) *Soil Erosion and Crop Productivity*, Madison, WI: American Society of Agronomy.

CRUMB STRUCTURE

The combination of individual **soil** particles into loose aggregates or crumbs. This is brought about in part by **humus** which helps to bind groups of mineral particles together. The resulting structure provides pore spaces that improve the **water**-holding capacity of the soil and allow for easy penetration by plant roots.

CRUST

The outermost solid part of the earth which together with the upper mantle makes up the **lithosphere**. It varies in thickness from about 6 km in the ocean basins to 35–70 km in the continents. Composition also varies from basaltic **igneous rocks** in the ocean basins to **sedimentary**, igneous and **metamorphic rocks** in the continental areas.

Further reading
Brown, G.C. and Mussett, A.E. (1993) *The Inaccessible Earth: An Integrated View of its Structure and Composition* (2nd edition), New York: Chapman and Hall.

CRYSTAL

A substance in which the **atoms** or **molecules** are arranged in an orderly pattern to produce a characteristic three-dimensional, geometrical form (e.g. cube, rhombus, hexagon). Solids which do not form crystals are described as amorphous.

CURBSIDE COLLECTION

See **blue box**.

CURIE

A unit of **radioactivity** formerly widely used but now replaced by the **becquerel**. It is based on the radioactivity of **radium** (Ra) – 3.7×10^{10} disintegrations per second – against which the radioactivity of other substances is compared. The unit was named after Marie Curie, the Polish-born physicist (1867–1934).

CYANOBACTERIA

Formerly classified as blue-green **algae**, cyanobacteria are single-celled microscopic organisms which contain **chlorophyll** and are therefore capable of **photosynthesis**. Currently widespread in **salt** and fresh **water** and in **soils**, **fossil** evidence suggests that cyanobacteria have been present on earth for more than 3 billion years, and through the release of **oxygen** (O) during **photosynthesis** may have contributed to the development of the current gaseous composition of the **atmosphere**.

Further reading
Miller, G.T. (1994) *Living in the Environment*, Belmont, CA: Wadsworth.

CYCLONE

An atmospheric low pressure **system**, generally circular in shape, with the airflow counterclockwise in the northern hemisphere and clockwise in the southern hemisphere and converging towards the centre of the system. Although the term may be used for mid-latitude frontal **depressions**, which display such characteristics, it is commonly used to refer to the intense tropical storms in the Indian Ocean which are the equivalent of Atlantic **hurricanes** or Pacific **typhoons**.

See also
Tropical cyclone.

Further reading
Henderson-Sellars, A. and Robinson, P.J. (1986) *Contemporary Climatology*, London/New York: Longman/Wiley.

D

DAMS

Structures designed to restrict the flow of surface **runoff**, usually to control flooding or to provide **water** for **irrigation** or the production of **hydroelectricity**. They range from relatively small earth-fill features, ponding back thousands of cubic metres of water, to massive reinforced concrete structures which create **reservoirs** containing several billion cubic metres of water. All dams, large or small, have an environmental impact on the area in which they are built. The obvious change is in the **hydrological cycle**, but environmental interrelationships ensure that the effects are felt in the local climatology and in the **flora** and **fauna** of the region.

See also
Reservoirs – environmental effects.

Further reading
Petts, G.E. (1984) *Impounded Rivers: Perspectives for Ecological Management*, Chichester/New York: Wiley.
Jansen, R.B. (ed.) (1988) *Advanced Dam Engineering for Design, Construction and Rehabilitation*, New York: Van Nostrand Reinhold.

DARWIN, C.R. (1808–1882)

Nineteenth-century biologist usually credited with the development of the theory of **evolution**, which he termed 'descent with modification'. Much of the data on which his theory was based were collected during his voyage on the *HMS Beagle* around the coasts of South America in the 1830s, but he continued his work for almost another 20 years before publishing the results in his classic volumes, *Origin of Species* (1859) and *The Descent of Man* (1871). The central themes of the theory were developed around the concept that all organisms were descended from common ancestors, and as they evolved they were subject to natural selection. The latter is commonly referred to as the 'survival of the fittest', in which **species** able to adapt to a particular **environment** or a change in the environment survive, whereas those unable to adapt ultimately became extinct. In developing his evolutionary theories, Darwin may have been influenced by the writings of the geologist Charles Lyell, who first recognized the role of gradual – as opposed to catastrophic – change in the physical environment, and by the work of Thomas **Malthus** on the relationship between **population** growth and food supply. At about the same time, ideas similar to those of Darwin were being examined by Alfred Russel Wallace. The two men corresponded and their theories were presented jointly at the Linnean Society of London in 1858, but the evolutionary concepts have come to be associated almost exclusively with Darwin, and this is reflected in their popular representation by the term 'Darwinism'. The work of Darwin and the other evolutionary scientists was controversial, and gave rise to a debate on the relative merits of creation and evolution which is yet to be resolved.

See also
Lamarck, J.B. de.

Further reading
Desmond, A. and Moore, S. (1992) *Darwin*, London: Michael Joseph.
Stiling, P.D. (1992) *Introductory Ecology*, Englewood Cliffs, NJ: Prentice-Hall.

'DAUGHTER' PRODUCT

A radioactive decay product. **Radon**, for example, is a decay product of ^{226}**radium** (Ra), and in turn produces its own daughters as it undergoes spontaneous decay.

DECIDUOUS

Describing an organism which sheds parts of itself at a particular time, season or stage of growth. In the **plant community**, deciduous trees shed their leaves annually to allow them to cope with the stress of cold or dry conditions. Without leaves, for example, **transpiration** is reduced, allowing the trees to survive when **soil** moisture is low or unavailable. Among animals, the loss of milk-teeth as infants mature or the annual shedding of horns by deer are examples of deciduous processes.

DECOMMISSIONING WASTES

Waste material created when a **nuclear reactor** site is closed. Such wastes create major disposal problems. Areas in and adjacent to the reactor chamber, for example, contain materials which remain radioactive for thousands of years. Reactors cannot therefore be demolished like normal structures. The buildings must remain isolated on site or perhaps dismantled and placed in more secure and permanent storage. As the nuclear power stations built in the 1960s and 1970s begin to age, the problem of dealing with decommissioning wastes will grow.

See also
Radioactivity.

Further reading
Pasqualetti, M.J. (1990) *Nuclear Decommissioning and Society: Public Links to a Technical Task*, New York: Routledge & Kegan Paul.

DECOMPOSERS

Organisms such as **bacteria** and **fungi** that break down dead organic material. Also called saprophytes, they release digestive **enzymes** that reduce complex organic **molecules** into simpler inorganic **compounds**. The chemicals are recycled back into the **environment** where they can be reused.

DEFLATION

The lifting and transportation of fine particles from loose, dry surfaces by **wind**. In an area where a mixture of particle sizes is present, deflation will tend to remove the smaller particles such as **clay**, **silt** and **sand** and leave behind gravel and pebbles. Where the process is concentrated in a particular area it may create a shallow depression called a deflation hollow or blowout. Although most common in **desert** or semi-desert areas, deflation can also occur on beaches and in dry areas where the vegetation cover has been destroyed locally – for example, by overgrazing. Deflation can be an important agent of environmental change in such areas, but it also brings about change in the areas where the the deflated material is redeposited.

DEFORESTATION

The clearing of forested areas as part of a commercial forestry enterprise or for some other economic purpose such as the expansion of settlement or the development of agriculture. Deforestation has traditionally been associated with economic development, but it is no longer characteristic of developed nations. In such areas, all the trees that are likely to be removed have already been removed. There are exceptions, however. In Canada, for example, large areas are cleared every year in the commercial logging of the **boreal forest**. There and in other areas of the developed world, where forestry continues to be an important element in the economy, good forest management requires that deforested areas be regenerated, but the process is not always effective. Currently, deforestation is most extensive in the world's less-developed nations, particularly those in the tropics. **Clear cutting** and burning of the tropical **rainforest** in South and Central America, Africa and South East

Table D-1 Estimated annual deforestation rates in selected low- to middle-income economies in the 1980s

	'000 SQ KM		'000 SQ KM		'000 SQ KM
Brazil	13.8	Nigeria	4.0	Indonesia	10.0
Mexico	10.0	Zaire	3.7	Malaysia	2.7
Colombia	8.9	Tanzania	3.0	Thailand	2.4
Paraguay	4.5	Cameroon	1.9	Philippines	1.4

Source: Compiled from data in World Resources Institute (1992) *World Resources 1992–93 A Guide to the Global Environment*, New York: Oxford University Press

Asia is proceeding rapidly, although the exact rate is difficult to assess accurately. Estimates for the Amazon region in Brazil, for example, have ranged from 13,800 to 21,000 sq km per annum. Whatever the true value, given the demographic and economic pressures on these areas, the regeneration of such a large area of rainforest is unlikely.

The impact of deforestation extends beyond the destruction of the trees. It destroys the **habitat** for other plants and for forest-dwelling animals. By opening up the **environment**, and altering such elements as **radiation, temperature, wind** and moisture, deforestation also produces significant micro-climatological changes. The higher levels of **solar radiation** reaching the surface cause temperatures to rise, accelerating the decomposition of organic matter. Temperature ranges and wind speeds increase following deforestation and the intensity of the rainfall reaching the surface is higher. The latter causes more rapid **leaching** of the **nutrients** released from the organic matter. **Runoff** also increases, creating a greater potential for **soil erosion**, particularly in hilly areas. On a larger scale, deforestation may contribute to the enhancement of the **greenhouse effect**. Much of the deforestation in tropical regions is achieved through burning which releases **carbon dioxide** (CO_2) directly into the atmosphere. In addition, growing forests are significant **carbon** (C) **sinks**, and once they are removed, the **recycling** of carbon dioxide is reduced, allowing it to remain in the **atmosphere** to enhance **global warming** (see Figures C-5 and C-12).

Further reading
Colchester, M. and Lohmann, L. (eds) (1992) *The Struggle for Land and the Fate of the Forests*, London/Atlantic Highlands, NJ: Zed Books.
Kummer, D.M. (1992) *Deforestation in the Postwar Philippines*, Chicago: University of Chicago Press.
Mather, A.S. (1990) *Global Forest Resources*, London: Benaven Press.
Monastersky, R. (1993) 'The deforestation debate', *Science News*, 10 July: 26–7.
Richards, J.F. and Tucker, R.P. (eds) (1988) *World Deforestation in the Twentieth Century*, Durham, NC: Duke University Press.

DEGRADATION-INDUCED DROUGHT

Drought promoted by environmental degradation usually initiated by human activity which disrupts the regional **energy** and **water** balances. Environmental degradation induced by overgrazing or wood cutting may have contributed to drought in the **Sahel** through a biogeophysical feedback mechanism.

Further reading
Hulme, M. (1989) 'Is environmental degradation causing drought in the Sahel? An assessment from recent empirical research', *Geography* 74: 38–46.

DEHYDRATION

The removal of **water** or its constituents from

a chemical **compound**. Also applied to excess water loss from the human body brought about by vomiting, dysentery and diarrhoea. Dehydration is common in Third World nations as a by-product of unhygienic living conditions.

DELTA

A depositional landform created when a river or stream flows into a body of standing **water**, such as a lake or sea. The sudden reduction in the **velocity** of the stream which results causes it to deposit the sediment it is carrying. **Flocculation** of **clay** particles in **salt** water also increases the rate of depositon when a river flows directly into the sea. Although deposition is generally rapid, it is not disorganized and deltaic sediments exhibit well-defined bedding patterns. Coarser sediments tend to be deposited first, forming the foreset beds, while finer clays are carried further to form the bottomset beds. The rapid deposition of sediments tends to block the main stream channel, causing it to divide and subdivide into secondary channels

Figure D-1 The structure and morphology of one of the many forms of delta

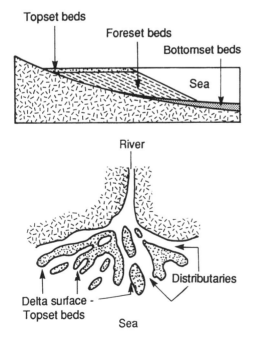

called distributaries. The settling out of sediments as the distributary channels change helps to produce the topset beds that form the delta surface. The classic fan-shaped plan from which the landform derives its name (the Greek letter Δ) is well illustrated by the delta of the River Nile, where it flows into the Mediterranean Sea. However, during and following deposition, such factors as the volume and size distribution of the sediments, the form of the lake or seabed offshore and the impact of **wind**, waves and **tides** will influence the shape of the delta. With their large areas of flat, easily worked land, kept fertile by a regular supply of **nutrients**, the deltas of the world's major rivers, such as the Nile in Egypt and the Ganges/Brahmaputra in Bangladesh, have supported large agriculturally based populations for thousands of years. Despite regular **floods**, which can destroy crops and homes and cause major loss of life, the advantages apparently outweigh the disadvantages, and habitation has persisted. Such long-term occupation of the deltas, however, has ensured that they are no longer completely natural features. The construction of flood protection structures and the creation of permanent shipping channels have altered both the form of deltas and the processes involved in their formation. In addition, the construction of **dams** or the alteration of the stream channel upstream from the delta can have an impact on the entire delta **environment** by changing the amount and nature of the sediments reaching it. For example, reduced sediment yields on many deltas have also meant a reduction in the availability of natural nutrients. Replacement of the latter with chemical **fertilizers** has created previously unknown environmental problems.

Further reading
Colella, A. and Prior, D.B. (eds) (1990) *Coarse-grained Deltas*, Oxford/Boston: Blackwell Science.
Ritter, D.F., Kochel, R.C. and Miller, J.R. (1995) *Process Geomorphology*, Dubuque, IA: Wm C. Brown.

DEMOGRAPHIC TRANSITION

A concept which combines **population** change with socioeconomic development. The demographic transition model reflects the stages

Figure D-2 A graphical representation of the various stages of the demographic transition

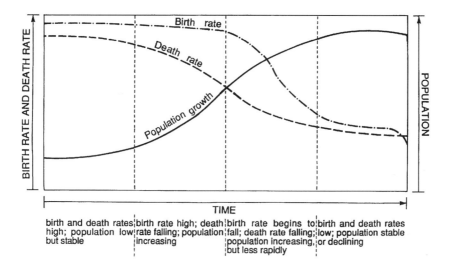

birth and death rates | birth rate high; death | birth rate begins to | birth and death rates
high; population low | rate falling; population | fall; death rate falling; | low; population stable
but stable | increasing | population increasing, | or declining
 | | but less rapidly |

through which most European and North American developed nations have passed. In Stage 1, birth and death rates are high and populations remain small but relatively stable. However, death rates can fluctuate over a considerable range – for example, because of **famine** and disease – and at times populations may decline. Improved economic and social conditions carry the population into Stage 2, in which death rates decline as a result of an increase in food supply or better disease control. Birth rates remain high and, as a result, rapid population growth begins. With continued technological and economic development, usually associated with industrialization, the transition enters Stage 3. Birth rates begin to fall and population growth rates decline until eventually birth and death rates become balanced at a low level, and in Stage 4 the population stabilizes. In a number of developed nations, an additional stage can be added in which birth rates actually fall below death rates and the population enters a declining phase. The demographic transition model suggests that technological and economic development will naturally cause population to stabilize. However, the conditions that allowed the developed nations to proceed through the transition – abundant **energy** and other natural **resources**, availability of land, improved food supply – are no longer available to less-

developed nations. Currently in Stage 2 of the transition, with rapidly growing populations, the latter do not have the economic and technological resources required to proceed to the next stage of the model.

See also
Population – environmental effects.

Further reading
Ehrlich, P.R. and Ehrlich, A.H. (1990) *The Population Explosion*, New York: Simon & Schuster.
Parnwell, M. (1993) *Population Movements and the Third World*, London: Routledge.
Simmons, O.G. (1990) *Perspectives on Development and Population Growth in the Third World*. New York: Plenum.

DEMOGRAPHY

(The study of) statistics on human populations, including such elements as growth rate, age and sex ratios, distribution and density, and their effects on socioeconomic and environmental conditions.

DENATURING

The addition of an unpleasant or noxious substance to a product in order to make it unpalatable. Grain, for example, may be treated so that it must be used for seed rather than direct human consumption. Similarly, methylated spirit, a mixture of **ethanol** and **methanol** capable of causing extreme intoxication, normally contains pyridine, an unpleasant, smelly **liquid**, to discourage human use.

DENDROCHRONOLOGY

A method of absolute dating based on the growth rings of trees. Since growing trees normally add one ring every year, the age of a tree can be calculated by counting its annual growth rings. In turn, consideration of the tree in the evironmental context of the area in which it is growing can provide information on the age of other elements in that area. Under certain environmental conditions, some trees may produce multiple rings in a given year or perhaps no rings at all. Any problems this might cause can be overcome by sampling a representative number of trees. The process is so accurate that it has been used to provide a correction factor for pre-3000 BP **radio-carbon dates**.

See also
Dendroclimatology.

Further reading
Schweingruber, F.H. (1988) *Tree Rings, Basics and Applications of Dendrochronology*, Dordrecht: Reidel.
Schweingruber, F.H. (1993) *Trees and Wood in Dendrochronology*, Berlin/New York: Springer-Verlag.

Figure D-3 The elements of tree-ring analysis

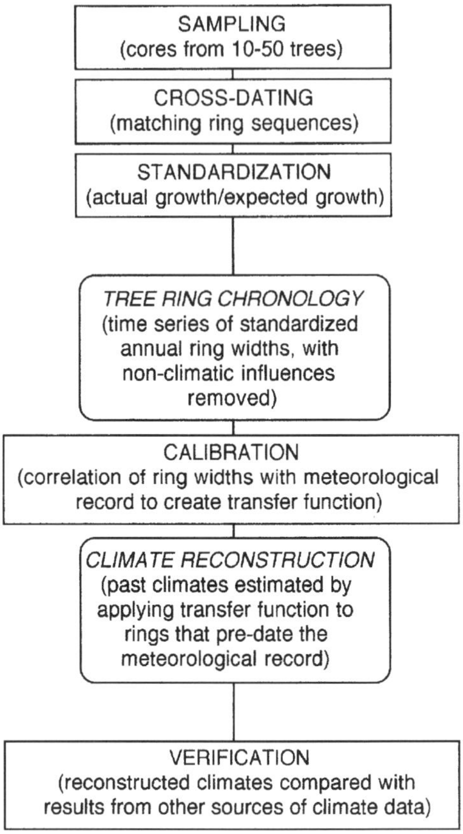

SAMPLING
(cores from 10-50 trees)

CROSS-DATING
(matching ring sequences)

STANDARDIZATION
(actual growth/expected growth)

TREE RING CHRONOLOGY
(time series of standardized annual ring widths, with non-climatic influences removed)

CALIBRATION
(correlation of ring widths with meteorological record to create transfer function)

CLIMATE RECONSTRUCTION
(past climates estimated by applying transfer function to rings that pre-date the meteorological record)

VERIFICATION
(reconstructed climates compared with results from other sources of climate data)

DENDROCLIMATOLOGY

The study of past **climates** as revealed by variations in the widths of the annual growth rings of trees. Such variations reflect changes in growing conditions from one year to the next, and although the rate of growth represents the combination of a variety of factors, the role of climate is usually sufficiently strong to establish a correlation between tree growth and such climatic elements as **temperature** and **precipitation**. The most obvious relationships have been identified in areas where the trees are growing under some form of environmental stress or close to their climatological limits. For example, the pioneering work in dendroclimatology at the University of Arizona was carried out in the south-western part of the United States where temperatures are normally sufficiently high for growth, but moisture is a significant limiting factor. Elsewhere, along the northern tree-line in Europe and North America, for example, moisture levels are adequate, and summer temperatures correlate well with ring widths. In some areas, however, it may not be possible to identify a relationship between growth and a specific climate element, while in the tropics the technique is limited by the absence of distinct annual rings in many species.

Figure D-4 Palaeoclimatic reconstruction in dendroclimatology: mean deviations of tree growth and mean reconstructed deviations in temperature and precipitation for the winters of 1861–1870

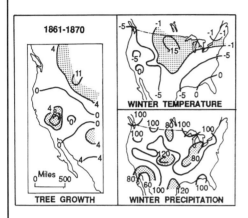

Source: After Fritts, H.C., Lofgren, C.R. and Gordon, G. (1980) 'Past climate reconstructed from tree rings', *The Journal of Interdisciplinary History* X (4): 773–95

Climate reconstruction using tree rings is now technically very advanced. The process of establishing a record involves a series of stages. Initially, the samples obtained in an area have to be treated statistically to remove the influence of non-climatic elements from the measured ring widths. As a tree ages, for example, the rings that it produces tend to become thinner. By comparing the actual width of a ring with the expected width for a tree of a particular age on the site being examined, standardized values which reflect only the influence of climate can be obtained. These standardized values can then be used to calibrate the sequence of rings. Using a period for which meteorological records are available, a statistical relationship is sought between tree ring width over the same period and a specific meteorological parameter such as temperature or precipitation. Once the relationship has been established, the other rings can be used to create a record of meteorological conditions outside the period for which numerical data are available. The results are then verified by comparing them with additional records from outside the calibration period or results from other sources. Using samples from bristlecone pines, which are among the oldest living trees in the world, it has been possible to reconstruct changing climates over the past 4000 years in the mountains of southern California. By overlapping samples from these living trees with ring sequences from dead trees, the record has been extended back more than 8000 years. Most records are much shorter, but climatic reconstructions based on tree rings are available for large areas of North America and Europe, providing information on such elements as temperature, precipitation and **atmospheric pressure**, both annually and seasonally.

Further reading
Bradley, R.S. and Jones, P.D. (1992) *Climate Since A.D. 1500*, London: Routledge.
Douglass, A.E. (1971) *Climatic Cycles and Tree Growth*, New York: Stechert-Hafner.
Fritts, H.C. (1976) *Tree Rings and Climate*, New York: Academic Press.
Schulman, E. (1956) *Dendroclimatic Changes in Semiarid America*, Tucson: University of Arizona Press.
Pearce, F. (1996) 'Lure of the rings', *New Scientist* 152 (2060): 38–42.

DENITRIFICATION

The breakdown of **soil nitrates** by **bacteria** to release free **nitrogen** (N). Most denitrification takes place in waterlogged soils where the low **oxygen** (O) environment favours the **anaerobic** bacteria involved.

See also
Nitrogen cycle.

DENSITY

The **mass** of a substance per unit **volume**. Usually expressed as grams per cubic centimetre. In environmental studies the term is often used to denote the number of **elements** in a given area – **population** density in persons per square kilometre, for example.

DENUDATION

The wearing away, or literally the 'laying bare' of the land surface by the combined effects of **weathering, mass movement** and **erosion**. The result is a net reduction in the relief of the landscape. The main agent of denudation is **water**. Glacial **ice** also contributes to the process in mountainous areas and in high latitudes, and during the **ice ages**, **glaciers** had a major role in wearing down the landscape. Denudation rates are usually calculated by measuring **sediment yield** from drainage basins, expressed as millimetres per 1000 years. Average values may be only a few millimetres over that time period, but rates vary considerably in time and place. They are greatest in areas of strong relief with steep slopes, little vegetation and abundant **precipitation**. Warm, semi-arid regions, with a sparse vegetation cover, also exhibit high denudation rates, and rates are generally lowest on moist, level plains, where gradients are low and there is sufficient **precipitation** to keep the landscape well vegetated. Changing climatic conditions over time also cause variations in denudation rates. Human intervention in the **environment** through mining and quarrying, forest **clear cutting** and inappropriate agricultural practices has accelerated the natural rates of denudation in many areas.

Further reading
Sparks, B.W. (1971) *Rocks and Relief*, London: Longman.
Summerfield, M.A. (1991) *Global Geomorphology*, London/New York: Longman/Wiley.
Walling, D.E. (1987) 'Rainfall, runoff and erosion of the land; a global view', in K.J. Gregory (ed.) *Energetics of the Physical Environment*, Chichester: Wiley.

DEOXYRIBONUCLEIC ACID (DNA)

A complex organic **polymer** consisting of a series of nucleotides linked together in the form of a double helix. Nucleotides consist of a nitrogenous base, plus a pentose **sugar** and a **phosphate** group, and in the DNA **molecule**, the main strands of the helix are sugar-phosphate chains, while the nitrogenous base provides the link between the two. DNA, located in the **chromosomes** of living **cells**, contains the genetic codes that are necessary for the development and function of a

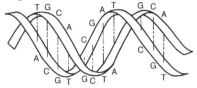

Figure D-5 The DNA double helix

Source: Faughn, J.E., Turk, J. and Turk, A. (1991) *Physical Science*, Philadelphia: Saunders

particular organism. The DNA molecule is self-replicating, capable of passing on perfectly matched genetic information from one generation of cells to the next. In this way, it determines the inherited characteristics of an organism. Damage to the structure of DNA, as a result of exposure to chemicals or **radiation**, for example, can lead to genetic disorders.

See also
Genes, Ultraviolet radiation.

Further reading
Drlica, K. (1984) *Understanding DNA and Gene Cloning*, New York: Wiley.
Hecht, S.M. (ed) (1996) *Bioorganic Chemistry: Nucleic Acids*, New York: Oxford University Press.
Hoagland, M.B. (1981) *Discovery, the Search for DNA's Secrets*, Boston, MA: Houghton Mifflin.

DEPRESSION

A mid-latitude low pressure system. Depressions are usually circular in form and travel from west to east in mid-latitudes in both hemispheres, bringing with them changeable **weather** conditions accompanied by strong **winds** and **rain**. They tend to follow well-developed storm tracks and are major contributors to the **climate** of mid-latitudes.

See also
Cyclone, Mid-latitude frontal model.

Further reading
Hidore, J.J. and Oliver, J.E. (1993) *Climatology: An Atmospheric Science*, New York: Macmillan.

DEPRESSION (THE)

A period of major economic decline in the 1930s associated with reduced industrial activity, high unemployment and the collapse of international trade. The impact of The

Depression in North America was intensified because it coincided with a period of **climate change** which caused **drought** and created the **Dustbowl** in the **Great Plains**.

Further reading
Garraty, J.A. (1986) *The Great Depression*, San Diego: Harcourt Brace Jovanovich.

DESALINATION

The removal of dissolved salts from sea water or saline **groundwater** to provide fresh **water** for domestic, industrial or agricultural use. **Distillation** and **reverse osmosis** are the most common processes employed. In distillation, **water vapour**, driven off as steam when salt water is boiled, condenses as fresh water, leaving the solid **salts** behind. **Reverse osmosis** involves the pumping of salt water at high pressure through a **semi-permeable membrane**. Dissolved salts are retained on the membrane while fresh water passes through. Other less frequently used methods which accomplish the separation of the salts from the fresh water include electrodialysis – the passage of an electric current through the saline solution – and freezing, which causes the formation of freshwater ice and leaves behind a concentrated salt solution. Distillation is most widely used, accounting for between 85 and 90 per cent of the freshwater produced by desalination. The main concerns associated with desalination centre on the high **energy** use involved, and the potential environmental problems created by the disposal of large volumes of salt or brine. Currently, desalination plants provide less than 0.1 per cent of the world's fresh water. They are most commonly found where energy costs are relatively low and the costs of supplying water from any source are high – as in Kuwait and Saudi Arabia – or where other sources are limited and the demand is high enough to meet the costs – as in Florida and California in the United States.

Further reading
Porteous, A. (1983) *Desalination Technology*, London: Applied Science Publishers.

DESERT

An area of permanent **aridity** in which **precipitation** is infrequent or irregular in its occurrence, and the resulting low annual rainfall totals are exceeded by high **evapotranspiration** rates. Such conditions are characteristic of sub-tropical regions between 25° and 30° north and south of the equator. There, the descending arms of the tropical **Hadley Cells** inhibit precipitation, and the high **temperatures** encourage high evapotranspiration rates which contribute to major **soil moisture deficits**. The Sahara Desert in North Africa – the world's biggest – is a product of such conditions. Elsewhere, deserts are created by the rain shadow effects of mountain systems that lie across the prevailing **wind** directions – as in North America, where the Mojave and Sonoran Deserts owe their aridity to the presence of the Sierra Nevada blocking the moist **air masses** from the Pacific Ocean. The rain shadow effect also applies to the Gobi and other deserts of Central Asia. The Himalayas prevent the moisture of the Indian **monsoon** from reaching the area, but the continentality of the region is also an important element. Its distance from the ocean means that the air masses reaching the area have lost most of their moisture before they arrive. The world's deserts cover some 5 million km², and an additional 40 million km² are arid or semi-arid. In total, some 33 per cent of the earth's land surface exhibits desert characteristics to some degree or other. The area covered by desert is not static. In the past it has been larger and also smaller. At various times during the **Pleistocene**, many areas now considered deserts were much wetter. In the Sahara, for example, the abandoned shorelines of Lake Chad indicate that it was once larger and deeper than it is now. At other times, the deserts have been even drier and more extensive. In North America in the nineteenth century, parts of the **Great Plains** were so dry that the area was referred to as 'The Great American Desert'. At present the main concern is that the deserts are

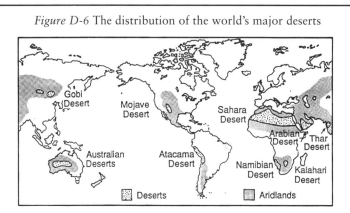

Figure D-6 The distribution of the world's major deserts

expanding again, with the **desertification** of sub-Saharan Africa having the potential to affect the lives and livelihood of millions of people. Despite their overall aridity, water may be available in deserts from time to time and from place to place. Occasional storms provide a short-term **water** supply and sometimes cause flash-floods, while rivers supplied from wetter areas outside the deserts provide a permanent source of water in some areas. The Nile in Egypt, the Tigris-Euphrates system in Iraq and the Colorado in the United States are examples of such rivers. In addition, there is often abundant **groundwater** beneath the desert. It may appear at the surface in the form of springs or at **oases**, but access to the groundwater supply in desert areas usually involves the drilling of wells. Since such subsurface water has taken thousands of years to accumulate, and the replenishment of **aquifers** in deserts is extremely slow, careful management of the groundwater resource is important. However, in most cases, even steady, controlled use leads to a progressive depletion of the groundwater supply. The typical desert landscape is often represented as a sea of **sand**, overlying a relatively subdued plain, with some relief provided by the presence of sand dunes. This is the so-called erg desert, which is well developed in the Sahara, although even there only 20 per cent of the surface is sand. Elsewhere, the combination of **climate**, rock type and relief has produced rocky or stony surfaces, bare rock pavements, deep valleys and steep slopes. Mass wasting processes

such as **exfoliation** help to steepen slopes and produce **scree**.

Although water is not common in desert areas, when it is available – for example, during a major storm – it can achieve spectacular amounts of **erosion**. This is made possible in part by the absence of plant cover, which encourages rapid **runoff**, and the availability of large quantities of easily eroded and transported unconsolidated sediments. The most obvious erosional landforms are the major, usually dry, river channels or wadis of the Middle East, but in areas of strong relief, smaller scale steepsided canyons are common. The materials eroded from such features are deposited rapidly, often in the form of fans, when water is no longer available. Wind is also a potent erosional and depositional force in desert areas. Lifting loose sand from a surface, it can create deflation hollows and the moving sand can erode adjacent rocks in a form of natural sand-blasting. The redeposited sand may take the form of sheets or sand dunes which move according to the prevailing wind speed and direction.

Limited precipitation and high evapotranspiration limit plant growth, and desert surfaces are characterized by sparse, **xerophytic vegetation**, or, in some areas, by the complete absence of plant life. In places, the high evapotranspiration rates create very salty soils which also inhibit plant growth. However, even those surfaces that are apparently completely devoid of vegetation may bloom immediately following precipitation, as seeds and dormant plants respond rapidly to the availability of moisture. Those plants which survive in the desert environment have adapted to the conditions, for example, by reducing their water needs, and the same applies to desert animals. Many have developed nocturnal habits to avoid the heat of the day, and all have adapted to living with the lack of water. Despite such adaptability, the **biomass** productivity of desert areas remains low.

In the past, the human occupation of deserts was restricted to those areas where water was readily available, such as at oases or along river systems, for example, in Egypt and Mesopotamia. More recently, technological advancements and changing economic factors have allowed the development of new areas. Improved **irrigation** techniques, for example, have allowed the Israelis to bring parts of the Negev Desert into agricultural production, while in the United States the demand for vacation and retirement facilities in California and Ari-

zona has made it economically feasible to bring water from the neighbouring mountains to supply new residential areas. The water is also used to irrigate high value crops, such as strawberries or out-of-season vegetables, and even golf-courses. Such situations are the exception, however, and the potential for further development of desert areas remains limited.

See also
Desertification.

Further reading
Abrahams, A. and Parsons, A. (1993) *Geomorphology of Desert Environments*, London: Chapman and Hall.
Cooke, R.U., Warren, A. and Goudie, A.S. (1993) *Desert Geomorphology*, London: Longman.
Louw, G.N. and Seely, M.K. (1982) *Ecology of Desert Organisms*, London/New York: Longman.
Reisner, M. and Bates, S. (1990) *Overtapped Oasis: Reform or Revolution for Western Water*, Covello, CA: Island Press.

Figure D-7 The rugged desert landscape of the south-western United States with its sparse covering of xerophytic vegetation

Photograph: Courtesy of Susan and Glenn Burton

DESERTIFICATION

The expansion of **desert** or desert-like conditions into adjacent areas, initiated by natural environmental change, by human degradation of marginal environments or a combination of both. Most modern approaches to the definition of desertification recognize the combined impact of adverse climatic conditions and the stress created by human activity (Verstraete 1986). Both have been accepted by the United Nations as the elements that must be considered in any working definition of the process (Glantz 1977). The **United Nations Environmental Program** (UNEP) has tended to emphasize the importance of the human impact over **drought,** but the relative importance of each of these elements remains very controversial. Some see drought as the primary element, with human intervention aggravating the situation to such an extent that the overall expansion of the desert is increased, and any recovery – for example, following a change in climatic conditions – is lengthier than normal. Others see direct human activities as instigating the process. In reality, there must be many causes that together bring desert-like conditions to perhaps as much as 60,000 sq km of the earth's surface every year and threaten up to a further 30 million sq km. The areas directly threatened are those adjacent to the deserts and semi-deserts on all continents. Africa is currently receiving much of the attention, but large sections of the Middle East, the central Asian republics of the former Soviet Union, China adjacent to the Gobi Desert, north-west India and Pakistan, along with parts of Australia, South America and the United States are also susceptible to desertification. Even areas not normally considered as threatened, such as Southern Europe from Spain to Greece, are not immune. At least 50 million people are directly at risk of losing life or livelihood in these regions. In a more graphic illustration of desertification, the United States Agency for International Development (USAID), at the height of the Sahelian drought in 1972, claimed that the Sahara was advancing southwards at a rate of as much as 30 miles (48 km) per year along a 2000 mile (3200 km) front (Pearce 1992b). Although these specific numbers are not universally accepted – for example, Nelson (1990) has suggested that all such data be treated with a healthy scepticism – they do give an indication of the magnitude of the problem, and remain one of the reasons why there has been increasing cause for concern in recent years (van Ypersele and Verstraete 1986).

Human nature being what it is, when drought strikes an area, there is a natural tendency to hope that it will be short and of limited intensity. The inhabitants of drought-prone areas, therefore, may not react immediately to the increased aridity. They may continue to cultivate the same crops, perhaps even increasing the area under cultivation to compensate for reduced yields, or they may try to retain flocks and herds which have expanded during the times of plenty. If the drought is prolonged in the arable areas, the crops die and the bare earth is exposed to the ravages of **soil erosion.** The **Dustbowl** in the **Great Plains** developed in this way. Once the available moisture had evaporated and the plants had died, the **wind** removed the topsoil – the most fertile part of the **soil profile** – leaving a barren landscape, which even the most drought-resistant desert plants found difficult to colonize (Borchert 1950).

Prolonged drought in pastoral areas is equally damaging. It reduces the forage supply, and, if no attempt is made to reduce the animal population, the land may fall victim to overgrazing. The retention of larger herds during the early years of the Sahelian drought, for example, allowed the vegetation to be overgrazed to such an extent that even the plant roots died. When the wind blew, it lifted the exposed, loose soil particles and carried them away, taking with them the ability of the land to support plant and animal life. In combination, these human and physical activities seemed to be pushing the boundaries of the Sahara Desert inexorably southwards. Out of this grew the image of the **shifting sands**, which came to represent desertification in the popular imagination. As an image, it was evocative, but the reality of such a representation has been increasingly questioned in the 1990s (Nelson 1990; Pearce 1992b).

Climatic variability clearly made a major contribution to desertification in both the **Sahel** and the Great Plains, and in concert with human activities created serious environmental problems. An alternative view sees human activity in itself capable of initiating desertification in the absence of increased **aridity** (Verstraete 1986). For example, human interference in areas where the environmental balance is delicate might be sufficient to set in motion a train of events leading eventually to desertification. The introduction of **arable agriculture** into areas more suited to grazing, or the removal of forest cover, to open up agricultural land or to provide fuelwood, may disturb the ecological balance to such an extent that the quality of the environment begins to decline, and if nothing is done, the soil becomes highly susceptible to erosion. In such cases, desertification is initiated by human activities with little or no contribution from nature.

Although human activities have been widely accepted as causing desertification, and the processes involved have been observed, there is increasing concern that the human contribution has been overestimated. Current academic and popular attitudes to desertification owe a lot to the findings of a **United Nations Conference on Desertification** (UNCOD) held in Nairobi, Kenya in 1977. At the Conference, the role of human activities in land degradation was considered to be firmly established, and the contribution of drought was seen as secondary at best. Since human action had caused the problem, it seemed to follow that human action could solve it. In keeping with this philosophy, the United Nations Environment Program (UNEP) was given the responsibility for taking global initiatives to introduce preventive measures which would alleviate the problem of desertification (Grove 1986). Fifteen years and US$6 billion later, few effective countermeasures have been taken, and the plan of action is widely seen as a failure (Pearce 1992a).

The data upon which the UNEP responses were based are now considered by many researchers to be unrepresentative of the real situation. Nelson (1990), for example, has suggested that the extent of irreversible

Figure D-8 Areas at risk from desertification

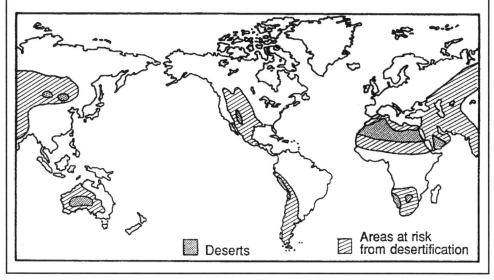

desertification has been over-estimated, although he does not deny that it remains a serious concern in many parts of the world. Problems arising from the timing and method of collection of the data were aggravated by the UNEP premise that human activity was the main cause of the land degradation that produced desertification. Natural causes such as short-term drought and longer term climatic change were ignored or given less attention than they deserved, yet both can produce desert-

Table D-2 Action required for the prevention and reversal of desertification

1 **Prevention**

   (a) Good land-use planning and management:

      e.g. cultivation only where and when precipitation is adequate

          animal population based on the carrying capacity of land in driest years

          maintenance of woodland where possible

   (b) Irrigation appropriately managed to minimize sedimentation, salinization and waterlogging

   (c) Plant breeding for increased drought resistance

   (d) Improved long-range drought forecasting

   (e) Weather modification:

      e.g. rainmaking

          snowpack augmentation

   (f) Social, cultural and economic controls:

      e.g. population planning

          planned regional economic development

          education

2 **Reversal**

   (a) Prevention of further soil erosion:

      e.g. by contour ploughing

          by gully infilling

          by planting or constructing windbreaks

   (b) Reforestation

   (c) Improved water use:

      e.g. storage of runoff

          well-managed irrigation

   (d) Stabilization of moving sand:

      e.g. using matting

          by re-establishment of plant cover

          using oil waste mulches and polymer coating

   (e) Social, cultural and economic controls:

      e.g. reduction of grazing animal herd size

          population resettlement

like conditions without input from society. The inclusion of areas suffering from short-term drought may well have inflated the final results in the UNEP accounting of land degradation. Failure to appreciate the various potential causes of desertification would also limit the response to the problem. Different causes would normally elicit different responses, and UNEP's application of the societal response to all areas, without distinguishing the cause, may in part explain the lack of success in dealing with the problem (Pearce 1992b). The debunking of some of the myths associated with desertification and the realization that even after more than 15 years of study its nature and extent are inadequately understood, does not mean that desertification should be ignored. There are undoubtedly major problems of land degradation in many of the earth's arid lands. Perhaps sensing an increased vulnerability as a result of the current controversy, and certainly fearful of being left behind in the rush to deal with the problems of the developed world, the nations occupying the land affected appeared at the Rio **Earth Summit** in 1992 and proposed a **Desertification Convention** to address their problems. Negotiations continued after the Summit, and in 1994, 110 governments signed the UN Convention to Combat Desertification. A much more comprehensive treaty than earlier efforts, the Convention will seek innovative solutions through national action programmes and partnership agreements. Particular attention will be paid to Africa, where desertification is most severe.

Although obscured by the current controversy and lost in the complexity of attempts to define the issue of desertification more accurately, two questions remain of supreme importance to the areas suffering land degradation. Can desertification be prevented? Can the desertification that has already happened be reversed? In the past, the answer to both has always been a qualified yes, and seems likely to remain so, although some researchers take a more pessimistic view (e.g. Nelson 1990). In theory, society could work with the

environment by developing a good understanding of environmental relationships in the threatened areas or by assessing the capability of the land to support certain activities and by working within the constraints that these would provide. In practice, non-environmental elements – such as politics and economics – may prevent the most ecologically appropriate use of the land. The most common approaches to the prevention and reversal of desertification are listed in Table D-2.

The fight against desertification has been marked by a distinct lack of success. Recent reassessments of the problem, beginning in the late 1980s, suggest that this may be the result of the misinterpretation of the evidence and a poor understanding of the mechanisms that cause and sustain the degradation of the land. The additional research required to resolve that situation will further slow direct action against desertification, but it may be the price which has to be paid to ensure future success.

References and further reading
Borchert, J.R. (1950) 'The climate of the central North American grassland', *Annals of the Association of American Geographers* 40: 1–39.
Dregne, H.E. (1983) *Desertification of Arid Lands*, New York: Harwood Academic Publishers.
Glantz, M.H. (ed.) (1977) *Desertification: Environmental Degradation in and around Arid Lands*, Boulder, CO: Westview Press.
Grove, A.T. (1986) 'The state of Africa in the 1980s', *The Geographical Journal* 152: 193–203.
Hulme, M. and Kelly, M. (1993) 'Exploring the links between desertification and climate change', *Environment* 35: 4–11 and 39–45.
Nelson, R. (1990) *Dryland Management: The 'Desertification' Problem*, World Bank Technical Paper No. 16, Washington, DC: World Bank.
Pearce, F. (1992a) 'Last chance to save the planet', *New Scientist* 134 (1823): 24–8.
Pearce, F. (1992b) 'Miracle of the shifting sands', *New Scientist* 136 (1851): 38–42.
Thomas, D.S.G. and Middleton, N.J. (1994) *Desertification: Exploding the Myths*, Chichester: Wiley.
van Ypersele, J.P. and Verstraete, M.M. (1986) 'Climate and desertification – editorial', *Climatic Change* 9: 1–4.
Verstraete, M.M. (1986) 'Defining desertification: a review', *Climatic Change* 9: 5–18.

DESERTIFICATION CONVENTION

See **United Nations Convention to Combat Desertification**.

DESERTIZATION

The term formerly used for **desertification**.

DESICCATION

The removal of moisture from a substance, often through the use of an agent that absorbs moisture. Silica gel, for example, is used to both dry out and maintain the dryness of pharmaceutical products. It is also a climatological concept, applied to an **environment** that is drying up, either as part of natural climatic change, as the result of human activity or as a combination of the two.

See also
Desertification.

DESTRUCTIVE DISTILLATION

The heating of solid organic substances in the absence of **air** (also called carbonization) to produce **gases**, volatile **liquids** and **charcoal**. The destructive distillation of wood, for example, is used to produce acetic **acid** (vinegar), but the most common substance used is **coal**. When heated to temperatures of between 900°C and 1200°C, coal releases coal gas and coal tar, with **coke** as a by-product. Prior to the large-scale use of **natural gas**, coal gas, consisting mainly of **hydrogen** (H) (50%) and **methane** (CH_4) (30%), was commonly used for heating and lighting. Coal tar can be refined to provide a variety of **hydrocarbons** including **benzene** (C_6H_6), and naphthalene ($C_{10}H_8$), and coke is an important raw material in the **iron** (Fe) and steel industry.

See also
Pyrolysis.

Further reading
McNeil, D. (1966) *Coal Carbonization Products*, Oxford/London: Pergamon Press.
Payne, K.R. (1987) *Chemicals from Coal: New Processes*, Chichester/New York: Wiley.

DETERGENTS

Synthetic cleansing products, usually derived from **petrochemicals**. They vary in composition depending upon their intended use, but all contain chemicals called surfactants, which lower the surface tension of **water**, enabling the detergent to dislodge dirt from the soiled material more easily. Builders are incorporated in detergents to deal with specific problems such as **hard water** or acidity. Other additives such as optical brighteners, bleaches and perfumes are included in most domestic laundry detergents. Effective as they are at providing cleaning power for domestic, industrial and institutional purposes, detergents have also contributed to some serious environmental problems. The early detergents which became popular following the Second World War did not break down rapidly in the **environment**, causing serious foam build-up in streams and **sewage treatment** plants. The production of biodegradable detergents in the mid-1960s reduced that problem. More serious was the

Figure D-9 The products of the destructive distillation of coal and their uses

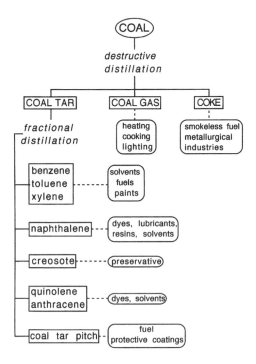

use of **phosphates** in laundry detergent as a builder to soften hard water and reduce its acidity. The phosphate-rich **waste** water that this produced provided extra **nutrients** for aquatic plants and contributed to the accelerated **eutrophication** of lakes and rivers. The solution lay in the banning or strict control of phosphate-based detergents and their replacement by phosphate-free products.

Further reading
Davidsohn, A. and Milwidsky, B.M. (1987) *Synthetic Detergents* (7th edition), Harlow/New York: Longman/Wiley.
McGucken, W. (1991) *Biodegradable Detergents and the Environment*, College Station, TX: A and M University Press.

DEUTERIUM

A non-radioactive, heavy **isotope** of **hydrogen**. With a **mass number** of 2, it is twice the mass of ordinary hydrogen and makes up about 0.015 per cent of the natural hydrogen in the **environment**. Deuterium is a component of **heavy water** (D_2O) used as a **moderator** in some **nuclear reactors**.

See also
CANDU.

DEWPOINT

The **temperature** at which an **air mass** becomes completely **saturated** as a result of cooling. Cooling beyond the dewpoint temperature will cause **condensation**. If the cooling is the result of contact between the saturated air and a cool surface, water droplets or dew will form on the surface. If cooling is the result of **adiabatic processes**, **clouds** will form.

DIALYSIS

The separation of dissolved substances using a **semipermeable membrane**. Normally the smaller **molecules** pass through the membrane while the larger remain behind. By applying an electric potential across the membrane it can be made **permeable** to positive or negative **ions**. The process is called

electrodialysis and is used in **desalination** plants. Natural dialysis allows the kidneys to remove nitrogenous **waste** from the body. If the kidneys fail, the wastes must be removed through artificial dialysis processes – haemo-dialysis.

See also
Osmosis.

Further reading
Drukker, W., Parsons, F.M. and Maher, J.F. (eds) (1983) *Replacement of Renal Function by Dialysis: A Textbook of Dialysis* (2nd edition), Boston, MA: M. Nijhoff.

DIATOMS

Microscopic, unicellular brown or green **algae**. They contain **chlorophyll** and are therefore capable of **photosynthesis**. Diatoms make a major contribution to **primary production** in **aquatic ecosystems**, and provide a base for a variety of fresh and saltwater **food chains**. The cell walls of diatoms are impregnated with **silica** (SiO_2), which is hard and insoluble. When the organisms die they sink, and the siliceous skeletons survive to form deposits of diatomaceous earth, which is used commercially in **detergents**, polishes and **fertilizers**. Some **petroleum** deposits may have their origins in major accumulations of dead diatoms. Diatoms are also used to provide information on past **environments** in **Quaternary** studies.

See also
Plankton.

Further reading
Round, F.G., Crawford, R.M. and Mann, D.G. (1990) *Diatoms*, Cambridge: Cambridge University Press.

DICHLORODIPHENYL-TRICHLOROETHANE (DDT) $(C_6H_4CL)_2.CH.CCL_3$

A **chlorinated hydrocarbon** once widely used as a broad-spectrum **insecticide**. Introduced during the Second World War as a delousing agent, it proved very effective against diseases such as **malaria**, yellow fever and typhus, which were spread by insects. Relatively low production costs, its effectiveness and its persistence in the **environment** – one application could continue to kill insects for up to a year – encouraged its worldwide acceptance. Over the longer term, however, it was recognized that these advantages came with serious side-effects. Being a broad-spectrum product it killed beneficial insects as well as pests, and having a relatively long **half-life** – perhaps 20 years – it tended to accumulate in the environment. Athough not soluble in **water** it was soluble in fat, which allowed it to migrate up the **food chain**, where it accumulated in the body tissue of the predators. In birds, it caused the thinning of eggshells, seriously reducing the breeding success of some species, as Rachel Carson pointed out in her book *Silent Spring*. By the mid-1960s, DDT was found to be widespread in the fatty tissue of the world's human population, passed on from mother to child through breast milk. Although the link between DDT concentration and human health was not clear, its potential to cause serious ecological disruption was recognized, and it was eventually banned or had its use severely restricted.

See also
Organochlorides, Pesticides.

Further reading
Davies, J.E. and Edmundson, W.F. (eds) (1972) *Epidemiology of DDT*, Mount Kisco, NY: Futura Publishing.
Wurster, C.F. (1969) 'Chlorinated hydrocarbon insecticides and the world ecosystem', *Biological Conservation* 1 (2): 123–9.

DIEBACK

See **tree dieback**.

DIFFRACTION

The bending, or change in direction, of **electromagnetic waves**. In the **natural environment**, diffraction is responsible for such phenomena as the coloured haloes visible when a light source is viewed through **fog**.

See also
Scattering of light.

Further reading
Born, M. and Wolf, E. (1993) *Principles of Optics: Electromagnetic Theory of Propogation, Interference and Diffraction of Light* (6th edition), Oxford/New York: Pergamon Press.

DIFFUSION

The free or random movement of a substance from a region in which it is highly concentrated into one in which it is less concentrated. In **gases** and **liquids**, it happens spontaneously at the molecular level, and continues until the concentration becomes uniform. Diffusion is a feature of many **pollution** problems, but since the rate of diffusion is related to measurable **elements** such as **temperature** and molecular weight in gases or the temperature and **viscosity** of liquids, the distribution and extent of the problem can often be predicted. Diffusion can also take place through barriers that are porous or permeable, for example, allowing substances to move in and out of **cells**. The diffusion of light is brought about by the scattering of a beam of light. In cultural geography, the term is used to describe the spread of ideas, information, commodities or diseases from a source region to adjacent areas.

Further reading
Alters. S. (1996) *Biology: Understanding Life*, St Louis: Mosby.

DIGESTION

The process by which animals convert food into a form in which it can be absorbed. Food products such as **carbohydrates**, fats and **proteins** are ingested through the mouth. In the stomach and the upper part of the small intestine, the food is broken down into its constituent **nutrients** which are then absorbed into the lower part of the small intestine. These nutrients provide the body with **energy** and contribute to **cell** and organ growth as well as the replacement of damaged tissue. Indigestible materials are passed through the system and expelled as **waste**. The term is also used to describe industrial processes in which organic materials are processed into a simpler usable form using mechanical and chemical methods. The manufacture of paper from trees, for example, involves the digestion of the original wood to separate out the fibres required to produce the paper.

See also
Metabolism.

Further reading
Mader, S.S. (1996) *Biology*, Dubuque, IA: Wm C. Brown.

DILUTION

The reduction in the concentration of a **solution** by the addition of a **solvent,** or by its addition to a larger volume of solvent. An example of the former is the dilution of a beverage, such as orange juice, by the addition of **water.** In environmental studies the latter is more common, when, for example, effluent is released into a body of water and is diluted as a result.

DIMETHYL SULPHIDE (DMS)

A **sulphur** (S) **compound** emitted by **phytoplankton** during their seasonal bloom, with an estimated annual production of 40×10^{12} gm. It is oxidized into **sulphur dioxide** (SO_2) and **methane sulphonic acid** (MSA), which provide a natural contribution to **acid rain.** The MSA is ultimately converted into sulphate **aerosols** which are very effective in reducing the inflow of **solar radiation.** They also have the potential to increase **cloud** cover through their ability to act as **condensation nuclei.** As a result of this, DMS may have a role in **global warming.** Higher **temperatures** will encourage greater **plankton** productivity, which in turn will bring about an increase in the production of DMS. The resulting increase in **atmospheric turbidity** and cloud cover will then act as a **negative feedback** and counter global warming by reducing incoming **radiation.**

Further reading
Charlson, R.J., Lovelock, J.E., Andreae, M.O. and Warren, S.G. (1987) 'Oceanic plankton, atmospheric sulphur, cloud albedo and climate', *Nature* 326: 655–61
Fell, N. and Liss, P. (1993) 'Can algae cool the planet?', *New Scientist* 139 (1887): 34–8.

DIODE

An electronic device that allows a current to flow in only one direction. The earliest diodes were **electron** tubes or valves, but these have now been superseded by semiconductors, which are smaller, cooler in operation and more rugged than the valves. Diodes are used to control power supply and regulate voltages in such devices as battery chargers, motors, radios and televisions, and they are essential components in most modern electronic equipment.

DIOXINS

A group of approximately seventy-five **chlorinated hydrocarbons** formed as by-products of chemical reactions involving chlorine (Cl) and hydrocarbons. Dioxins appear as manufacturing impurities in some **herbicides,** wood preservatives and disinfectants, and are released into the **environment** during the **incineration** of chlorine-based **plastics** or as a result of the chlorine bleaching process in pulp and paper mills. The most toxic of the group is TCDD, 2,3,7,8-tetrachlorodibenzo-*p*-dioxin ($C_{12}H_4Cl_4O_2$). It is present in the herbicide 2,4,5-T, and was a contaminant in **Agent Orange,** the defoliant used by the US Forces during the Vietnam War. Substantial amounts of TCDD were also released as a result of an industrial accident at Seveso in Italy in 1976. Dioxins are persistent chemicals, accumulating in **soil** and human fatty tissue. Health effects are varied and complex, ranging from skin problems, such as chloracne, to **cancers,** birth defects and serious immunological, neurological and behavioural problems. A reduction of dioxin levels in the environment will require a steady decline in the use of chlorine in manufacturing, and a ban on the incineration of plastics and other chlorine-based products.

Further reading
Gough, M. (1986) *Dioxin, Agent Orange: The Facts*, New York: Plenum Press.

DISPERSAL/DISPERSION

The spread and subsequent **dilution** of pollutants in **water** and **air.** The effectiveness

Figure D-10 The release of pollutants from a hospital incinerator. The plume is coning, allowing effective and relatively rapid dilution of the pollutants.

Photograph: The author

of the dilution will depend on such factors as the concentration and release rate of the **effluent** and the prevailing water and atmospheric conditions. The term dispersal is also used to refer to the dissemination of plants and animals from their source regions to other parts of the world.

See also
Diffusion, Oxygen sag curve, Plume.

Further reading
Williamson, S.J. (1973) *Fundamentals of Air Pollution*, Reading, MA: Addison-Wesley.
Stiling, P.D. (1992) *Introductory Ecology*, Englewood Cliffs, NJ: Prentice-Hall.

DISPOSAL WELLS

Specially drilled deep wells or dry **petroleum** exploration or production wells into which fluids are pumped as a means of **waste disposal**. Any introduction of **liquid** waste

into the **lithosphere** has the potential to cause **groundwater pollution**, but if the wells are sufficiently deep, and the rock structures such that they restrict the flow of the wastes, pollution is less likely. The injection of liquids into the earth's crust in certain areas has been observed to cause an increase in earth tremors. Although the seismic activity produced has been relatively minor, there is some concern that it might act as a trigger for a larger **earthquake**.

Further reading
Healy, J.H., Rubey, W.W., Griggs, D.T. and Raleigh, C.B. (1968) 'The Denver earthquakes', *Science* 161 (3848): 1301–10.

DISSOCIATION

The breakdown of a substance through the decomposition of its **molecules**, either by the addition of heat – thermal dissociation – or the passage of an electric current – electrolytic dissociation. The reaction is usually temporary and reversible. Ammonium chloride (NH_4Cl), for example, readily dissociates into ammonia (NH_3) and hydrochloric acid (HCl) when heated, but the two products recombine again on cooling.

$$NH_4Cl \underset{cooling}{\overset{heating}{\rightleftharpoons}} NH_3 + HCl$$

In electrolytic dissociation, a current of electricity passed between **electrodes** through an aqueous **solution** will cause the constituents to be deposited at the electrodes, with the reaction taking place at the ionic level.

DISTILLATION

A process in which a **liquid** is vapourized and the **vapour** subsequently condensed to produce a purified form of the liquid or one of its constituents. Distillation is a natural part of the **hydrological cycle**. **Energy** supplied by the sun causes surface **water** to evaporate. The vapour is carried up into the **atmosphere** until it is cool enough to condense as water droplets. These droplets are pure water, although the original source may have been salty or polluted. Distillation is the principal method of purifying **liquids**. It is the most common process used in **desalination**, and has been used increasingly by domestic consumers to provide fresh drinking-water. Domestic distillers remove toxic **metals**, non-volatile **organic compounds** and kill **bacteria** through the boiling process, but do not remove **volatile** organics which simply condense and become incorporated in the distilled water. Distillation is used in the production of **alcohol**, for domestic and industrial use. In the **petroleum** industry, it is employed to separate out the different products contained in crude **oil**. Since the various compounds contained in petroleum vapourize and condense at different **temperatures**, it is possible to separate and then collect the individual constituents of the original liquid oil. The process is called fractional distillation.

See also
Condensation, Destructive distillation, Drinking-water standards.

Further reading
Kister, H.Z. (1990) *Distillation Operation*, New York: McGraw-Hill.
World Bank (1980) *Alcohol Production from Biomass in the Developing Nations*, Washington, DC: World Bank.

DIVERGENCE

The spreading of moving **air** in the **atmosphere**. It may involve both vertical and horizontal components. Air descending vertically, for example, will spread horizontally once it reaches the surface. This type of divergence is associated with **anticyclones**, creating a net outflow of air from the centre of the **system**. Because the descending air is also being warmed adiabatically, it normally creates dry, fair weather conditions. Divergence may also occur when air is forced to rise. In equatorial regions, the rising air over the **intertropical convergence zone** (ITCZ) gradually cools and becomes less buoyant. Unable to return to the suface because of the warm air rising beneath it, it is forced to spread (or diverge) north and south. Divergence has also been recognized in horizontal airflows in cities. The speed of the air flowing over urban areas is reduced by the

greater surface roughness and the increase in frictional drag that it causes. This causes **convergence** of the airflow within cities. Once the air has passed into the adjacent rural area the surface roughness is reduced and divergence of the airflow takes place. Divergence also occurs when air exits from the confines of an urban canyon into a more open area such as a square or plaza.

See also
Anticyclone, Convergence.

Further reading
Barry, R.G. and Chorley, R.J. (1992) *Atmosphere, Weather and Climate*, (6th edition), London: Routledge.
Oke, T. R. (1987) *Boundary Layer Climates* (2nd edition), London: Methuen.

DOBSON UNIT (DU)

The unit used to represent the thickness of the **ozone layer** above a given location at standard (sea-level) **temperature** and **pressure**. If brought to normal pressure at sea-level, all the existing atmospheric ozone (O_3) would form a band no more than 3 mm thick. One Dobson Unit is equivalent to 0.01 mm.

DOMESTIC WASTE

See **garbage**.

DOPPLER EFFECT

The apparent change in pitch of a **sound** due to a change in the relative positions of the sound source and the listener. The pitch of a train whistle, for example, will appear to increase as it approaches an observer and decrease once it has passed. The effect is caused by the wave-like motion of sound, and applies to all types of waves. The Doppler effect has been developed for navigational purposes, and Doppler radar is an important tool in tracking storm systems for weather forecasting.

Further reading
Griffith, W.T. (1992) *The Physics of Everyday Phenomena: A Conceptual Introduction*, Dubuque, IA: Wm C. Brown.

DROUGHT

Drought is a rather imprecise term with both popular and technical usage. To some, it indicates a long, dry spell, usually associated with lack of **precipitation**, when crops shrivel and **reservoirs** shrink. To others, it is a complex combination of meteorological **elements**, expressed in some form of **moisture index**. There is no widely accepted definition of drought. It is, however, very much a human concept, and many current approaches to the study of drought deal with moisture deficiency in terms of its impact on human activities, particularly those involving agriculture. Agricultural drought is defined in terms of the retardation of crop growth by reduced **soil** moisture levels. This, in turn, may lead to economic definitions of drought when, for example, dry conditions reduce yield or cause crop failure, leading to a reduction in income. It is also possible to define drought in purely meteorological terms, where moisture deficiency is measured against normal or average conditions which have been established through long-term observation (Katz and Glantz 1977).

The establishment of normal moisture levels also allows a distinction to be made between **aridity** and drought. Aridity is usually considered to be the result of low average rainfall, and is a permanent feature of the climatology of a region. In contrast, drought is a temporary feature, occurring when precipitation falls below normal or when near normal rainfall is made less effective by other weather conditions such as high **temperature**, low **humidity** and strong **winds** (Felch 1978).

Aridity is not a prerequisite for drought. Even areas normally considered to be humid may suffer from time to time, but some of the worst droughts ever experienced have occurred in areas that include some degree of aridity in their climatological make-up. Along the desert margins in Africa, for example, annual precipitation is low, ranging between 100 and 400 mm, but, under normal conditions, this would allow sufficient vegetation growth to support

pastoral agriculture. Some arable activity might also be possible if **dry-farming** techniques were employed, yet drought occurs with considerable regularity in these areas (Le Houerou 1977). The problem lies not in the small amount of precipitation, but rather in its variability. Mean values of 100 to 400 mm are based on long-term observations and effectively mask totals in individual years, which may range well above or below the values quoted. Weather records at Beijing, in drought-prone northern China, show that the city receives close to 600 mm of precipitation in an average year. However, the amount falling in the wetter years can be six to nine times that of the drier years. Only 148 mm were recorded in 1891, for example, and 256 mm in 1921, compared to a maximum of 1405 mm in 1956 (NCGCC 1990). Rainfall variability is now recognized as a major factor in the occurrence of drought (Oguntoyinbo 1986), and a number of writers have questioned the use of 'normal' values in such circumstances. In areas of major rainfall variability, the nature of the **environment** reflects that variability rather than the so-called normal conditions, and any response to the problems which arise from drought conditions must take that into account (Katz and Glantz 1977).

Drought is perhaps the most ubiquitous climatological problem society has to face. Its greatest impact is in areas that experience **seasonal drought** or **contingent drought**. The former is common in the subtropics, where the year includes distinct dry and wet seasons. Drought intensifies when the dry season is extended beyond its normal duration or the wet season provides less precipitation than usual. The extension of such conditions over a period of several years has produced some of the most catastrophic droughts on record, in areas such as the **Sahel**, East Africa, India, China and Australia, all of which are subject to seasonal drought. Contingent drought is characterized by irregular and variable precipitation in areas that normally have an adequate supply of moisture to meet their needs. The droughts of 1975–1976 and 1988–1992 in the normally humid UK were contingent droughts. The **Great Plains** of North America have suffered contingent drought for thousands of years, from pre-historic times up to the present (Phillips 1982; Rosenberg 1978; Van Royen 1937). Arguably the most important of these, in terms of its extent, duration and impact on government policies and agricultural prac-tices, was the drought of the 1930s. It created the **Dustbowl** and caused widespread hard-ship and misery for the inhabitants of the area, but it also brought about the realization that drought was an integral component of the environment of the plains and had to be treated as such.

Modern views of drought vary with time and place and with the nature of the event itself. It may be seen as a technological problem, an economic problem, a political problem, a cultural problem, or sometimes a multifaceted problem involving all of these. Whatever else it may be, however, it is always an environmental problem, and basic to any understanding of the situation is the relationship between society and environment in drought-prone areas.

Over thousands of years, certain plants and animals have adapted to life with the limited moisture. Their needs are met, therefore no drought exists. This is the theoretical situation in most arid areas. In reality it is much more complex, for although the **flora** and **fauna** may exist in a state of equilibrium with other elements in the environment, it is a **dynamic equili-brium**, and the balance can be disturbed. Fluctuations in weather patterns, for example, might reduce the amount of precipitation available, thereby changing the whole relationship. If plants and animals can no longer cope with the reduced water supply, they will suffer the effects of drought. Depending upon the extent of the change, plants may die from lack of moisture, they may be forced out of the area as a result of competition with **species** more suited to the new conditions, or they may survive, but at a reduced level

of productivity. The situation is more complex for animals, but the response is often easier. In addition to requiring water, they also depend upon the plants for food, and their fate will therefore be influenced by that of the plants. They have one major advantage over plants, however. Being capable of movement, they can respond to changing conditions by migrating to areas where their needs can be met. Some degree of balance will eventually be attained again, although certain areas – such as the world's desert margins – can be in a continual state of flux for long periods of time.

The human animal, like other species, is forced to respond to such changing environmental conditions. In earlier times this often involved migration, which was relatively easy for small primitive com-munities, living by hunting and gathering, in areas where the overall **population** was small. As societies changed, however, this response was often no longer possible. In areas of permanent or even semi-permanent agricultural settlement, with their associated physical and socioeconomic structures, migration was certainly not an option – indeed, it was almost a last resort. The establishment of political boundaries, which took no account of environmental patterns, also restricted migration in certain areas. As a result, in those regions susceptible to drought, the tendency, perhaps even the necessity, to challenge the environment grew. If sufficient water was not available from precipitation, either it had to be supplied in other ways – for example, by well and aqueduct – or different farming techniques had to be adopted to reduce the moisture need in the first place. The success of these approaches depended very much on such elements as the nature, intensity and duration of the drought, as well as on various human factors, which included the numbers, stage of cultural development and technological level of the peoples involved.

Technology has helped those living in drought-prone areas to cope with drought. In some cases **drought prediction** has allowed responses to be planned and the impacts mitigated, but there are concerns that technology will create greater problems in the future, either directly through the rapid depletion of **groundwater** supplies or indirectly through the activities that lead to **global warming** (IPCC 1996; NCGCC 1990).

References and further reading
Felch, R.E. (1978) 'Drought: characteristics and assessment', in N.J. Rosenberg (ed.) *North American Droughts*, Boulder, CO: Westview Press.
Glantz, M.H. (ed.) (1994) *Drought Follows the Plough, Cultivating Marginal Areas*, Cambridge: Cambridge University Press.
IPCC (1996) *Climate Change 1995*, Cambridge: Cambridge University Press.
Katz, R.W. and Glantz, M.H. (1977) 'Rainfall statistics, droughts and desertification in the Sahel', in M.H. Glantz (ed.) *Desertification:*
Environmental Deterioration in and around Arid Lands, Boulder, CO: Westview Press.
Le Houerou, H.N. (1977) 'The nature and causes of desertification', in M.H. Glantz (ed.) *Desertification: Environmental Deterioration in and around Arid Lands*, Boulder, CO: Westview Press.
NCGCC (1990) *An Assessment of the Impact of Climate Change Caused by Human Activities on China's Environment*, Beijing: National Coordinating Group on Climate Change.
Oguntoyinbo, J. (1986) 'Drought prediction', *Climatic Change* 9: 79–90.
Phillips, D.W. (1982) *Climate Anomalies and Unusual Weather in Canada during 1981*, Downsview, Ont: Atmospheric Environment Service.
Rosenberg, N.J. (ed.) (1978) *North American Droughts*, Boulder, CO: Westview Press.
Van Royen, W. (1937) 'Prehistoric droughts in the central Great Plains', *Geographical Review* 27: 637–50.

DROUGHT PREDICTION

The attempt to forecast the occurrence of **drought** so that responses can be planned and consequences much reduced. The simplest approach to drought prediction is the **actuarial forecast**, which estimates the probability of drought based on past occurrences. Problems with the length and homogeneity of meteorological records often reduce the reliability of actuarial predictions. Links between meteorological variables and other physical or environmental variables have been assessed as possible predictors. There is, for example, a relationship between **sunspot** activity and **precipitation**. Drought on the **Great Plains** has been correlated with the minimum of the 22-year double sunspot cycle. The correlation has no physical theory to explain it, however, and the relationship does not appear to apply outside the western United States. Most modern attempts at drought prediction involve **teleconnection**. Global **sea-surface temperatures** (SSTs) have been correlated with drought in the **Sahel,** for example, and **ENSO** events have been recognized as precursors of drought in Brazil, Australia, Indonesia and India. The time lag between the occurrence of a specific SST or an ENSO event and the onset of drought allows the prediction to be made.

Further reading
Schneider, S.H. (1978) 'Forecasting future droughts: is it possible?', in N.J. Rosenberg (ed.) *North American Droughts*, Boulder, CO: Westview Press.
Oguntoyinbo, J. (1986) 'Drought prediction', *Climatic Change* 9: 79–90.

DRY ADIABATIC LAPSE RATE

See **adiabatic processes**.

DRY DEPOSITION

A form of **acid precipitation** consisting of dry acidic particles, which usually fall out close to their emission sources. **Sulphur dioxide** (SO_2) is the most common constituent of dry deposition. The particles are converted into **acids** when dissolved in surface **water**, at which time their environmental impact is similar to that of **wet deposition**. They also contribute to health problems when they come into contact with the moist tissues of the human respiratory system.

DRY FARMING

A technique which involves the preservation of several years of **precipitation** to be used for the production of one crop. It includes the use of deep ploughing, to provide a **reservoir** for the **rain** that falls, plus a combination of techniques to reduce losses by **evapotranspiration**. The ratio in dry-farming areas is typically one crop year for every three or four years of **fallow**. Although only about a quarter of the rainfall total over the fallow period may be available to the crops, it can produce a doubling of the yield.

Further reading
Moore, R.J. (1969) 'Water and crops', in R.J. Chorley (ed.) *Water, Earth and Man*, London: Methuen.

DRY SEDIMENTATION

The **fallout** of dry **particulate matter** from the **atmosphere** under the effects of **gravity**. Deposition usually takes place close to the source of the material, but **dust** produced by volcanic eruptions and pushed higher into the atmosphere may be carried further. Volcanic eruptions also provide the largest amounts of dustfall. The eruption of **Mount St Helens** in 1980, for example, deposited between 0.2 and 0.4 km³ of fine ash over the western United States. Dry sedimentation is common in arid regions where there is insufficient moisture for other forms of atmospheric cleansing such as **precipitation scavenging**. Dustfall regularly occurs in northern China in the spring, for example, when the prevailing **winds** bring dust from the Gobi Desert. Less frequent occurrences can be significant locally. In Australia in 1983, a major dust storm deposited 106 kg of dust per hectare over the city of Melbourne. Such amounts pale in comparison with the 350 m of **loess** deposits in China formed by dry sedimentation during the **Pleistocene**.

See also
Volcano.

Further reading
Burroughs, W.J. (1981) 'Mount St Helens: a review', *Weather* 36: 238–40.
Lourenz, R.S. and Abe, K. (1983) 'A dust storm over Melbourne', *Weather* 38: 272–4.

DRYLAND SALINITY

See **salinization**.

DUPONT

The major producer of **chlorofluorocarbons** (CFCs) under the trade name **Freon**. Although apparently unwilling to respond in the 1970s to problems associated with CFC production and use, following the **Montreal Protocol**, the company worked to produce less damaging alternatives such as **hydrochlorofluorocarbons** (HCFCs) and pledged to reduce its output by 95 per cent by the year 2000.

DUST

A general term for solid particles of varying materials, sizes and origins, carried in suspension in the **atmosphere**. Most dust particles are considered to have diameters greater than 1 μm, but some volcanic dust may be finer than that. Larger particles remain in the **troposphere**, but volcanic eruptions sometimes push smaller particles into the **stratosphere**. Dust originates naturally as a result of volcanic activity, fires and the **deflation** of particles from **deserts** and other arid areas. Human activities such as mining, quarrying and various agricultural practices contribute dust directly, while activities which increase **desiccation** or **desertification** contribute indirectly by increasing the area susceptible to deflation or **wind** erosion. Dust in the atmosphere has a role in **climate change** through its ability to scatter **radiation**, and at the local level may present an environmental hazard.

See also
Aerosols, Atmospheric turbidity, Dry sedimentation, Dust veil index.

Further reading
Kellogg, W.W. (1980) 'Aerosols and climate', in W. Bach, J. Pankrath and J. Williams (eds), *Interactions of Energy and Climate*, Dordrecht: Reidel.
Shaw, R.W. (1987) 'Air pollution by particles', *Scientific American* 257: 96–103.

DUST VEIL INDEX (DVI)

A rating system developed by climatologist **H.H. Lamb** to provide an assessment of the impact of volcanic eruptions on **atmospheric turbidity** and hence on global **weather** and **climate**. It was derived from such parameters as radiation depletion, the estimated lowering of average **temperatures**, the volume of **dust** ejected and the extent and duration of the veil. The 1883 eruption of **Krakatoa**, with a DVI of 1000, provided a base against which all other eruptions were measured. The 1963 eruption of **Mount Agung** was rated at 800, for example, whereas the DVI for **Tambora** in 1815 was 3000. Other indices have been introduced to refine and modify the DVI, but none has been so widely used.

See also
Glaciological volcanic index, Volcanic explosivity index.

Further reading
Lamb, H.H. (1970) 'Volcanic dust in the atmosphere; with a chronology and assessment of its meteorological significance', *Philosophical Transactions of the Royal Society*, A. 266: 435–533.

DUSTBOWL (THE)

An area of the **Great Plains**, stretching from Texas in the south to the Canadian prairies in the north, which suffered the effects of **desertification** in the 1930s. A combination of **drought** and inappropriate farming practices caused the destruction of the **topsoil** and allowed it to be carried away by the **wind**. The social and economic impacts of the Dustbowl helped to bring about a reassessment of agricultural development and land use in semi-arid areas such as the Plains.

Further reading
Rosenberg, N.J. (ed.) (1978) *North American Droughts*, Boulder, CO: Westview Press.
Worster, D. (1979) *DustBowl: The Southern Plains in the 1930s*, New York: Oxford University Press.

DYNAMIC EQUILIBRIUM

A concept originally used in **geomorphology**, but now generally accepted as having much wider implications. It considers the components of the **environment** to be in, or attempting to achieve, some degree of equilibrium. The balance is never complete, however, but requires a continuing series of mutual adjustments among the **elements** that make up the environment. The rate, nature and extent of the adjustments required will vary with the amount of disequilibrium introduced into the system, but in every environment there will be periods when relative stability can be maintained with only minor adjustments. The environment is then said to be in a **steady state**.

See also
Chaos theory, Ecological balance, Gaia hypothesis.

DYSTROPHIC (LAKES)

Acidic **water** bodies that are deficient in **calcium** (Ca) and have low **nutrient** levels. As a result, they tend to be unproductive. Dystrophic lakes are characteristic of acid **peat** areas. Decomposition rates are low and the lake bottoms are often covered with undecomposed vegetable matter.

See also
Eutrophication, Oligotrophic lakes.

E

EARTH FIRST

A radical international **environmental movement** involved in direct action against individuals, companies and organizations it sees as threatening the **environment**, Earth First undertakes grassroots organizing, litigation and civil disobedience. In North America, Earth Firsters have been involved in the disruption of lumbering operations to protect forests from chainsaws, and in Britain they have campaigned against the construction of motorways by occupying construction sites. They have also instigated boycotts against companies such as Shell and McDonalds as a protest against their perceived disregard for the environment.

EARTH RESOURCES TECHNOLOGY SATELLITES (ERTS)

A series of earth-observing satellites launched by the United States between 1972 and 1984. Referred to as the Landsat series, they evolved from the TIROS and NIMBUS series of meteorological satellites launched in the 1960s. By observing the earth's surface in different sectors of the **electromagnetic spectrum**, Landsat was able to provide information on crop production forecasts, soil and forestry management, **energy** and mineral resource exploration and the assessment of urban population densities. Descendants of these satellites monitor all aspects of the **environment** from the earth's **radiation** balance to **desertification** and **pollution**, and a new series of twenty-six satellites is planned to provide for better monitoring of natural hazards and disasters.

Further reading
Lawler, A. (1995) 'NASA mission gets down to Earth', *Science* 269: 1208–10.
Park, C.C. (1997) *The Environment: Principles and Applications*, London: Routledge.

EARTH SUMMIT

Popular name for the **United Nations Conference on Environment and Development (UNCED)** held in Rio de Janeiro in 1992.

EARTHQUAKE

A series of earth movements brought about by a sudden release of **energy** during tectonic activity in the earth's crust. Earthquakes frequently accompany volcanic eruptions, but the most severe examples are associated with movements along fault lines. Movement along most active faults is infinitesimal and smooth. The rocks on either side of the fault line will move vertically or horizontally in relation to each other and the small amounts of energy involved will be dissipated into the **environment**. If for some reason no movement takes place, despite the addition of energy, the fault is said to be locked, and energy continues to accumulate in the fault system until it is sufficient to break the lock. The sudden release of accumulated energy causes the earthquake. Waves of energy – seismic waves – spread out in all directions from the origin or focus of the earthquake, but those which reach the surface immediately above the focus – the epicentre of the quake – and spread from there cause the most visible damage. Earthquakes are classified according to their magnitude and intensity. Magnitude is a measure of the energy released and is recorded on the Richter Scale, an open-ended logarithmic scale starting at zero, but with no upper limit, since there is no evidence that the values of past large earthquakes are the greatest possible. Maximum observed intensities range between 8.0 and 8.6 – the 1906 San Francisco earthquake, for example, had an estimated magnitude of 8.3 on the Richter Scale – but thousands of earthquakes of lesser magnitude occur every year. The intensity of an earthquake is reflected in the damage that it does to property and is measured on the Modified Mercalli Scale. The scale was developed in the early part of the twentieth century and modified in 1931. It includes twelve values

ranging from Negligible (detected by instruments only) to Catastrophic (total destruction) based on the damage caused to property. Earthquakes of the same magnitude and intensity may cause different levels of destruction depending upon local conditions. Structures built on a **sand** or **clay** substrate, or land reclaimed from the sea, for example, suffer greater damage than those built directly upon rock. Areas in which the buildings have been constructed or renovated to withstand earthquake shocks usually suffer less damage than those where the buildings have been constructed using local materials and techniques. Factors such as these often lie behind the great differences in the death toll associated with earthquakes of similar magnitude and intensity. In Tangshan, China in 1976, buildings of local brick using traditional methods simply disintegrated under the effects of an earthquake measuring 7.6 on the Richter Scale and more than 200,000 people died. Nearly 10 years later in Mexico City an

Figure E-1 The world's earthquake zones

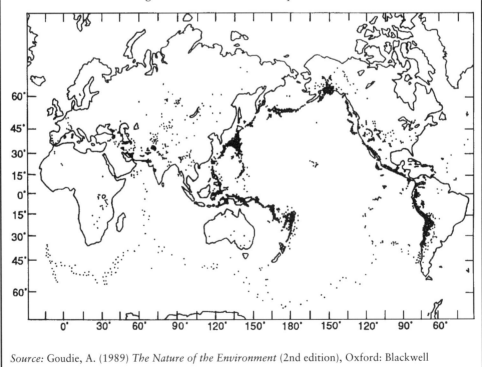

Source: Goudie, A. (1989) *The Nature of the Environment* (2nd edition), Oxford: Blackwell

		Table E-1 Recent Japanese earthquakes		
EARTHQUAKE	DATE	MAGNITUDE	DAMAGED HOUSES	DEATHS
Great Kanto	1923	7.9	576,262	142,807
Kita-Tajima	1925	6.8	3475	428
Kita-Tango	1927	7.3	16,295	2925
Kito-izu	1930	7.3	2240	272
Sanriku Offshore	1933	8.1	7479	8008
Tottori	1943	7.2	7736	1083
Tonankai	1944	7.9	29,189	998
Mikawa	1945	6.8	12,142	2306
Nankai	1946	8.0	15,640	1432
Fukui	1948	7.1	39,111	3848
Chile Tidal Wave	1960	8.5	2830	139
Niigata	1964	7.5	2250	26
Tocachi Offshore	1968	7.9	691	52
Miyagi Offshore	1978	7.4	1383	28
Central Sea of Japan	1983	7.7	1584	104
SW Hokkaido Offshore	1993	7.8	601	230
S. Hyogo (Kobe)	1995	7.2	100,209	5504

Source: Asahi Shimbun, *Japan Almanac*, 1996

earthquake of similar magnitude produced a much lower death-toll of between 9000 and 10,000, despite the city's high population density and the fact that it is built mainly on the bed of a dried-up lake. Most of the damage and the highest death-toll was in those areas where the sub-surface materials were unstable or the houses were of traditional construction. Earthquake-proofed buildings and those on more stable ground had a much greater survival rate. Indirect effects of major earthquakes include **fire** – caused by the rupturing of **gas** lines and the availability of large amounts of combustible materials – which is difficult to fight because the **water** supply system has usually been ruined. The destruction of the water system may also allow disease to spread among the survivors of the earthquake. Earthquake occurrence is not evenly distributed around the earth, but is concentrated in those areas that are most tectonically active. Such areas include the boundary zones between tectonic plates and zones of sea floor spreading. As a result, the areas that experience most earthquakes are located in a belt around the Pacific. This circum-Pacific belt accounts for more than 80 per cent of the world's earthquakes, among them the most devastating in terms of both life and property damage. Japan, for example, has experienced seventeen earthquakes with a magnitude greater than 7 this century, resulting in a death toll of more than 165,000 and damage to nearly 720,000 buildings. On the other side of the Pacific in California, the San Andreas Fault is an earthquake zone which has the potential to devastate the large population centres of southern and central California.

It has formed where the Pacific and North American tectonic plates grind together, and sudden movement along it was responsible for the 1906 San Francisco earthquake. Current concern is with the southern end of the fault which is locked and under great stress in a number of places. The sudden release of the energy represented by that stress would have a catastrophic effect on most of southern California. Other earthquake concentrations occur in a belt stretching through the Mediterranean to the Himalayas and then south-east through Indonesia to meet up with the circum-Pacific belt in the area of the Philippines. Mid-oceanic ridge development in the Atlantic Ocean and the Indian Ocean are also regions of increased earthquake activity. Outside these areas earthquakes may be less common, but they do occur in even the most stable parts of the earth's crust, such as the shield areas of Canada and Australia. Since they are unexpected, earthquakes in these areas often cause a disproportionate amount of damage.

Earthquakes cause major environmental change. They disrupt drainage patterns, both on the surface and underground; in mountainous areas they initiate landslides that change the **geomorphology** of these areas and alter vegetation patterns; coastal areas may be altered catastrophically by the tidal waves or **tsunamis** which follow major offshore earthquakes Attempts at earthquake prediction with the aim of reducing loss of life have met with little success. Potential precursors of damaging earthquakes have included swarms of minor earth tremors, emissions of gases from cracked bedrock or unusual animal behaviour. These may apply to individual earthquakes, but as yet there is no universal signal of an impending earthquake.

Further reading
Chandler, A.M. (1986) 'Building damage in Mexico City earthquake', *Nature* 320 (6062): 497–501.
Davidson, K. (1995) 'Waiting for the Big One', *New Scientist* 141 (1918): 24–8.
deBlij, H.J. and Muller, P.O. (1993) *Physical Geography of the Global Environment*, New York: Wiley.
Normile, D. (1995) 'Cracking up', *New Scientist* 147 (1988): 26–31.

EARTHWATCH

A non-profit, privately funded organization, founded in 1972, Earthwatch is one of the world's largest sponsors of scientific field research. Originating in the United States, it now has affiliates in Australia, Britain and Japan. Its mission is to improve understanding of the planet, the diversity of its inhabitants and the processes that affect the quality of life on earth. Since 1972, Earthwatch has been involved in nearly 2000 projects in 111 countries, with more than 40,000 volunteers participating in their activities. In 1997, funds have been allotted to support twenty-five new projects that include wildlife management, archaeology, **rainforest** ecology, art and architecture.

EARTHWATCH (UNEP)

A programme for monitoring and co-ordinating environmental information collected through the **Global Environment Monitoring System** (GEMS) of the **UN Environment Program** (UNEP).

EASTERLY TROPICAL JET

A rapidly moving easterly airstream encountered in the upper **atmosphere** in equatorial latitudes. It is less persistent than **jet streams** in higher latitudes, being a summer phenomenon best developed over India and Africa, and generally absent over the oceans. The jet is associated with the strong lateral **temperature** gradient that develops in the upper atmosphere at the time of the southwest **monsoon** over India. A similar but separate jet is present over West Africa at about the same time. In both areas, the zones of maximum **precipitation** lie mainly to the north of the jet axis and in India the movement of monsoon **depressions** from east to west up the Ganges valley appears to be controlled by the jet and the associated upper easterlies.

Further reading
Barry, R.G. and Chorley, R.J. (1992) *Atmosphere, Weather and Climate* (6th edition), London: Routledge.

ECOCATASTROPHE

A massive deterioration of the **environment** which threatens the **flora, fauna** and other ecological attributes of an area. It may take place suddenly, for example, following an **oil tanker** shipwreck, or it may reflect the cumulative effects of environmental damage over a longer period of time. Western Siberia provides an example of the latter situation. There the rapid and uncontrolled exploita- tion of forests, mineral deposits and **petroleum** reserves has created ecological disaster in an area that until the 1960s and 1970s was one of the world's largest and least altered **wilderness** areas.

See also
Nuclear winter, Oil tankers.

Further reading
Buchholz, R.A. (1992) 'The big spill: oil and water still don't mix', in R.A. Buchholz, A.A. Marcus and J.E. Post (eds) *Managing Environmental Issues: A Casebook*, Englewood Cliffs, NJ: Prentice-Hall.
Pearce, F. (1993) 'The scandal of Siberia', *New Scientist* 140 (1901): 28–33.

ECOLOGICAL BALANCE

Stability in an **ecosystem** achieved through the development of **equilibrium** among its various components. This does not imply that the community is static. It is subject to natural variations associated with ecological **succession** and other influences such as **fire**, disease and **climate change**, but the system is normally sufficiently elastic to make the necessary adjustments without major displacement of the balance. Human inter- vention that includes the introduction or removal of plants and animals, **pollution** of the **environment** and destruction of **habitat** is now a main cause of imbalance in many ecosystems.

See also
Dynamic equilibrium, Steady state system.

Further reading
Lieth, H., Esser, G. and Overdieck, D. (eds) (1991) *Modern Ecology: Basic and Applied Aspects*, New York: Elsevier.

ECOLOGICAL ENERGETICS

The study of the flow of **energy** within or through **ecosystems**, involving the flow of **solar energy** into the system, its conversion into chemical energy via **photosynthesis** and its ultimate dissipation as heat energy. The process is highly inefficient – on average only 1 per cent of the **solar radiation** is converted during photosynthesis and 90 per cent of the stored chemical energy is lost to the **environment** as heat at each **trophic level** in a **food chain** – but it provides and redistributes sufficient energy to allow the various com- ponents of the **biosphere** to function.

Further reading
Wiegert, R.G. (ed.) (1976) *Ecological Energetics*, Stroudsburg, PA: Dowden, Hutchinson & Ross.

ECOLOGICAL EXPLOSION/ POPULATION EXPLOSION

A very rapid increase in the numbers of an organism, typical of **r-strategists** such as grass- hoppers and mice which are usually small, have a short life-span and produce large numbers of offspring. The rapid spread of infectious diseases such as influenza and plague is often the result of explosive growth among the **viruses** and **bacteria** that cause them. The specific causes of ecological explosions vary, but most are associated with the removal of elements that normally act as constraints on growth. A limited food supply, for example, will keep a **population** in check. If the food supply suddenly becomes abundant, however, the population will expand rapidly. Such is the case with the **algal blooms** which follow local nutrient enrichment. Alternatively, if the predators that keep a population in check are removed, or the organism involved is moved to a location where no natural predators are present, a population explosion will follow. When rabbits were introduced into Australia, for example, the lack of natural predators allowed the population to balloon and become a major environmental problem.

Further reading
Bolen, E.G. and Robinson, W.L. (1995) *Wildlife Ecology and Management* (3rd edition), Englewood Cliffs, NJ: Prentice-Hall.

ECOLOGICAL INTRODUCTIONS

The introduction of an organism into an area in which it is not normally resident, usually as a result of human activity. The process may be deliberate, as with the importation of grazing animals and new plant **species** into areas of European colonization in the eighteenth and nineteenth centuries, or it may be accidental, such as the introduction of lamprey and zebra mussels into the Great Lakes in the second half of the twentieth century. Should the introduced species be unable to find its **ecological niche** in the new **ecosystem**, it will decline and eventually become extinct. Rabbits, which had been so successful on their introduction to Australia, failed to become established on the island of South Georgia, for example, when introduced there in 1872. In most cases, however, the introduction of species into a new **environment** is detrimental to that environment. Selective grazing by non-native cattle and sheep has caused a change in the grass cover in many areas in the southern hemisphere and overgrazing by goats has caused **soil erosion**. Rats that have escaped from visiting ships have decimated the indigenous bird population in Tristan da Cunha in the south Atlantic and feral cats – domestic cats which have gone wild – have caused the extinction of at least five species of birds native to New Zealand and found nowhere else. The net effect of ecological introductions is to disrupt the existing **ecological balance**. Ultimately a new balance may be established, often with the introduced species as a dominant element, but in the process, the original environment may be destroyed or at least radically changed.

See also
Extinction.

Further reading
Drake, J.A., Mooney, H.A., di Castri, F., Groves, R.H., Kruger, F.J., Rejmanek, M. and Williamson, M. (eds) (1989) *Biological Invasions: A Global Perspective*, Chichester: Wiley.
Holdgate, M.W. and Wace, N.M. (1971) 'The influence of man on the floras and faunas of southern islands', in T.R. Detwyler (ed.) *Man's Impact on Environment*, New York: McGraw-Hill.
Turk, J. and Turk, A. (1988) *Environmental Science*, Philadelphia, PA: Saunders.

ECOLOGICAL NICHE

See **niche**.

ECOLOGY

Originally defined by Ernst Haeckel in 1866, ecology is the study of the relationships that develop among living organisms and between these organisms and their **environment**. Investigation may take place at different levels, involving, for example, the study of the relationships of individual **species** of plants or animals – **autecology** – or the study of community patterns – synecology. Subdivision into functional ecology, which deals with such topics as **population** size, behaviour, competition and predation, and historical ecology, which includes the investigation of past distribution patterns and **evolution**, is also common. Other classifications are possible – terrestrial ecology, freshwater ecology and marine ecology, for example. Popular usage of the term tends to be loose, and is commonly applied to environmental problems that have a human origin.

See also
Carrying capacity, Green Parties.

Further reading
Ehrlich, P.R., Ehrlich, A.H. and Holdern, J.P. (1977) *Ecoscience: Population, Resources and Environment*, New York: W.H. Freeman.
Krebs, C.J. (1985) *Ecology*, San Francisco: Harper & Row.
Odum, E.P. (1995) *Ecology and our Endangered Life Support Systems* (2nd edition), Sunderland, MA: Sinauer Associates.
Tudge, C. (1991) *Global Ecology*, New York: Oxford University Press.

ECOSPHERE

See **biosphere**.

ECOSYSTEM

A term coined in 1935 by the British ecologist A.G. Tansley to refer to a **community** of interdependent organisms and the physical **environment** they inhabit. Although there is no accepted ecosystem hierarchy, the eco-system concept can be applied at a variety of scales, from the microscopic to the whole earth. The individual organisms interact with each other and with their environment in a series of relationships made possible by the flow of matter and **energy** within and through the **system**. The relationships are dynamic and routinely respond to change, without altering the basic characteristics of the ecosystem. Cyclical changes in animal **populations**, for example, are an integral part of most ecosystems. Despite the large fluctu-ations that may be involved, the functional relationships among organisms – such as those between predator and prey – continue, and ultimately some degree of balance is restored. Major environmental disruption such as that caused by **climate change** or **fire**, however, may alter a specific ecosystem irreversibly, and bring about its replacement by a system with different characteristics. Regular fires, for example, may create a grassland ecosystem where once a forest ecosystem existed. Natural ecosystems are theoretically self-sustaining, but increasing human interference is threatening their sustainability in many parts of the world, and various methods of ecosystem management have been introduced in an attempt to preserve and protect characteristic natural ecosystems.

See also
Carrying capacity, Trophic levels.

Further reading
Stiling, P.D. (1992) *Introductory Ecology*, Engle-wood Cliffs, NJ: Prentice-Hall.
Odum, E.P. (1971) *Fundamentals of Ecology* (3rd edition), Philadelphia, PA: Saunders.
Owen, O.S. and Chiras, D.D. (1995) *Natural Resource Conservation*, Englewood Cliffs, NJ: Prentice-Hall.
Tansley, A.G. (1935) 'The use and abuse of vegetation concepts and terms', *Ecology* 16: 248–307.

ECOTONE

A transition zone between two **ecosystems**. Although usually most obvious through the changing distribution of vegetation, it also includes other organisms from the two communities. The transition zone may be many kilometres wide and include the characteristics of the two ecosystems, as in the parkland area between the forest and **savanna grasslands** of the sub-tropics, or it may be relatively narrow, with the change from one ecosystem to the other taking place within as little as a few hundred metres, as at some altitudinal tree lines. Where the zone is broad, the ecotone will include animals from both adjacent ecosystems and perhaps others that are characteristic of the ecotone itself, producing a **community** that may be more diverse than in the ecosystems. The location of particular ecotones is usually associated with changes in such environmental elements as **climate**, geology and **soils**, but the nature of the transition may also be influenced by such factors as **fire**, the number of grazing animals present in an area or the extent of the competition between the two communities it separates.

Further reading
Goudie, A. (1989) *The Nature of the Environment* (2nd edition), Oxford: Blackwell.

ECOTOURISM

In its simplest form, the recreational use of the landscape and the **flora** and **fauna** that it contains. Although it may involve hunting and fishing, ecotourism has come to be identified with the non-destructive use of the **environment**, through, for example, bird-watching, wildlife safaris or the appreciation of special plant communities such as the Californian Redwoods. The wildlife elements in ecotourism are major sources of income in many parts of the world, generating millions of dollars every year, for example, in less developed countries such as Kenya and Tanzania in East Africa. By introducing large numbers of people to the environment, ecotourism is also a very effective educational tool, providing participants with a greater appreciation of the environment, while encouraging them to contribute financially to the alleviation of environmental problems. Paradoxically, the popularity of ecotourism has the potential to threaten the very environmental attributes it is designed to protect. In some areas, for example, sensitive **ecosystems** might be harmed when excessive numbers of tourists disturb wildlife, damage vegetation or generate extra **waste** and **garbage**. Thus, while ecotourism can be a financial boon for many areas and can contribute to the preservation of environments under threat, it is not without its potential problems and must be carefully managed.

Further reading
Boo, E. (1990) *Ecotourism: The Potential and Pitfalls* (vols 1 and 2), Washington, DC: WWF.
Pearce, F. (1995) 'Selling wildlife short', *New Scientist* 147 (1993): 28–31.
Whelan, T. (1991) *Nature Tourism: Managing for the Environment*, Covelo, CA: Island Press.

EDAPHIC FACTORS

The physical, chemical and biological properties of **soils** that contribute to the characteristics of **ecosystems**. These contributions may not always be obvious, because they work indirectly through other factors – **soil texture** making its influence felt through drainage – but in some cases the effects are strong and clear. Acidity or **pH**, for example, will influence the types of organisms that occupy the soil, the types of plants that will grow and consequently the types of **fauna** that will inhabit an area. Thus, **limestone** areas develop their own peculiar ecosystems in large part because of the very high pH of their soils.

EDDY DIFFUSION

A process by which a concentrated **gas** or **liquid** becomes diluted as it mixes with the adjacent **air** or **water**. Mixing is brought about by the development of relatively small-scale circular or semi-circular currents or eddies, generated by such conditions as **convection**, changes in surface roughness and **friction**. Eddies may be generated in the

vertical or horizontal plane, but the net effect is to create turbulence, and reduce the concentration by incorporating small amounts of the concentrate into the larger **air mass** or water body. Eddy diffusion is one of the most common processes by which air and water pollutants are diluted and therefore made less harmful.

See also
Diffusion, Pollution, Turbulent flow.

Further reading
Oke, T.R. (1987) *Boundary Layer Climates* (2nd edition), London: Routledge.
Williamson, S.J. (1973) *Fundamentals of Air Pollution*, Reading, MA: Addison-Wesley.

EFFECTIVE RAINFALL

That part of the total rainfall that is available to meet a particular need. In agriculture, for example, the effective rainfall might include that which is retained in the crop root zone, after **evaporation**, surface **runoff** and drainage have taken place. For other purposes – for example, hydroelectric power production – the effective rainfall might be the proportion of the **precipitation** that becomes runoff.

See also
Evapotranspiration.

EFFLUENT

Technically any fluid emanating from a source, but commonly used to refer to gaseous or liquid **waste** discharged treated, partially treated, or untreated into the **environment** from residential, industrial or agricultural sources. Domestic **sewage** is perhaps the most ubiquitous effluent.

ELECTRODE

A conductor through which an electric current enters or leaves a **system** or device. The flow of **electrons** is from the negatively charged **cathode** to the positively charged **anode**.

See also
Electrolysis, Electrolyte.

ELECTROLYSIS

The chemical decomposition of an **electrolyte** brought about by passing an electric current through it. The electrolyte is first dissolved or melted to allow its **dissociation** into positively and negatively charged **ions**. When the electric current is applied, the positively charged ions (**cations**) migrate to the **cathode** and the negatively charged ions (**anions**) to the **anode**, where they are liberated or deposited. The electrolysis of **water**, for example, produces positively charged **hydrogen** ions and negatively charged **oxygen** ions that can be drawn off at the cathode and anode of the system respectively. Industrial electrolysis requires large amounts of electrical **energy**, but it is widely used for the production and purification of metals such as **sodium** (Na), **aluminum** (Al) and **copper** (Cu). It is also the main process for the commercial production of **chlorine** (Cl) from brine.

$$\underset{brine}{2NaCl} + 2H_2O \xrightarrow{electrolysis} \underset{\substack{sodium \\ hydroxide}}{2NaOH} + \underset{chlorine}{\overset{anode}{Cl_2}} + \underset{hydrogen}{\overset{cathode}{H_2}}$$

Further reading
Lawrence, C., Rodger, A. and Compton, R.G. (1996) *Foundations of Physical Chemistry*, New York: Oxford University Press.

ELECTROLYTE

A **compound** that, when dissolved in a **solvent**, or melted by the addition of heat, is capable of conducting an electric current. In the process, the compound is decomposed.

See also
Electrolysis.

ELECTROMAGNETIC SPECTRUM

The range of electromagnetic **radiation** classified according to wavelength, and extending from long radio and television waves to short **gamma rays** and **cosmic radiation**. Each electromagnetic wave consists of an electric field and magnetic field vibrating at right angles to each other. The frequency of

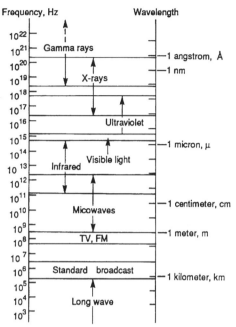

Figure E-2 The electromagnetic spectrum – frequencies, wavelengths and forms of radiation

the vibrations is related to the **energy** level of the radiation source, and indicated in the wavelength. **Longwave radiation** (**infrared**, for example) is low in energy, whereas **short-wave radiation** (**ultraviolet**, for example) has a high energy content. The energy available in shortwave radiation is sufficient to bring about **ionization** in some **atoms**, through the displacement of **electrons**. The resulting highly reactive products may cause damage to living **cells** and contribute to the onset of various types of **cancer**. In contrast, the non-ionizing radiation of low-energy waves is generally considered harmless, although there are claims that long-term exposure to radiation from overhead electric transmission lines, video display terminals and other electric appliances may also cause cell damage. The bulk of the electromagnetic radiation in the earth/atmosphere system originates in the sun. Being a high energy body, the sun emits mainly shortwave radiation, although it also produces a wide range of frequencies, including **visible light**, the only type of electromagnetic radiation detectable by the human eye. When **solar radiation** is absorbed

– for example, by the **ozone layer** or the earth's surface – it is re-emitted at a lower energy level and therefore at a longer wavelength. The redistribution of solar electromagnetic radiation through **scattering**, **reflection** and **absorption** in the **environment** is the main determinant of the **energy budget** in the earth/ atmosphere system.

See also
Atmospheric turbidity, Terrestrial radiation.

Further reading
Carter, R.G. (1990) *Electromagnetic Waves: Microwave Components and Devices*, London/ New York: Chapman and Hall.
Faughn, J.S., Turk, J. and Turk, A. (1991) *Physical Science*, Philadelphia, PA: Saunders.

ELECTRON

A negatively charged elementary particle occupying an orbit or shell around the **nucleus** of an **atom**. The negative charge of the electrons balances the positive charge of the **protons** in the nucleus of the atom. Although much smaller than the other atomic particles, electrons determine the chemical character-istics of atoms, since chemical reactions involve the sharing or transferring of electrons

between the atoms of the **elements** involved.

ELEMENT

A substance that cannot be separated into simpler materials by chemical or physical means. Each element is distinguished by its **atomic number**. Since all **atoms** with the same atomic number have the same chemical pro-perties, the atomic structure of the element will determine its chemical characteristics. Each element is identified by a symbol, which may be the first letter of its English name plus another letter if necessary to avoid confusion (e.g. C for **carbon** or Ca for **calcium**) or the first letter of its Latin name (e.g. K (kalium) for **potassium**). Although there are wide differences between elements, there are also sufficient similarities to allow them to be arranged in a **periodic table** in which they are grouped according to their chemical and physical characteristics.

EL NIÑO

A flow of abnormally warm water across the eastern Pacific Ocean towards the coast of Peru. It is associated with changing pressure patterns and a reversal of airflow in the

Figure E-3 Sea surface temperature anomalies in the Pacific Ocean, September 1997

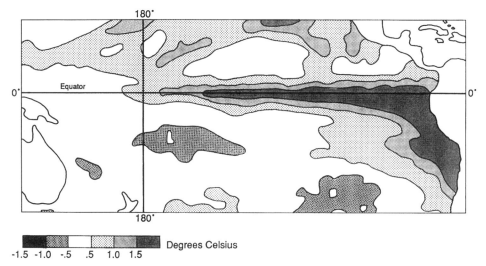

Source: Courtesy NOAA Climate Diagnostics Center

equatorial Pacific, a phenomenon referred to as the **Southern Oscillation**. An El Niño event is preceded by a strong east to west airflow which pushes warm surface **water** to the western Pacific basin, allowing cold water to well up and take its place in the east. When the Southern Oscillation brings about a reversal of the airflow, the warm water ponded up in the western Pacific flows east again to replace the cold and create an El Niño. The name originally referred to a warm current that appeared off the coast of Peru close to Christmas – hence El Niño, the (Christ) Child. Although originally considered a regional event, when strongly developed, El Niño can keep areas in the eastern Pacific warmer than normal for periods of a year or more, and it is now recognized as being of global significance, capable of disturbing weather patterns in many parts of the world.

See also
Drought prediction, ENSO, La Niña.

Further reading
Glantz, M.H. (1996) *Currents of Change: El Niño's Impact on Climate and Society*, Cambridge/New York: Cambridge University Press.
Quinn, W.H. and Neal, V.T. (1992) 'The historical record of El Niño events', in R.S. Bradley and P.D. Jones (eds) *Climate Since AD 1500*, London: Routledge.
Rasmussen, E.M. and Hall, J.M. (1983) 'El Niño', *Weatherwise* 36: 166–75.
UNEP (1992) *The El Niño Phenomenon*, Nairobi: United Nations Environment Program.

ELUVIATION

The removal of colloidal-sized particles such as **clay** and organic matter from upper **soil horizons** by percolating **water** usually followed by their re-deposition or illuviation in lower horizons. Although eluviation is a natural part of **soil** development, if the particles are removed more rapidly than they can be replaced, the fertility of the soil will be reduced. Similarly, the illuviation of the particles deeper in the **soil profile** may affect the quality of the soil by causing drainage problems.

See also
Colloids, Leaching.

EMBRYO

A multicellular organism in its early stages of development, following the fertilization of an ovum or egg-cell. Although both plants and animals pass through an embryonic stage, the term is most commonly applied to animals, which are considered to be in the embryonic stage from the time of fertilization to birth, in the case of mammals, or hatching in the case of birds. The term foetus (fetus) is used for a human embryo which is more than eight weeks old, by which time the organ systems have formed.

EMERSON, R.W. (1803–1882)

A US philosopher and writer who embraced the transcendentalist philosophy of the divinity and unity of man and nature. Such beliefs foreshadowed later environmentalist concepts and ideas.

See also
Thoreau, H.D.

Further reading
Cox, G.W. (1993) *Conservation Ecology: Biosphere and Biosurvival*, Dubuque, IA: Wm C. Brown.

EMISSION STANDARDS

Permissible limits of emissions from **pollution** sources. They are commonly incorporated in clean **air** or **water** quality legislation, which not only sets out the standards but also incorporates mechanisms by which they can be enforced. Primary standards have been developed to protect human health, while secondary standards are aimed at other **elements** including crops, materials and structures that have implications for human welfare and personal comfort. Specific maximum allowable emission levels are based on the known detrimental impacts of certain pollutants or upon the laboratory-based estimates of such impacts. They generally assume that emissions below a certain level will have no adverse effects, and do not always take into account the cumulative effects of long-term low level emissions.

Standards have changed with time, as the technology to reduce emission levels has been developed and as the understanding of the impacts of specific pollutants has improved.

See also
Air quality standards, Clean air legislation, Emissions trading, Environmental Protection Agency (EPH), Safe Drinking Water Act, Water quality standards.

Further reading
Holmes, G., Theodore, L. and Singh, B. (1993) *Handbook of Environmental Management and Technology*, New York: Wiley.

EMISSIONS TRADING

A scheme developed by the United States **Environmental Protection Agency** (EPA) to encourage companies to reduce **pollution**. Various approaches are involved, but in essence, if a company is able to reduce its output of pollutants below the EPA emission levels, it receives credit for the difference. That credit can be used to offset excess pollution at another of the company's plants, it can be banked for future use or sold to another company. The trade is not on a direct 1:1 ratio, however. The credit allowed is less than the original saving so that ultimately there will be a reduction in total emissions. Evidence indicates that emissions trading has not had a major impact on pollution reduction, and environmentalists suggest that unless the transfer credit limits are reduced significantly, the scheme will produce only a shifting of emissions rather than the desired reduction.

Further reading
Heggelund, M. (1991) *Emissions Permit Trading: A Policy Tool to Reduce the Atmospheric Concentration of Greenhouse Gases*, Calgary: Canadian Energy Research Institute.
Rapaport, R. (1986) 'Trading dollars for dirty air', **Science** 86 (7): 75.
Smith, Z. A. (1995) *The Environmental Policy Paradox*, Englewood Cliffs, NJ: Prentice-Hall.

ENDANGERED SPECIES

Species of plants and animals threatened with **extinction** because their numbers have declined to a critical level as a result of overharvesting or because their **habitat** has been drastically changed. That critical level is the **minimum viable population** (MVP), and represents the smallest number of breeding pairs required to maintain the viability of the species. The numbers involved are difficult to determine, but vary with the species. Large carnivores, for example, tend to have a lower MVP than small mammals such as rabbits or mice. The total number of endangered species is not easily estimated, but the **Convention on International Trade in Endangered Species** (CITES) lists some 3000 animals and about 24,000 species of plants sufficiently endangered that trade in them is either completely prohibited or strictly regulated. CITES has been signed by 125 nations, many of which have also passed legislation directed at reducing pressure on endangered species, through restrictions on

Figure E-4 Numbers and distribution of tigers in India

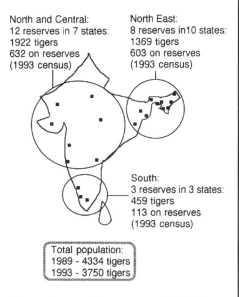

North and Central:
12 reserves in 7 states:
1922 tigers
632 on reserves
(1993 census)

North East:
8 reserves in10 states:
1369 tigers
603 on reserves
(1993 census)

South:
3 reserves in 3 states:
459 tigers
113 on reserves
(1993 census)

Total population:
1989 - 4334 tigers
1993 - 3750 tigers

Source: Courtesy Government of India Web Pages

hunting, wildlife management and habitat preservation. The movement to retain **bio-diversity** also favours the **conservation** of endangered species. Large mammals such as elephants, whales, gorillas and tigers have received most attention, but smaller organisms from migratory birds to minnows to spiders have also been identified as endangered. Exotic plants, such as tropical orchids or the cacti of the south-western United States, also make headlines from time to time when dealers, collectors or poachers are apprehended, but thousands of other plant species are endangered by everyday human activities such as agriculture or forestry. Some endangered species have been brought back from the edge of extinction – for example, the whooping crane, the American bison, the peregrine falcon – while others are being preserved in zoos or wildlife refuges. However, the number of endangered species or threatened species (those likely to become endangered in the foreseeable future) is growing rapidly and according to some environmentalists the annual extinction rate of all species has reached epidemic proportions. Arguments for the preservation of endangered species commonly cite the economic, medical, aesthetic or recreational importance of particular species, but on a broader scale all species may be considered as integral components of the environmental system, and any loss of even one species must have an effect on the integrity of that system.

See also
Blue List, Red Data Books.

Further reading
Cohn, J.P. (1990) 'Elephants: remarkable and endangered', *Bioscience* 40 (1): 10–14.
Leigh, J. , Boden, R. and Briggs, J. (1984) *Extinct and Endangered Plants of Australia*, Melbourne: Macmillan.
Miller, G.T. (1994) *Living in the Environment*, Belmont, CA: Wadsworth.

ENERGY

The capacity to do work. Energy takes a variety of forms and can be converted from one form to another to meet specific needs. There is no simple universally accepted classification of energy forms, but most classifications include the following:

- *kinetic* – the energy possessed by an object in motion.
- *potential* – the energy possessed by an object as a result of its position. A bag of flour on a shelf, for example, retains the energy expended to place it in that position. If the bag falls off the shelf the potential energy will become kinetic energy.
- *thermal* – heat energy.
- *electrical* – the energy associated with an electric charge in an electric field.
- *chemical* – the energy released during a chemical reaction.
- *nuclear* – the energy released during a nuclear reaction.
- *radiant* – energy transmitted in the form of **radiation**.

Energy can be converted from one form to another. The chemical energy in **coal**, for example, is converted into thermal energy through **combustion**, which in turn can be converted into electrical energy in a **thermal electric power station**. Such conversions are not perfect, however. All involve some energy loss. Even a natural process such as **photosynthesis**, in which radiant energy from the sun is converted and stored in plants as chemical energy, may convert as little as 2 per cent of the **solar radiation** falling on a plant. A division into renewable and non-renewable energy resources is often made in environmental studies. **Renewable energy** can easily be replaced once it has been used because of regular regeneration at source. Solar energy used in the earth/atmosphere system, for example, is continuously replaced by new energy created by **fusion** processes in the sun; the energy obtained from plants is replaced by the growth of new plants; the energy in flowing **water** is renewed through the **hydrological cycle**. In contrast, when non-renewable energy is used it cannot be replaced, or can be replaced only on a time-scale that lies outside the human frame of reference. When coal is burned, for example, the energy released cannot be renewed, and although there are places on the earth's surface where new coal is being formed, the thousands of years required for its formation take it out of even the long-term energy plans of society. Until the mid-eighteenth century, society depended mainly upon **renewable energy** resources, and its net impact on the environment was very limited. The impact of energy use only began to be global with the increased use of coal from the 1750s on. Coupled with the **steam engine** and used in a variety of industrial processes, coal was responsible for initiating the increasingly rapid environmental change that continued with the development of **petroleum** energy resources. In the late twentieth century, the developed industrial societies depend on petroleum and electricity as their main sources of energy. Coal is generally less important than it once was, although it retains its predominant position in some less developed nations, such as China, for example, where it is abundant and cheap. Energy consumption has increased six-fold since 1800, and accompanying that increase

there has been a major deterioration in environmental conditions. Global environmental issues such as **acid rain**, increased **atmospheric turbidity** and the enhancement of the **greenhouse effect** all have an energy component, and the ever-growing demand for energy is at the root of the serious deterioration of **air**, water and **soil** in many parts of the world. Many environmentalists advocate the return to **renewable energy** resources, but while that has been successful in some areas, it is unlikely that such sources as solar energy, **biomass** energy or the energy from **wind** and waves would be sufficient to meet modern demands.

Further reading
Berry, R.S. (1991) *Understanding Energy:Energy Entropy and Thermodynamics for Everyman*, River Edge, NJ: World Scientific.
Byrne, J. and Rich, D. (1992) *Energy and Environment: The Policy Challenge*, New Brunswick, NJ: Transaction Publishers.
Howes, R.H. and Fainberg, A. (eds) (1991) *The Energy Sourcebook*, New York: American Institute of Physics.
Long, R. (1989) *Energy and Conservation*, New York: H.W. Wilson.
Nathwani, J.S., Siddall, E. and Lind, N.C. (1992) *Energy for 300 Years: Benefits and Risks*, Waterloo, Ont: Institute for Risk Research, University of Waterloo.
Priest, J. (1984) *Energy: Principles, Problems, Alternatives*, Reading, MA: Addison-Wesley.

Figure E-5 Changing technology and energy consumption

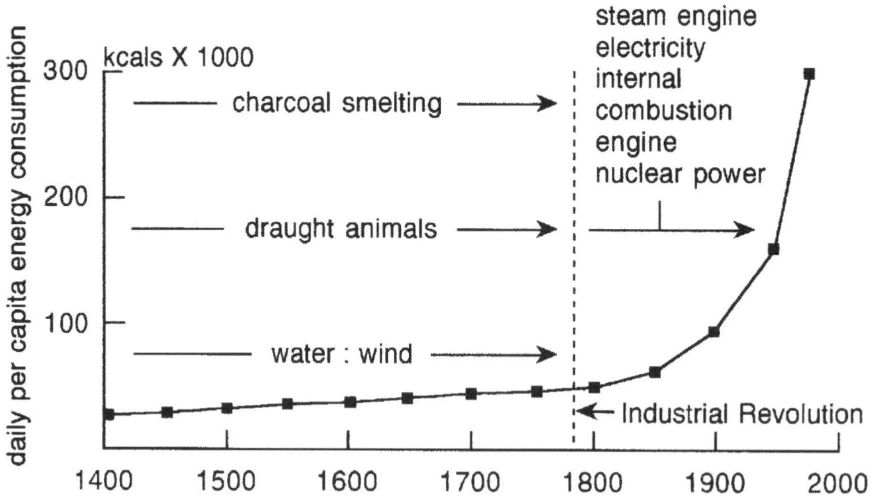

ENERGY BUDGET

An accounting of the flow of **energy** through a **system**. Originally applied by ecologists to **ecosystems**, the approach is also useful in industry to check the **energy efficiency** of industrial processes. The concept is also applied by climatologists to the relationship between the amount of **solar radiation** entering the earth/atmosphere system and the amount of **terrestrial radiation** leaving. In theory, these energy fluxes should balance. In practice, although the concept of balance is a useful one, it applies only in general terms to the earth as a whole over an extended time period. It is not applicable to any specific area over a short period of time.   .

See also
Heat budget.

ENERGY CONSERVATION

Using less **energy** to achieve the same amount of work or decreasing the amount of **fuel** used to produce the same energy output. By reducing demand and improving **energy efficiency**, energy **resources** can be conserved. Energy conservation has both environmental and economic implications. By using less energy and using it more efficiently, the output of **combustion** products is reduced, to the benefit of the **atmospheric environment**, and the necessity to mine and drill for new energy resources is lessened, to the benefit of the **terrestrial environment**. Conservation receives most attention when energy costs are high. Greater efficiency and reduced demand will offset the impact of such economic conditions. During the period of rising **petroleum** prices in the 1970s and 1980s, improved **insulation** in residential buildings, the development of more efficient **internal combustion engines** and changes to **renewable energy** sources all contributed to energy conservation and helped to diminish the impact of the rising costs.

Further reading
Aubrecht, G.J. (1989) *Energy*, Columbus, OH: Merrill.
Ricketts, J. (1995) *Competitive Energy Management and Environmental Technologies*, Lilburn, GA: Fairmont Press.

ENERGY EFFICIENCY

The ratio of the work done by a process or mechanism to the total **energy** consumed, normally expressed as a percentage. It is commonly applied to the conversion of energy from one form to another, with the amount of energy put in to a convertor compared to the amount of energy available following the conversion process. At each stage in a **food chain**, for example, the efficiency is only about 10 per cent. The remaining 90 per cent is lost in the form of heat or used to allow the converting organisms to function. In industry, energy efficiency values vary considerably. **Thermal electric power stations** are usually less than 40 per cent efficient in their conversion of chemical energy to electrical energy, whereas hydroelectric power stations, in which mechanical energy is converted to electrical energy, can be as much as 95 per cent efficient.

With the growing concern for declining energy **resources**, rising energy costs and the impact of large-scale energy consumption on the **environment**, the term has assumed a less technical meaning. It refers, for example, to the willingness of society to change its user habits so that less energy is wasted, and may involve increased government intervention in the form of incentive programmes or greater education on energy use.

Table E-2 Energy efficiency in a typical thermal electric power plant

Chemical energy in coal	100 units
Heat lost in stack gases	10 units
Heat lost in cooling water	50 units
Electrical transmission losses	**3 units**
Total losses	**63 units**
Electrical energy delivered	37 units
Energy efficiency of the plant: 37%	

Further reading
Chapman, J.D. (1989) *Geography and Energy: Commercial Energy Systems and National Policies*, London: Longman.
Kleinbach, M.H. and Salvagin, C.E. (1986) *Energy Technologies and Conversion Systems*, Englewood Cliffs, NJ: Prentice-Hall.
Natural Resources Canada (1996) *Energy Efficiency Trends in Canada: Energy Efficiency Indicators*, Ottawa: Natural Resources Canada.

ENHANCED GREENHOUSE EFFECT

See greenhouse effect.

ENSO

An acronym for El Niño – Southern Oscillation.

ENTROPY

A thermodynamic measure of the degree of disorder within a system. In a well-ordered system, for example, there will be a distinct separation between warm bodies and cool bodies. With time, however, the heat is gradually redistributed so that all bodies

Figure E-6 The impacts of an ENSO event

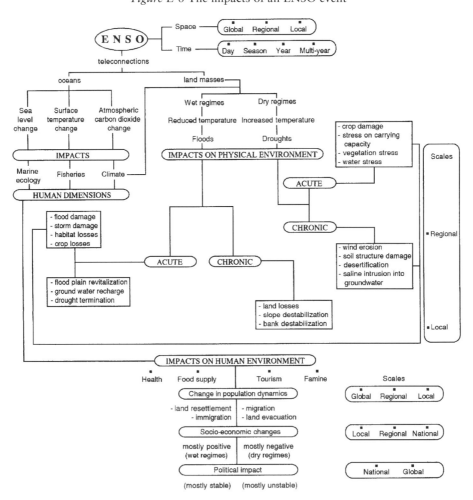

Source: Nkemderim, L.C. and Budikova, D. (1996) 'The El Niño-Southern Oscillation has a truly global impact', *IGU Bulletin* 46: 30

within the system have the same **temperature**. Such a system is considered to be much less ordered than the original and therefore possesses greater entropy. In any closed system, the level of entropy always increases with time, and as it does so its **energy** level decreases. As a result, entropy can also be considered as a measure of a system's inability to do work.

Further reading
Berry, R.S. (1991) *Understanding Energy: Energy, Entropy and Thermodynamics for Everyman*, River Edge, NJ: World Scientific.
Glassby, G.P. (1988) 'Entropy pollution and environmental degradation', *Ambio* 17 (5): 330–5.

ENVIRONMENT

A combination of the various physical and biological **elements** that affect the life of an organism. Although it is common to refer to 'the' environment, there are in fact many environments, all capable of change in time and place, but all intimately linked and in combination constituting the whole earth/atmosphere system. They vary in scale from microscopic to global and may be subdivided according to their attributes. The **aquatic environment**, for example, is that of rivers, lakes and oceans, the **terrestrial environment** that of the land surface. The term 'built environment' has been applied to areas, such as cities, created by human activity. The human element has a dominant role in modern environmental studies, and as a result environmental science includes not only the traditional sciences such as chemistry, physics and biology, but also engineering, economics, sociology, politics and law. The study of the environment is thus very much interdisciplinary in nature.

Further reading
Anderson, S.H., Beiswinger, R.E. and Purdom, P.W. (1993) *Environmental Science*, New York: Macmillan.
Holland, H.D. and Peterson, U. (1995) *Living Dangerously: The Earth, its Resources and the Environment*, Princeton, NJ: Princeton University Press.
Simmons, I.G. (1993) *Interpreting Nature: Cultural Constructions of the Environment*, London: Routledge.

ENVIRONMENTAL CHANGE RESEARCH CENTRE (ECRC)

An organization based at University College London involved in the monitoring and modelling of change in aquatic and terrestrial **ecosystems**. Its activities include investigation of **acid precipitation**, the **eutrophication** of lakes, **climate change** and nature **conservation**, in the UK and internationally.

ENVIRONMENTAL CONSEQUENCES OF NUCLEAR WAR (ENUWAR)

An investigation into the impact of large-scale nuclear war on the **environment**, sponsored by the **International Council of Scientific Unions** (ICSU) through the **Scientific Committee on Problems of the Environment** (SCOPE). It grew out of the concerns raised by the concept of **nuclear winter** postulated in the early 1980s. Studies by the SCOPE-ENUWAR researchers identified the direct and indirect biological effects of nuclear war on **ecology** and agriculture as a major threat to society.

See also
TTAPS scenario.

Further reading
Harwell, M.A. and Hutchinson, T.C. (1985) *Environmental Consequences of Nuclear War, Volume II: Ecological and Agricultural Effects, SCOPE 28*, New York: Wiley.
SCOPE-ENUWAR (1987) 'Environmental consequences of nuclear war: an update', *Environment* 29: 4–5 and 45.

ENVIRONMENTAL EQUILIBRIUM

The concept of dynamic balance among the constituents of the **environment**. Changes in one of the components, tending to produce instability, are countered by charges in others which attempt to restore the balance. The balance is never complete, however. It is a dynamic process that includes a continuing series of mutual adjustments among the **elements** involved. The rate, nature and extent of the adjustments required will vary with the amount of disequilibrium introduced into the system, but in every environment there will be periods when relative stability can be maintained with only minor adjustments. This inherent stability of the environment tends to dampen the impact of changes even as they happen, and any detrimental effects that they produce may go unnoticed. At other times the equilibrium is so disturbed that stability is lost, and major responses are required to restore the balance. Many environmentalists view modern environmental deterioration as the result of human interference in the system at a level that has pushed the stabilizing mechanisms to their limits, and perhaps beyond.

See also
Dynamic equilibrium, Gaia.

Further reading
Kemp, D.D. (1994) *Global Environmental Issues: A Climatological Approach* (2nd edition), London/New York: Routledge.
Lovelock, J.E. (1979) *Gaia – A New Look at Life on Earth*, Oxford: Oxford University Press.

ENVIRONMENTAL IMPACT ASSESSMENT (EIA)

Assessment and analysis of the potential impact of various forms of human activities on the **environment**. Formal studies of existing environmental conditions and expected changes are commonly required before major development projects such as **water** diversion or mineral extraction are allowed. The results of the assessment are presented in the form of an environmental impact statement which is made available for discussion among the various levels of government, with additional input from environmental organizations and the general public. Such discussions may lead to the abandonment of the project, or its modification to reduce the perceived impact. In theory, a project should only be allowed to proceed if the impact statement indicates minimal environmental disruption, but in some cases socioeconomic considerations are allowed to override strict environmental factors. Environmental impact assessment can be made to work well in the consideration of short-term, direct impacts, but it is generally less successful in dealing with the cumulative impacts of longer term, less direct impacts, such as climate change, for example.

Further reading
Gilpin, A. (1995) *Environmental Impact Assessment (EIA): Cutting Edge for the Twenty-First Century*, Cambridge/New York: Cambridge University Press.
Glasson, J., Therivel, R.S. and Chadwick, A. (1994) *Introduction to Environmental Impact Assessment*, London: UCL Press.
Wood, C. (1995) *Environmental Impact Assessment: A Comparative Review*, London: Longman.

ENVIRONMENTAL LAPSE RATE

The rate at which **temperature** declines with increasing altitude in the **troposphere**. Although the lapse rate varies with time and place it is normally considered to average −6.4°C per 1000 m. Since the **atmosphere** is heated from below, abnormal heating or cooling can increase or reduce the lapse rate and in some cases cause it to be reversed. The relationship between the environmental lapse rate and the dry and saturated **adiabatic lapse rates** determines **atmospheric stability** in the **tropopause**.

Further reading
McIlveen, J.F.R. (1991) *Fundamentals of Weather and Climate*, London: Chapman and Hall.

ENVIRONMENTAL LEGISLATION

Laws, acts, by-laws, protocols and conventions aimed at preserving and protecting the **environment**. At one end of the scale are international agreements, such as the **Montreal Protocol** or **CITES**, that require further legislative action by the signatories. At the other are municipal by-laws, such as **waste disposal** regulations, designed to meet local needs or conditions. Current environmental legislation is less effective than it might be because of inadequate enforcement. This is a particular problem at the international level, but often applies to national legislation where several jurisdictions are involved.

See also
Clean air legislation, Environmental Protection Act (UK), National Environmental Policy Act,

Ozone Protection Act, Resource Conservation and Recovery Act, Safe Drinking Water Act, Superfund, Tall stacks policy, 30 per cent club, US Forest Service Multiple Use Sustained Yield Act, US Global Climate Protection Act, US Wilderness Act.

ENVIRONMENTAL MOVEMENT

The environmental movement has its roots in the growing concern for nature which characterized all sectors of society – from literature to science – in the nineteenth century. Interest was mainly in the **conservation** of natural conditions that manifested itself in the creation of national parks, forests reserves and game preserves in North America and Europe. Between the wars, particularly in North America, where **drought** devastated large areas of agricultural land in the 1930s, more attention was paid to **soil conservation**. By the 1950s and 1960s, **pollution** had become the central environmental issue. After a decline in the 1970s, when concerns over **energy** replaced the **environment** in the public interest, the environmental movement rebounded, reflecting an increasing level of concern with society's ever-increasing ability to disrupt environmental **systems** on a large scale. A new environmentalism has emerged, characterized by a broad global outlook, increased politicization and a growing environmental consciousness that takes the form of **waste** reduction, prudent use of **resources** and the development of environmentally safe products. There is also a growing appreciation of the economic and political components in environmental issues, particularly as they apply to the problems arising out of the economic disparity between rich and poor nations. The modern environmental movement is aggresive, with certain organizations using direct action in addition to debate and discussion to draw attention to the issues.

See also
Earth First, Earthwatch, Emerson, R.W., Friends of the Earth, Greenpeace, Green parties, Muir, J., National Audubon Society, Nature Conservancy, Nature Conservancy Council, Rainforest Action Network, Sierra Club, Thoreau, H.D., Wilderness Society, Worldwatch Institute, Worldwide Fund for Nature.

Further reading
Cox, G.W. (1993) *Conservation Ecology: Biosphere and Biosurvival*, Dubuque, IA: Wm C. Brown.

Marcus, A.A. and Rands, G.P. (1992) 'Changing perspectives on the environment', in R.A. Buchholz, A.A. Marcus and J.E. Post (eds) *Managing Environmental Issues: A Casebook*, Englewood Cliffs: Prentice-Hall.

Shabecoff, P. (1993) *A Fierce Green Fire: The American Environmental Movement*, New York: Farrar, Straus & Giroux.

Wall, D. (ed.) (1993) *Green History: A Reader in Environmental Literature, Philosophy and Politics*, London: Routledge.

White, L. (1987) 'The historical roots of our ecologic crisis', *Science* 155: 1203–7.

Worster, D. (1995) *History of Ecological Thought*, (2nd edition), Cambridge: Cambridge University Press.

ENVIRONMENTAL POLLUTION

The contamination of the physical and biological components of the earth/atmosphere system to such an extent that normal environmental processes are adversely affected. Although some writers and researchers regard pollution as anthropogenic in origin, natural sources can also provide sufficient extraneous material to disrupt normal environmental processes. Volcanic activity, for example, can cause major **air pollution** or **water pollution** and destroy **flora** and **fauna** by covering them with volcanic ash or **lava**. With time, the **environment** adjusts to take such changes into account, since the earth/atmosphere system includes mechanisms capable of dealing with almost any form of contamination or disruption. Larger air or water pollutants, for example, fall out of the air or water stream as a result of **gravity**, **gases** may be absorbed by plants, neutralized by **oxidation** or dissolved in water, and organic materials may be destroyed by **bacteria**. The capacity of the **environment** to deal with pollutants in this way is not unlimited, but it might be argued that pollution occurs only when the environment's capacity for dealing with additional material is surpassed. Such levels are difficult to establish, however, since the impact of a given amount of pollutant will vary with such factors as the physical nature of the environment, the age and mix of organisms and the timing or duration of the pollution event. Emissions from 100 cars in a restricted **airshed** would have the potential to cause a greater environmental impact than the same emissions in an open, well-ventilated location. Similarly, 1000 barrels of **oil** spilled on a rocky, sub-Arctic shoreline would not have the same impact as 1000 barrels spilled on a sandy, tropical beach. For some contaminants, the introduction of a human element allows specific pollution levels to be established. It is possible, for example, to calculate maximum levels of air pollutants that can be tolerated by normal, healthy individuals without harm. Permissible levels of chemicals in drinking water can be developed also. They are not ideal, however, since like the wider environment individual humans also react differently to the same levels of contamination. Young children, older adults or individuals with existing health problems commonly suffer more than healthy, mature adults. Furthermore, the tests used to estimate these levels are not necessarily conducted on human subjects, but on animals under laboratory conditions. Thus, attempts at establishing the levels of pollution that the environment or **elements** of the environment can withstand have met with only limited success. It is increasingly clear, however, that human activities are adding contaminants to the environment at rates which easily exceed its ability to cope. Pollution takes many forms, most of which started as local problems, but which are now global in extent. Even in the Arctic, far removed from industrial activity, air pollution is present in the form of **Arctic Haze**, and **oil slicks** can be found in the middle of the world's oceans. Some of the pollutants, such as human **sewage** or **smoke**, are obvious, whereas others, such as radioactive emissions, toxic gases or chemicals dissolved in water are invisible and may not evoke a response until they have caused major environmental damage. **Noise** pollution is similar, in that it is not visible. However, along with more traditional forms of pollution, it makes a serious contribution to the quality of the environment in cities and has begun to create problems in rural areas also as power boats, snow-machines and all-terrain vehicles increase in popularity. To be treated effectively, pollution must be controlled at source. There is little point, for example, in liming a lake to reduce the acidity of water, if industry continues to emit acid gases into the **atmosphere**. In the case of the nuclear industry and certain branches of the chemicals industry, potential pollutants may be so hazardous that there must be complete control, with a no-emissions policy and permanent or long-term management of **waste** products. Although the technology exists to handle most, if not all, pollution problems, socioeconomic or political factors may prevent its adoption. There have been some successes, such as the reduction in water pollution by **detergent phosphates** and control of the production and emission of **CFCs**, but other pollutants such as **sewage** and automobile exhaust gases remain serious and growing problems in some areas. Since the 1940s and 1950s, governments have introduced significant volumes of anti-pollution legislation and set standards for air and water quality. Success has been mixed, however, and as pollution problems become increasingly global in scope, international co-operation becomes essential. The successful implementation of the provisions of the **Montreal Protocol** shows that co-operation can work, but differences in national economic and political agendas create the potential for ongoing problems.

See also

Acid rain, Air quality, Atmospheric turbidity, Bhopal, Chernobyl, Clean air legislation, Dust veil index, Eutrophication, Fumigation, Greenhouse gases, Incineration, Leachate, Livestock waste, Love Canal, Noise abatement, Nuclear waste, Oil pollution, Oil tanker accidents, Oxygen sag curve, Photochemical smog, Precipitation scavenging, Scrubbers, Soot, Tall stacks policy.

Further reading
Costello, M.J. and Gamble, J.C. (1992) 'Effects of sewage sludge on marine fish embryos and larvae', *Marine Environmental Research* 33: 49–74.
Freedman, B. (1995) *Environmental Ecology: The Ecological Effects of Pollution, Disturbance and other Stresses* (2nd edition), San Diego, CA: Academic Press.

Howard, R and Perley, M. (1991) *Poisoned Skies*, Toronto: Stoddart.
Newson, M. (1992) 'The geography of pollution', in M. Newson (ed.) *Managing the Human Impact on the Natural Environment: Patterns and Processes*, London/New York: Belhaven Press.
Turco, R. (1997) *Earth Under Siege: From Air Pollution to Global Change*, New York: Oxford University Press.
Vatavuk, W. (1990) *Estimating Costs of Air Pollution Control*, Chelsea, MI: Lewis.

ENVIRONMENTAL PROTECTION ACT (UK)

Legislation enacted in Britain in 1990 to improve the control over the **pollution** of **air**, land and **water**. It promoted an integrated approach to pollution in which the environmental impacts of major emissions to air, land and water are considered in combination rather than individually. It requires the use of the best available techniques not entailing excessive costs to be used in controlling emissions.

ENVIRONMENTAL PROTECTION AGENCY (EPA)

US agency established in 1970 to co-ordinate government action on environmental issues, and thus it is involved in the establishment, monitoring and enforcement of environmental standards. It engages in research and supports environmental protection activities at the state and local level. In the past the agency has been criticized for lax enforcement of regulations and being politically motivated in its decision making.

See also
Superfund.

ENZYME

A **protein** produced by living **cells** that acts as a **catalyst** in organic chemical reactions. Enzymes tend to be reaction specific, being involved in only one type of reaction or at most only a limited range of reactions. They have an important role in **metabolic processes** in living organisms, through, for example, their ability to initiate reactions, that normally require high **temperatures**, at temperatures that can be tolerated by the organisms. Enzymes also assist in the microbiological reactions associated with such processes as decay and **fermentation**, and have been developed commercially for use in the food and drug industries. **Pollution** can destroy or impair the functions of enzymes by causing chemical or physical changes in their protein. The enzymes are then no longer able to initiate essential reactions, cells die and living organisms are damaged or destroyed.

Further reading
Alters, S. (1996) *Biolog: Understanding Life*, St Louis: Mosby.
Dugas, H. (1995) *Bioorganic Chemistry: A Chemical Approach to Enzyme Action* (3rd edition), New York: Springer-Verlag.

EPICENTRE

See **earthquake**.

EPILIMNION

The surface layer of warmer **water** lying above the **thermocline** in a lake. It is best developed in the summer months when **solar radiation** levels are higher and **wind** and wave action help to distribute the available **energy**. **Temperature** changes and the consequent vertical movement or overturning of the lake water in the autumn destroys the layer, but it is re-established following the overturning of the lake waters again in the spring.

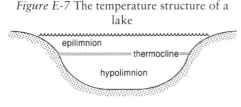

Figure E-7 The temperature structure of a lake

See also
Hypolimnion, Lake stratification.

EPIPHYTE

A plant which is not rooted in the ground, but lives upon or is attached to another plant. The host plant provides no **nutrients**, only support. Epiphytes are common in the tropical **rainforest** where plants such as orchids are found perched in trees where they can have access to more sunlight than at the surface and receive nutrients and moisture from the **canopy** above. The host tree appears to obtain no benefits from the relationship, but is unharmed by the presence of the epiphyte.

Further reading
Fitter, A. and Hay, R. (1987) *Environmental Physiology of Plants* (2nd edition), London: Academic Press.

EPOCH

A subdivision of one of the major periods into which the geological history of the earth is divided. The **Pleistocene** and **Holocene** epochs, for example, are subdivisions of the **Quaternary** period.

EQUATORIAL RAINFOREST

One of the world's major **biomes**. Equatorial rainforests are located in a zone some 10°N and S of the equator, mainly in Amazonia, equatorial west and central Africa and south-east Asia, all of which experience heavy **precipitation** (1750–2500 mm) and high, fairly constant **temperatures** (25–28°C) through the year. Such hot, moist conditions encourage rapid and abundant plant growth, with the tallest trees exceeding 30 m in height, and supporting a **canopy** sufficiently

thick that it allows little light to reach ground level. Many of the trees, shallow rooted in thin **soils**, are underpinned by buttresses. The trees in turn support a great variety of climbing plants and **epiphytes** that use the trees to reach the sunnier conditions in the upper levels of the forest. The rainforest vegetation is typically stratified, consisting of as many as five layers from the forest canopy to the forest floor. The greatest amount of **biomass** is present in the tree layers that have access to light, whereas, on the forest floor, where little light penetrates, the vegetation is poorly developed and widely spaced. With the absence of seasons, the trees do not shed their leaves at any particular time of year, and, as a result, the forest is considered to be evergreen. Because of the ideal growing conditions and the rapid **recycling** of **nutrients** through the **system**, the equatorial rainforest is biologically very productive and is claimed (although not without dispute) to have the greatest abundance and diversity of plants and animals of any of the terrestrial biomes. Many commercially important tree **species** such as ebony, teak, mahogany and rubber grow there, while forest plants also provide a wide range of raw materials that can be processed into pharmaceuticals. Like all tropical forests, the equatorial rainforest is under increasing pressure from development. In Asia, for example, high value hardwoods such as teak are being overharvested, while in Brazil the forest is being cleared at a rate of 10,000 km² per year to provide land for agricultural or industrial development. According to **Friends of the Earth**, such destruction is responsible for the **extinction** of more than 8000 rainforest species every year. Diverse as the rainforest **flora** and **fauna** may be, they cannot continue to undergo such stress without serious environmental effects. Additional impacts include changes to temperature and moisture regimes, reduced soil fertility and **soil erosion** which ensure that once the forest is destroyed it is very difficult to re-establish. On a global scale, rainforest destruction has been linked to **global warming**. Clearing the forest by burning introduces additional **carbon dioxide** (CO_2) into the **atmosphere**, while the removal of the vegetation reduces **photosynthesis** and therefore the **recycling** of the **gas.** As a result, destruction of the equatorial rainforest is the second largest cause of increased atmospheric CO_2, which contributes to the progressive enhancement of the **greenhouse effect.** Clearly the threat to the rainforest is serious and has implications that extend beyond the tropics, but attempts to improve the situation through such approaches as **sustainable development,** bans on logging and mining and the creation of forest reserves have as yet had only limited impact.

See also
Extinction, Statement of Forest Principles.

Further reading
Cunningham, W.P. and Saigo, B.W. (1990) *Environmental Science: A Global Concern,* Dubuque, IA: Wm C. Brown.
Friends of the Earth (1989) *Rainforests: Protect Them,* London: Friends of the Earth.
Park, C.C. (1992) *Tropical Rainforests,* London: Routledge.
Place, S.E. (1993) *Tropical Rainforests: Latin American Nature and Society in Transition,* Wilmington, DE: Scholarly Resources.
Whitmore, T.C. (1990) *An Introduction to Tropical Rainforests,* Oxford: Oxford University Press.

EQUILIBRIUM

A concept often applied in environmental studies, it implies a state of balance among the individual components of a particular **environment.** It may also be applied to the whole earth/atmosphere **system.** If equilibrium is achieved, the system is said to be in a **steady state.** This does not mean that the system is static. The balance is achieved through mutual adjustments, reflected in **positive** or **negative feedbacks,** for example, among the various **elements** in the environment.

See also
Dynamic equilibrium, Environmental equilibrium.

EQUILIBRIUM MODELS

A form of **general circulation model** (GCM). Change is introduced into a **model** representing existing climate conditions, and the

model is allowed to run until a new equilibrium is established. The new model climate can then be compared with the original to establish the overall impact of the change.

See also
Transient models.

EQUINOX

The time at which the sun is directly overhead at the equator. It occurs twice a year, on 21 March and 23 September, as the sun (apparently) travels between the Tropics of Cancer and Capricorn. Although the earth's axis is tilted at 23.5° from the vertical, at the equinox it is neither tilted towards the sun nor away from it, and the circle of illumination passes through both poles. As a result, day and night are of equal length at all **latitudes.**

Figure E-8 The relative positions and orientations of the earth and sun at the solstices and equinoxes

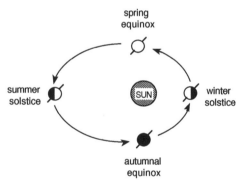

See also
Precession of the equinoxes, Solstice.

EROSION

The wearing away of the earth's land surface by such processes as **weathering, solution,** abrasion, **corrosion** and transportation. The mechanisms involved include **gravity, wind, water** and **ice.** Each creates characteristic landforms and varies in importance with time and place. During the **Pleistocene** epoch, for example, ice (glacial erosion) replaced running water (fluvial erosion) as the dominant form of erosion in many northern areas. Rock type can also be important. Unconsolidated **sands** and gravel are more easily eroded than solid granite, while rocks such as **limestone** are worn down by chemical processes such as **corrosion** and **solution** rather than by the direct physical impact of abrasion. Although erosion is entirely natural, it can be deliberately or inadvertently enhanced or reduced by human activities. Poor cultivation practices or the removal of vegetation in hilly areas contribute to increased **soil erosion**; interference with the **hydrology** of an area – by building a **dam**, for example – can alter the rate and nature of fluvial erosion. Less directly, **acid rain** can contribute to increased erosion in limestone areas. Measures for combating erosion vary with the mechanisms involved, but include such activities as slope stabilization, stream channel modification and coastal protection using seawalls and breakwaters.

See also
Denudation.

Further reading
Ritter, D.F., Kochel, R.C. and Miller, J.R. (1995) *Process Geomorphology*, Dubuque, IA: Wm C. Brown.
Thornes, J.B. (ed) (1995) *Vegetation and Erosion: Processes and Environments*, Chichester/New York: Wiley.

ESTER

An **organic compound** that is the equivalent of the **salt** of an inorganic **acid**, produced when the **hydrogen** in the organic acid is replaced by an organic group. Acetic acid (CH_3COOH), for example, becomes the ethyl ester, ethyl acetate ($CH_3COOC_2H_5$) when the **hydrogen** (H) in its carboxyl group (COOH) is replaced by the ethyl group (C_2H_5). Esters are important commercial chemicals, being present in many animal fats and oils, and some are used as flavouring essences because of their pleasant smells.

Further reading
Solomons, T.W.G.(1992) *Organic Chemistry* (5th edition), New York: Wiley.

ETHANOL

Ethyl alcohol (C_2H_5OH). One of the most common **alcohols**, traditionally produced by the **fermentation** of the natural **sugars** in grain and fruit, and the base for many alcoholic beverages. In industry, where it is in demand as a **solvent**, most ethanol is produced through the **hydrolysis** of ethene (ethylene), a **gas** derived from **petroleum**. In Brazil and the United States, ethanol produced by the fermentation of sugar cane waste and grain respectively has been used as a vehicle **fuel**, usually in the form of **gasohol**, a combination of **gasoline** and ethanol.

Further reading
World Bank (1980) *Alcohol Production from Biomass in the Developing Countries*, Washington, DC: World Bank.

EUROPEAN ARCTIC STRATOSPHERIC OZONE EXPERIMENT (EASOE)

An experiment undertaken during the northern winter of 1991–1992, using ground measurements; balloons, aircraft and a variety of modelling techniques to establish the nature and extent of **ozone depletion** over the Arctic.

Further reading
Pyle, J. (1991) 'Closing in on Arctic ozone', *New Scientist* 132 (1794): 49–52.

EUSTASY

The consideration of **sea-level change** on a worldwide scale. In recent geological time the major eustatic changes have been associated with the glaciations of the **Quaternary** era. These glacio-eustatic changes involved a lowering of **sea level** during glacial periods, as more **water** was retained on land in the form of **ice**, and a rising sea level during **interglacials** as the ice melted and the water was returned to the ocean basins. Estimates of the decline in sea level at various glacial maxima range from 85 m to 120 m below current sea level, whereas during the interglacials sea levels rose again, perhaps exceeding current levels by a few metres. In postglacial times, during the **climatic optimum** (hypsithermal), sea level was perhaps 6 m higher than at present. In northern latitudes, the question of eustatic sea-level changes during glaciations, interglacials and in postglacial times is complicated by isostatic movements of the earth's crust. It is difficult to obtain accurate values of eustatic change under such conditions, and most estimates have been obtained from observations in lower latitudes well away from the direct effects of ice. Although glacial variations have caused the greatest eustatic changes, other factors, such as tectonic changes, which alter the shape and capacity of the ocean basins, also have the potential to contribute. A change in **temperature** of sea water could also cause sea level to rise. An increase in 1°C of the mean temperature of the oceans would cause sufficient expansion of the water to bring about a rise in sea level of perhaps as much as 60 cm. Recent concern with **global warming** has drawn attention to this fact. Glacio-eustatic change is also a potential consequence of global warming. The increase in the melting of **glaciers** and ice-caps following an estimated increase of 8°C in high latitudes has the potential to cause a measurable rise in sea level. In the extreme case of temperature increases sufficiently great to cause complete melting of the Greenland and Antarctic ice sheets, sea level might increase by some 60 m.

See also
Isostasy.

Further reading
Dott, R.H. (ed.) (1992) *Eustasy: The Historical Ups and Downs of a Major Geological Concept*, Boulder, CO: Geological Society of America.
Flint, R.F. (1971) *Glacial and Quaternary Geology*, New York: Wiley.
IPCC (1996) *Climate Change 1995: The Science of Climate Change*, Cambridge: Cambridge University Press.
Morner, N.A. (1980) *Earth Rheology, Isostasy and Eustasy*, New York: Wiley.
Ritter, D.F., Kochel, R.C. and Miller, J.R. (1995) *Process Geomorphology*, Dubuque, IA: Wm C. Brown

EUTROPHIC (LAKES)

Water bodies that have a high concentration of **nutrients** and are high in organic productivity. High organic levels also tend to make eutrophic lakes cloudy in nature. Lakes usually become eutrophic with age as nutrients accumulate, but the addition of pollutants such as **sewage, phosphates** and agricultural **fertilizers** can speed up the process.

See also
Eutrophication, Oligotrophic lakes.

EUTROPHICATION

The natural ageing of a **water** body characterized by increasing levels of dissolved **nutrients** in the water. These in turn encourage the growth of aquatic plants often in the form of **algal blooms**. Eutrophication also occurs in coastal sea areas, where 'red tides' are an indication of a rapid increase in local nutrient levels. When the plants begin to die, the increase in bacterial activity raises the **biochemical oxygen demand** (BOD), and so the water becomes deoxygenated. Eutrophication is a natural process, but it has been accelerated by human activities which add such products as **sewage**, agricultural **fertilizers** and **detergents** to the water bodies. The additional **nitrogen** (N) and **phosphorus** (P) from such **effluents** encourages plant growth. The introduction of phosphate-free detergents and the improvement of **sewage treatment** facilities has slowed eutrophication in some areas. In the late 1960s and 1970s, for example, Lake Erie in North America was in a state of advanced eutrophication that many environmentalists considered to be irreversible. In places it was covered by great mats of floating **algae**, and stretches of shoreline were contaminated by masses of rotting plant material. Since then, effluent control has reduced the problem significantly. Eutrophication remains a major concern in many areas, however, including enclosed seas such as the Mediterranean, which are more susceptible to the problem than the open ocean.

Further reading
Ashworth, W. (1986) *The Late, Great Lakes: An Environmental History*, New York: Alfred A. Knopf.
Pearce, F. (1995) 'Dead in the water', *New Scientist* 145 (1963): 26–31.

EVAPORATION

The process of vaporization by which a **liquid** is converted into a **gas** or **vapour**. This change in state requires the absorption of **energy** normally supplied by the adjacent **environment** – for example, from a hot surface, or from the surrounding **air**. Once the change has taken place, the absorbed energy is retained by the vapour in the form of **latent heat**. When the process is reversed and the vapour is condensed back into a liquid, the latent heat is released into the environment in the form of **sensible heat**. Evaporation has a role in several major processes in the earth/atmosphere system. It is an integral part of the **hydrological cycle**, for example, initiating the **recycling** process which involves both the physical redistribution and the cleansing of the system's water supply.

Evaporation also contributes to the earth's **energy budget**, by helping to redistribute energy within the **atmosphere**. The energy from **solar radiation** used to evaporate water in the tropics is retained by water vapour in the form of latent heat. If that vapour is carried into higher latitudes, it carries the latent heat with it, to be released as sensible heat if **condensation** takes place. Similar conditions also develop in frontal systems, and the net result is a redistribution of energy in the environment. The rate at which evaporation takes place depends upon moisture availability, **temperature**, **humidity** and air movement. If no moisture is available, no evaporation will take place. If moisture is available, however, a combination of high temperatures, low **relative humidities** and strong **winds** will produce maximum evaporation. Under calm conditions, the evaporation rate will decline with time, as the input of water vapour into the atmosphere increases the relative humidity. Air movement helps to maintain a high evaporation by replacing the saturated air with drier air that can absorb more moisture. The greatest annual evaporation rates occur over the tropical oceans. Rates are also high over the continents in tropical regions, where moisture is available. More evaporation takes place over the **rainforest**, for example, than over the tropical **deserts**. In the latter case, although high temperatures and low relative humidities encourage high evaporation rates there is insufficient moisture and actual evaporation is restricted. Rates are also high over the Gulf Stream in the western Atlantic Ocean and over the Kuro Siwo current in the Pacific, particularly when dry continental air moves out over the **oceans**. The accurate measurement of evaporation is not easy. Instruments such as evaporating pans, which provide evaporation rates from an open water surface, and atmometers, which provide evaporation rates from a saturated, porous surface, present information on potential rather than actual evaporation, since the supply of moisture is not limited. Comparisons with other variables such as **precipitation** and soil moisture availability allow actual evaporation rates to be calculated. Formulae have been developed for estimating evaporation from such variables as precipitation, drainage, **runoff** and **soil moisture storage** (water budget approach) or from duration of sunshine, air temperature, air humidity and wind speed (energy budget/aerodynamic approach), but all contain potential sources of error, usually in the form of data reliability.

Further reading
Barry, R.G. and Chorley, R.J. (1992) *Atmosphere, Weather and Climate* (6th edition), London: Routledge.
Monteith, J.L. (1981) 'Evaporation and surface temperature', *Quarterly Journal of the Royal Meteorological Society* 107: 1–27.

EVAPOTRANSPIRATION

The transfer of **water** from the terrestrial **environment** into the **atmosphere**, combining **evaporation** from the land surface with **transpiration** from plants. A division is commonly made into actual and **potential evapotranspiration**. The former represents measurable evapotranspiration, and is limited by the availability of water – once the available water has been used up no more evapotranspiration can take place – whereas the latter is an artificial value based on the assumption that there are no restrictions on the availability of water. It represents the environment's capacity for evapotranspiration.

The difference between actual and potential evapotranspiration is a measure of the water deficit in an area. Evapotranspiration is an important element in water budget studies, where, for example, it is used in the calculation of **irrigation** requirements.

See also
Moisture deficit, Moisture surplus, Thornthwaite, C.W.

Further reading
Kovacs, G. (1987) 'Estimation of average areal evapotranspiration', *Journal of Hydrology* 95: 227-40.
Oke, T. (1987) *Boundary Layer Climates* (2nd edition), London: Routledge.

EVOLUTION

The concept of gradual, cumulative change. It can be applied to organisms, objects and ideas. The concept is commonly associated with **Charles Darwin**, but it owes its initial development to the work of **Jean Baptiste de Lamarck** on changes in **species** in the late eighteenth and early nineteenth centuries. His ideas were further developed by the geologist Charles Lyell, who first recognized the role of gradual change (as opposed to catastrophic change) in the physical environment. It continues to have a role in explaining the development of landforms, such as hillslopes, or the patterns associated with river systems, for example, but it is recognized that rapid change initiated by tectonic activity or severe storms cannot be ignored. In the **atmospheric environment**, many of the **models** developed to study and predict change are based on evolutionary concepts, in that they involve the introduction of relatively small changes to the **system** which are allowed to accumulate over a given period of time or until a specific condition has been realized. The concept has achieved its widest development in the biological sciences, based mainly on the ideas of Darwin and Alfred Russel Wallace in the mid-nineteenth century. In their theory of evolution, all organisms were descended from common ancestors, with the resultant variations brought about by the survival of species or groups of individuals within species that were best suited to a particular environment. In this natural selection process, only the fittest survived. Those that could not adapt to a particular environment or to a change in the environment faced ultimate **extinction**. In some cases, co-evolution, in which the evolution of one organism was intimately tied to the evolution of another, was recognized. Evolutionary concepts have become widely accepted and supported by findings in palaeontology and genetics research which were not available to Darwin. In some quarters, however, the controversy between evolution and creationism continues.

Further reading
Cherfas, J. (ed.) (1982) *Darwin Up to Date*, London: New Science Publications.
Darwin, C. (1859) *On the Origin of Species by Means of Natural Selection*, London: John Murray.
Gish, D.T. (1979) *Evolution, The Fossils Say No* (3rd edition), San Diego: Creation-Life Publishers.
Margulis, L. and Olendzenski, L. (eds) (1992) *Environmental Evolution: Effects of the Origin and Evolution of Life on Planet Earth*, Cambridge, MA: MIT Press.
Schneider, S.H. and Londer, R.S. (1984) *The Coevolution of Climate and Life*, San Francisco, CA: Sierra Club.

EXFOLIATION

A form of mechanical **weathering** in which the surface layers of a rock are peeled off in layers, sometimes referred to as 'onion weathering', since, on spheroidal rocks, the peeling takes the form of concentric shells. Exfoliation may be caused by repeated expansion and contraction of the rock as a result of heating and cooling or by pressure release following the removal of the overlying material by **erosion**. In some cases, exfoliation may be the result of chemical weathering when the **absorption** of **water** causes minerals to swell and creates enough pressure to loosen the layers.

EXPONENTIAL GROWTH

Change in which growth is cumulative, as illustrated by the concept of **compound interest**. In that case, the interest which accrues over a fixed period of time, such as a year, is added to the original sum (principal)

Figure E-9 A graph to illustrate the concept of exponential growth

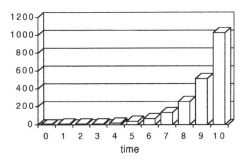

and interest during the second year is calculated on the principal plus the first year's interest. Although the rate of increase (as indicated by the interest rate) may remain the same, the amount received as interest increases every year. In environmental studies, exponential growth applies particularly to **population**. If the offspring in a population produce at the same rate as their parents, the population will grow exponentially. If a woman gives birth to two daughters, for example, and each of these daughters gives birth to two daughters, at the third generation there will be four daughters descended from the original mother. After four generations, her female descendants will number eight, and after five generations will number sixteen. Such a sequence of numbers – 1, 2, 4, 8, 16, 32, etc. – illustrates exponential growth (or geometric growth as it was once called) and when plotted on a graph, forms a curve that becomes steeper as the sequence progresses. Exponential growth of this type is common at some time in the development of all organisms, and continues until the population reaches some degree of stability as it approaches the **carrying capacity** of the **environment**. In some cases, however, especially among small organisms such as **bacteria**, the population exceeds the carrying capacity of the environment and crashes. **Malthus** compared the exponential (geometric) growth – 1, 2, 4, 8, 16, 32, etc. – of the world's human population with the arithmetic growth – 1, 2, 3, 4, 5 , 6, etc. – of its food **resources** and predicted that its population would crash.

See also
Overpopulation.

EXTINCTION

The elimination of all individuals in a particular **species**. The species is effectively removed from the earth/atmosphere system and cannot be replaced. Extinction is a natural process brought about, for example, by the inability of a species to cope with changing environmental conditions or with increased competition from another species. It is also part of the Darwinian concept of the 'survival of the fittest', in which organisms that adapt to change survive, whereas those that do not adapt become extinct. Perhaps as many as 98 per cent of the species that ever existed on earth are now extinct, although many of these are now represented by their descendants. The modern horse (Equus), for example, is a descendant of the 'dawn horse' (Eohippus) which became extinct some 40 million years ago. Mass extinctions punctuate the environmental history of the earth, with some, such as the elimination of the dinosaurs at the end of the Cretaceous, giving rise to much speculation. Despite these periods of more rapid extinction, natural extinction rates are much slower than modern figures which are inflated as a result of human activities. The current extinction rate is estimated to be as much as 1000 times that of the average natural rate, and the total pool of some 29 to 30 million species is being depleted at a rate that may exceed 40,000 per year. The accelerated extinction of plant and animal species is brought about by such activities as **habitat** alteration, commercial and sports hunting and fishing, introduction of new species and attempts at predator and pest control. As a result, many species are now endangered and face imminent extinction. Animals such as whales, big cats and pandas receive most attention, but many insects and plants are also threatened. Since plants are at the base of all **food chains**, their increasingly rapid extinction may in turn threaten other organisms. Not all human-induced extinctions are necessarily bad – the elimination of the smallpox **virus**, for example, was beneficial – but the current rate is increasingly considered unacceptable. Much more attention to the **conservation** of species and the maintenance of **biodiversity** is required, however, if the trend is to be reversed.

See also
Darwin, C.R., Endangered species, Evolution, Extirpation.

Further reading
Chiras, D.D. (1991) *Environmental Science*, Redwood City, CA: Benjamin/Cummings.
Stanley, S.M. (ed.) (1987) *Extinction*, New York: Scientific American.
Stiling, P.D. (1992) *Introductory Ecology*, Englewood Cliffs, NJ: Prentice-Hall.

EXTIRPATION

The elimination of a **species** from a specific area as a result of such factors as environmental change, over-predation or disease. It is less serious than **extinction**, since members of the species survive in other areas and may be reintroduced naturally or by human intervention if the conditions that led to their initial elimination change. For example, wolves that had been wiped out in Yellowstone National Park in the western United States were reintroduced in the form of breeding pairs captured in a similar habitat in Canada. In some cases, extirpation may be seen as bringing a species one step closer to extinction.

EXXON VALDEZ

A tanker which ran aground on a reef in Prince William Sound, Alaska in March 1989, spilling some 38,000 tonnes of Alaskan crude oil, and ultimately contaminating nearly 1000 km of coastline. Parks, wildlife refuges, critical habitat areas and a game sanctuary were among the affected areas. Initial responses involved containing the spill, skimming **oil** from the **water,** rescuing oiled wildlife and protecting salmon hatcheries in the area. Exxon, owner of the tanker, employed several thousand workers to clear the shoreline, using a variety of techniques, including manual removal of the oil, cleaning of rocks by hand, high pressure, hot water washing and **bioremediation**. Large amounts of oil remained along the coast at the end of the 1989 season, and clean-up work was necessary in some areas as late as 1994. A major survey of the area in 1995 found residual contamination in a number of places, but also found wildlife in all stages of recovery. Bald eagles appeared to have

Figure E-10 The spread of oil following the Exxon Valdez grounding

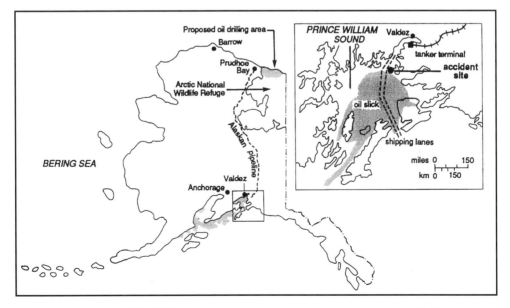

Source: After Miller, G.T. (1994) *Living in the Environment* (8th edition), Belmont, CA: Wadsworth

recovered fully, some waterfowl were on the way to recovery, but had not reached pre-spill populations, while others like the Pacific herring were in significant decline. The monitoring and research, which is likely to continue for some years, has already provided a better understanding of the most appropriate techniques for the clean-up of **oil spills** in a fragile northern environment.

See also
Oil tankers, Trans-Alaska pipeline.

Further reading
Davidson, A. (1990) *In the Wake of the Exxon Valdez: The Devastating Impact of the Alaska Oil Spill*, San Francisco: Sierra Club Books.
Keeble, J. (1991) *Out of the Channel: The Exxon Valdez Oil Spill in Prince William Sound*, New York: HarperCollins.

F

FAHRENHEIT SCALE

A scale developed by Gabriel Daniel Fahrenheit (1686–1736) in which the melting point of **ice** is set at 32° and the boiling point of **water** under standard **atmospheric pressure** is 212°. Although still used in some countries, such as the United States, in popular **weather** reporting and forecasting, it is generally no longer in scientific use, having been replaced by the **Celsius scale**. Conversion between the two scales can be achieved using the following formulae:

$$°F = 32 + 9/5°C$$

$$°C = 5/9(°F - 32)$$

See also
Kelvin.

FALLOUT

The deposition of **particulate matter** from the **atmosphere** on to the earth's surface. Fallout may take place rapidly and near the source of the particles, particularly if the material is large and released close to the surface. Finer particles released or pushed high into the atmosphere may remain suspended for several years before returning to the surface as fallout. The particle size and quantity of materials involved is quite variable. The **dry deposition** of **acid** particles, for example, includes minute chemical particles that are unlikely to accumulate in sufficient quantity to be obvious at the earth's surface, whereas major volcanic eruptions, such as those of **Mt St Helens** and **Pinatubo**, are often accompanied by fallout of large ash fragments that accumulate locally to depths of tens of centimetres. The term 'radioactive fallout' is used in reference to radioactive particles released into the atmosphere by nuclear explosions.

See also
Atmospheric turbidity, Atomic bomb, Thermonuclear device.

Further reading
Chester, D.K. (1993) *Volcanoes and Society*, London/New York: E. Arnold/Routledge, Chapman and Hall.
OECD, Nuclear Energy Agency (1987) *The Radiological Impact of the Chernobyl Accident on OECD Countries*, Paris: Nuclear Energy Agency (OECD).

FALLOW

Arable land left untilled or tilled, but unsown, for a season. Fallow is a normal component of **crop rotations** in **arable agriculture,** allowing the land to rest and recover from the effects of cropping. It is also common in areas where **dry farming** is practised. There the land may be tilled, but left without a crop for several years to provide a reservoir for **precipitation,** allowing **soil moisture** levels to rise.

FAMINE

Acute food shortage leading to widespread starvation. It is usually associated with large scale natural disasters such as **drought, flood** or plant disease which produce crop failure and disruption of food supply. Famine has been a recurring problem in Africa in the second half of the twentieth century. Drought-induced famine killed thousands of

Figure F-1 The distribution of famine in
Africa

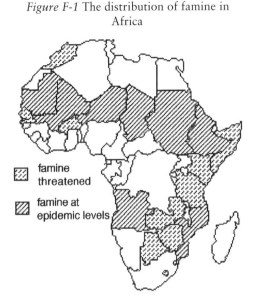

▨ famine
   threatened

▧ famine at
   epidemic levels

people in the **Sahel** between 1968 and 1973,
for example, and in Ethiopia in the early
1980s, a food supply already decimated by
drought and locusts, was further reduced by
civil war and political dissension, putting
more than 3 million people at risk of
starvation. Perhaps the greatest famine of
recent times was that in China between 1958
and 1962, when a combination of drought
and problems associated with the collectiv-
ization of agriculture led to widespread
harvest failure. Over that four-year period,
China suffered 25-30 million more deaths
than might have been expected under normal
conditions. Famine also appears with some
regularity in the history of Europe. The Irish
potato famine of the 1840s, which led to
death or emigration for millions of people,
was caused by plant disease. Even earlier, in
the seventeenth and eighteenth centuries,
famines were recorded in Scotland, Scandinavia and Iceland, apparently brought
about by the deteriorating climatic conditions
associated with the **Little Ice Age**. Despite
improved technology, conditions such as
drought, disease and social unrest, which have
contributed to famine in the past, remain, and
famine is therefore likely to continue to recur.

See also
Desertification, Natural hazard.

Further reading
Canadian International Development Agency
(1985) *Food Crisis in Africa*, Hull, Quebec: CIDA.
Jowett, J. (1989) 'China: the population of the
People's Republic', *SAGT Journal* 18: 38–49.
Kemp, D.D. (1994), *Global Environmental Issues:
A Climatological Approach*, (2nd edition) London/
New York: Routledge.
Parry, M.L. (1978) *Climate Change, Agriculture
and Settlement*, Folkestone: Dawson.

FAMINE EARLY WARNING
SYSTEM (FEWS)

An information system funded by USAID
aimed at forecasting **famine** in sub-Saharan
Africa. Using remotely sensed and ground-
based data on such factors as **drought** and
crop health, specialists provide forecasts of
potential food availability. The provision of
pre-harvest crop assessments allows warnings
of potential famine to be issued where necessary. FEWS warnings are used in assessing aid
to countries at risk from famine.

FAUNA

The animal life characteristic of a particular
biome. The **savanna** biome, for example, supports large populations of **herbivores**, such as
wildebeest, antelope and kangaroo, and
predators in the form of lions, cheetahs,
hyenas or dingoes that prey on them. Any
change in a biome, whether natural or
human-induced, has the potential to alter the
associated fauna.

Further reading
Cole, M.M. (1986) *The Savannas*, San Diego:
Academic Press.

FEEDBACK

Occurs in integrated systems where change in
one part of a **system** will initiate change
elsewhere in the system. The feedback may be
direct, involving only two **elements**, or it may
be looped involving one or more additional
variables. The change may be fed back into
the system in such a way as to diminish
(negative feedback) or augment (positive
feedback) the effects of the original change.
Negative feedbacks are common in the

Figure F-2 Feedback and the maintenance of homeostasis in humans

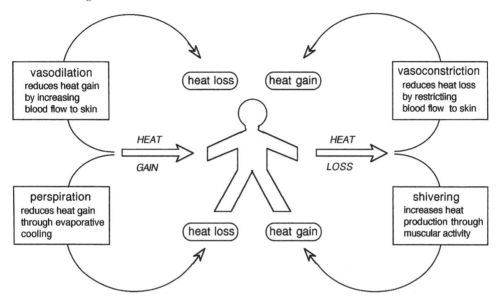

environment, where they act as a form of self-regulation. The ability of mammals to maintain a near constant or steady body **temperature** – a condition called homeostasis – for example, depends upon negative feedback. Positive feedback is illustrated by the relationship between temperature and **albedo** represented by **snow** cover. A lowering of temperatures at the earth's surface would allow the persistence of snow cover beyond the normal season. This, in turn, would increase the amount of **solar radiation** reflected back into space, causing surface temperatures to fall even more and encouraging snow to remain even longer. The relationship between temperature and albedo through solar radiation is an example of a looped feedback. Many feedbacks are much more complex than represented by these examples, and this causes difficulties when environmental relationships are examined through modelling. In the study of **climate change**, for example, the feedbacks may be so intimately interwoven that their ultimate environmental impact can be difficult to assess. The higher temperatures associated with an intensified **greenhouse effect** would bring about more **evaporation** from the earth's surface. Since **water vapour** is a very effective **greenhouse gas**, this would create a

positive feedback to augment the initial rise in temperature. With time, however, the rising water vapour would condense, leading to increased cloudiness. The **clouds** would reduce the amount of **radiation** reaching the earth's surface and therefore cause a temperature reduction – a negative feedback – which might moderate the initial increase. Such complexities add to the difficulties of creating **models** that represent the workings of the earth/atmosphere system accurately.

See also
Autovariations, General circulation models.

Further reading
Chorley, R.J. and Kennedy, B.A. (1971) *Physical Geography: A Systems Approach*, London: Prentice-Hall International.
Washington, W.M. and Parkinson, C.L. (1986) *Introduction to Three Dimensional Climate Modelling*, Mill Valley, CA: University Science Books.
Cusbach, U. and Cess, R.D. (1990) 'Processes and modelling', in J.T. Houghton, G.J. Jenkins and J.J. Ephraums (eds) *Climate Change: The IPCC Scientific Assessment*, Cambridge: Cambridge University Press.

FEN

See **peat/peatlands**.

FERMENTATION

The chemical breakdown of **carbohydrates** by **yeasts** and **bacteria** through the action of **enzymes**, usually under **anaerobic** conditions. It includes lactic fermentation, which leads to the formation of lactic acid – a common process in the production and preservation of food – but it is most often applied to alcoholic fermentation in which the action of enzymes, such as zymase, on certain **sugars** produces **ethanol** and releases **carbon dioxide** (CO_2).

Lactic fermentation

$$C_6H_{12}O_6 \rightarrow 2CH_3CH(OH)COOH$$
sugars   lactic acid

Alcoholic fermentation

$$C_6H_{12}O_6 = 2C_2H_5OH + 2CO_2$$
sugars   ethanol   carbon
                                      dioxide

The production of ethanol through the fermentation of a variety of high carbohydrate crops such as sugar cane, sugar beets, sorghum and corn (maize) has been suggested as an environmentally appropriate means of providing a substitute for petroleum-based **fuels**. Projects have been developed in Brazil and the United States using sugar cane and corn respectively to produce automotive fuel. The technology is relatively simple, but production costs remain higher than for **gasoline**, and most of the **alcohol** produced through fermentation continues to be used in alcoholic beverages and as a **solvent** or chemical raw material in industry.

See also
Gasohol.

Further reading
Ward, O.P. (1989) *Fermentation Biotechnology: Principles, Processes and Products*, Englewood Cliffs, NJ: Prentice-Hall.

FERNAU GLACIATION

See **Little Ice Age**.

FERREL'S LAW

Deduced by William Ferrel in the mid-nineteenth century, it states that moving **air** in the northern hemisphere is deflected to the right of its path as a result of the **Coriolis effect**. In the southern hemisphere, the deflection is to the left.

FERTILE CRESCENT

The name given to an arc of land stretching from the Nile Valley in Egypt, north along the coast of the Mediterranean sea, east to the valleys of the Tigris and Euphrates and south to the head of the Persian Gulf. It includes the modern nations of Egypt, Israel, Lebanon, Syria, Iraq and part of western Iran. The climate is characterized by hot, dry summers and mild, moderately wet winters. **Precipitation** is unreliable, however, and **drought** is not uncommon. Although arguably not fertile by current standards, the availability of **water** in the major rivers of the region supported permanent agriculture and allowed the development of the world's first agriculturally based civilizations.

See also
Agrarian civilizations.

Further reading
Haberman, A. and Hundey, I. (1994) *Civilizations: A Cultural Atlas*, Agincourt, Ont: Gage Educational Publishing.
Hughes, J.D. (1975) *Ecology in Ancient Civilizations*, Albuquerque: University of New Mexico Press.

FERTILIZER

Any substance added to **soil** to provide a source of **nutrients** for plants. The active components in all fertilizers are **nitrogen** (N), **phosphorus** (P) and **potassium** (K) combined in proportions that vary according to such factors as their source or the requirements of particular crops. Other chemicals such as **lime** ($CaCO_3$) may be added to the **soil** to improve its chemical composition or texture, but they are not usually regarded as fertilizers. Fertilizers are grouped according to origin into organic and inorganic forms, with the latter sometimes referred to as chemical or artificial fertilizers. Organic forms include farmyard manure (the combination of animal **wastes** and straw), green manure (growing crops such as grass and clover ploughed back into the soil), **compost** (decayed or decaying plant waste) and sewage sludge (treated urban **sewage**). In China, raw human sewage – night soil – is added directly to the soil as a readily available and effective fertilizer. Commercial inorganic fertilizers are produced with the constituent chemicals (N-P-K) combined in varying proportions according to the needs of particular crops. A broad spectrum fertilizer might be coded 10-10-10, for example, meaning that it contained equal parts of all three chemicals, whereas one coded 5-20-20 would be used specifically for crops requiring less nitrogen, but a higher proportion of both phosphorus and potassium. The use of chemical fertilizers such as these has often been criticized by producers and consumers of natural food products. They do provide essential **nutrients**, however, and without them worldwide food production would decline substantially. Indeed, the sharp increase in food production that accompanied the so-called **green revolution** was directly linked to the increased availability of nitrogen-rich fertilizers. Chemical fertilizers are often easier to use and faster-acting than their organic equivalents, but they do have the disadvantage that they do not add organic matter to the soil. In contrast, farmyard manure, compost and green manure not only fertilize the soil but also help to improve its structure by adding organic materials such as **humus**, which increase the water-holding capacity of the soil and reduce the potential for **erosion**. The excessive or uncontrolled use of fertilizers, and the consequent washout of soluble nitrates and phosphates from the soil, can lead to the contamination of **groundwater** supplies or the **eutrophication** of water bodies.

See also
Inorganic matter, Organic compounds.

Further reading
Bacon, P.E. (1995) *Nitrogen Fertilization in the*

Environment, New York: M. Dekker.
Colwell, J.D. (1994) *Estimating Fertilizer Requirements: A Quantitative Approach*, Wallingford: CAB International.
Smith, S.R. (1996) *Agricultural Recycling of Sewage Sludge and the Environment*, Tucson, AZ: CAB International.

FIELD CAPACITY

A measure of the amount of **water** retained in a **soil** after excess moisture has drained away. When this condition is attained, the voids and pore spaces in the soil are holding the maximum volume of water possible.

See also
Soil moisture storage.

FILTRATION

The process by which unwanted substances are removed from a fluid medium – **solids** from a **liquid** or **gas**, for example. The results are achieved by passing the fluid through a filter consisting of a porous material such as paper, woven cloth, ceramics or **sand** in which the pores are sufficiently small that only fluids can pass through and larger materials are retained on the filter. In addition to such mechanical devices, filtering can be accomplished very effectively by **adsorption**, in which particles become attached to the filter by physical or chemical bonds. The adsorptive qualities of **charcoal** make it a very efficient filter, for example. In some cases filtration is achieved at the expense of the filter. A crushed **lime** ($CaCO_3$) filter, for example, will remove **acid** gases such as **sulphur dioxide** (SO_2) from flue gases, but in the process it is converted to calcium sulphate ($CaSO_4$). Filtration is one of the most common and effective approaches to controlling the emission of pollutants into the **environment**.

See also
Scrubbers.

Further reading
Cheremisinoff, P.N. (1995) *Solids/Liquids Separation*, Lancaster, PA: Technomic Publishing.
Orr, C. (ed.) (1979) *Filtration: Principles and Practices*, New York: M. Dekker.

FIRE

One of the results of **combustion**, and a major force for change in the **environment** Most natural fires are caused by **lightning**. Although seen in human terms as destructive events, fires are an integral part of the **ecology** of many areas. Forest fires, for example, help to regenerate the forest **community** by destroying dead or dying trees and those infested with disease or insects. The resulting ash provides a ready supply of **nutrients** for the regrowth of new vegetation after the fire. Some trees such as jackpine, Douglas fir and lodgepole pine actually benefit from fires, because their seed cones will only open under the high **temperatures** produced by the fires. The ecology of tropical and temperate **grasslands** also depends upon the regular occurrence of fire. The fires burn off the dead grass that accumulates at the end of the growing season, encouraging the growth of new grass, and providing a renewed food supply for the local **fauna**. Because fires tend to be more destructive to trees than to grass, persistent fires will help to prevent the colonization of grasslands by trees, and may even allow grass to move into areas which would normally support tree **species**. Plant communities that are shaped by periodic fires in this way are known as fire-climax or pyroclimax communities, because they are prevented by fire from achieving the mix of species associated with a true **climax community** based on such factors as **soil** and **climate**. The natural vegetation of Mediterranean climate regions such as southern France and California is an example of a pyroclimax community. Fire has been used by society for millions of years for lighting, heating and cooking. Primitive societies used fire to clear land for agriculture or to encourage the growth of new grass to provide food for their animals. Fire was also used directly in hunting to drive game animals into traps or into areas where they could be more easily slaughtered. Used in this way, fire was one of the first mechanisms by which human beings brought about major (if local) change to the environment. Fire continues to be used to clear forests – in the 1987 burning season in Brazil, some 80,000 km² of tropical **rainforest** were destroyed by fire – and in some agricultural practices such as burning grain straw or stubble. In modern society, however, the tendency is to prevent fires wherever possible because of the damage that they do to the **terrestrial environment** and because of the **gases** and **particulate matter** they add to the **atmosphere**. Running contrary to this is the view among some ecologists that fires are natural, and unless there is danger to life or property they should be allowed to run their course. This 'let-it-burn' approach was adopted by the US Forest Service in 1972, and received its greatest test in 1988 when major fires destroyed thousands of hectares of forest in Yellowstone National Park. Although the Park Service was severely criticized in some quarters, surveys soon after the event showed that loss of wildlife and damage to soil were minimal, and the long-term impact of the fire would be a healthier, more resilient forest, although lacking some of the aesthetic qualities of the pre-fire landscape. In the built environment, fires continue to cause millions of dollars'-worth of damage annually. Their impact on urban morphology is most obvious, but fires involving chemicals, **plastics** and rubber can add large amounts of noxious substances to the local and regional environment.

Further reading
Carey, A. and Carey, S. (1989) *Yellowstone's Red Summer*, Flagstaff, AZ: Northland Publishing.
Fuller, M. (1991) *Forest Fires: An Introduction to Wildland Fire Behaviour, Management, Firefighting and Prevention*, New York: Wiley.
Rossotti, H. (1993) *Fire*, New York: Oxford University Press.
Whelan, P. (1995) *The Ecology of Fire*, Cambridge: Cambridge University Press.

FIRST WORLD CLIMATE CONFERENCE (GENEVA 1979)

Sponsored by the **WMO**, the Conference examined the impact of **climate change** on agriculture, fishing, forestry, **hydrology** and urban development. To meet the need for co-ordinated scientific research into the causes and implications of climate change, the conference endorsed a WMO proposal to set up the **World Climate Program** (WCP).

FISCHER-TROPSCH PROCESS

One of the processes used in the liquification of **coal**. When coal is burned in the presence of **oxygen** (O) and steam, it produces a mixture of **carbon monoxide** (CO) and **hydrogen** (H). With the addition of a **catalyst** such as **nickel** (Ni) or **cobalt** (Co), **hydrogenation** of the carbon monoxide takes place, and **methanol** (CH_3OH), plus various waxes and oils, is

Figure F-3 The stages and products of the Fischer-Tropsch process

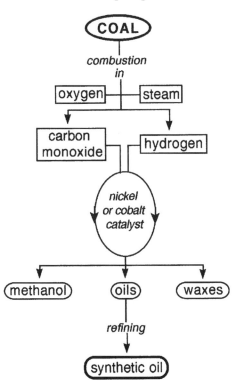

produced. The oils can then be refined for specific uses. The process was first developed in the early twentieth century and was used in Germany during the Second World War as a source of synthetic **oil**. It is currently one of the processes used in South Africa, the world's leading producer of synthetic oil, yielding about two barrels of oil and as much as 1000 m³ of **gas** per ton of coal.

See also
Coal gasification, Coal liquefaction.

Further reading
Aubrecht, G (1989) *Energy*, Columbus, OH: Merrill Publishing.
Rylander, P.N. (1985) *Hydrogenation Methods*, London: Academic Press.

FISSION

See **nuclear fission**.

FIXATION

See **nitrogen fixation**.

FLASH-POINT

The lowest **temperature** at which a flammable substance (usually a **liquid**) gives off sufficient **vapour** to catch **fire** when ignited.

FLOCCULATION

The aggregation of fine sediments, such as **clay**, held in suspension in a **liquid**. The resulting particles are still small, but may be too large to continue in suspension and therefore settle out of the liquid. In the littoral **environment**, flocculation is an important process in the formation of **deltas**. Clay **colloids** carried into the sea by a stream or river begin to flocculate as soon as they meet the salt **water**. The latter is an **electrolyte** which causes the electrically charged clay particles to bind together and sink to the sea bed where they contribute material to the delta-building process. Flocculation is also part of the **sewage treatment** process where it aids the removal of fine solids from sewage effluent.

FLOOD

The inundation of normally dry land by **water**. Flooding causes millions of dollars'-worth of property damage and takes hundreds of lives every year (see Table F-1). It is most common in river valleys or along the coastal areas of lakes, seas and oceans. River floods are caused when a river channel is incapable of carrying the volume of water added to it, and the excess spills over on to the adjacent **floodplain**. Heavy and prolonged **precipitation**, snowmelt, channel constrictions, dam failures and alterations to drainage basins may produce or contribute to flooding either singly or in combination. Intense, prolonged precipitation will cause flooding, particularly if the ground is already saturated, but the flooding may be aggravated by urban development or **deforestation** within a drainage basin which leads to an increase in the rate of **runoff**.

Similarly, in many areas snowmelt causes annual spring flooding, but the extent and duration of the flooding can be increased when ice accumulation in the river channel constricts the flow. This is common in Canada and Siberia, where **snow** in the headwaters of north-flowing rivers begins to melt, while the lower reaches remain ice-bound and unable to cope with the increased flow. Some degree of coastal flooding frequently accompanies high **tide** levels, but the area affected may be increased significantly when high tides are accompanied by river floods or by storm surges in which strong onshore **winds** drive the water ashore. The 1953 North Sea floods were caused by a combination of high spring tides and strong north winds that pushed large volumes of water into the southern end of the basin to inundate the low-lying coastal areas of the Netherlands

Table F-1 The causes and consequences of some major floods

YEAR	LOCATION	DEATHS/DAMAGE	CAUSE
1997	South China/Hong Kong	?	Monsoon rains/ Typhoons
1997	River Oder, Germany/Poland	100+/$3 billion	Heavy rain/Poor dyke maintenance
1997	Red River, USA/Canada	<50/c. $1 billion	Spring snowmelt
1993	Mississippi River, USA	50/$10 billion	Rain/Snowmelt
1991	Bangladesh	125,000/?	Cyclone/Storm surge
1988	Bangladesh	2000/?	Monsoon rains
1988	Sudan	?	Torrential rains
1982	Peru	2500/?	Torrential rains/El Niño
1973	Mississippi River, USA	11/$1.2 billion	Rain/Snowmelt
1972	Black Hills, S. Dakota, USA	242/$163 million	Torrential rains/Flash flood
1970	Bangladesh	2000,000+/?	Cyclone/Storm surge
1963	Northern Italy	2000+/?	Dam overtopped
1953	Northern Europe	2000+/?	Storm surge/North Sea
1938	Yellow River, China	1000,000/?	Destruction of dykes by military action
1928	Florida, USA	2400/?	Hurricane
1911	Yangtze River, China	100,000/?	Monsoon rains
1889	Johnstown, Pennsylvania, USA	2000+/?	Dam burst

Sources: Various

and eastern England. Similar storm surges accompany the **hurricanes** that pass through the Caribbean or the **cyclones** in the Bay of Bengal that regularly devastate Bangladesh. Seismic sea waves or **tsunamis** are less common, but following **earthquakes** or volcanic eruptions, they too can cause very rapid and destructive flooding in coastal areas. **Global warming** through the increased melting of **glaciers** and **ice** sheets, and the subsequent rise in **sea level**, has the potential to increase the frequency and extent of coastal flooding.

Floods are a natural part of the **hydrological cycle** and contribute to both **erosion** and deposition. In human terms, however, they are seen as a serious hazard to life and property, and human responses to flooding reflect that. The most obvious response is to provide protection against it by building embankments or barriers, such as the dykes which protect the coast of the Netherlands, or the **levées** which line major rivers such as the Mississippi. The diversion of rivers and the straightening or deepening of their channels allow them to carry more water and therefore reduce the amount that spills over on to the floodplain. At the other end of the scale, the response may include acceptance of flooding plus some form of adjustment to minimize the impact. Zoning by-laws, for example, may prevent the use of a flood-prone area for residential development, or require floodproofing of buildings or structures so that flood damage is minimized. Acceptance of flooding in this way is usually combined with flood frequency analysis and emergency measures procedures to warn people and secure structures when flooding is imminent. A modern approach to flood problems is to consider the integrated nature of the entire drainage basin rather than only the area prone to flooding. This recognizes that activities allowed in one part of the basin may have serious consequences elsewhere in the system. The channelization of one section of a river, for example, might cause flooding downstream where the existing channel may be unable to accommodate the new flow regime. Similarly, a change in land use in the upper reaches of a drainage basin may well have consequences far downstream. Deforestation of the upper reaches of the Ganges and Bhramaputra rivers in the foothills of the Himalayas has been blamed by some researchers for major flooding problems in Bangladesh, where the rivers combine to flow into the Bay of Bengal. Such circumstances suggest that flood prevention requires the management of an entire drainage basin rather than just the areas that are obviously prone to flooding.

Further reading
Coch, N.K. (1995) *Geohazards: Natural and Human*, Englewood Cliffs, NJ: Prentice-Hall.
Handmer, J. (ed.) (1987) *Flood Hazard Management: British and International Perspectives*, Norwich: Geo.
Ives, J.D. and Messerli, B. (1989) *The Himalayan Dilemma: Reconciling Development and Conservation*, London: Routledge.
Mayer, L. and Nash, D. (eds) (1987) *Catastrophic Flooding*, Boston/London: Allen & Unwin.
Rasid, H. and Pramanik, M.A.H. (1993) 'Areal extent of the 1988 flood in Bangladesh: how much did the satellite imagery show?', *Natural Hazards* 8: 189–200.

FLOODPLAIN

An area of limited relief bordering a river inundated when the river overflows its banks during a *flood*. Floodplains are generally low and flat, but possess some relief in the form of the **levées** which border the main channel and the abandoned channels which indicate the former course of the river as it meandered across the floodplain. Poor drainage on the floodplain encourages the persistence of wetlands, such as the backswamp areas found towards the landward edge of the plain. The floodplain surface is composed of fine sediments deposited during flooding. Regular deposition of these sediments maintains the **fertility** of the floodplain **soils** and makes them attractive to settlement, despite the hazards posed by regular flooding. The **agrarian civilizations** of Egypt and Mesopotamia depended upon this regular renewal of soil quality, and the large **populations** that inhabit the floodplains of Bangladesh and

China attest to their continued attraction. The availability of flat land and proximity to the river providing **water** for industry and transportation also made floodplains attractive for urban development. As a result, many of the world's largest cities, such as London, Paris, New Orleans, Shanghai and Calcutta, have grown up on floodplains or on **deltas** where the floodplains reach the sea. Modern environmental planners, however, tend to regard all floodplains as hazard land, best suited for recreational and certain types of agricultural land use rather than for urban residential or industrial development.

Further reading
Ritter, D.F., Kochel, R.C. and Miller, J.R. (1995) *Process Geomorphology*, Dubuque, IA: Wm C. Brown.
Ward, R.C. (1978) *Floods – a Geographical Perspective*, London: Macmillan.

FLORA

The combination of plants in a particular area. Each **biome** has a characteristic flora. The term also refers to the friendly **bacteria** which help to protect the human body against invasion by **pathogens**.

FLUE GAS

A mixture of hot waste gases released during **combustion**. The constituents are predominantly **nitrogen** (N), **carbon dioxide** (CO_2), **carbon monoxide** (CO) and steam, but depending upon the **fuel** being used and the combustion process involved, it may also contain **sulphur dioxide** (SO_2) and **oxides of nitrogen** (NO_x). **Particulate matter** such as fly ash, a by-product of the combustion of the very fine pulverized **coal** used in modern furnaces, may also be exhausted along with the gases.

See also
Acid rain, Flue gas desulphurization.

FLUE GAS DESULPHURIZATION (FGD)

The process by which **sulphur dioxide** (SO_2)

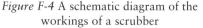

Figure F-4 A schematic diagram of the workings of a scrubber

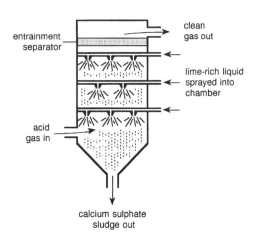

is removed from the exhaust **gases** produced by the burning of **coal**. FGD is commonly used in coal-fired thermal generating stations to reduce the output of acid gases which lead to the formation of **acid precipitation**. The process takes several forms, but usually involves the exposure of the acid gases to an alkaline or basic substance, such as **limestone** or a lime-rich liquid, in a **scrubber**. The sulphur dioxide in the **flue gases** combines with the **calcium carbonate** ($CaCO_3$) in the limestone to produce calcium sulphate or gypsum ($CaSO_4.2H_2O$). FGD is currently the most popular form of acid emission control, in part because it is capable of reducing the output of sulphur dioxide in exhaust gases by between 80 and 95 per cent, but also because the equipment required is technically quite simple and can be added to existing power plants relatively easily. Problems remain with the disposal of the large amounts of gypsum generated – some 1.5 m tonnes per annum from a large plant – and the environmental impact of quarrying required to meet the increasing demand for high-grade limestone. Most industrial nations have sulphur dioxide reduction programmes in place and these are likely to be met in large part by the installation of FGD equipment.

Further reading
Ellis, E.C., Erbes, R.E. and Grott, J.K. (1990) 'Abatement of atmospheric emissions in North

America: progress to date and promise for the future', in S.E. Lindberg, A.L. Page and S.A. Norton (eds) *Acidic Precipitation, Volume 3, Sources, Deposition and Canopy Interactions,* New York: Springer-Verlag.

Kyte, W.S. (1986) 'Some aspects of possible control technologies for coal-fired power stations in the United Kingdom', *Mine and Quarry* 15: 26–9.

FLUIDIZED BED COMBUSTION (FBC)

The burning of a mixture of crushed **coal**, **limestone** and **sand** in the presence of high-pressure **air**. The air is blown in from the base of the **combustion** chamber through a bed of sand, until the sand begins to resemble a boiling **liquid**. At that point the coal and limestone are added and the combustion process begins. Continual mixing ensures that combustion is very efficient, allowing the use of relatively low-grade **fuels** if necessary. Acid gas emission control is also very efficient, with up to 90 per cent of the **sulphur** (S) in the fuel being absorbed by the limestone. Since furnace **temperatures** remain relatively low, FBC also leads to reduced emissions of **oxides of nitrogen** (NO$_x$). As yet FBC technology is not widely used, being restricted to low-pressure boilers and a number of pilot **thermal electric generating**

Figure F-5 A schematic diagram of the workings of a fluidized bed combustion system

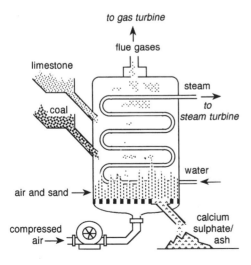

plants, but with the ability to produce 20 per cent more electricity than conventional systems for the same coal consumption, it has the potential to expand.

See also
Acid rain, LIMB, Scrubbers.

Further reading
Ellis, E.C., Erbes, R.E. and Grott, J.K. (1990) 'Abatement of atmospheric emissions in North America: progress to date and promise for the future', in S.E. Lindberg, A.L. Page and S.A. Norton (eds) *Acidic Precipitation, Volume 3, Sources, Deposition and Canopy Interactions,* New York: Springer-Verlag.
Ramage, J. (1983) *Energy: A Guidebook,* Oxford: Oxford University Press.

FLUORINE (F)

A highly toxic, greenish-yellow **gas** with a pungent odour. Along with **bromine** (Br), **chlorine** (Cl) and **iodine** (I), it is a member of the **halogen group** of **elements**. Being the most reactive of all elements, it is not found free in the **environment**, but must be manufactured through the **electrolysis** of potassium hydrogendifluoride (KHF$_2$). It occurs mainly in the form of fluorspar (CaF$_2$), used in the chemical and ceramics industries, and cryolite (Na$_3$AlF$_6$), used extensively in **aluminum** (Al) smelting. Demand for the element increased rapidly with the development of **fluorocarbon**-based refrigerants and propellants such as **Freon** and the tetra-fluoroethylene **polymer**, Teflon, marketed as a corrosion resistant and non-stick coating. Fluorine compounds are also used in **uranium** (U) processing, the production of **detergents** and the refining of some forms of **gasoline**. As sodium fluoride (NaF), it has been added to drinking-water and toothpaste as an aid in the prevention of tooth decay.

See also
Chlorofluorocarbons.

Further reading
Chambers, R.D. (1973) *Fluorine in Organic Chemistry,* New York: Wiley.
Waldbott, G.L., Burgstahler, A.W. and McKinney, H.L. (1978) *Fluoridation: The Great Dilemma,* Lawrence, KA.: Coronado.

Figure F-6 The chemical structure of typical fluorocarbons

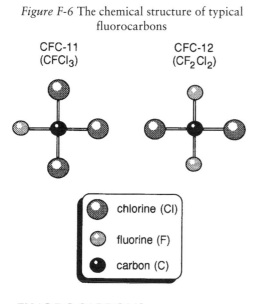

CFC-11
(CFCl₃)

CFC-12
(CF₂Cl₂)

chlorine (Cl)

fluorine (F)

carbon (C)

FLUOROCARBONS

One of a group of synthetic **organic compounds** called **halocarbons** in which some or all of the **hydrogen atoms** have been replaced by **fluorine atoms**. First developed in the United States in the 1930s as refrigerants, it was soon recognized that being for the most part nonflammable, noncorrosive, nontoxic and chemically inert, they had much wider applications. Through their use as **solvents**, propellants in aerosol spray cans and as coating materials such as Teflon (polytetrafluoroethene), their use expanded rapidly in the 1960s and 1970s. Like other halocarbons, however, fluorocarbons have been implicated in the destruction of the **ozone layer**, and their production is being phased out.

See also
Bromofluorocarbons, Chlorofluorocarbons, Ozone depletion.

Further reading
Downing, R.C. (1988) *Fluorocarbon Refrigerants Handbook*, Englewood Cliffs, NJ: Prentice-Hall.

FLUVIAL PROCESSES

Processes associated with flowing **water** in rivers or streams in the **runoff** sector of the hydrological cycle. Including the **erosion**, transportation and deposition of sediments, they are the most important group of geomorphic processes at work in the terrestrial environment.

Further reading
Morisawa, M. (1985) *Rivers*, New York: Longman.
Richards, K.S. (ed.) (1987) *River Channels: Environment and Process*, New York: Blackwell.

FLUX

(1) The rate of flow of **mass** or **energy** per unit area. The horizontal flux of moisture is represented as kilograms per square metre per second (kg m^{-2}s^{-1}), for example, and the flow of **solar energy** as **calories** per square centimetre per day (cal cm^{-2}day^{-1}).
(2) A substance used in the **smelting** of metallic **ores**, that combines with **waste** materials in the ores to form scum or slag which can be removed from the surface of the molten **metal**. Fluxes are also used in the brazing, soldering and welding of metals to reduce **oxidation** and allow for better fusion of the metals being joined.

FLY ASH

Finely divided particulates carried into the **atmosphere** by **flue gases** following **combustion**. The ash may include unburned **fuel** as well as combustion products. Once a common source of **pollution** from **thermal electric power stations** and industrial plants, fly ash is now commonly filtered out of the gases before they are released into the environment.

FLYWAYS

The routes followed by migratory birds as they move between their wintering and breeding grounds in spring and autumn. In North America, four major flyways may carry as many as 100 million birds, mainly waterfowl, in one season. Knowledge of these migratory routes and the resting places used by the birds *en route* provided hunting societies with a reliable source of food at least

Figure F-7 The flyways of North America

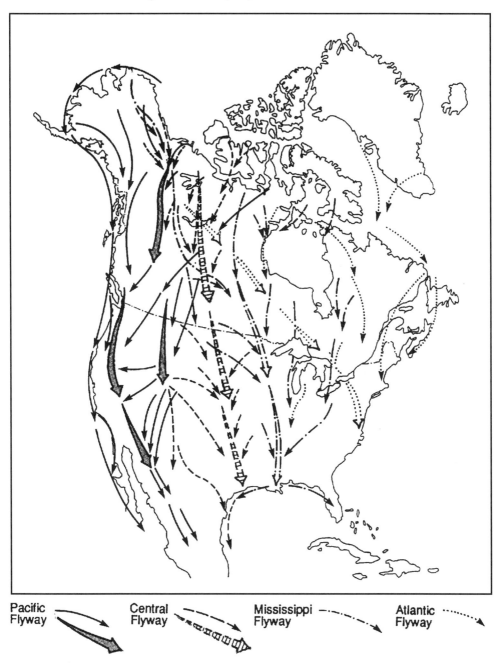

Pacific Flyway   Central Flyway   Mississippi Flyway   Atlantic Flyway

Source: After Enger, E.D. and Smith, B.F. (1995) *Environmental Science: A Study of Interrelationships,* Dubuque, IA: Wm C. Brown

twice a year. The aboriginal inhabitants of the Hudson Bay area in northern Canada, for example, participated in a spring and autumn goose hunt, made possible by the presence of major flyways along the west and east coasts of the bay. Similarly, in Europe, a variety of migratory **species** from songbirds to wildfowl were regularly netted, snared and shot on their way between Africa and northern Europe in the spring and on the return journey in the autumn. An estimated 240 million songbirds continue to be harvested in Italy every year. Modern sports hunting also takes advantage of the presence of these flyways. Now strictly controlled, hunting using modern firearms once decimated such unlikely prey species as hawks and cranes. The numbers of birds using the flyways has always varied, depending upon such factors as reproductive success and **weather** conditions, but in recent years human activities have had a significant impact. In both North America and Europe, the draining of **wetlands** and the destruction of forests has reduced the availability of resting places for the migrating birds, and the building of tall structures such as radio and television towers, lighthouses and refinery smokestacks causes the death of perhaps as many as half-a-million birds annually in North America alone. The maintenance of existing wetlands or the provision of wildlife refuges and the development of various treaties dealing with migratory bird populations attempt to deal with such problems.

Further reading
Cox, G.W. (1993) *Conservation Ecology: Biosphere and Biosurvival*, Dubuque, IA: Wm C. Brown.

FOAM

An aggregation of gaseous bubbles in **water** or other **liquids**, produced by vigorous mixing, **fermentation** or some chemical action which causes effervescence. With sufficient aeration, foam will form naturally in the **aquatic environment**, but the widespread adoption of synthetic **detergents** in the 1950s and 1960s led to the serious **pollution** by foam of rivers and lakes. Not being biodegrad-

able, the detergents retained their foaming ability when released in **waste** water, and the natural agitation of waves or flowing water reactivated the foam. Although mainly an aesthetic problem, the foam indicated the likely presence of other pollutants. Under public and governmental pressure in the late 1960s, the detergent industry began to introduce biodegradable products that were more easily removed through **sewage treatment** and less likely to foam in rivers and streams. The production of solid **plastic** foams has been implicated in the destruction of the **ozone layer**, because of the escape of **CFCs** used as blowing or foaming agents in the process.

Further reading
Garrett, P.R. (ed) (1993) *Defoaming: Theory and Industrial Applications*, New York: M. Dekker.

FODDER

Animal food which has been grown or collected specifically for animals. Examples include grass, shrubs, hay, straw, various grains and turnips.

See also
Forage.

FOG

A suspension of small **water** droplets in the lower **atmosphere** which causes visibility to be reduced to less than 1 km. It is caused by **condensation** when moist **air** is cooled below the **dewpoint**. The most common types of fog are **radiation** fog and advection fog. Radiation fog is formed on clear, calm nights when **terrestrial radiation** easily escapes into the atmosphere causing the **temperature** of the surface to fall. If outward radiation continues and the **air** is sufficiently moist, the atmosphere close to the surface may be cooled below its dewpoint, leading to condensation and the formation of fog. Advection fog forms when relatively warm, moist air is cooled as it moves over a colder surface. When maritime air flows over a cold land surface during winter, for example, advection fog will form and persist as long as the **wind** blows. The natural formation of fog may be aggravated by human activities which, for example, add **condensation nuclei** or abundant **water vapour** to the atmosphere.

See also
Clouds, Haze, Smog.

Further reading
Anthes, R.A., Cahir, J.J., Fraser, A.B. and Panofsky, H.A. (1981) *The Atmosphere* (3rd edition), Columbus, OH: Charles E. Merrill. Murray, W.A. and Kurtz, J. (1976) *A Study of Ice Fog and Low Temperature Water Fog Occurrence at Mildred Lake, Alberta*, Edmonton, Alta: Syncrude Canada.

FOOD CHAIN

A group of organisms linked to each other through their production and consumption of food and **energy**. Most food chains are short (perhaps four or five links) and linear, with some form of green plant at one end and a carnivore or omnivore at the other. A simple food chain could be made up as follows:

Grass   → Antelope   → Lion
                 (herbivore)   (carnivore)

(producer)   (primary   (secondary
                 consumer)   consumer)

The grass produces the food and energy through **photosynthesis**. It is then consumed by the antelope, which in turn is consumed by the lion. In this way food and energy are passed along the chain, with each stage referred to as a **trophic level**. This is an example of a grazing food chain. Ultimately the energy reaches the decomposers. They convert dead organic matter into its constituent parts, releasing nutrients and initiating a detrital food chain, which includes a variety of detrivores and other detritus consumers. The conversion process in any food chain is relatively inefficient, with as much as 90 per cent of the useful energy being lost during the conversion from one level to another, usually in the form of heat. Thus only 1 per cent of the energy available from the grass would be stored in the body of the lion. Aquatic food chains tend to be longer than terrestrial chains, and as a result the amount of the original **solar energy** that

reaches the ultimate consumer is even less. The numbers of organisms in a food chain also tend to decline as the levels increase, creating the so-called pyramid of numbers. Large herds of antelope are required to support only a few lions, for example. Linkages among the **elements** in the chain are so strong that disruption at one level will be felt along the entire chain. Removal of the producer, for example, will reduce the food supply for the primary consumers, perhaps causing a decline in their numbers by starvation, which in turn reduces the food supply for the predators. At the other end of the chain, a reduction in the numbers of predators will allow the population of the grazing primary consumers to increase, thus placing stress on their food supply. The introduction of rabbits into Australia, for example, created a short food chain with only a limited number of predators. This allowed the rabbit **population** to increase to such an extent that it competed with other primary

Figure F-8 The transfer of energy in a food chain or trophic chain

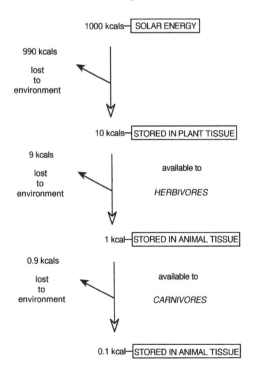

consumers, both natural and domesticated, for food. In some places, the abundance of rabbits caused the complete destruction of the vegetation and the initiation of **soil erosion**. The introduction of the viral disease **myxomatosis** in the 1950s was so effective in killing off the rabbits that predators such as foxes and eagles turned to other native species, but eventually their populations also slumped. A similar disruption of the food chain occurred following the accidental escape of the experimental rabbit calicivirus in 1995.

Humans are at the upper end of many food chains, and can adapt to the relative inefficiency of the chains. In the developed nations, for example, farmers grow grain to be fed to cattle or pigs. The beef or pork is then eaten by humans. Being omnivorous, however, they could take a more energy-efficient approach, and eat the grain before it is processed by the animals. The resulting food savings could then be made available to the less developed nations where under-nourishment and even **malnutrition** are major problems. Harvesting the lower levels of a food chain has also been considered in the **aquatic environment**, where experimental fishing of large **zooplankton**, such as Antarctic krill, has been carried out. Lack of knowledge of the impact of such harvesting on the oceanic food chains, which depend upon krill, plus negative economic factors have ensured that it has not gone beyond the experimental stage.

See also
Ecological introductions.

Further reading
Cox, G.W. (1993) *Conservation Ecology: Biosphere and Biosurvival*, Dubuque, IA: Wm C. Brown.
Kaufman, D.G. and Franz, C.M. (1993) *Biosphere 2000: Protecting our Global Environment*, New York: HarperCollins.

FOOD WEB

Although **food chains** are basically linear, individual chains are commonly interlocked at various levels to form a web. Different predators, for example, may share the same prey, and consumers may alter their eating habits – and become involved in a new food chain – if their preferred food source is no longer available. The presence of **decomposers** at all **trophic levels** also helps to link individual chains into a web.

FORAGE

Animal food in its natural state available for grazing by animals. It includes almost any type of plant material, even those from which human beings can extract little of food value. Excessive foraging can lead to **overgrazing** and environmental deterioration.

FOSSIL

Remains of organisms or parts of organisms that have been preserved in the rocks (usually **sedimentary rocks**) of the earth's **crust**. For fossilization to occur, the organism involved must possess a skeleton or hard structure and must be buried beneath sediments soon after death. In some cases the entire organism is preserved, but more commonly the soft parts decay and only the skeleton remains. The original composition of the skeleton may also be changed, **calcium carbonate** ($CaCO_3$) being replaced by **silica** (SiO_2), for example. Occasionally whole organisms or parts of organisms may survive. Fossil mammoths have been found in frozen ground in Siberia, for example, and the fossil **pollen** used in **palynology** consists essentially of the unaltered exoskeletons of pollen grains.

See also
Palaeontology.

Further reading
Paul, C.R.C. (1980) *The Natural History of Fossils*, London: Weidenfeld & Nicolson.

FOSSIL FUELS

Fuels that are the residues of organic material. Like fossils, plants and animals were buried by sediments or drowned, and in the resulting **anaerobic environment** retained much of their stored **energy**. Coal, for example, contains chemical energy converted from **solar energy** by **photosynthesis** when the plants were growing. It is that energy which is released when fossil fuels are burned. Fossil fuels include coal, **oil** and **natural gas**, the main energy sources in advanced industrial societies. They are non-renewable sources of energy and finite in quantity, but world supplies remain sufficient for several centuries. Fossil fuel consumption is a major contributor to a number of current environmental issues including **acid rain**, **atmospheric turbidity** and **global warming**.

Further reading

Kraushaar, J.J. and Ristinen, R.A. (1993) *Energy and the Problems of a Technological Society* (2nd edition), New York: Wiley.

FRAMEWORK CONVENTION ON CLIMATE CHANGE (FCCC)

One of the conventions signed at the **Earth Summit** in Rio de Janeiro in 1992. The FCCC grew out of concern for **global warming**, but was signed only after much controversy, and ended up as a relatively weak document lacking even specific emission reduction targets and deadlines. Its main aim is to bring about the stabilization of **greenhouse gas** emissions at a level which would prevent dangerous anthropogenic interference with the climate system. Following the Summit, the **IPCC** was reorganized to provide information needed to support the implementation of the Convention, and in early 1996 it presented its report to the 159 countries that had ratified the document. A Conference of the Parties (COP) was set up to deal with the transfer of scientific and technical knowledge among the parties and the implementation of the provisions of the Convention.

Further reading

Adger, W.N. (1995) 'Compliance with the climate change convention', *Atmospheric Environment* 29: 1905–15.
IPCC (1996) *Climate Change 1995: The Science of Climate Change*, Cambridge: Cambridge University Press.
Kauppi, P.E. (1995) 'The United Nations Climate Convention: unattainable or irrelevant', *Science* 270: 1454.
Pitt, D. and Nilsson, S. (1994) *Protecting the Atmosphere: The Climate Change Convention and its Context*, London: Earthscan.

FREE RADICAL

A group of **atoms** which usually exists in combination with other atoms, but may exist independently under special circumstances. Possessing one or more unpaired **electrons**, free radicals are highly reactive. They have an important role in the formation of **photochemical smog**. **Hydrocarbon** radicals such as the ethyl radical (C_2H_5), for example, contribute to the formation of **peroxyacyl nitrate** (PAN), an **oxidant** which is one of the major contributors to the toxicity of photochemical smog.

Further reading

Williamson, S.J. (1973) *Fundamentals of Air Pollution*, Reading, MA: Addison-Wesley.

FREON

The commercial name for **chlorofluorocarbons** (CFC), developed by the **DuPont** Company.

FRICTION

A force which prevents or resists relative motion between surfaces in contact. Before one body can move over another, enough **energy** must be applied to overcome the static friction between the two. This is associated with the surface roughness of the two bodies. Generally, the rougher the two surfaces, the greater will be the coefficient of static friction. The concept is perhaps more easily seen when two solid bodies are involved, but friction also applies when a fluid moves over a solid surface, for example. **Wind** moving over the surface of the earth or **water** flowing over the bed of a river both experience the effects of friction, which slows the movement of the **air** or the water. If the earth's surface or the bed of the stream varies in roughness, the resulting variations in the coefficients of friction cause turbulence, which may have implications for the mixing of pollutants in the **atmosphere** or the extent and distribution of **erosion** and deposition in river beds.

See also
Turbulent flow.

Further reading
Rabinowicz, E. (1965) *Friction and Wear of Materials*, New York: Wiley.

FRIENDS OF THE EARTH (FOE)

A federation of autonomous environmental groups founded in 1971. FOE is represented in fifty-two countries and claims to be the largest international network of environmental groups. It is concerned with social and economic factors in relation to the **environment**, and operates at all scales from international, where it is involved in the preservation of the **rainforest**, to local, where groups attempt to control inappropriate **waste disposal** or oppose the construction of roads through environmentally fragile areas.

FUEL

A substance used to produce **thermal energy**, either through **combustion** - as in the case of the **fossil fuels** – or through changes in its physical chemistry – as in the case of nuclear fuels.

FUEL DESULPHURIZATION

The reduction in the **sulphur** content of **fuels** such as **coal** and **oil** prior to **combustion**. Methods vary from simple, cost-effective processes such as the crushing and washing of coal to more complex chemical cleaning methods such as coal **gasification** or **liquefaction**.

FUEL SWITCHING

One of the simplest approaches to the control of acid **gas** emissions, involving the replacement of high-sulphur **fuels** with low-sulphur alternatives. The most common form of fuel switching is the replacement of high-sulphur coal with a low-sulphur coal. **Coal** may also be replaced entirely by **oil** or **natural gas**.

FUELWOOD

Wood products harvested for use as **fuel**. In well-wooded mid-latitudes, fuelwood commonly takes the form of large logs. Elsewhere, fuelwood may be no more than the branches and twigs of scrub woodland, as is the case in the arid or semi-arid parts of Africa and Asia. The cutting of fuelwood makes a major contribution to **desertification** in such areas.

Further reading
De Montalembert, M.R. and Clement, J. (1983) *Fuelwood Supplies in the Developing Countries*, Rome: Food and Agricultural Organization of the United Nations.

FUMIGATION

(1) The destruction of **bacteria**, insects, rodents and other pests by exposure to poisonous **gases**, **smoke** or **vapours**. It is commonly used in enclosed spaces such as greenhouses or grain stores.
(2) The rapid build-up of **air** pollutants, often to dangerous levels, in the atmospheric boundary layer. Fumigation is frequently produced when a **temperature inversion** prevents the upward dispersion of pollutants. Frequently a night-time phenomenon in urban areas, early morning heating usually causes sufficient turbulence to mix the pollutants with cleaner air and therefore dilute them.

See also
Turbulent mixing.

Further reading
Oke, T.R. (1987) *Boundary Layer Climates* (2nd edition), London: Routledge.

FUNGI

A group of relatively simple organisms which range from easily visible mushrooms and toadstools to microscopic **yeasts** and mildews. Fungi may consist of a single **cell** or a mass of very fine threads or filaments (hyphae), and are characterized by their lack of **chlorophyll**. Since they cannot participate in **photosynthesis**, they are considered to be heterotrophic – they obtain their **nutrients** directly from the **environment**. Reproduction takes place by means of **spores**. Numbering perhaps as many as 47,000 **species,** fungi have an important role in the **recycling** of plant and animal matter in the **soil,** but the same qualities that make them good **decomposers** also allow them to cause problems of decay in food, fabrics and other organic materials important to society. The fungal infestation of timber causes dry rot, for example. Some fungi are also parasitic on plants and animals. Beneficial forms of fungi include various varieties of edible mushroom plus the yeasts used in baking, brewing and the preparation of other foodstuffs. **Antibiotics** such as penicillin can be produced by culturing the appropriate fungi.

See also
Fungicide.

Further reading
Carlile, M.J. and Watkinson, S.C. (1994) *The Fungi,* Boston: Academic Press.

FUNGICIDE

A chemical used to destroy harmful **fungi.**

FUSION

See **nuclear fusion.**

G

GAIA HYPOTHESIS

First developed in 1972 by James Lovelock, and named after an ancient Greek earth goddess, the Gaia hypothesis views the earth as a single organism in which the individual **elements** coexist in a symbiotic relationship. Internal homeostatic control mechanisms, involving positive and negative **feedbacks**, maintain an appropriate level of stability. It has much in common with the concept of **environmental equilibrium**, but goes further in presenting the view that the living components of the **environment** are capable of working together actively to provide and retain optimum conditions for their own survival. In the simplest case, animals take up **oxygen** (O_2) during **respiration** and return **carbon dioxide** (CO_2) to the **atmosphere**. The process is reversed in plants, carbon dioxide being absorbed and oxygen being released. Thus the **waste** product from each group becomes a **resource** for the other. Working together over millions of years, these living organisms have combined to maintain oxygen and carbon dioxide at levels capable of supporting their particular forms of life and, through carbon dioxide, maintain the **greenhouse effect** at a level which can provide a **temperature** range appropriate for that life. This is one of the more controversial aspects of Gaia, flying in the face of conventional scientific opinion, which since at least the time of **Darwin** has seen life responding to environmental conditions rather than initiating them. Some interesting and possibly dangerous corollaries emerge from this. It would seem to follow, for example, that existing environmental problems which threaten current forms of life and life

processes – for example, **global warming, ozone depletion** – are transitory, and will eventually be brought under control again by the environment itself. Some scientists view the acceptance of this aspect of Gaia as irresponsible, since it also requires the acceptance of the efficacy of natural regulatory **systems** which are as yet unproven, particularly in their ability to deal with large-scale human interference. Lovelock himself has allowed that Gaia's regulatory mechanisms may well have been weakened by human activity. Systems cope with change most effectively when they have a number of options by which they can take appropriate action, and this was considered to be one of the main strengths of Gaia. It is possible, however, that the earth's growing population has created so much stress on the environment that the options are much reduced, and the regulatory mechanisms may no longer be able to nullify the threats to balance in the system. This reduction in the variety of responses available to Gaia may even have cumulative effects that could threaten the survival of the human **species**. Although the idea of the earth as a living organism is a basic concept in Gaia, the hypothesis is not anthropocentric. Humans are simply one of the many forms of life in the **biosphere**, and, whatever happens, life will continue to exist, but it may not be human life. For example, Gaia includes mechanisms capable of bringing about the **extinction** of those organisms that adversely affect the system. Since the human species is at present the source of most environmental deterioration, the partial or complete removal of mankind might be Gaia's natural answer to the earth's current problems.

See also
Symbiosis.

Further reading
Joseph, L.E. (1990) *Gaia: The Growth of an Idea*, New York: St Martin's Press.
Lovelock, J.E. (1972) 'Gaia as seen through the atmosphere', *Atmospheric Environment* 6: 579–80.
Lovelock, J.E. (1986) 'Gaia: the world as living organism', *New Scientist* 112 (1539): 25–8.
Lovelock, J.E. (1988) *The Ages of Gaia; A Biography of our Living Earth*, New York: Norton.
Lovelock, J.E. (1995) *Gaia: A New Look at Life on Earth* (2nd edition), Oxford: Oxford University Press.
Schneider, S.H. and Boston, P.J. (eds) (1991) *Scientists on Gaia*, San Francisco: Sierra Club Books.

GAMMA RAYS

High-energy **electromagnetic radiation** with a wavelength of less than 10^{-4} μm, capable of causing **ionization**. They are similar to **X-rays**, but have greater penetrating power, allowing at least partial penetration of substances such as concrete and **lead** (Pb) which normally stop X-rays. Gamma radiation reaches the earth as **cosmic radiation** from space, but most is absorbed in the **atmosphere**. Other sources of gamma rays include the decay of certain radioactive minerals such as **uranium** (U) or **radium** (Ra), either naturally or as the result of controlled **nuclear reactions** and the detonation of **nuclear weapons**. With their high penetrative power, gamma rays easily enter the human body, where their ionizing properties can lead to major **cell** damage and the initiation of various types of **cancer**.

See also
Ionizing radiation.

Further reading
Miller, E.W. and Miller, R.M. (1990) *Environmental Hazards: Radioactive Materials and Wastes: A Reference Handbook*, Santa Barbara, CA: ABC-Clio.

GARBAGE

Domestic refuse or municipal solid **waste**, including both organic and inorganic materials. The bulk of the waste is paper and cardboard (41 per cent in North America; 38 per cent in the UK), followed by metals, plastics and glass (c. 23 per cent in both North America and the UK) and food wastes or vegetable matter (c. 26 per cent in North America; 20 per cent in UK). Most garbage is disposed of in **sanitary landfill** sites or burned in incinerators, but both types of disposal are being re-examined. Appropriate landfill sites are increasingly difficult to find and **incineration** causes problems in the form of **air pollution** and toxic ash. As a result, greater attention is being given to **recycling** the waste, although not all garbage can be recycled and some other form of disposal will continue to be required.

See also
Ocean dumping.

Further reading
Jones, B.F. and Tinzmann, M. (1990) *Too Much Trash?*, Columbus, OH: Zaner-Bloser.
Pfeffer, J.T. (1992) *Solid Waste Management Engineering*, Englewood Cliffs, NJ: Prentice-Hall.

GAS

A substance which has the form of a completely elastic fluid in which the **atoms** and **molecules** move freely in random patterns.

See also
Gas laws, Liquid, Solid.

GAS LAWS

Thermodynamic laws that deal with the relationship between the **temperature, pressure** and **volume** of gases. The most important of these are Boyle's law and Charles' law. According to Boyle's law, the volume of a given **mass** of **gas** at constant **temperature** is inversely proportional to its pressure (i.e. $pV = a\ constant$). Charles' law states that the volume of a given mass of gas, at constant pressure, is directly proportional to its absolute temperature (i.e. V/T = a constant). The two laws may be combined in the gas equation – $pV = RT$, where p = pressure; V = volume; R = the gas constant and T = temperature K. Although these are theoretical relationships that apply only to the so-called perfect gas, they can be proven experimentally and have a role in weather forecasting using atmospheric **models**.

Further reading
Cutnell, J.D. and Johnson, K.W. (1995) *Physics* (3rd edition), New York/Toronto: Wiley.

GASOHOL

A mixture of **gasoline** and **ethanol** or **methanol** used as a **fuel** for gasoline-powered motors. Since ethanol and methanol can be produced from **waste** agricultural and wood products, the production of gasohol has been seen as a means of reducing the **energy** loss caused when these materials are discarded, while at the same time reducing the demand for gasoline. It is used regionally in the United States, in areas where the abundance of corn provides the raw material for ethanol production, but even there it is not economically competitive with gasoline and requires government subsidy.

See also
Alcohol.

GASOLINE

Petrol. A mixture of **volatile hydrocarbons** used mainly as a **fuel** in the **internal combustion engine**. Most gasoline is produced through the refining of **petroleum**, but it can also be obtained from **coal** through **destructive distillation** and subsequent refining. Different qualities of gasoline can be produced by mixing the appropriate **volatile** components. Aircraft and automobiles require different mixtures, for example. In addition to the **hydrocarbons**, gasolines commonly contain other chemicals such as antioxidants, anti-icing agents and **detergents** in proportions that depend upon their intended use.

Further reading
Society of Automotive Engineers (1994) *Gasoline: Composition and Additives to Meet the Performance and Emission Requirements of the Nineties*, Warrendale, PA: Society of Automotive Engineers.

GAS PHASE REACTION

Chemical reactions that take place with the reactants in a gaseous state. In the conversion of **sulphur dioxide** (SO_2) and **oxides of nitrogen** (NO_x) into sulphuric and nitric acid, all the reactions take place with the various **compounds** remaining in a gaseous state. Gas phase reactions are less efficient than **liquid phase reactions**.

See also
Acid precipitation.

GEIGER COUNTER

Geiger-Muller counter. An instrument used to detect the presence of **radiation** by measuring the **ionization** caused by radioactive particles. Ionization of a **gas** such as argon in the sensor of the instrument disrupts its electrical field and the resulting voltage pulse is recorded by a counter or amplified to produce an audible signal.

GENE

Genes consist of segments of **DNA** which control the characteristics of specific organisms or parts of organisms, and allow these characteristics to be transmitted from generation to generation. Located on **chromosomes**, they contain the chemically coded instructions for the assembly of particular **proteins**, which in turn determine the form and function of the organism involved. The human body contains some 100,000 genes located on twenty-three chromosomes. They control such obvious attributes as eye and skin colour, but the presence or absence of specific genes can be a contributing factor to such diseases as haemophilia or cystic fibrosis. Genes that produce alternative characteristics are called alleles. For example, alleles of the gene for eye colour can produce brown or blue eyes in an individual. Any change in the nature of a particular gene will lead to a change in the the organism. This can happen through natural **mutation**, which produces a new gene slightly or significantly different from the original, or through accidental exposure to chemicals or **radiation**. Through **genetic engineering**, organisms can be given new characteristics by manipulating the genes they possess.

See also
Evolution, Genome.

Further reading
Lewin, B. (1987) *Genes*, New York: Wiley.
Suzuki, D.T. (1989) *An Introduction to Genetic Analysis*, New York: Freeman.

GENE POOL

The sum of all the different genes, including alleles, that are contained in the total **population** of a **species**. The gene pool therefore contains all the genetic or hereditary information about that species. The make-up of the gene pool changes with time because of differential reproduction – not all individuals in a species reproduce at the same rate. As a result, certain alleles from the more prolific individuals will become more common, whereas those from individuals which do not reproduce may ultimately be lost from the population. Selective breeding of plants and animals to produce particular traits also tends to reduce the variety of genes in the gene pool. While this may be beneficial in the short term, in the longer term it may make a species less able to respond to change and therefore less likely to survive if the conditions for which it was selectively bred no longer apply.

Further reading
Cook, L.M. (1991) *Genetic and Ecological Diversity*, New York: Chapman and Hall.

GENERAL CIRCULATION MODELS (GCMs)

Climate models take various forms, and involve various levels of complexity depending upon the application for which they are designed. GCMs are three-dimensional models that provide full spatial analysis of the **atmosphere**. They incorporate major atmospheric processes plus local climate features predicted through the process of **parameterization**. The simulations of current and future climates provided by these models require powerful computers capable of processing as many as 200,000 equations at tens of thousands of points in a three-dimensional grid covering the earth's surface, and reaching through two to fifteen levels as high as 30 km in to the atmosphere. In addition to these **grid-point models**, **spectral models** have been developed. In these, the emphasis is on the representation of atmospheric disturbances or waves by a finite number of mathematical functions. Many of the more advanced models incorporate this approach.

GCMs can be programmed to recognize the role of land and sea in the development of global climates and climate change. Their complex representation of atmospheric processes allows the inclusion of important **feedback** mechanisms, and they can deal with the progressive change set in motion when one or more of the components of the atmosphere is altered. In an attempt to emulate the integrated nature of the earth/atmosphere system, atmospheric GCMs have been coupled with other environmental models. Recognizing the major contribution of the **oceans** to world climatology, the most common coupling is with ocean models. In theory, such models combining the atmospheric and oceanic circulations should provide a more accurate representation of the earth's climate. This is not always the case, however. The coupling of the models leads to the coupling of any errors included in the individual models. The so-called 'model drift' which occurs can be treated, but it remains a constraint for **coupled models**. Another major problem is the difference in time-scales over which atmospheric and oceanic phenomena develop and respond to change. The atmosphere generally responds within days, weeks or months, while parts of the oceans – for example, the ocean deeps – may take centuries or even millennia to respond. As a result, running a completely interactive **coupled ocean-atmosphere model**, until all elements reach equilibrium, is time-consuming and costly. Because of this, the oceanic element in most coupled models is much less comprehensive than the atmospheric element. The ocean is commonly modelled as a slab which represents only the uppermost layer of **water** in which the **temperature** is relatively uniform with depth. Oceanic heat storage is calculated only from the chosen depth of the layer and other elements such as oceanic heat transport and exchanges with the deeper parts of the ocean are neglected or calculated only indirectly. Thus the accuracy of the results is limited. **Sea-ice models, carbon cycle models** and chemical models have also been recognized as having the portential to contribute to climate simulation when coupled to existing GCMs. Carbon cycle models, particularly important in studies of **global warming**, have already been coupled to ocean models, and chemical models have been developed to investigate the influence of other trace **gases** on the general circulation of the atmosphere.

Most of the effort in the development of GCMs has gone into producing **equilibrium models**. In these, change is introduced into a model which represents existing climate conditions and the model is allowed to run until a new equilibrium is reached. The new model climate can then be compared with the original to establish the overall impact of the change. Most of the numerous GCMs used to study the impact of a doubling of atmospheric **carbon dioxide** (CO_2) on world climates have been of this type. They make no attempt to estimate changing conditions during the transient phase of the model run, although these conditions may well have important environmental impacts long before equilibrium is reached. The development of transient or time-dependent models which would provide the interim information lagged behind that of equilibrium models, but that discrepancy is now being addressed. In the IPCC second assessment of climate change compiled in 1995, for example, ten experiments were run using **transient models** compared with only one available for the first assessment five years earlier.

Despite the growing sophistication of general circulation models, problems remain, often associated with their inability to deal adequately with elements that are integral to the functioning of the earth/atmosphere system. These include the roles of **clouds**, oceans and feedback mechanisms. Clouds are important because of the significant influence which they exert on the earth's **heat budget** through the **reflection** and **absorption** of **radiation**, but they are difficult to simulate, in part because they develop at the regional level, whereas GCMs are global in scale. Parameterization provides only a partial solution, but progress is being made in the provision of regional scale conditions to GCMs – for example, through nested models – which may improve the representation of clouds. Despite lengthy experimentation with coupled ocean-atmosphere models, representation of the oceans remains a source of uncertainty. Part of the problem is the paucity of observational data, and this reflected to some extent in the relatively low resolution of the ocean components in coupled models. Attempts must be made to address the inadequacy of the data through the development of an oceanic observing network, if the ocean component in models is to be improved. Both clouds and oceans are also involved in feedback mechanisms. The representation of feedbacks is a highly complex but important element in the consideration of climate-related processes. Feedbacks are incorporated in some form in most GCMs, with their number and complexity varying from model to model. The **IPCC** investigators in 1995 identified feedbacks involving clouds, surface radiation budgets and the carbon cycle in the terrestrial **biosphere** as particularly important sources of uncertainty. The constraints that this imposes on GCMs must be recognized, and appropriate allowances made, when analysing their predictions or incorporating them in policy-making decisions.

All models represent a compromise between the complexities of the earth/atmosphere system and the technical constraints of computer hardware and software. Despite the tens of thousands of hours of computing time that went into the production of the IPCC 1995 reports, for example, the results are far from definitive. Progress in the development of GCMs since 1990 is impressive, however, and with further development and refinement, the courses now being pursued by modellers have considerable potential for narrowing the gap between climate simulation and reality.

Further reading
Cusbach, U. and Cess, R.D. (1990) 'Processes and modelling', in IPCC *Climate Change: the IPCC Scientific Assessment*, Cambridge: Cambridge University Press.
Henderson-Sellers, A. (1991) 'Global climate change: the difficulties of assessing impacts', *Australian Geographical Studies* 29: 202–25.
Hengeveld, H. G.(1991) *Understanding Atmospheric Change*, SOE Report 91-2, Ottawa: Environment Canada.
IPCC (1990) *Climate Change: The IPCC Scientific Assessment*, Cambridge: Cambridge University Press.
IPCC (1996) *Climate Change 1995: The Science of Climate Change*, Cambridge: Cambridge University Press.
Kemp, D.D. (1997) 'As the world warms: climate change 1955', *Progress in Physical Geography* 21 (2): 310–14.
Ramanathan, V., Pitcher, G.J., Malone, R.C. and Blackmon, M.L. (1983) 'The response of a spectral GCM to refinements in radiative processes', *Journal of Atmospheric Science* 40: 605–30.
Washington, W.M. and Parkinson, C.L. (1986) *An Introduction to Three Dimensional Modeling*, Mill Valley, CA: University Science Books.

GENETIC ENGINEERING

The manipulation of the genetic make-up of an organism to produce some desired – usually beneficial – effect. A **gene** removed from a **chromosome** in one organism may be spliced to the chromosome of another to produce a specific condition in the second organism. Alternatively, they may be attached to microscopic organisms such as **bacteria** and introduced into the **cells** of the second organism that way. Genetic engineering has been used to produce desirable traits in domesticated plants and animals, and is a well-established process for the production of **antibiotics** and **hormones**. On the negative side, the release of genetically engineered organisms into a new **environment** which has no particular controls or restraints for that organism could have disastrous results.

Further reading
Fincham, J.R.S. and Ravetz, J.R. (1991) *Genetically Engineered Organisms: Benefits and Risks*. New York: Wiley.

GENETIC DRIFT

Random change in the genetic make-up of a **population**. It is most common in small populations, where it reduces genetic variation and may contribute ultimately to **extinction** of a **species**.

See also
Gene pool.

GENETICS

The scientific study of heredity.

GENOME

The complete set of **genes** characteristic of a particular organism. The Human Genome Mapping Project is in the process of establishing the genetic complement of the human **species**, by identifying all 60,000 to 80,000 human genes and mapping their location on specific **chromosomes**.

GENUS

A grouping of **species** that share common or similar characteristics.

GEOGRAPHIC INFORMATION SYSTEMS (GIS)

Systems developed to collect, store and analyse data that include a spatial element. They are usually computer-based and include several components designed to allow:

1 The acquisition of data from census reports, maps or satellite imagery, to create an

inventory of appropriate information.

2 The processing and management of the data base.

3 Analysis of data to produce new information. By comparing individual elements in the data bank, GIS can establish relationships among them. For example, comparison of the **hydrology, floodplain** topography and **soil** type of an area might allow the establishment of criteria for delineating hazard land. Additional comparison with elements from the built **environment** would then allow the identification of properties at risk.

4 The output of the results of the analyses, often in the form of a series of computer-generated maps.

Geographic information systems have become important tools in environmental planning, in large part because they can be used to collect, integrate and analyse large volumes of data accurately and more rapidly than is possible with traditional methods of analysis.

Further reading

Bonham-Carter, G. (1994) *Geographic Information Systems for Geoscientists: Modelling with GIS*, New York: Pergamon Press.

Martin, D. (1991) *Geographic Information Systems and their Socioeconomic Applications*, London: Routledge.

Maguire, D.J., Goodchild, M.F. and Rhind, D.W. (1991) *Geographical information systems*, Harlow, Essex: Longman Scientific.

GEOMORPHOLOGY

The study of the form of the earth's surface. It grew up in the late nineteenth century as a branch of geology dealing with the morphology or form of the earth's surface. Although it continues as a sub-discipline of geology, particularly in the United States, it is now more often considered an integral part of physical geography. Early studies tended to be descriptive in nature with at best a qualitative assessment of the processes involved. In contrast, modern geomorphological studies are highly quantitative, commonly involving detailed measurement and analysis of landforms and geomorpho-

logical processes. They investigate the actions of such agents of **erosion**, transportation and deposition as **weathering**, running **water**, waves, glacial **ice** and **wind**, which individually and collectively contribute to the formation of landforms.

Further reading

Chorley, R.J., Dunn, A.J. and Beckinsale, R.P. (1973) *The History of the Study of Landforms, Vol 2: The Development of Geomorphology*, London: Methuen.

Goudie, A. (1995) *The Changing Earth: Rates of Geomorphological Processes*, Oxford/Cambridge, MA: Blackwell.

Ritter, D.F., Kochel, R.C. and Miller, J.R. (1995) *Process Geomorphology* (3rd edition), Dubuque, IA: Wm C. Brown.

Strahler, A.H. and Strahler, A.N. (1992) *Modern Physical Geography* (4th edition), New York: Wiley.

Summerfield, M.A. (1991) *Global Geomorphology*, London/New York: Longman/Wiley.

GEOPHYSICAL FLUID DYNAMICS LABORATORY (GFDL)

The GFDL is one of the research units of the US National Oceanic and Atmospheric Administration, based in Princeton, New Jersey. Its goal is to understand and predict the earth's **climate** and **weather**, including consideration of the human elements involved. Researchers at GFDL are currently investigating weather and **hurricane** forecasting, **El Niño** predictions, stratospheric **ozone depletion** and **global warming**.

GEOSTROPHIC WIND

A **wind** in the free **atmosphere** (i.e. at an altitude of 500 to 1000 m and therefore not affected by surface **friction**) in which direction and velocity represent a balance between the pressure gradient and the **Coriolis effect**. It flows parallel to the isobars, unlike surface winds, which are affected by friction and therefore cross the isobars at an angle. In the northern hemisphere the geostrophic wind flows with low pressure to the left and high pressure to the right. In the southern hemisphere the situation is reversed. The **velocity** of the geostrophic wind is directly dependent upon the magnitude of the

Figure G-1 The dynamics of the geostrophic
wind

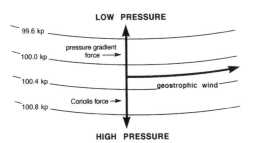

pressure gradient and is inversely proportional to the **latitude**, being less at higher
latitudes than at lower latitudes for the same
pressure gradient.

See also
Isopleth.

GEOTHERMAL ENERGY

Energy available in the molten and semi-
molten rocks beneath the earth's **crust**. The
high **temperatures** that this creates in
adjacent solid rocks in certain areas causes
sub-surface **water** to be superheated or
converted into steam, which can be used for
direct space heating or converted into
electricity in a conventional **power** plant.
Most existing geothermal plants depend
upon the production of naturally heated
water, but the expansion of production is
technically possible through the injection of
cold surface water into hot, dry rocks 3 to 5
km beneath the earth's surface. The injected
water would be under pressure and become
superheated. When returned to the surface
and normal pressure it would convert
instantly into steam, which would provide
the power for the generation of electricity.
After **condensation** the water would be
returned again beneath the surface for re-
heating. Despite having the advantage of
supplying apparently free and **renewable
energy**, geothermal zones are being developed
only slowly. Development and production
costs remain sufficiently high that geothermal
energy cannot readily compete with
conventional energy sources. Technical
problems also exist. For example, the rapid

extraction of energy from the geothermal
zone may allow the rocks to cool down and
reduce the efficiency of the process. In
addition, the dissolved **gases** and minerals
present in geothermal fluids can lead to rapid
corrosion of the system. The environmental
impacts of geothermal power plants are much
less than those of conventional plants, but
they include the primary problems of **noise**,
odour and **groundwater** contamination, plus
secondary concerns such as the disruption of
local hot springs or geysers and even induced
seismic activity. Geothermal fields in the
USA, Iceland, Italy and New Zealand already
contribute to power production, but their
impact is largely local and the contribution of
geothermal zones to global energy requirements is limited.

Figure G-2 Tapping the geothermal energy
in the earth's crust

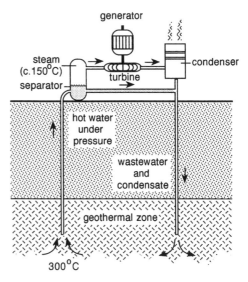

Further reading
Kleinbach, M.H and Salvagin, C.E. (1986) *Energy
Technologies and Conversion Systems*. Englewood
Cliffs, NJ: Prentice-Hall.
Wohletz, K. and Heiken, G. (1992) *Volcanology
and Geothermal Energy*, Berkeley: University of
California Press.

GLACIATION

A period of global cooling which caused
glaciers to advance and **ice** sheets to expand.

See also
Ice ages.

GLACIER

A body of **snow** and **ice** lying wholly or
mostly on land and capable of movement.
Glaciers occur when the accumulation of winter
snow exceeds summer melting or ablation.
When the accumulation is sufficiently great,
the base of the snowpack is converted into ice
through a process of compaction and recrystallization. The amount of snow that must
collect before ice is formed depends upon
such factors as local **temperature** and the
density of the snow, and may vary between
13 and 80 m, but once formed, it is the
deformation of the basal ice that causes
glaciers to move. Once moving they are capable
of **erosion**, transportation and deposition,

Figure G-3 The distribution of glaciers in
Scotland: c. 10,500–10,000 BP

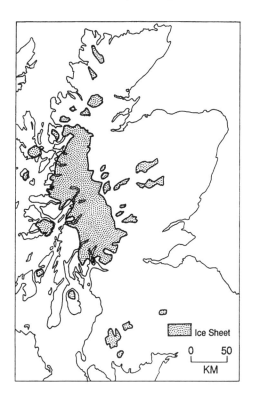

Ice Sheet

0   50
KM

Source: After Sissons, J.B. (1967) *The Evolution
of Scotland's Scenery*, Edinburgh: Oliver and Boyd

producing characteristic glacial landforms.
Glaciers are commonly classified according
to their morphological (e.g. valley glaciers,
ice sheets), dynamic (e.g. active – fast moving,
passive – slow moving) or thermal (e.g.
temperate, polar) characteristics. Currently,
glaciers cover some 10 per cent of the earth's
land surface, with the ice sheets of Greenland
and Antarctica accounting for most of that
total, but during the **Pleistocene**, ice sheets
covered much of the northern hemisphere,
reaching as far south as the Great Lakes in
North America, and southern Britain,
Germany and Poland in Europe. The ice
retreated from these areas some 10- or
12,000 years ago but they retain typically
glaciated landscapes formed at that time.

Further reading
Hart, J. and Martinez, K. (1997) *Glacial Analysis*
(CD-Rom), London: Routledge.
Menzies, J. (ed.) (1995) *Modern Glacial Environ-
ments: Processes, Dynamics and Sediments*,
Oxford: Butterworth-Heinemann.
Sharp, R.P. (1988) *Living Ice: Understanding
Glaciers and Glaciation*, New York: Cambridge
University Press.
Sugden, D.E. and John, B.S. (1976) *Glaciers and
Landscape*, London: Edward Arnold.

GLACIOLOGICAL VOLCANIC
INDEX (GVI)

An index based on acidity levels in glacial **ice**
as revealed by **ice cores**. This gives an
indication of **sulphur dioxide** (SO_2) levels
associated with past volcanic eruptions, an
element absent from both the **dust veil index**
and the **volcanic explosivity index**.

GLOBAL ATMOSPHERE WATCH
(GAW)

A programme established by the **World
Meteorological Organization (WMO)** in 1989
to monitor the composition of the **atmosphere**. It is involved in research aimed at
developing a better understanding of the
atmosphere and predicting the evolution of
such elements as **climate change**, **ozone
depletion** and atmospheric **pollution**. The GAW
provides the scientific assessment of the state
of the **atmospheric environment** for **GEMS**.

GLOBAL CLIMATE OBSERVING SYSTEM (GCOS)

Created by the **WMO, IOC, ICSU** and **UNEP** to develop a dedicated observing system to meet requirements for monitoring **climate**, detecting **climate change** and predicting climate variations and change. The work of the GCOS is intended to reduce current uncertainties associated with inadequate observational data in these areas. The data collected are expected to have an important impact on the ability of **general circulation models** (GCMs) to improve the quality of their predictions.

GLOBAL ENERGY AND WATER CYCLE EXPERIMENT (GEWEX)

An interdisciplinary project of the **WCRP**, aimed at providing a better understanding of the **hydrological cycle** and global **energy** fluxes. It uses direct observation as well as information provided by model predictions.

GLOBAL ENVIRONMENT MONITORING SYSTEM (GEMS)

A **system** co-ordinated by **UNEP** as part of its **Earthwatch** effort to monitor the **environment** worldwide. It involves more than 140 nations who provide data on **climate change**, **pollution**, **deforestation**, **ozone depletion**, and **greenhouse gas** changes. By analysing such information GEMS makes periodic assessments of the world's environmental health.

Further reading
McCormick, J. (1991) *UNEP/WHO GEMS Assessment of Urban Air Quality: Urban Air Pollution*, Nairobi: UNEP.

GLOBAL FORUM

A conference of non-government organizations (NGOs) held at the same time as the **United Nations Conference on Environment and Development** (UNCED) in Rio de Janeiro, 1992. It included a wide range of topics – from the preservation of **biodiversity** to **sustainable development** – that generally paralleled those included in the main UNCED event.

Further reading
Parson, E.A., Hass, P.M. and Levy, M.A. (1992) 'A summary of the major documents signed at the earth summit and the global forum', *Environment* 34: 12–15 and 34–6.

GLOBAL OZONE OBSERVING SYSTEM (GO₃OS)

A network of some 140 **ozone** (O$_3$) monitoring

Table G-1 Global Forum – treaties and other documents prepared by NGOs

---

Earth Charter

A short statement of eight principles for sustainable development intended to parallel the Rio Declaration

Treaty groupings

NGO Co-operation and Institution-Building Cluster: includes treaties on technology, sharing of resources, poverty, communications, global decision making and proposals for NGO action.

Alternative Economic Issues Cluster: includes treaties on alternative economic models, trade, debt, consumption and lifestyles.

Major Environmental Issues Cluster: includes treaties on climate, forests, biodiversity, energy, oceans, toxic waste and nuclear waste.

Food Production Cluster: includes treaties on sustainable agriculture, food security, fisheries.

Cross-Sectorial Issues Clusters: includes treaties on racism, militarism, women's issues, population, youth, environmental education, urbanization and indigenous peoples.

---

stations worldwide which use ground monitoring and remote sensing techniques to obtain information on the horizontal and vertical distribution of ozone as well as the total atmospheric concentration, all of which is provided to the **Global Environmental Monitoring System** (GEMS).

GLOBAL WARMING

Since the beginning of the century, there has been a rise in global mean **temperatures** of about 0.5°C. The basic cause of this warming is seen as the enhancement of the **greenhouse effect** over the same period brought on by rising levels of anthropogenically produced **greenhouse gases**. **Carbon dioxide** (CO_2) emissions have received most attention but **methane** (CH_4), **nitrous oxide** (N_2O) and **CFCs** have also contributed through their ability to retain **terrestrial radiation** in the **atmosphere**, and thus produce warming. The observed warming is what might be expected from the rising level of greenhouse gases, but the change has not been consistent. The main increase took place between 1910 and 1940 and again after 1975. Between 1940 and 1975, despite rising greenhouse levels, global temperatures declined. Such variations are not uncommon in the climate record, and

the changes that have taken place are well within the range of normal natural variations in global temperatures. However, the warming between the 1960s and 1980s was more rapid than that between the 1880s and 1940, and James Hansen of the **Goddard Institute for Space Studies** (GISS) claimed in 1988 that the global greenhouse signal is sufficiently strong for a cause-and-effect relationship between the anthropogenically produced carbon dioxide increase and global warming to be inferred. His conclusion was not widely accepted at the time, but in 1996 the **IPCC** also claimed to recognize a human influence in the current global warming.

Estimates of warming are commonly obtained by employing **general circulation models**, set to provide information on the temperature impact of a set increase in carbon dioxide. By the mid-1980s it was

Figure G-4 Measured globally averaged (i.e. land and ocean) surface air temperature in the twentieth century

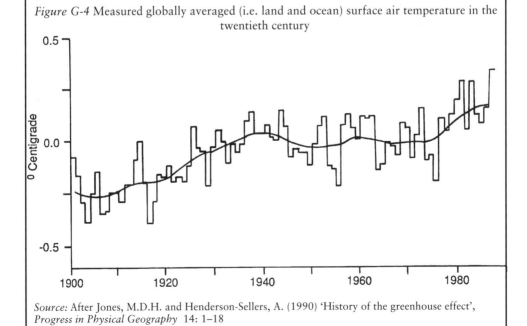

Source: After Jones, M.D.H. and Henderson-Sellers, A. (1990) 'History of the greenhouse effect', *Progress in Physical Geography* 14: 1–18

Figure G-5 Projected change in global surface temperature following a doubling of carbon dioxide (a) December, January, February (b) June, July, August

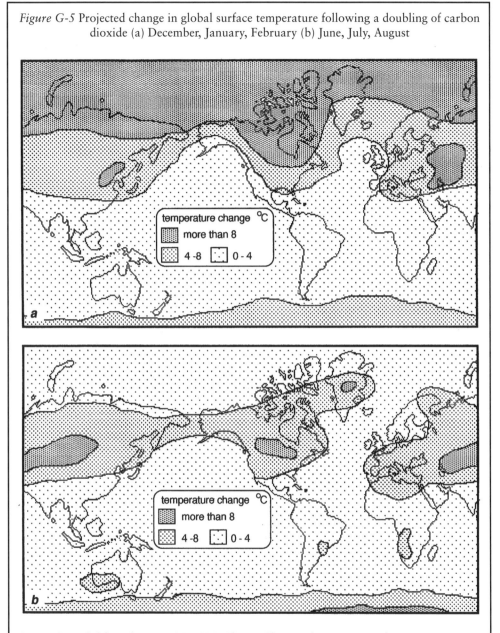

Source: Compiled from data in IPCC (1990) *Climate Change: The IPCC Scientific Assessment,* Cambridge: Cambridge University Press

estimated that a doubling of CO_2 would cause an average warming of 1.3–4°C. In 1990, the IPCC assessment produced values of 1.5–4.5°C with a best estimate of 2.5°C. By the time of their second assessment in 1995, their best estimate had been reduced to 2°C with a maximum of 3.5°C, to be reached by 2100.

Although the estimated temperatures are not particularly high – mainly because they are global averages – evidence from past world temperature changes indicates that they are of a magnitude which could lead to significant changes in climate and climate-related activities. During the **Climatic Optimum** some 5000 to 7000 years ago, temperatures in North America and Europe were only 2–3°C higher than the present average, but they produced major environmental changes. Evidence from that time period and from another warm spell in the early Middle Ages 800 to 1000 years ago also suggests that the greatest impact of any change will be felt in mid- to high latitudes in the northern hemisphere. Northern Canada, for example, might experience an increase of 8–12°C in mean temperatures in the winter, whereas in lower latitudes in Southern Europe, the **Sahel** and south-east Asia the predicted increases generally fall between 1–2°C.

Because the various **elements** in the **atmospheric environment** are closely inter-related, it is only to be expected that if temperature increases, changes in other elements will also occur. Moisture patterns are likely to be altered, for example. Semi-arid **grassland** areas would likely experience

a reduction in **precipitation**, but some projections indicate that more rainfall is possible in parts of Africa and south-east Asia. Confidence in the ability of computerized models to predict precipitation changes is low, however.

Given the range of possibilities presented in the estimates of future global warming it is difficult to predict its environmental and socioeconomic impacts. However, using a combination of investigative techniques – ranging from laboratory experiments with plants to the creation of computer-generated models of the atmosphere and the analysis of past climate anomalies – researchers have produced results which provide a general indication of what the consequences might be in certain key sectors. Warming would bring about changes in the length and intensity of the **growing season** which, coupled with new moisture regimes would disrupt existing vegetation patterns, both natural and cultivated. Across the northern regions of Canada, Scandinavia and Russia, the trees of the **boreal forest** would begin to colonize

Figure G-6 Impact of global warming on central and eastern Canada

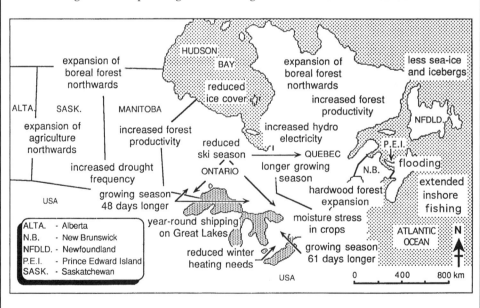

Source: After Kemp, D.D. (1994) *Global Environment Issues: A Climatological Approach*, London/ New York: Routledge

the **tundra** as they have done during warmer spells in the past. Along its southern margin, however, they would come under pressure from the poleward spread of deciduous trees and grassland.

Higher temperatures might threaten the forests indirectly through an increase in the frequency of forest fires and insect infestations, and if the rate of change takes place more rapidly than the forest can respond, rapid die-off of large numbers of trees is a possibility. Whatever the final outcome, it appears likely that global warming in the northern forest would disrupt the northern **ecosystems**, and in turn have a significant impact on those countries, such as Canada, Sweden, Finland and Russia, where national and regional economies depend very much on the harvesting of softwoods from the boreal forest.

In lower latitudes where the temperature element is less dominant, the impact of global warming would be felt through changes in the amount and distribution of moisture. In northern Australia, for example, the increased poleward penetration of **monsoon** rains would allow the expansion of tropical and subtropical vegetation. In China, agricultural output is expected to rise as a result of the northward spread of crops such as rice, corn and cotton, and the increase in the area in which tropical and subtropical fruits can be cultivated. Of major concern to many agriculturalists is the expected decrease in precipitation in the world's grain-producing areas, which along with higher temperatures, could increase the frequency and severity of **drought** in the grain belts of North America, Russia and Ukraine, leading to more frequent crop failures. Martin Parry – a leading analyst of the agricultural implications of global warming – has predicted a decline in grain yields of 15–20 per cent in Africa, tropical Latin America and much of India and south-east Asia by the middle of the twenty-first century.

Another major concern associated with **water** is a rising **sea level** brought about by the thermal expansion of the **oceans** and the additional water added to the sea by the melting of **snow** and **ice** as warming progresses. The IPCC has predicted a rise of 38–55 cm by the year 2100, but other estimates suggest that an increase of as much as a metre is possible. Even with the lower estimates, coastal regions would experience increased flooding, greater coastal **erosion** and the disruption of economic activities. For some nations occupying islands that are currently only a few metres above sea level (the **Alliance of Small Island States**, for example), the predicted rise would be disastrous. Some, such as the Maldive Islands in the Indian Ocean, would become uninhabitable, particularly if the increased storminess predicted by some models came to pass.

The range of potential impacts is great, but in many cases the reality of the situation may only become apparent when the changes have occurred, for there are many variables involved in the predictions. The human factors, as always, are particularly unpredictable. Technology, politics, socio-economic conditions and even **demography** can influence warming, for example, through their contribution to changes in greenhouse gas concentrations, yet the nature and magnitude of the variations in such elements is almost impossible to forecast.

Despite such uncertainties and a certain degree of dissension among scientists and politicians, the international view as expressed by the **Framework Convention on Climate Change** is to see global warming as a serious threat to the environment and society in the future, which must be dealt with now. Attempts to arrest continued warming have centred mainly on the control and reduction of greenhouse gas emissions, but the degree of success is as yet limited.

Further reading

Flavin, C. (1989) *Slowing Global Warming: A Worldwide Strategy* (Worldwatch Paper 91), Washington, DC: Worldwatch Institute.
Hansen, J. and Lebedeff, S. (1988) 'Global surface air temperatures: update through 1987', *Geophysical Research Letters* 15: 323–6.
Henderson-Sellers, A. (1990) 'Greenhouse guessing: when should scientists speak out?',

Climatic Change 16: 5–8.

Hulme, M. (1993) 'Global warming', *Progress in Physical Geography* 17: 81–91.

IPCC (1990) *Climate Change: The IPCC Scientific Assessment,* Cambridge: Cambridge University Press.

IPCC (1996a) *Climate Change 1995: The Science of Climate Change,* Cambridge: Cambridge University Press.

IPCC (1996b) *Climate Change 1995: Impacts, Adaptations and Mitigation of Climate Change* Cambridge: Cambridge University Press.

IPCC (1996c) *Climate Change 1995: Economic and Social Dimensions of Climate Change,* Cambridge: Cambridge University Press.

Kellogg, W.W. (1987) 'Mankind's impact on climate: the evolution of an awareness', *Climatic Change* 10: 113–36.

Kemp, D.D. (1991) 'The greenhouse effect: a Canadian perspective', *Geography* 76: 121–31.

Lindzen, R.S. (1994) 'On the scientific basis for global warming scenarios', *Environmental Pollution* 83: 125–34.

NCGCC (1990) *An Assessment of the Impact of Climate Change Caused by Human Activities on China's Environment* Beijing: National Coordinating Group on Climate Change.

Parry, M. and Duncan, R. (1994) *The Economic Implications of Climate Change in Britain* London: Earthscan.

Pearce, F. (1992) 'Grain yields tumble in a greenhouse world', *New Scientist* 134 (1817): 4.

Pittock, A.B. and Salinger, M.J. (1991) 'Southern hemisphere climate scenarios', *Climatic Change* 18: 205–22.

Simons, P. (1992) 'Why global warming would take Britain by storm', *New Scientist* 136 (1846): 35–8.

Titus, J.G. (1986) 'Greenhouse effect, sea level rise and coastal zone management', *Coastal Zone Management Journal* 14: 147–72.

Warrick, R.A. (1993) 'Slowing global warming and sea level rise: the rough road from Rio', *Transactions of the Institute of British Geographers* NS 18:140–8.

GLUCOSE

One of the simplest **carbohydrates**, glucose or dextrose $(C_6H_{12}O_6)$ is a hexose (six-carbon **atom**) monosaccharide (simple **sugar**). Glucose is produced during **photosynthesis** and is the basic **energy** source for metabolic processes in plants and animals, other sugars and carbohydrates being converted into glucose before being utilized to provide energy. Glucose occurs naturally in honey and sweet fruits, and is prepared commercially from **starch** and other carbohydrates by **hydrolysis**. It is used in the brewing and confectionery industries.

GODDARD INSTITUTE FOR SPACE STUDIES (GISS)

Established in 1961, GISS is a US National Aeronautics and Space Administration research institute located in New York City. It is primarily concerned with studies of global **climate change**, both past and future, but researchers at GISS are also involved in the investigation of other planetary atmospheres, atmospheric chemistry and astrophysics. Much of the research involves the analysis of global data sets using **models** of the atmospheric, terrestrial and oceanic environments. The research approach also includes the study of other planets as an aid to forecasting the future evolution of the earth.

GOLD (Au)

A soft, yellow, precious **metal**. It is eminently malleable and ductile, capable of being beaten into sheets as thin as 0.0001 mm. Gold is very resistant to **corrosion** and does not combine readily with other chemicals. It is found free in nature, and extracted from its **ore** using **mercury** (Hg) or cyanide. Gold is used for coins and jewellery and in the electronics industry. Where increased hardness is required it is alloyed with **copper** (Cu) or **silver** (Ag).

Further reading
Boyle, R.W. (1987) *Gold: History and Genesis of Deposits*, New York: Van Nostrand Reinhold.

GRASSLANDS

Regions in which the climax vegetation is dominated by grasses, but may contain a variety of annual and perennial plants. Grasslands form one of the earth's major **biomes**. They occur in both temperate and subtropical regions, and are naturally associated with areas where the available moisture is insufficient to support trees or other woody

Figure G-7 The distribution of the world's grasslands

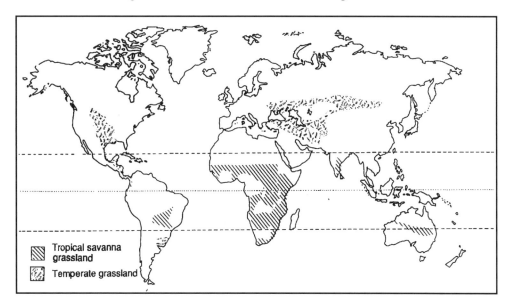

Tropical savanna grassland

Temperate grassland

Source: After Park (1992)

vegetation. The prairies of North America, the steppes of central Asia, the veld of South Africa and the pampas of South America are examples of temperate grassland. All experience high **evapotranspiration** rates that restrict the availability of moisture at certain times of the year. A similar effect is produced by periods of **seasonal drought** in the subtropical **savanna** grasslands of the **Sahel** and of the East African plains. Even where sufficient moisture may be available for the growth of woody plants, for example, close to the grassland margins, the presence of grazing animals or natural **fires** tends to favour the survival of the grasses. Large grazing animals are the main **fauna** in grassland areas, from the bison of North America to the various kinds of antelope in Africa and the kangaroos in Australia. Associated with these are the predators such as lions, cheetahs, wolves and hyenas which prey on the grazing animals. Smaller burrowing mammals such as rabbits and prairie dogs, which can easily work the grassland **soil** are also characteristic of the world's grasslands. In both temperate and subtropical grassland the return of **nutrients** and **humus** to the soil when the grass dies and the

enrichment provided by the droppings of grazing animals tends to produce fertile soils. Given the low levels of moisture characteristic of such areas, however, **overgrazing** or cultivation can make them susceptible to **soil erosion** and even **desertification**. Most grassland areas are no longer natural. Native grasses have been replaced by cultivated **species** to improve grazing conditions, or by grains such as wheat or maize. In many areas, grasslands have been created by clearing forests or improving wasteland – for example, draining swamps – where the potential agricultural return warrants it.

See also
Great Plains.

Further reading
Bourlière, F. (ed.) (1983) *Tropical Savannas*, Amsterdam: Elsevier.
Coupland, R.T. (ed.) (1979) *Grassland Ecosystems of the World*, Cambridge/New York: Cambridge University Press.

GRAVITY

The force of attraction between the earth and a body on its surface. All objects have a gravitational attraction for each other, but it is only when large masses such as the earth are involved that the implications of that attraction become evident. The force that causes a body released into the **atmosphere** to fall to the earth's surface, for example, is gravity, and the weight of that body is a measure of the force of gravity. Having such a large **mass** the earth's gravitational influence spreads beyond its boundaries, allowing it to retain the Moon as its satellite.

GRAY

The **SI unit** of **radiation** absorbed dose (**rad**). One gray (Gy) is equivalent to an energy absorption of 1 **joule** per kilogram.

GREAT AMERICAN DESERT

A term applied to the western interior plains of North America in the nineteenth century. The image grew out of the reports of exploratory expeditions which visited the plains during one of the periods of **drought** common to the area, and observed the results of natural **desertification**.

Further reading
Bowman, I. (1935) 'Our expanding and contracting desert', *Geographical Review* 25: 43–61.
Lawson, M.P. (1976) *The Climate of the Great American Desert*, Lincoln: University of Nebraska Press.

GREAT PLAINS

An area of temperate **grassland** with a semi-arid **climate** in the interior of North America. They stretch for some 2500 km from western Texas in the south along the flanks of the Rocky Mountains to the Canadian prairie provinces in the north. Although the Plains are a major producer of agricultural products such as cattle and grain, they suffer from periodic **droughts** of great severity. These have included the droughts that produced the **Dustbowl** of the 1930s, the so-called **Great American Desert** of the nineteenth century and the **Pueblo Drought** of the thirteenth century.

Figure G-8 The Great Plains of North America

Further reading
Bamforth, D.B. (1988) *Ecology and Human Organization on the Great Plains*, New York: Plenum Press.

GREEN DATA BOOK

A companion volume to the **Red Data Book**, the Green Data Book lists plants that are rare, endangered and under threat of extinction.

GREEN PARTIES

Political organizations which aim to protect the **environment** through the use of established parliamentary procedures. Although individual parties have specific goals, in general the so-called Greens favour self-sufficiency, **sustainable development** and the use of **appropriate technology**, while opposing the further development of **nuclear energy** and challenging existing **pollution** control and abatement programmes. The largest and most powerful green party is **Die Grünen** in Germany, and the European Parliament includes a coalition of Greens. In 1990 Alaska introduced green politics to North America by becoming the first state to give a Green Party official standing.

Further reading
Dobson, A. (1990) *Green Political Thought*, London/New York: Routledge.
Spretnak, C., Capra, F. and Lutz, W.R. (1986) *Green Politics*, Santa Fe, NM: Bear.

GREEN REVOLUTION

The name given to the rapid increase in crop production brought about in the late 1950s and 1960s by a combination of increased **fertilizer** use and the introduction of new higher yielding varieties of grain. The main grains involved were wheat and rice, introduced particularly to south-east Asia, where they helped food production to keep pace with the rapidly growing population. These new varieties were much more demanding than indigenous grains, requiring large quantities of artificial fertilizers and abundant **water** for best results. When the cost of fertilizers rose rapidly in the 1970s, many of the nations that had benefited initially could no longer afford the necessary quantities, and the advantages provided by the new crop varieties could not be fully realized. Furthermore, there is some concern that the replacement of a range of local plant varieties with only one or two high yielding types increases the vulnerability of the crops to pests and disease. Although that situation can be kept in check using **pesticides**, economic and environmental costs can be high and success is not guaranteed. In its time, the Green Revolution showed that agricultural technology appropriately applied

Figure G-9 Increasing grain yields made possible by the Green Revolution

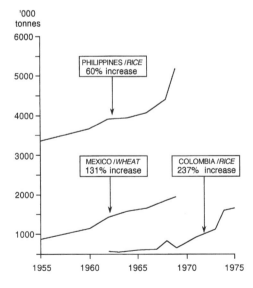

Source: Derived from data in Foin, T.C. (1976) *Ecological Systems and the Environment,* Boston: Houghton Mifflin

could improve food production significantly. However, problems with increasing costs, potential environmental impacts and the predicted agricultural impacts of **global warming** suggest that the benefits will be difficult to sustain.

Further reading
Brown, L.R. (1970) *Seeds of Change: The Green Revolution and Development in the 1970s*, New York: Praeger/Overseas Development Council.
Day, R.H. and Singh, I. (1977) *Economic Development as an Adaptive Process: The Green Revolution in the Indian Punjab*, Cambridge/New York: Cambridge University Press.
Gordon, R.C. and Barbier, E.R. (1990) *After the Green Revolution: Sustainable Agriculture for Development*, East Haven, CT: Earthscan.

GREEN REFRIGERATOR

A refrigerator developed in Germany in which the normal **CFC** refrigerant has been replaced by a propane/butane mixture. Although such a mixture has a greater cooling capacity than CFCs, it is also inflammable and therefore banned in some countries.

Further reading
Toro, T. (1992) 'German industry freezes out green fridge', *New Scientist* 135 (1935): 16.

GREENHOUSE EFFECT

The name given to the ability of the **atmosphere** to be selective in its response to different types of radiation. Incoming short wave **solar radiation** is transmitted unaltered to heat the earth's surface. The returning long wave **terrestrial radiation** is unable to penetrate the atmosphere, however. It is absorbed by the so-called **greenhouse gases**, causing the **temperature** of the atmosphere to rise. Some of the **energy** absorbed is returned to the earth's surface, and the net effect is to maintain the average temperature of the earth/atmosphere system some 30°C higher than it would be without the greenhouse effect. The process has been likened to the way in which a greenhouse works – allowing sunlight in, but trapping the resulting heat inside – hence the name. Although the

Figure G-10 Radiation fluxes and the greenhouse effect

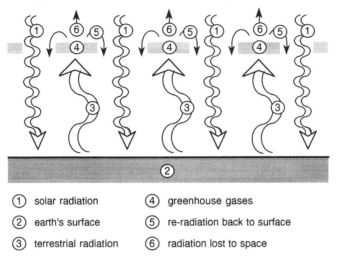

① solar radiation   ④ greenhouse gases

② earth's surface   ⑤ re-radiation back to surface

③ terrestrial radiation   ⑥ radiation lost to space

analogy is not perfect, the term continues to be widely used for descriptive purposes. Enhancement of the greenhouse effect as a result of rising levels of greenhouse gases has contributed to **global warming**.

Further reading
Bolin, B., Doos, B.R., Jager, J. and Warrick, R.A. (eds) (1986) *The Greenhouse Effect, Climatic Change and Ecosystems*, SCOPE 29, New York: Wiley.
Pickering, G.T. and Owen, L.A. (1994) *An Introduction to Global Environmental Issues*, London/New York: Routledge.

GREENHOUSE GASES

The group of about twenty **gases** responsible for the **greenhouse effect** through their ability to absorb long wave **terrestrial radiation**. They are all minor gases and together make up less than 1 per cent of the total volume of the **atmosphere**. **Carbon dioxide** (CO_2) is the most abundant, but **methane** (CH_4), **nitrous oxide** (N_2O), the **chlorofluorocarbons** (CFCs) and tropospheric **ozone** (O_3) also make significant contributions to the **greenhouse effect**. **Water vapour** also exhibits greenhouse properties, but has received less attention than the others. Since the beginning of the twentieth century, rising levels of these gases in the atmosphere, associated with increasing **fossil fuel** use, industrial development and agricultural activity, have brought about an

enhancement of the greenhouse effect. The amount of carbon dioxide recycled during **photosynthesis** is also being reduced because of worldwide **deforestation**, which allows the extra gas to remain in the atmosphere. The net result has been a gradual **global warming**, projected to continue as long as the volumes of greenhouse gases in the atmosphere

Figure G-11 Contribution of greenhouse gases to global warming: (a) 1880–1980; (b) 1980s

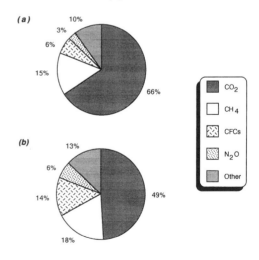

Source: After Kemp, D.D. (1994) *Global Environment Issues: A Climatological Approach*, London/New York: Routledge

Table G-2 Sources of greenhouse gas emissions

SECTORS	ACTIVITIES	GASES	PERCENTAGE OF GLOBAL TOTAL
Energy	Fossil fuel combustion	CO_2, CH_4, N_2O, O_3	54
	Natural gas leakages		
	Industrial activities		
	Biomass burning		3
Forest	Harvesting	CO_2, CH_4, N_2O	
	Clearing		
	Burning		8
Agriculture	Rice production (paddies)	CO_2, CH_4, N_2O	4.5
	Animal husbandry (ruminants)		3
	Fertilizer use		1.5
Waste management	Sanitary landfill waste	CO_2, CH_4, N_2, O_3, CFC	5
	Disposal		
	Incineration		
	Biomass decay		
Other	Cement production	CO_2, N_2O, CFC	1
	CFC production/use		11.5
	Miscellaneous		8.5

Source: After Green, O. and Salt, J. (1992) 'Limiting climate change: verifying national commitments', *Ecodecision*, December: 9–13

continue to rise. Plans to slow the warming include a reduction in greenhouse gas emissions and a slowing down of deforestation, but the results are, as yet, insignificant.

Further reading
Boag, S., White, D.H. and Howden, S.M. (1994) 'Monitoring and reducing greenhouse gas emissions from agricultural, forestry and other human activities', *Climatic Change* 27: 5–11.
Bolle, H.J., Seiler, W. and Bolin, B. (1986) 'Other greenhouse gases and aerosols', in B. Bolin, B.R. Doos, J. Jager and R.A. Warrick (eds) *The Greenhouse Effect, Climatic Change and Ecosystems, SCOPE 29*, New York: Wiley.

GREENPEACE

An international, independent environmental organization founded in Vancouver, Canada in 1971. Its non-violent but confrontational approach to global environmental issues has created major publicity and won the organization a large following worldwide. It has tackled such issues as commercial whaling, the harvesting of seal pups, dumping of **waste** on land and at sea, clear cutting by lumber companies, and the emission of toxic pollutants into the **air** and **water** by industry. Following a policy of supporting peace and disarmament, Greenpeace has made several attempts to prevent the continued testing of nuclear weapons, most recently in 1996 by confronting the French government in the South Pacific. All these activities are carried out in pursuit of Greenpeace's goal of ensuring the continuing ability of the earth to nurture life in all its diversity.

Further reading
Hunter, R. (1979) *Warriors of the Rainbow: A Chronicle of Greenpeace*, New York: Holt, Reinhart & Winston.

GREY WATER

Waste water which does not contain the products of bodily functions, being mainly the product of bathing, showering, dishwashing and similar activities. It is generally considered suitable for lawn and garden **irrigation**, and in areas such as the US south-west where water is scarce, it is seen as a simple way of increasing the efficiency of water use.

GRID-POINT MODELS

Climate **models** that provide full spatial analysis of the **atmosphere** by means of a three-dimensional grid covering the earth's surface and reaching an altitude of as much as 30 km. The progressive solution of thousands of equations at each of these points allows powerful computers to provide simulations of current and future climates.

See also
General circulation models.

GROSS NATIONAL PRODUCT (GNP)

The total value of all goods and services produced by a nation's economy in a given period of time, usually a year. It is a convenient but crude indicator of the level of economic activity in a country, and is commonly used to differentiate between more developed countries (MDCs) and less developed countries (LDCs). In environmental terms it can be misleading. The costs of **pollution** clean-up, for example, add to the GNP, although in real environmental and economic terms the impact is negative. Similarly, positive environmental developments such as **energy**-efficient appliances, which benefit society by reducing **pollution** and conserving **resources**, may actually reduce the GNP through the savings they generate.

Further reading
Hanink, D.M. (1994) *The International Economy: A Geographical Perspective*, New York: Wiley.
Meadows, D.H., Meadows, D.L., Randers, J. and Behrens, W.W. (1972) *The Limits to Growth*, New York: Universe Books.

GROUND CONTROL

The use of observation and measurement at the earth's surface to verify information provided by remote sensing from satellites or aircraft.

GROUNDWATER

The **water** that accumulates in the pore spaces and cracks in rocks beneath the earth's surface. It originates as **precipitation** and percolates down into sub-surface **aquifers**. The upper limit of groundwater saturation is the **water table**. Groundwater moves under the influence of **gravity**, although usually only slowly, and may return to the surface naturally – for example, through springs. Increasingly it is pumped from wells and **boreholes** for human use. The rate of withdrawal commonly exceeds the rate of recharge, and in many areas the groundwater supply is declining rapidly. In the United States, most of the major **aquifers** are not being recharged by precipitation as rapidly as they are being exploited and consequently are overdrawn. The withdrawal rate from the Ogallala Aquifer beneath the **Great Plains** in the United States is estimated to be 100 times the recharge rate, and the rapidly dropping water table means that water is being pumped to the surface from depths in excess of 1800 m. Similar problems of aquifer depletion are occurring in other areas such as Saudi Arabia, China and India, where surface water is in short supply, and groundwater is essential to meet local needs. In Mexico City, the demand is so high that the water-table is declining at a rate of more than 3 m per year. Another threat to groundwater supplies comes from contamination. **Petroleum** and chemical spills frequently seep into groundwater systems, and use of excess **fertilizers** and **pesticides** also contributes to **pollution** in some areas. The slow through-flow in most aquifers ensures that the impact of these pollutants

Figure G-12 The nature and sources of threats to groundwater

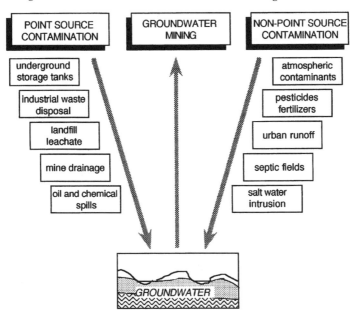

continues for many years. In coastal areas, the excess removal of freshwater from aquifers may allow the incursion of salt water, which effectively prevents further use of the system for most domestic and industrial purposes.

Further reading
Brown, A.G. (ed.) (1995) *Geomorphology and Groundwater*, Chichester/New York: Wiley.
Miller, G.T. (1994) *Living in the Environment* (8th edition), Belmont, CA: Wadsworth.
Sloggett, G. and Dickason, C. (1986) *Groundwater Mining in the United States*, Washington, DC: Government Printing Office.

GROWING SEASON

The period of the year when mean daily temperatures exceed the **temperature** at which plant growth takes place. Since different plants mature at different rates, the length of the growing season will determine the mix of natural vegetation and the types of crop that will grow in a particular area. In mid- to high latitudes, frost in spring and autumn will damage plants at the beginning and end of the growing season, when daily temperatures exceed the growth threshold but night-time values fall below zero.

GULF STREAM

A warm ocean current, originating in the eastern Caribbean, which flows north along the eastern seaboard of the United States before swinging north-eastwards into the Atlantic Ocean, where it becomes the weaker and cooler North Atlantic Drift. The latter is responsible for warming Western Europe particularly in the winter, when **temperatures** as far north as northern Norway are abnormally high for that latitude at that time of year.

Further reading
Neumann, G. (1968) *Ocean Currents*, Amsterdam/New York: Elsevier.

GYRE

The roughly circular patterns assumed by the major surface currents in the world's **oceans**. Centred on the subtropical high pressure **cells**, the **water** in these gyres circulates clockwise in the northern hemisphere and anticlockwise in the south, under the influence of the atmospheric circulation.

See also
Oceanic circulation.

H

HABITAT

The specific **environment** in which an organism lives. Although the term may be linked with a particular **species** – for example, the habitat best suited for elephants – any given environment will be shared by a variety of organisms that have requirements in common, or depend upon other organisms in the habitat. Habitats range in scale, from worldwide to continental, to regional or local and even microscopic. The nature of any habitat is determined by a large number of variables, but most can be grouped into climatic, topographic, edaphic and **biotic** factors. Of these, the climatic factors – for example, light, heat, moisture and wind – are generally considered to be of most importance, particularly at larger scales. The others become more important at a regional or local scale, sometimes in combination with climate. The combination of **topography** and **climate**, for example, leads to the zonation of vegetation with altitude, with a consequent zonation in habitats. **Edaphic factors** include **elements** of **soil structure**, **texture**, chemistry and **moisture** characteristics. Biotic factors include the impact of animal grazing and, increasingly, human interference. The combination of all these elements determines the resource availability in a specific habitat and therefore its **carrying capacity**. Habitats tend to include a natural patchiness, brought about by local variations in topographical or edaphic factors, for example, and they change naturally as a result of normal environmental dynamics. However, the response to change may differ from one part of a habitat to another and as a result a specific habitat such as a forest may include a mosaic of areas in different stages of development. Changes may be minor or temporary – for example, the result of a forest fire – or much more dramatic and long lasting – for example, following the major climate change which produced the **ice ages**. The impact on organisms also varies, from minor to catastrophic, with the complete loss of habitat perhaps bringing about the **extinction** of a particular species. Increasingly, habitat change or loss is being caused by human activities such as **deforestation**, development of agriculture, industrialization and urbanization. In many parts of the world the nature of the local or regional habitat is anthropogenically determined, and even in areas generally considered natural, such as the Arctic, the human imprint is present.

Further reading
Cox, G.W. (1993) *Conservation Ecology: Biosphere and Biosurvival*, Dubuque, IA: Wm C. Brown.
Newman, A. (1990) *The Tropical Rainforest: A World Survey of Our Most Valuable Endangered Habitats*, New York: Facts on File.
Woodward, F.T. (1987) *Climate and Plant Distribution*, Cambridge: Cambridge University Press.
World Resources Institute (1992) 'Wildlife and habitat', in *World Resources 1992-93: A Guide to the Global Environment*, New York: Oxford University Press.

HADLEY CELLS

Convection cells that form in the tropical **atmosphere** north and south of the equator. Named after George Hadley who, in the eighteenth century, developed the classic model of the general circulation of the atmosphere based on convection. High

Figure H-1 The location and structure of the tropical Hadley Cells

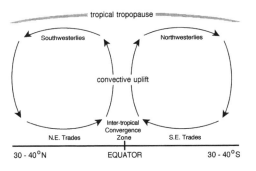

temperatures cause the surface **air** in equatorial regions to warm and become buoyant. Rising away from the source of heat and under the effects of adiabatic cooling, the temperature of the air falls and it becomes less buoyant. Unable to sink back to the surface because of the warm air rising behind it, it is diverted north and south at the altitude of the **tropopause** and reaches between 20 and 30°N and S before beginning to descend. The world's great hot **deserts** are located at these latitudes beneath the descending arms of the Hadley Cells where the subsiding air creates large, warm, sub-tropical **anticyclones**. The stability of these high-pressure systems is so effective in counteracting the processes required to cause **precipitation** that **aridity** is normal. From these anticyclones some of the air returns towards the equator as the north-east and south-east trade winds to complete the circulation. Some of the air moving pole-wards in the upper section of the Hadley Cell is incorporated in the subtropical **jet stream**, which flows in the vicinity of the tropopause above the sub-tropical high-pressure cells.

See also
Adiabatic process.

Further reading
Barry, R.G. and Chorley, R.J. (1992) *Atmosphere, Weather and Climate* (6th edition), London: Routledge.

HADLEY, G.

See **Hadley Cells.**

HALF-LIFE

The time taken for a **radionuclide** to lose one half of its **radioactivity**. For example, a **nuclide** with a half-life of ten years will lose 50 per cent of its original radioactivity after the first ten years, 50 per cent of the remaining radioactivity over the second ten years, and so on. Thus at the end of thirty years, the substance will still contain 12.5 per cent of its original radioactivity. The process involves progressive radioactive decay, in which an unstable **element** releases **ionizing radiation**, while producing a unique sequence of decay products until a stable element remains. ^{238}**Uranium**, for example, ultimately decays into the stable product ^{206}**lead**, but intermediate steps in the process include such elements as ^{234}thorium and ^{214}polonium. The half-life of different nuclides varies from seconds (^{214}polonium at 1.64×10^{-4} seconds) to billions of years (^{238}uranium at 4.5×10^{9} years). With each succeeding half-life, the hazard presented by a substance decreases, but with highly radioactive materials which have half-lives of millions or even billions of years, the hazard persists beyond any human time-scale. Although the concept is most commonly used with reference to radioactive substances, it is also sometimes used to indicate the time taken for pollutants to be expelled from biological systems. Organic

Table H-1 The half-life of selected radionuclides

RADIONUCLIDE		HALF-LIFE
Polonium	^{214}Po	1.64×10^{-4}s
Krypton	^{89}Kr	3.16 min
Radon	^{222}Rn	3.83 days
Strontium	^{90}Sr	28.5 years
Radium	^{226}Ra	1.6×10^{3} years
Carbon	^{14}C	5.73×10^{3} years
Uranium	^{238}U	4.47×10^{9} years
Indium	^{115}In	4.41×10^{14} years

Source: After Cutnell and Johnson (1995)

mercury (Hg), for example, has a half-life of seventy days in the human body.

Further reading
Cutnell, J.D. and Johnson, K.W. (1995) *Physics* (3rd edition), New York/Toronto: Wiley.
Das, A. and Ferbel, T. (1994) *Introduction to Nuclear and Particle Physics*, New York: Wiley.

HALOGENATED HYDROCARBONS

Hydrocarbons into which **chlorine** (Cl), **bromine** (Br) and/or **fluorine** (F) – the halogen elements – have been introduced. In some cases (for example, **chlorinated hydrocarbons**) the **halogens** are added to the hydrocarbons, whereas in others (**chlorofluorocarbons**, for example), the halogens replace the **hydrogen** (H) **atoms** in hydrocarbons such as ethane and **methane** (CH_4). Halogenated hydrocarbons such as organochloride **pesticides** have caused major problems in the terrestrial and aquatic environments and chlorofluorocarbons have been responsible for **ozone depletion** and the environmental problems associated with it.

See also
Halon, Organochlorides.

HALOGENS

A group of highly reactive, toxic **elements** – **fluorine** (F), **chlorine** (Cl), **bromine** (Br), **iodine** (I) – which occupy Group 7A in the **periodic table**. They react readily with **metals** to form **salts** (for example, sodium chloride) and with organic compounds to form a variety of complex chemicals (for example, **organochlorides, chlorofluorocarbons**). The salts, or halides, are highly soluble in **water** and are common constituents of sea water.

Further reading
Gutmann, V. (ed.) (1967–1968) *Halogen Chemistry* (3 vols), New York: Academic Press.
Zumdahl, S.S. (1993) *Chemistry* (3rd edition), Lexington, MA: D.C. Heath.

HALONS

Synthetic **organic compounds** containing bromine (Br), commonly used in fire extinguishers. With **ozone depletion potentials** between 3 and 10, they are more effective in destroying the **ozone layer** than **chlorofluorocarbons**. A ban on halons was implemented in 1994.

See also
Bromofluorocarbons.

HAMBURGER CONNECTION

The link between the clearing of the **equatorial rainforest** and fast food production in the developed nations, particularly the United States. The forests are cleared to provide grazing land for cattle and the meat produced is exported to earn revenue. The direct environmental damage caused by **deforestation** is compounded by the loss of the **carbon** (C) **sink** provided by the trees, and an increase in **methane** (CH_4) production from the cattle. The term applies to beef from Central America, but not that from Amazonia, which is prevented from entering the United States for health reasons. The 'hamburger connection' has been opposed, with some success, by environmental groups such as the **Rainforest Action Network**.

Further reading
Hecht, S. (1990) 'The sacred cow in the green hell', *The Ecologist* 20: 229–35.
Myers, N. (1981) 'The hamburger connection: how Central American forests become North America's hamburgers', *Ambio* 10: 3–8.

HARD WATER

See **water quality**.

HARAPPAN CIVILIZATION

A **water**-based civilization which developed in the Indus valley some 5000 years ago. Like the Egyptian and Mesopotamian civilizations, it consisted of a number of city states supported by agriculture on the fertile river **floodplains**. Its decline around 1800 BC was once attributed to invasion by adjacent nomadic tribes, but current evidence suggests that **drought** was a major contributing factor.

See also
Agrarian civilizations.

Further reading
Calder, N. (1974) *The Weather Machine and the Threat of Ice*, London: BBC Publications.
Singh, G., Joshi, R.D., Chopra, S.R. and Singh, A.B. (1974) 'Late-Quaternary history of the vegetation and climate of the Rajasthan Desert, India', *Philosophical Transactions of the Royal Society*, B 267: 467–501.

HARMATTAN

A hot, dry, dusty **wind** which blows out of the Sahara Desert over the **Sahel** and much of West Africa during the northern winter. It brings **continental tropical air** southwards which contributes to the **seasonal drought** characteristic of the area. The Harmattan (or the Doctor) brings a welcome respite from the hot, humid, uncomfortable and unhealthy conditions associated with the **maritime tropical air mass** that covers the region for the remainder of the year, but the **dust** which accompanies it causes transport and communications problems.

Further reading
Adetunji, J., McGregor, J. and Ong, C.K. (1979) 'Harmattan haze', *Weather* 34: 430–6.

HAZARDOUS WASTE

See **waste classification.**

HAZE

Fine **aerosols**, $< 1\mu m$ in diameter, held in suspension in the **atmosphere**. Haze includes **dust** and **salt** particles produced by natural and human activities as well as **hydrocarbons** such as **terpenes** released by vegetation. Although individually the constituents of haze are invisible, together they can reduce visibility, particularly if they act as **condensation nuclei** and attract moisture. Haze aerosols scatter light and lead to variations in the colour of the atmosphere, particularly at sunrise and sunset.

See also
Arctic Haze.

HEAT BUDGET

A measure of the net transfer of **energy** from one body or **system** to another as a result of **temperature** differences between them. The transfer takes place through **conduction, convection, radiation** and **evaporation**. It applies to both animate and inanimate objects with the ideal situation being a balanced budget in which the heat energy entering a body or system is balanced by the amount leaving. In homeothermic animals such as humans, for example, a balanced heat budget is important for the maintenance of their characteristic stable body temperature. The relationship may be represented as follows:

$$S = (M - W) + (\pm R \pm C - E)$$

where   S   = rate of heat storage in body

M   = rate of total metabolic energy production

W   = rate of external work performed, expressed as its heat equivalent

R   = net rate of heat exchange by radiation between the body and its environment

C   = net rate of heat exchange by conduction-convection between the body and its environment.

E   = rate of heat loss by evaporation-convection to the environment

(Mather 1974)

To maintain equilibrium, S should be zero. It may vary from that by small amounts for limited periods of time, but a positive budget will ultimately lead to heat stroke, whereas a negative budget, in which heat loss exceeds heat gain, may lead to **hypothermia**. On a larger scale, the heat budget concept can be applied to the entire earth/atmosphere system. The budget is determined by comparing the input of **solar radiation** with the output of **terrestrial radiation**, which can be represented by the equation:

$$Q^* = K^* + L^*$$

where Q^* = net all-wave radiation flux
density

K^* = net short wave radiation

L^* = net long wave radiation

(Oke 1987)

Q^* and L^* may be positive or negative, but K^* is either positive or zero (for example, at night), never negative. Balance is achieved when $Q^* = 0$. Any imbalance would cause the earth/atmosphere system to become either hotter or colder than normal. In practice, the earth/atmosphere system's heat budget is extremely complex – for example, it includes storage elements which slow the heat transfer – and seldom balances over the short term. The complexity in part reflects the great variety of sub-systems in the **environment**. The heat budget of tropical regions differs from that in high latitudes, for example; the

budgets of **oceans** and continents are also quite different and a specific area can have variations in its heat budget from season to season. Geological and palaeoclimatic evidence suggests that even over the longer term the system remains in a state of flux. The disruption of the environment by human activities also interferes with the heat budget. At a local scale, **deforestation**, urbanization and the creation of **reservoirs**, for example, cause a change in the local heat flow. On a larger scale, increased **atmospheric turbidity** and the depletion of the **ozone layer** alter the flow of incoming solar radiation while the enhancement of the **greenhouse effect** alters the outward flow of terrestrial radiation, all of which have the potential to disrupt the heat budget.

See also
Energy budget.

Further reading
Barry, R.J. and Chorley, R.G. (1992) *Atmosphere, Weather and Climate*, (6th edition), London: Routledge.
Mather, J.R. (1974) *Climatology: Fundamentals and Applications*, New York: McGraw-Hill.
Oke, T.R. (1987) *Boundary Layer Climates* (2nd edition), London: Routledge.

HEAT ISLAND

See **urban heat island**.

HEAVY METALS

Metals such as **mercury** (Hg), **lead** (Pb), **tin** (Sn) and **cadmium** (Cd) that have a relatively high atomic weight. Problems arise when they are converted into a soluble organic form or concentrated by hydrological or biological processes until they become hazardous to natural **ecosystems** and human health. The mobilization of heavy metals in areas suffering from **acid rain** may contribute to the decline of aquatic biota. Methyl mercury, a soluble mercury **compound**, has been linked to nervous disorders and lead poisoning is a health problem of long standing. Although heavy metals are found naturally in the **environment**, the main **pollution** problems arise from their release from mineral and metal processing and industrial activities. In Brazil, for

Figure H-2 The components of a human scale heat budget

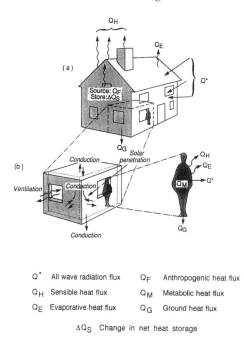

Q^*	All wave radiation flux	Q_F	Anthropogenic heat flux
Q_H	Sensible heat flux	Q_M	Metabolic heat flux
Q_E	Evaporative heat flux	Q_G	Ground heat flux

ΔQ_S Change in net heat storage

Source: After Oke, T. (1978) *Boundary Layer Climates*, London: Methuen

example, rivers have been polluted with mercury compounds used in **gold** (Au) mining and in Japan in the 1950s, Minamata Bay was polluted by mercury compounds released in the effluents from **plastics** factories.

See also
Minamata disease.

Further reading
Fergusson, T.G. (1990) **The Heavy Metals: Chemistry, Environmental Impact and Health Effects**, Oxford/New York: Pergamon Press.

HEAVY WATER

A form of **water** in which the **hydrogen** (H) **atoms** have been replaced by **deuterium**. Heavy water, or deuterium oxide (D_2O), is present in natural water in a ratio of about one part in 6500 and is separated from it by fractional **distillation** or **electrolysis**. Heavy water is used in the laboratory as a tracer in chemical reactions, and on an industrial scale is commonly used as a moderator and **coolant** in **nuclear reactors**.

See also
CANDU.

Further reading
Miller, A.E. and van Alstyne, H.M. (1994) *Heavy Water: A Distinctive and Essential Component of CANDU*, Chalk River, Ont: Chalk River Nuclear Laboratories, Atomic Energy of Canada Limited.
Ramage, J. (1983) *Energy: A Guidebook*, Oxford: Oxford University Press.

HERBICIDES

Chemicals used to kill plants or inhibit their growth. They are sprayed on the foliage or introduced into the root systems. General or broad-spectrum herbicides are toxic to all plants, whereas selective herbicides target only specific plants. Herbicides take many forms, that differ widely in their effectiveness and environmental impact. Inorganic chemicals such as common salt (**sodium chloride** (NaCl)) are simple and safe general herbicides, while the more complex chlorinated phenoxyacetic acids, such as 2,4-D and 2,4,5-T, which mimic the structure of plant growth **hormones**, are very effective

selective herbicides. Phenoxy herbicides are routinely used in forest regeneration to kill off the unwanted **species** that compete with the young seedings, and for many years thousands of tonnes were used in North America during the growing season to rid lawns of unwanted dandelions and other weeds. Although very effective, these synthetic herbicides come with an environmental price. Many are now recognized as **carcinogens**, and 2,4,5-T has been banned for most uses in the United States since 1979. Both 2,4-D and 2,4,5-T were constituents of **Agent Orange**, a defoliant used by the US forces in Vietnam. Their contamination by **dioxin**, a potent carcinogen at very low concentrations, made them even more hazardous than ususal. Environmental problems have led to the restricted use of many herbicides and the development of less hazardous replacements. New urea-based herbicides, for example, which work by inhibiting **photosynthesis**, are less toxic to animals and less persistent in the **environment**. Continued research into new herbicides is also required because of the development of resistance in many weeds to existing herbicides.

See also
Carbamates, Organochlorides.

Further reading
Grower, R. (ed.) (1988) *Environmental Chemistry of Herbicides*, Boca Raton, FL: CRC Press.

HERBIVORE

A plant-eating organism. The more obvious herbivores are mammals, but the group includes a wide range from insects such as aphids through small mammals such as rabbits and grazing animals such as deer, to the largest of the land mammals, the elephant. Herbivores are considered primary consumers, directly dependent on the plant **species** that occupy the base of all **food chains**. Some insect herbivores have developed a dependence upon particular plant species, and adaptations among grazing mammals include specialized teeth and complex digestive systems necessary to enable them to digest plant material. Domesticated herbivores such as sheep and cattle are an important component of agriculture in many parts of the world. Through **overgrazing**, the production of **waste** and the release of **methane** (CH_4) from their digestive systems they contribute to a number of environmental problems.

Further reading
Crawley, M.J. (1983) *Herbivores: The Dynamics of Animal–Plant Interactions*, Berkeley: University of California Press.

HETEROGENEOUS CHEMICAL REACTIONS

Chemical reactions which take place on the surface of the **ice** particles that make up **polar stratospheric clouds**. These clouds contain nitric and hydrochloric acid particles which through a complex series of reactions ultimately release **chlorine** (Cl) into the **stratosphere**. In the form of **chlorine monoxide** (ClO) it then attacks the **ozone layer**. Similar reactions have been identified on the surface of stratospheric **sulphate particles** such as those released during the eruption of **Mount Pinatubo** in 1991.

Figure H-3 Heterogeneous chemical reactions in the Antarctic stratosphere

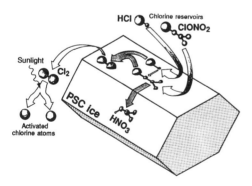

Source: Turco, R.P. (1997) *Earth under Siege*, Oxford/New York: Oxford University Press

Further reading
Keys, J.G., Johnston, P.V., Blatherwick, R.D. and Murcray, F.J. (1993) 'Evidence for heterogeneous reactions in the Antarctic autumn stratosphere', *Nature* 361: 49–51.

Shine, K. (1988) 'Antarctic ozone – an extended meeting report', *Weather* 43: 208–10.

HIGH-LEVEL RADIOACTIVE WASTE

Waste material produced by the nuclear industry, it includes spent **fuel** and reprocessing waste. High-level **nuclear waste** is hot and intensely radioactive when produced, and its disposal must take both of these factors into account. Much of the heat is lost during the first decade of storage, but many of the **fission** products present in the waste have long **half-lives** and it will remain radioactive for tens of thousands of years. The bulk of the high-level waste currently being produced by **nuclear reactors** is stored on site in **water**-filled pools, steel drums or concrete silos. As the amount of waste accumulates – by the year 2000, the United States alone will have close to 40,000 tonnes in temporary storage – some more permanent form of disposal will be required. Land-based disposal in deep mines or **boreholes** has been examined by a number of governments, while others have considered deep ocean burial as a possibility. Whatever method is chosen, the challenge will be to provide protection for the **environment** over the thousands of years required for the waste to lose its **radioactivity**.

See also
Actinides, Half-life, Nuclear waste.

Further reading
Brookins, D.G. (ed.) (1987) *Geological Disposal of High-Level Radioactive Wastes*, Athens: Theophrastus Publications.
Edwards, R. (1996) 'Sellafield's Trojan Horse', *New Scientist* 149 (2011): 11–12.
Murray, R.L. and Powell, J.A. (1988) *Understanding Nuclear Waste*, Columbus, OH: Battelle Press.

HOLOCENE

The most recent or postglacial part of the **Quaternary** period which began some 11,000 years ago, following the decay of the **Pleistocene** ice sheets. It has been characterized by considerable environmental change brought about mainly by climatic variability. The disappearance of the ice at the end of the Pleistocene was followed by the continued amelioration of climate that peaked about 5000 years ago in the Climatic Optimum, allowing animals and vegetation to migrate back to the areas recently vacated by the ice. At about the same time, the return of water to the oceans from the melted ice sheets caused a major marine transgression. Since then, climate has included a considerable number of fluctuations, the best developed being another warm spell in the early Middle Ages between c. AD 850 and 1200 – the Little Optimum – and a period of deterioration, the so-called Little Ice Age, between AD 1550 and 1850. Each of these fluctuations was accompanied by measurable variations in such environmental elements as the distribution of flora and fauna, particularly in higher latitudes in the northern hemisphere. Human activities also felt the impact of these variations – for example, through agriculture. In the latest part of the Holocene, the disruption of the environment by human activities appears to have the potential to bring about climate change, through, for example, the enhancement of the greenhouse effect.

Further reading
Roberts, N. (1989) *The Holocene: An Environmental History*, Oxford: Blackwell.
Mannion, A.M. (1991) *Global Environmental Change: A Natural and Cultural Environmental History*, Harlow, Essex: Longman.

HORMONE

An **organic compound** which controls specific metabolic functions in an organism. Produced in only small quantities at a limited number of sites within an organism, hormones are transported to target areas elsewhere in the organism as required. In animals, for example, hormones are produced by ductless endocrine glands, such as the thyroid and pituitary glands, released into the bloodstream and carried to the appropriate parts of the body. In plants, the hormones are translocated via the vascular tissue. Hormones control the growth, reproduction and development of plants and animals, and their use has become common in

agriculture and industry. Hormones are used to manage the growth and reproduction of domesticated animals, for example, and synthetic **herbicides**, which imitate natural hormones, are commonly sprayed on crops to inhibit the growth of weeds. Hormones are also widely used in medicine. Insulin, for example, is essential for the control of diabetes, while oestrogen, progesterone and testosterone have been used in various quantities and combinations for birth control since the 1960s.

See also
Metabolism.

Further reading
Moore, T.G. (1989) *Biochemistry and Physiology of Plant Hormones* (2nd edition), New York: Springer-Verlag.

HUMAN DIMENSIONS OF GLOBAL ENVIRONMENTAL CHANGE PROGRAM (HDP)

A programme set up by the International Social Science Council (ISSC) in 1988 to examine the dynamics of the inter-relationships between society and the global environment. A better understanding of such relationships will allow the identification and initiation of broad social strategies aimed at preventing or mitigating the undesirable impacts of global change. The HDP has developed a research programme to address the relationships between the environment and such human factors as population size, density, growth rate and social structure, production/consumption patterns and technological developments. Data from such studies will permit the analysis of policy options available for dealing with global environmental change while pursuing the goal of sustainable development.

HUMIDITY

A measure of the amount of **water vapour** in the **atmosphere**. **Absolute humidity** refers to the total **mass** of water in a given volume of air; **specific humidity** is the ratio of the mass of water vapour in the air to the combined mass of the water vapour and the air; **relative humidity** is the amount of water vapour in the air compared to the amount of water vapour the air can hold at that **temperature**. In popular usage the terms 'humidity' and 'relative humidity' are commonly considered as synonymous.

Further reading
Barry, R.G. and Chorley, R.J. (1992) *Atmosphere, Weather and Climate* (6th edition), London: Routledge.

HUMUS

Partially decomposed organic matter that is an essential component of fertile **soil**. Humus is dark brown in colour, colloidal in structure and rich in **bacteria** and **fungi**. It aids the retention of **soil moisture**, helps to bind soil particles together and bonds with **nutrients**, keeping them in the upper layers of the soil where they are easily accessible to growing plants. The humus-rich, black-earth soils (chernozems) of temperate **grassland** areas such as the midwestern United States were among the most fertile in the world when they were first brought under cultivation.

See also
Clay-humus complex, Colloids.

Further reading
Stevenson, F.J. (1994) *Humus Chemistry: Genesis, Composition, Reactions* (2nd edition), New York: Wiley.

HUNTING AND GATHERING

A pre-agricultural form of subsistence in which societies obtained food from the natural **environment** by hunting animals, catching fish and collecting fruits, nuts and seeds. Hunting and gathering societies continue to exist, but few, if any, are untouched by modern developments. They may use **iron** tools or firearms, for example, as well as modern methods of transportation, and participate in the money economy by selling the products of the hunt and buying manufactured goods. The impact of hunting and gathering societies on the environment was limited, and they are often seen as living

in harmony with the environment, taking from it only what they needed. That suggests some knowledge of the working of the environment and some degree of control over it. In reality, although these primitive societies understood the environment well enough to survive, they were most likely dominated by it, and their development was very much limited by what it allowed them to do. However, being few in number, and operating at low **energy** levels with only basic tools, they did little to alter their environment.

Further reading
Bettinger, R.L. (1991) *Hunter-Gatherers: Archaeology and Evolutionary Theory*, New York: Plenum Press.
Burch, E.S. and Ellanna, L.J. (eds) (1994) *Key Issues in Hunter-Gatherer Research*, Oxford/Providence: Berg.

HURRICANE

A cyclonic storm which forms over the tropical North Atlantic Ocean with sustained **wind** speeds in excess of 64 knots (120 km/hr).

See also
Tropical cyclones.

Further reading
Robinson, A. (1993) *Earthshock: Climate Complexity and the Forces of Nature*, London: Thames and Hudson.

HYDRATION

The addition of **water** or its constituents to a chemical compound.

See also
Anhydride, Dehydration.

HYDROCARBONS

Organic compounds composed of **hydrogen** (H) and **carbon** (C) bound together in chains or rings. They may be **solid**, **liquid** or **gas**. Hydrocarbons, such as **petroleum** and **natural gas**, are used mainly as **fuels**, but are also suitable for use as lubricants and as feedstocks for a variety of industrial materials.

Figure H-4 The chemical structure of simple and complex hydrocarbon

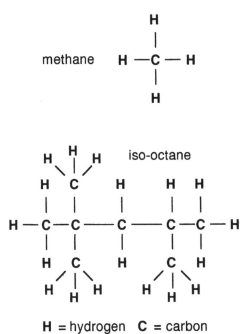

H = hydrogen   C = carbon

The **plastics** and **fertilizer** industries, for example, depend very much on products derived from hydrocarbons. As fuels they contribute to a number of environmental concerns. When burned in the presence of **oxygen** (O), hydrocarbons release thermal energy plus **water vapour** and **carbon dioxide** (CO_2), both of which are **greenhouse gases**:

e.g. $C_5H_{12} + 8O_2 \longrightarrow 5CO_2 + 6H_2O$
        (pentane)

If **combustion** is incomplete or inefficient, poisonous **carbon monoxide** (CO) will also be produced. Most of the liquid hydrocarbons contain few impurities, but if **sulphur** (S) is present combustion will result in the release of **sulphur dioxide** (SO_2) into the **atmosphere** to contribute to **acid rain**. Impurities such as sulphur can be removed from the liquid and gaseous hydrocarbons through refining, but the problem with carbon dioxide remains. Some relief can be obtained by using hydrocarbons in which the proportion of **carbon** (C) to **hydrogen** (H) is low. In that respect, natural gas is the best

fuel with a C:H ratio of 1:4 (e.g. CH_4 – **methane**), compared to petroleum with a ratio of *c.*1:2 (e.g. C_8H_{18} – octane), and **coal** which has a ratio of no more than 1:1.

See also
Carbon tax, Fossil fuels.

Further reading
Schobert, H.H. (1991) *The Chemistry of Hydrocarbon Fuels*, London: Newnes.

HYDROCHLOROFLUOROCARBONS (HCFCs)

A widely used substitute for **chlorofluorocarbons** (CFCs) consisting of **hydrogen** (H), **fluorine** (F), **chlorine** (Cl) and **carbon** (C). Being less stable than CFCs, hydrochlorofluorocarbons begin to break down in the **troposphere** before they can diffuse into the **ozone layer**. HCFCs have **ozone depletion potentials** (ODP) which range between 0.16 and 0.016 and limited atmospheric lifetimes of one to twenty years. As a result, they are about 95 per cent less damaging than normal CFCs, but still have a negative impact on the ozone layer. They are considered transitional chemicals between the major ozone destroyers and ozone-friendly replacements, and are to be phased out progressively by 2030 under the terms of a series of agreements arising out of the **Montreal Protocol**. Continuing reassessment of the ozone issue, however, makes it likely that deadlines will be subject to regular revision.

See also
Hydrofluorocarbons.

Further reading
Kanakidou, M., Dentener, F.J. and Crutzen, P.J. (1995) 'A global three-dimensional study of the fate of HCFCs and HFC-134a in the troposphere', *Journal of Geophysical Research* 100: 18781–801. Turco, R.P. (1997) *Earth Under Siege: From Air Pollution to Global Change*, Oxford/New York: Oxford University Press.

HYDROELECTRICITY

Electricity produced by using the **kinetic energy** available in flowing **water**. Where the gradient of a stream is steep or a natural waterfall exists, the water can be directed through a turbine to drive an electric generator. Damming the stream to create a **reservoir** allows the continued generation of electricity, even when the natural streamflow is low. Hydroelectric **systems** are characterized by their efficiency, which can reach as high as 90 per cent in some plants compared with as little as 30–40 per cent in equivalent thermal or steam-driven plants. Hydro systems can be brought on stream much more rapidly than thermal plants, and are therefore able to respond more effectively to changes in demand. Modern developments in the provision of hydroelectricity have involved the construction of large integrated systems, producing in excess of one gigawatt (1×10^9 **watts**) of power (see Table H-2) and providing **irrigation, flood** control and recreational facilities. Small-scale plants are also being brought into production. These microhydro units with outputs of as little as 100 kilowatts (100×10^3 watts) are being installed to provide electrical energy for local or on-site use with any excess being sold to the regional utility. Hydroelectricity has a number of environmental advantages over electricity produced thermally. Since no **fossil fuels** are burned, there is no **pollution** associated with the production of hydro-

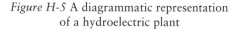

Figure H-5 A diagrammatic representation of a hydroelectric plant

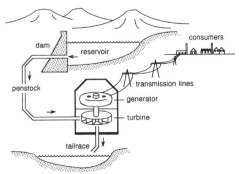

electricity. It is also much more efficient, since it uses a **renewable energy** source, with no loss of energy to the **environment** through the heating and cooling of water. It is not completely problem-free, however. The building of **dams** disrupts the local **hydrology**, for example, and the reservoirs created may take agricultural land out of production, or displace people. Because many of the best hydro sites are also areas of scenic beauty there may be conflict between the needs of tourists and the developers. The James Bay Project in Canada, the Itaipu Dam in Brazil and the Three Gorges development under construction in China have been criticized because of such problems.

Table H-2 Examples of large scale hydroelectric power schemes

PROJECT			CAPACITY (Gigawatts)
Cabora Bassa	Zambezi River	Mozambique	1.2
Kariba	Zambezi River	Zimbabwe/Zambia	1.4
Churchill Falls	Churchill River	Canada	5.25
Tucurui	Tocantins River	Brazil	8.0
James Bay	La Grande River	Canada	10.0
Guri	Caroni River	Venezuela	10.0
Itaipu	Parana River	Brazil/Paraguay	12.6
Three Gorges (under construction)	Yangtse River	China	13.0
Altamira	Xingu River	Brazil	18.0

Source: Various

Hydropower currently provides between 20 and 25 per cent of the world's electrical energy, and there is considerable potential for further development worldwide, but particularly in Africa, Asia and South America.

See also
Dams, Reservoirs – environmental effects.

Further reading
Deudney, D. (1981) *Rivers of Energy: The Hydropower Potential*, Washington DC: Worldwatch Institute.
Fritz, J.J. (ed.) (1984) *Small and Mini Hydropower Systems: Resource Assessment and Project Feasibility*, New York: McGraw-Hill.
McCutcheon, S. (1992) *Electric Rivers: The Story of the James Bay Project*, New York: Paul.

HYDROFLUOROCARBONS (HFCS)

Organic chemicals containing **hydrogen** (H), **fluorine** (F) and **carbon** (C), which are seen as appropriate substitutes for **CFCs** and **HCFCs** since they contain no **chlorine** (Cl) and have an **ozone depletion potential** (ODP) of zero. Hydrofluorocarbons are very effective **greenhouse gases**, however, with the potential to contribute to **global warming** as their use increases.

HYDROGEN (H)

A colourless, odourless, highly flammable **gas**. With a **mass number** of 1 it is the lightest **element**, normally existing in a diatomic molecular form (H_2). It does not exist free in the **environment**, but is commonly found in combination with other elements such as **oxygen** (O) and **carbon** (C) as **water, carbohydrates** or **hydrocarbons**. Two other **isotopes** of hydrogen are known – **deuterium** with a mass number of 2 and **tritium** with a mass number of 3. These 'heavy' hydrogens occur in only small amounts in the environment, but both have been developed for use in the nuclear industry. Free hydrogen is manufactured by the action of steam on water gas – a mixture of **carbon monoxide** (CO) and hydrogen – or **natural gas**, in the presence of a **catalyst**. It can also be produced by the **electrolysis** of water. Hydrogen is used in the synthesis of several industrial chemicals including **ammonia** (NH_3), hydrochloric acid (HCl) and **methanol** (CH_3OH). With two to three times the **energy** content of **gasoline** by weight, hydrogen has the potential to be a major **fuel** source. It has the added benefit that it is non-polluting when burned, producing only water. Being highly flammable, however, and susceptible to explosion under certain conditions, its widespread use would be accompanied by safety concerns. Some of these would be obviated by storing the hydrogen in the form of metal hydride fuel **cells**, from which the gas could be released by heating as required. At present the use of hydrogen as a fuel is not feasible because of the high energy cost – the energy used to produce the hydrogen is greater than that available from the gas itself. Further development seems unlikely unless a cheap supply of the electricity needed for electrolysis can be found.

Further reading
Dinga, G.P. (1988) 'Hydrogen: the ultimate fuel and energy carrier', *Journal of Chemical Education*, 65(8): 688–91.

HYDROGEN BOMB

See **thermonuclear device.**

HYDROGEN ION CONCENTRATION

The number of **hydrogen** (H) **ions** in a litre of a substance. It is used to measure the acidity of a **solution**, and usually represented in the form of a **pH** value.

See also
Acid.

HYDROGEN OXIDES (HO$_X$)

A group of naturally occurring compounds derived from **water vapour** (H_2O), **methane** (CH_4) and molecular **hydrogen** (H_2). They include atomic hydrogen (H), the **hydroxyl radical** (OH) and the perhydroxyl radical (HO_2), referred to collectively as **odd hydrogens**. Although relatively low in volume, they

Figure H-6 The destruction of ozone by hydrogen oxides

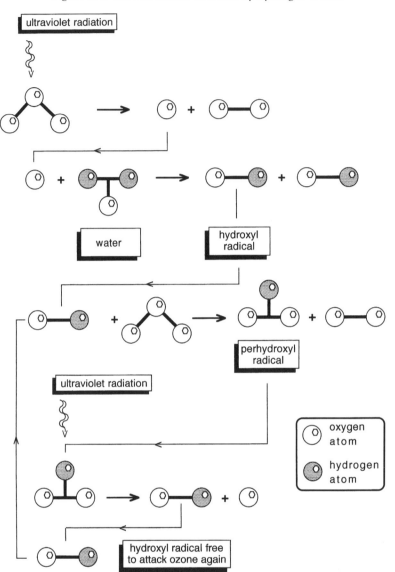

are very effective **ozone** destroyers, because of their participation in **catalytic chain reactions**. The HO_x group may be responsible for about 11 per cent of the natural destruction of ozone in the **stratosphere**. Hydrogen oxides lose their catalytic capabilities when they are converted to **water vapour**.

Further reading
Hammond, A.L. and Maugh T.H. (1974) 'Stratos-
pheric pollution: multiple threats to earth's ozone', *Science* 186: 335–8.

HYDROGENATION

The chemical combination of **hydrogen** (H) with another substance. The process normally requires the addition of heat and pressure plus the presence of a **catalyst**. The hydrogenation of **coal** causes the combination of

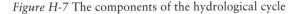

Figure H-7 The components of the hydrological cycle

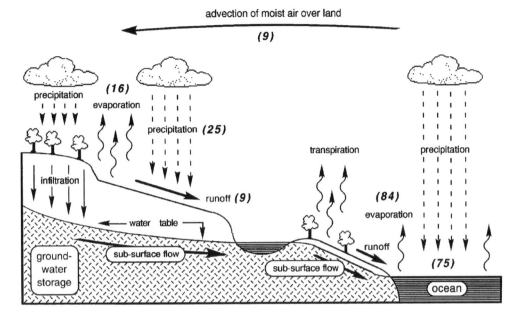

the **carbon** (C) in the coal with **hydrogen** (H) to produce a synthetic **oil** containing a variety of **hydrocarbons**. The process (named the **Fischer-Tropsch process** after its inventors) was developed in Germany in the 1930s, and can produce two barrels of oil and almost 300 m³ of **gas** per ton of coal. During the Second World War, the German war effort depended on the hydrogenation of coal to supply much of the oil it needed. More recently, South African plants have been hydrogenating some 3500 tons of coal per day by this method. Hydrogenation is also used in the food industry to harden **liquid** fats and oils. The production of margarine from animal and vegetable oils, for example, requires hydrogenation.

See also
Destructive distillation.

Further reading
Kleinbach, M.H. and Salvagin, C.E. (1986) *Energy Technologies and Conversion Systems*, Englewood Cliffs, NJ: Prentice-Hall.
Rylander, P.N. (1985) *Hydrogenation Methods*, London: Academic Press.

HYDROLOGICAL CYCLE

A complex group of processes by which **water** in its various forms is circulated through the earth/atmosphere system. It is powered by **solar radiation** which provides the **energy** to maintain the flow through such processes as **evaporation, transpiration, precipitation** and **runoff**. Short- and long-term storage of water in lakes, **oceans**, **ice** sheets and the **ground-water** reservoir is also part of the cycle. Although it is usual to consider the hydro-logical cycle as one all-encompassing system, there are in fact many regional and seasonal variations that may have environmental or societal implications. The relationship between the various **elements** in the cycle in a hot, semi-arid area, for example, is very different from that in a cool, moist location. Similarly, in higher latitudes the difference in the cycle between summer and winter is quite distinct. The human impact on the cycle is mainly in the runoff sector, with water being diverted for domestic, industrial and agricultural uses, but modern society interferes with almost all aspects of the cycle. Evaporation and trans-piration are disrupted by agricultural and forestry practices, **boreholes** and wells allow access to the groundwater system and the construction of **dams** and **reservoirs** creates additional storage. Since the cycle is a **closed system** in material terms, human activities do not deplete the entire system, but excess withdrawal from the runoff or groundwater sectors can create local shortages of water. Most human uses involve only short-term withdrawal from the system, but when the water is returned its quality is often much impaired by a variety of pollutants. The cleansing of such polluted water, mainly through evaporation and **precipitation**, is an important but less well-acknowledged feature of the hydrological cycle.

Further reading
Berner, E.K. and Berner, R.A. (1987) *The Global Watercycle: Geochemistry and Environment*, Englewood Cliffs, NJ: Prentice-Hall.
Briggs, D. and Smithson, P. (1997) *Fundamentals of Physical Geography* (2nd edition), London: Routledge.
Strahler, A.H. and Strahler, A.N. (1992) *Modern Physical Geography* (4th edition), New York: Wiley.

HYDROLOGY

The scientific study of **water** in the earth/atmosphere **system**. It includes not only surface water, but also water in the **atmosphere** and in the **groundwater** system. Physical hydrology focuses on the distribution and circulation of water, while applied hydrology is more concerned with water and human activities, and includes consideration of **water quality, irrigation**, drainage, **erosion** and **flood** control.

See also
Hydrological cycle.

Further reading
Ward, A.D. and Elliot, W.J. (eds) (1995) *Environmental Hydrology*, Boca Raton, FL: CRC/Lewis.
Ward, R.C. and Robinson, M. (1990) *Principles of Hydrology*, Maidenhead: McGraw-Hill.

HYDROLYSIS

The decomposition of a **salt** in **water** to form an **acid** and a **base**. The process is important in the chemical **weathering** of rocks, particularly in the breakup of silicate minerals,

such as the feldspars. The hydrolysis of orthoclase to kaolinite is an example of the process in action.

$$2KAlSi_3O_8 + 2H^+ + 9H_2O \longrightarrow H_4Al_2Si_2O_9 +$$
(orthoclase)   (kaolinite)

$$4H_4SiO_4 + 2K^+$$

Organic compounds can also be decomposed by hydrolysis. **Esters**, for example can be hydrolized into **alcohol** and acid, while the hydrolysis of the **sugars** and **starches** in **carbohydrates** is the starting point for many processes in the food industry. The hydrolysis of organic **waste** has the potential to reduce **waste disposal** problems and recover substances such as **ethanol** which could be recycled by the chemical industry.

Further reading
White, A.F. and Brantley, S.L. (eds) (1995) *Chemical Weathering Rates of Silicate Minerals*, Washington, DC: Mineralogical Society of America.

HYDROMETEOROLOGY

The science that combines **hydrology** and **meteorology**, with its main emphasis on the atmospheric sector of the **hydrological cycle**. The analysis of meteorological elements and events, for example, can be applied to hydrological issues such as flooding, **irrigation** and domestic **water** supply.

HYDROSPHERE

That part of the earth's **crust** covered by **water**, both salt and fresh. It includes the **oceans**, seas, rivers and lakes, that together cover about 74 per cent of the earth's surface. More than 95 per cent of the volume of water in the hydrosphere is in the oceans, with maximum depths in the deep ocean trenches in excess of 10 km. On land, the bulk of the freshwater is located in a relatively small number of large lakes such as Lake Baikal in Siberia and the Great Lakes in North America. Much of the hydrosphere, even that in the middle of the oceans, suffers from some degree of **pollution**.

HYDROXIDE

A **compound** formed when one of the **hydrogen** (H) atoms in **water** is replaced by another **atom** or group. Hydrogen is released in the process,

e.g. $2Na + 2H_2O \longrightarrow 2NaOH + H_2$
      (sodium)   (sodium
                           hydroxide)

Metal hydroxides such as **sodium hydroxide** are **alkalis**.

HYDROXYL RADICAL

See **hydrogen oxides**.

HYPOLIMNION

The colder, denser layer of **water** lying beneath the **thermocline** in a lake.

See also
Epilimnion.

HYPOTHERMIA

The decline of the core body **temperature** (37°C in humans) to dangerously low levels as a result of prolonged exposure to low temperatures and inadequate control of heat loss. This cooling causes reduced metabolic and heart rates and impairs mental functions because of reduced blood flow to the brain. When hypothermia threatens, the body responds by restricting blood flow to the extremities where heat loss is greatest, which may lead to tissue damage in the form of frostbite. Treatment for hypothermia involves the slow raising of the body's temperature, but in severe cases, (for example, core temperature below 26–27°C) this may be insufficient to prevent death.

See also
Heat budget.

Further reading
Pozos, R.S. and Wittmers, L.G. (eds) (1983) *The Nature and Treatment of Hypothermia*, Minneapolis: University of Minnesota Press.

HYPSITHERMAL

See **Climatic Optimum**.

I

ICE

The solid state of **water**. It occurs normally when the **temperature** of the water falls below its freezing point (0°C), but is also formed by the compaction and recrystallization of **snow**. Ice crystals form in the **atmosphere** through the **condensation** and freezing of **water vapour** around **condensation nuclei**.

ICE AGES

Periods in the geological history of the earth when **glaciers** and ice sheets covered large areas of the earth's surface. Ice ages occurred in series, separated by periods of temperate conditions called **interglacials**. Geological evidence indicates that major ice ages occurred in Precambrian times (>570 million years

Figure I-1 The distribution of ice during the most recent ice age: 20,000 to 18,000 years ago

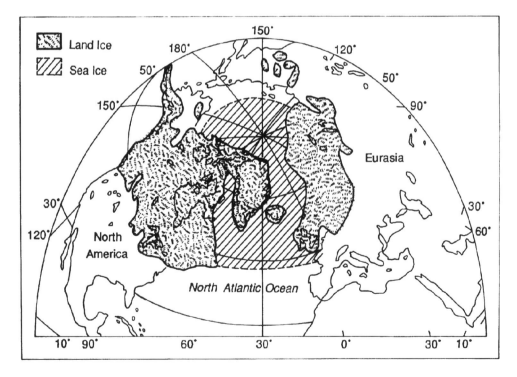

Source: Moore, P.D., Chaloner, B. and Stott, P. (1996) *Global Environmental Change*, Oxford: Blackwell Science

ago), and during the Cambrian, Ordovician and Permo-Carboniferous periods of the geological time-scale. The most recent series, the **Quaternary** glaciations, began some 2.5 million years ago and persisted until 10,000 years ago causing major disruption of land-forms, drainage, animal communities and vegetation. During that time period there may have been as many as twenty separate ice ages and some observers consider current conditions to represent an **interglacial** rather than a postglacial period. Although ice ages are brought about by changing climatic conditions, the exact mechanisms involved remain a matter of controversy. Theories put forward to explain the causes of the ice ages range from **sunspot cycles** and changing **solar energy** output to volcanic activity and changes in the earth's orbit (such as those postulated by **Milankovitch**), all of which have the ability to disrupt the earth's **energy budget** and the potential to initiate cooling.

See also

Interstadial, Little Ice Age, Milankovitch cycles.

Further reading

Dawson, A.G. (1992) *Ice Age Earth*, London/New York: Routledge.
Flint, R.F. (1971) *Glacial and Quaternary Geology*, New York: Wiley.
Harvey, L.D. (1988) 'Climatic impact of ice-age aerosols', *Nature* 334: 333–5.
Mannion, A.M. (1991) *Global Environmental Change: A Natural and Cultural Environmental History*, Harlow, Essex: Longman.
Sharp, R.P. (1988) *Living Ice: Understanding Glaciers and Glaciation,* New York: Cambridge University Press.

ICE CORES

Columns of **ice**, usually up to 100 m long, removed from ice sheets or glaciers and analysed to provide evidence of past environ-mental conditions. Cores have been taken from mountain glaciers and ice sheets around the world, but the most significant inform-ation has been obtained from the thick ice sheets of Greenland and Antarctica. The latter provide information on the mass balance of the world's major ice sheets and the changing chemistry of the **atmosphere**. The presence of **dust** layers indicates periods of increased

atmospheric turbidity associated with volcanic activity. Chemical analysis reveals the level of atmospheric acidity in the past and variations in the **oxygen** (O) **isotope** content of the ice allows past climates to be reconstructed as far back as 150,000 years.

Further reading

Delmas, R.J. (1991) 'Environmental information from ice cores', *Reviews of Geophysics* 30 (1): 1–21.
Mosely-Thompson, E. and Thompson, L.G. (1994) 'Dust in polar ice sheets', *Analysis* 22 (8): 44–6.
Robock, A. and Free, M.P. (1995) 'Ice cores as an index of global volcanism', *Journal of Geophysical Research* 100: 11549–67.

IGNEOUS ROCK

One of the three fundamental groups of rock that make up the earth's **crust**. Igneous rocks are formed by the solidification of molten **magma** at the earth's surface (extrusive origin) or within the crust (plutonic origin). They are generally crystalline in texture, larger **crystals** being more common in rock such as granite which has cooled slowly within the crust, whereas smaller crystals are typical of rock such as basalt which has cooled more rapidly on the surface. Because of their acidic composition, areas underlain by igneous rocks such as granite are particularly susceptible to **acid rain**.

See also

Lava, Metamorphic rock, Sedimentary rock, Volcano.

Further reading

McBirney, A.R. (1993) *Igneous Petrology* (2nd edition), Boston, MA: Jones and Bartlett.

IGNITION

The initiation of the process of **combustion**. It may take place slowly, by the application of an open flame, for example, or rapidly through the production of a high **energy** spark.

See also

Ignition temperature.

IGNITION TEMPERATURE

The **temperature** to which a substance must be heated to produce **combustion**.

IMPERMEABLE

Having a structure that does not allow the passage of **liquids** or **gases**. In the **natural environment**, the presence of impermeable materials has implications for **water** supply, **pollution** control and **petroleum** production. The presence of an impermeable layer may prevent or restrict the normal flow of **groundwater**, for example, and reduce the recovery rate of a water well. Natural or constructed layers of impermeable materials such as **clay** or **plastic** sheeting are commonly used to prevent the migration of pollutants from disposal sites into adjacent streams, lakes or groundwater systems. Because of their ability to restrict the sub-surface flow of liquids and gases, the nature and distribution of impermeable rocks exerts an important influence on the size and location of petroleum deposits. Industrially produced impermeable materials such as **metals, plastics** and rubber products are important for the provision of containers, pollution control devices, protective clothing and health products.

See also
Permeability, Porosity.

IMPERVIOUS

Impermeable.

See also
Pervious.

INCINERATION

The disposal of **waste** material by burning. Incineration was one of the earliest forms of **garbage** disposal, and small backyard incinerators in the form of old, perforated oil drums are still common in many parts of the United States. Incineration fell into disfavour in the 1970s and 1980s because of the **gases** and **particulate matter** that it released into the **atmosphere**. However, it does have certain advantages, and the provision of systems to prevent the emission of acid gases, toxic materials and particulate matter has led to its reconsideration as a **waste disposal** system. Municipal waste is frequently buried in **sanitary landfill** sites, but as the availability of suitable locations declines, incineration is increasingly being considered as an alternative. Since municipal waste commonly includes as much as 40 per cent combustible material, incineration would considerably reduce the bulk of the waste. In addition, the heat released can be used to produce steam and assist the generation of electricity, although the calorific value of waste is much lower than that of **coal**. The old disadvantages associated with the emissions of acid gases and particulate matter are now under control, but problems with toxic gases, produced as a result of incomplete combustion, still remain. The ash produced can also create problems, since it may contain **heavy metals** and other toxic products that require special storage or additional treatment before disposal. Incineration is particularly effective in dealing with hazardous organic wastes such as those produced by hospitals or by industry. Special incinerators have been designed to meet the requirements of specific waste products, and even the most hazardous products – for example, **polychlorinated biphenyls** (PCBs) and **dioxin** – can be disposed of safely by incineration. The United States **Environmental Protection Agency** (EPA) requires a 99.9999 per cent efficiency before granting permits for the operation of such in–cinerators. Such specialized devices also have the advantage that they are usually mobile and can be taken to the waste source, thus reducing the dangers associated with the transportation of **hazardous waste**.

See also
Recycling.

Further reading
Hester, R.E. and Harrison, R.M. (eds) (1994) *Waste Incineration and the Environment*, Cambridge: Royal Society of Chemistry.
Miller, G.T. (1994) *Living in the Environment: Principles, Connections and Solutions* (8th edition), Belmont, CA: Wadsworth.

INDEX CYCLE

See **zonal index.**

INDUSTRIAL REVOLUTION

A period of rapid transition from an agricultural to an industrial society, beginning in Britain in the mid-eighteenth century and spreading to other parts of the world in the next two hundred years. It was characterized by a major expansion in the use of **coal** as a **fuel,** in the **steam engine** and in the **iron** industry. Together, these encouraged the growth of new industrial cities incorporating heavy industries based on coal, iron and steel, as well as an expanding textile industry powered by the new steam engines. Railways and steamships linked these cities with their sources of raw materials and their markets. **Population** grew rapidly, providing the necessary labour force and also

Figure I-2 Improved metal working and engineering techniques associated with the Industrial Revolution allowed the creation of such structures as this late nineteenth-century railway bridge over the River Forth in central Scotland, still used by mainline traffic more than 100 years later

Photograph: The author

creating a growing consumer demand. Although the Industrial Revolution was a period of great technical achievement and economic development, it also marked the beginning of increasingly serious environmental deterioration. The rapidly growing population, new urbanization and industrialization created local and regional environmental stress through such elements as **sewage** disposal, mineral extraction, **energy** conversion and the sprawl of urban-industrial activities over the adjacent rural land. Current environmental issues such as **acid rain, global warming, atmospheric turbidity** and water **pollution** have their roots in activities initiated or expanded during the Industrial Revolution. Industrialization based on coal continues in China and India, creating the potential for further environmental deterioration in these areas.

See also
Acid mine drainage, Pollution, Smog.

Further reading
Alfrey, J. and Clark, C. (1993) *The Landscape of Industry,* London: Routledge.
Ashton, T.S. (1967) *The Industrial Revolution – 1760–1830,* London: Oxford University Press.
Hudson, P. (1992) *Industrial Revolution,* London: Edward Arnold.
Morris, R.J. and Langton, J. (1986) *Atlas of Industrializing Britain,* London: Routledge.

INERT (GASES)

Substances that are inert do not participate readily in chemical reactions. Few substances are completely inactive. Even some of the **noble gases** – argon, neon, helium, krypton, radon and xenon – once thought to be completely inert are now known to produce **compounds** with other chemicals. The stability of a substance may change with environmental conditions. **Nitrogen** (N), the commonest gas in the **atmosphere,** is effectively inert under normal atmospheric conditions. It acts as a dilutant for atmospheric **oxygen** (O), but does not become chemically involved with it. At high **energy** levels, however, following a flash of lightning or the spark in an **internal combustion engine,** the **nitrogen** (N) becomes active and combines with the **oxygen** (O) to produce **oxides of nitrogen** (NO_x). Similarly, **CFCs** which are inert under normal conditions of **temperature** and **pressure** in the **troposphere** become unstable when exposed to different conditions in the upper atmosphere. As a result they break down, creating by-products that initiate chemical reactions destructive to the **ozone layer**.

INFILTRATION

The penetration of the pore spaces or cracks in a permeable body by a fluid. The percolation of rainwater into **soil** is an example of the infiltration process.

See also
Infiltration capacity.

INFILTRATION CAPACITY

The rate at which **water** percolates into the **soil** from the surface. Infiltration capacity tends to be greater when a soil is dry and diminishes as the soil gets wetter. If the pore spaces in a soil are filled or water is added at a rate that exceeds the ability of the soil to absorb it – for example, during a heavy thunderstorm – the excess water will flow over the surface. Infiltration capacity varies with **soil structure**, vegetation cover and slope, and any activity that alters these factors may have a local environmental impact. The removal of vegetation, by allowing **precipitation** to reach the soil surface more rapidly than it can be absorbed, will effectively reduce infiltration capacity. This encourages greater **runoff** which can lead to accelerated **erosion**.

INFRARED RADIATION

Low **energy**, **long wave radiation**, sometimes referred to as heat radiation, with wavelengths between 0.7 μm and 1000 μm in the **electromagnetic spectrum**. **Terrestrial radiation** is infrared. It is captured by the atmospheric **greenhouse gases** and as a result is responsible for the heating of the earth/atmosphere system. Infrared radiation is not visible to the human eye, but instruments which sense the production of infrared radiation have been developed for a variety of purposes. For example, they are used in weather forecasting to measure variations in the **temperature** of **clouds**, in forestry to distinguish between healthy and diseased trees, in engineering to identify areas of heat loss in buildings, and in medicine to locate hot spots in the body which may indicate the presence of malignant growths. Because infrared radiation can penetrate **fog** and **haze**, infrared sensors may provide detailed information not available from visual sensors. Permanent records can be produced by using photographic film sensitive to radiation in the infrared section of the spectrum.

Further reading
Faughn, J.S., Turk, J. and Turk, A. (1991) *Physical Science*, Philadelphia: Saunders.
Vincent, J.D. (1990) *Fundamentals of Infrared Detector Operation and Testing*, New York: Wiley.

INORGANIC MATTER

Material which is of mineral origin, and does not contain **carbon** (C) compounds except in the form of carbonates.

INSECTICIDES

Chemicals used to kill insects. They may be natural or synthetic in origin. The former include nicotine sulphate, obtained from tobacco leaves, pyrethrum, extracted from the heads of chrysanthemums, and rotenone from the root of the tropical derris plant. In addition to these botanicals, micro-organisms also provide natural control of insects. The bacterium Bacillus thuringensis, for example, has proved very effective against the spruce budworm outbreaks that cyclically threaten the coniferous forests of North America. Natural insecticides do not persist in the **environment**, are effective at low doses, do not accumulate in organisms, are not biologically amplified and have low to moderate toxicity for humans and other animals. However, they tend to be expensive to produce. Synthetic insecticides include **chlorinated hydrocarbons**, **organophosphates** and **carbamates**. They are generally very effective in killing insects, but come with certain drawbacks. The chlorinated hydrocarbon

DDT, for example, played a major part in controlling malaria in tropical regions such as Sri Lanka (Ceylon) in the late 1940s and 1950s. However, because of its persistence in the environment, its **bioaccumulation** and biological amplification in **food chains**, it presented a major threat to wildlife and has since been banned or severely restricted in its use. Although organophosphates have only a low persistence and do not accumulate in the environment, they are highly toxic to mammals and account for most human deaths from insecticide poisoning. Carbamates are generally considered to be the least dangerous of the synthetic insecticides, but they too are highly toxic for fish and beneficial insects such as honey bees.

See also
Organochlorides, Pesticides.

Further reading
Hodgson, E. and Kuhr R.J. (eds) (1990) *Safer Insecticides: Development and Use*, New York: M. Dekker.
Laird, M., Lacey, L.R. and Davidson, E.W. (eds) (1990) *Safety of Microbial Insecticides*, Boca Raton, FL: CRC Press.

INSOLATION

A measure of the **solar radiation** arriving at the earth's surface. In detail, its value may represent the intensity of either direct or global (direct plus diffuse) radiation on a unit area at a specific time on a specific surface. It varies according to such **elements** as the **solar constant**, the time of year, latitude, slope and aspect of the receiving surface and the transparency of the **atmosphere**.

INSULATION

The reduction of heat transfer by **conduction**. Insulation can be provided by a variety of materials ranging from natural substances such as feathers or wood shavings to completely synthetic products such as **polystyrene** sheeting. Since **air** is a poor conductor of heat, insulating materials incorporate air spaces or are arranged in such a way as to produce and maintain air spaces. Natural goose or duck down in clothing, for example, and fibreglass batts in walls owe their efficiency to the air

spaces contained within them. The efficiency of insulating material is indicated by its RSI-value – a numerical representation of its resistence to heat transfer. Insulation is important for comfort in clothing and buildings and for safety in the home and in the workplace, but it also has economic and environmental implications. During the **energy** crisis in the 1970s, additional insulation was added to residential, office and industrial buildings to reduce heat loss and as a consequence save **fuel** and reduce costs. Greater attention to insulation would help to bring about a decline in the demand for energy, reduce the direct environmental impact of **fossil fuel** extraction and have a positive effect on the emission of **greenhouse gases** and acid gases. Insulation also refers to the prevention of the flow of an electric current. Electrical insulators are essential for the safe use of electricity. They include non-metallic substances such as glass, **plastic** and rubber.

Table I-1 Insulation efficiency

	*R-VALUE PER INCH
Loose fill insulation	
glassfibre	2.4–3.5
cellulose fibre	3.7
vermiculite	2.1–2.5
loose polystyrene	3.0–3.3
wood shavings	2.4
Batt or blanket insulation	
glassfibre batts	2.9–4.0
Rigid board insulation	
extruded polystyrene (blue)	5.0
expanded polystyrene (white)	3.4–4.2
phenolic foam board	4.2
polyurethane slabs	5.0–6.0
Foamed-in-place insulation	
ureaformaldehyde	4.3–4.9
polyurethane foam	4.7–5.0

*R-value is a measure of the resistance of insulation to heat transfer

Further reading
Argue, R. (1980) *The Well-Tempered House: Energy-Efficient Building for Cold Climates*, Toronto: Renewable Energy in Canada.

INTEGRATED PEST MANAGEMENT (IPM)

An ecological approach to pest management, which involves a combination of biological, chemical and cultivation techniques of pest control. It is not aimed at the complete eradication of a specific pest **population**, but is designed to maintain the pest population below the level at which it begins to cause economically unacceptable losses to plants and animals. The methods used tend to rely on low-level technology and may include, for example, the physical removal of certain bugs, the cultivation of crops that provide some mutual protection from pests or the use of botanical **insecticides** and natural predators. When used in the proper sequence and with appropriate timing, IPM techniques can reduce **pesticide** use and pest control costs by 50 to 90 per cent. With fewer pesticides being used, the risks to wildlife and to human health are also reduced. Although IPM has clear long-term benefits for the **environment**, its widespread adoption has been hampered by certain short-term disadvantages. Initially it tends to be costlier, more labour-intensive and slower to produce results than conventional pesticides, for example. In addition, it requires a better understanding of crop/pest relationships than other pest control systems and inadequate training has undoubtedly slowed the implementation of IPM techniques. Where it has been introduced – in Indonesia, China and Brazil – it has met with considerable success.

Further reading
Horn, D.J. (1988) *Ecological Approach to Pest Management*, New York: Guilford.
Van den Bosch, R. and Flint, M.L. (1981) *Introduction to Integrated Pest Management*, New York: Plenum.

INTEGRATED SOLID WASTE SYSTEM

A system based on the combination of a variety of methods of **solid waste disposal** aimed at causing the least damage to the **environment**. It includes all forms of disposal – **sanitary landfill, incineration, recycling** and composting – with the relative proportions varying from place to place and from time to time. Ideally the waste stream would be managed in such a way that the proportion of solid waste disposed of in landfill or by incineration would be minimal. In reality, perhaps 20 to 40 per cent of waste would still have to be buried or burned. Estimates in the United States see the proportion remaining at about 75 per cent until the end of the twentieth century, but individual communities, through a combination of recycling and composting, may already be approaching the lower rates.

See also
Compost.

Further reading
Pfeffer, J.T. (1992) *Solid Waste Management Engineering*, Englewood Cliffs, NJ: Prentice-Hall.

INTERACTIVE MODELS

General circulation models which are programmed to deal with the progressive change set in motion when one or more of the components of the **atmosphere** is altered. In the original **nuclear winter** studies, for example, the model used assumed that the **smoke** produced by the nuclear explosions would remain a passive constituent of the atmosphere, to be redistributed by the existing circulation pattern. However, further studies using an interactive model indicated a more complex situation. It showed that the ability of the smoke to absorb **solar radiation** would cause the **temperature** structure of the atmosphere to change, and this in turn would eventually alter the circulation pattern. The smoke would then be dispersed by this new circulation which it had helped to produce.

Further reading
Thompson, S.L. (1985) 'Global interactive transport simulations of nuclear war smoke', *Nature* 317: 35–9.

Figure I-3 Proposed interbasin transfer schemes in North America

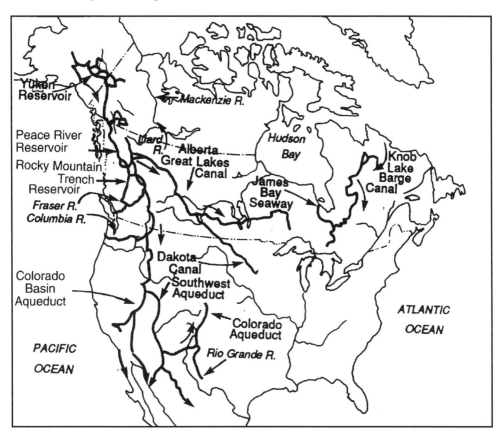

Source: Mungal, C. and McLaren, D.J. (1990) *Planet under Stress*, Don Mills, Ont: Oxford University Press

INTERBASIN TRANSFER

The movement of water from one drainage basin to another, usually to meet the needs of **irrigation** or the production of **hydro-electricity**. Associated with the basic hydrological changes that this produces, there are impacts on specific **elements** such as the regional **water balance**, stream morphology and aquatic biology of the basins involved. Existing environmental checks and balances appear to be able to cope with current relatively small-scale transfers, but as the demand for water continues to increase, it is expected that the need for larger scale interbasin transfers will be greater. Should that happen, significant changes in **evapor-** ation rates would likely follow, stream morphology would be altered because of changes in the rates of **erosion** and deposition, and **species** transfer and **habitat** change would have a substantial impact on the aquatic community. In the 1960s, major geographical engineering schemes were proposed to improve the distribution of continental water supplies through massive interbasin transfers. The **North American Water and Power Alliance** (NAWAPA), for example, was a scheme that envisioned the southward diversion of most of the major streams flowing into the Arctic, to provide domestic and irrigation water for the US south-west, and to improve water supply for navigation and **pollution** control in the Great

Lakes basin. A similar scheme in what was then the USSR proposed the diversion of the northward flowing rivers of Siberia southwards into the Caspian and **Aral Sea** basins and in Africa the diversion of the River Congo northwards into the Lake Chad basin was seen as a means of dealing with the chronic water shortage in the central Sahara Desert. Such grandiose schemes have the potential for major environmental disruption and, although none came to pass, future demands for water may see them resurrected.

Further reading
Griggs, N.S. (1996) *Water Resources Management: Principles, Regulations and Cases*, New York: McGraw-Hill.
Simons, M. (1971) 'Long-term trends in water use', in R.J. Chorley (ed.) *Introduction to Geographical Hydrology*, London: Methuen.

INTERGLACIAL

A warm phase between the glacials of an **ice age**. The improved climatic conditions supported major environmental changes in **soils**, vegetation and wildlife, which followed a recognizable pattern through the duration of the interglacial. Of these the vegetation changes were probably the most obvious. Although complex in detail, varying in time and place as the **plant communities** assembled and disassembled, the changes probably followed a similar sequence during each interglacial. In northern latitudes, for example, the **tundra** which colonized the open ground exposed by the retreating **ice** was replaced in sequence by **grassland** and woodland with the climax being represented by temperate **deciduous** forest. Subsequent climatic deterioration and the return of the ice reversed these improvements and ultimately brought the interglacial to an end. Although the present period of relative warmth – the **Holocene** – which began some 10,000 years ago, is often referred to as postglacial, it can also be considered as an interglacial. Peak ecological conditions occurred some 5000 years ago during the **Climatic Optimum**, and since then there has been an overall general deterioration with variations such as the **Little Optimum** and the **Little Ice Age**. Comparison with **ice core** and ocean sediment evidence of past interglacials suggests that the current conditions represent the later stages of an interglacial cycle. There are some indications, however, that **global warming** induced by human activities may be sufficient to slow or perhaps reverse the deterioration.

See also
Interstadial.

Further reading
Dawson, A. (1991) *Ice Age Earth*, London/New York: Routledge.
Deynoux, M. (ed.) (1994) *Earth's Glacial Record (International Geological Correlation Project 260)*, Cambridge/New York: Cambridge University Press.
Mannion, A.M. (1991) *Global Environmental Change: A Natural and Cultural Environmental History*, Harlow, Essex: Longman.

INTERGOVERNMENTAL OCEANOGRAPHIC COMMISSION (IOC)

Established as part of the **UN Educational, Scientific and Cultural Organization** (UNESCO) in the early 1960s to co-ordinate global ocean science programmes, the IOC promotes the study of a wide range of **elements** in the oceanic **environment**, including ocean dynamics, marine **pollution**, ocean mapping and biological resources. As a co-sponsor of the **World Climate Program** (WCP), it supports investigation into ocean/atmosphere interactions with a view to improving knowledge of weather and climate processes.

See also
TOGA, WOCE.

INTERGOVERNMENTAL PANEL ON CLIMATE CHANGE (IPCC)

A group of eminent scientists brought together in 1988 by the **World Meteorological Organization** (WMO) and the **United Nations Environment Program** (UNEP). It was charged with assessing the overall state of research on **climate change**, so that potential environmental and socioeconomic impacts might be evaluated, and appropriate response strategies developed. This involved three working groups charged with the following:

- WG I – to assess available scientific information on climate change

- WG II – to assess environmental and socio-economic impacts of climate change

- WG III – to formulate response strategies

Their reports – a scientific overview, an impact assessment and response strategies – were produced by 1991. A supplementary report was issued in 1992, generally confirming the original assessments, and by mid-1994 a second supplementary report, focusing on **radiative forcing** of **climate**, had been completed. Because of the rapid accumulation of data in the field of climate change, a second comprehensive report was considered necessary. Most of the contributors to the second assessment were university or government scientists, with limited numbers from private research agencies, companies and non governmental organizations (NGOs). Representatives from environmental advocacy groups such as **Greenpeace** were among those who reviewed the original documents. Completed in December 1995, it was presented to the signatories of the **UNCED Framework Convention on Climate Change** (FCCC) prior to publication in mid-1996. Central to the report was the recognition of the human contribution in current climate change. It also recommended that action should be taken to halt **global warming** and because of the time-lags involved it concluded that action could no longer be delayed. The 1995 IPCC report also provided background scientific data for subsequent FCCC negotiations on **greenhouse gas** emission targets.

Further reading

IPCC (1990a) *Climate Change: The IPCC Scientific Assessment*, Cambridge: Cambridge University Press.
IPCC (1990b) *Climate Change: The IPCC Impacts Assessment*, Canberra: Australian Government Publishing Service.
IPCC (1990c) *Climate Change: The IPCC*

Response Strategies, Washington, DC: Island Press.
IPCC (1992a) *Climate Change 1992: The Supplementary Report to the IPCC Scientific Assessment*, Cambridge: Cambridge University Press.
IPCC (1992b) *Climate Change 1992: The Supplementary Report to the IPCC Impacts Assessment*, Canberra: Australian Government Publishing Service.
IPCC (1995) *Climate Change 1994: Radiative Forcing of Climate Change and an Evaluation of the IPCC IS92 Emission Scenarios*, Cambridge: Cambridge University Press.
IPCC (1996a) *Climate Change 1995: The Science of Climate Change*, Cambridge: Cambridge University Press.
IPCC (1996b) *Climate Change 1995: Impacts, Adaptations and Mitigation of Climate Change: Scientific and Technical Analysis*, Cambridge: Cambridge University Press.
IPCC (1996c) *Climate Change 1995: Economic and Social Dimensions of Climate Change*, Cambridge: Cambridge University Press.
Kemp, D.D. (1997) 'As the world warms: climate change 1995', *Progress in Physical Geography* 21 (2): 310–14.
Masood, E. (1996) 'Climate report subject to scientific cleansing', *Nature* 381 (6583): 546.

INTERNAL COMBUSTION ENGINE

A heat engine powered by the controlled combustion of a **fuel** in an enclosed cylinder. The **thermal energy** is converted into mechanical **energy** by means of a moving piston. The most common fuel used in the internal combustion engine is **petroleum**, and at the high **temperatures** and pressures of the **combustion** cycle, pollutants such as **oxides of nitrogen** (NO_x) are produced and exhausted to the **atmosphere**. **Carbon dioxide** (CO_2) and **water** are also produced and inefficient combustion allows the formation of **carbon monoxide** (CO) and unburned **hydrocarbons**. In most urban areas, the internal combustion engine makes a major contribution to atmospheric **pollution**.

See also
Catalytic converter, Photochemical smog.

INTERNATIONAL ATOMIC ENERGY AGENCY (IAEA)

A United Nations agency concerned with the commercial and scientific uses of atomic **energy** and **radioisotopes**. It provides for the exchange of information on radioactive materials, for example, and monitors levels of exposure to **radioactivity**. Recommendations for treatment of victims of the **Chernobyl** disaster were based in part on information provided by the IAEA.

INTERNATIONAL BANK FOR RECONSTRUCTION AND DEVELOPMENT (WORLD BANK)

A financial institution founded in 1945 under the auspices of the United Nations. The initial impetus for the organization arose from the need to support reconstruction and development in war-torn Europe. Since then, it has expanded its operations to facilitate trade and development worldwide, and is particularly active in the **Third World** countries of Africa, Asia and Latin America. It contracts research and publishes technical papers on a wide range of monetary and development issues and sponsors other non-governmental agencies or institutes such as the **International Institute for Applied Systems Analysis** (IIASA). Through such activities, the World Bank has the potential to contribute to the understanding and amelioration of environmental issues.

Further reading
Nelson, R. (1990) *Dryland Management: The 'Desertification' Problem, World Bank Technical Paper No. 16*, Washington, DC: World Bank.
World Bank (1988) *The World Bank's Support for the Alleviation of Poverty*, Washington, DC: World Bank.

INTERNATIONAL COUNCIL OF SCIENTIFIC UNIONS (ICSU)

A non-governmental organization founded in 1931 with the aim of promoting all branches of science and encouraging the exchange of scientific data through international scientific co-operation. It initiates and co-ordinates research projects at the international level, encourages interdisciplinary research and monitors such elements as the rights, freedoms and responsibilities of scientists. The ICSU is particularly active in the environmental field, having been instrumental in the formation of groups such as the **Advisory** Group on Greenhouse Gases (AGGG) and comprehensive programmes such as the **International Geosphere-Biosphere Program** (IGBP).

INTERNATIONAL DEVELOPMENT ASSOCIATION (IDA)

An agency that, in association with the World Bank, makes interest-free loans to the world's poorest countries. Some seventy countries are eligible, but are required to prepare comprehensive environmental action plans as a prerequisite for IDA assistance.

INTERNATIONAL GEOGRAPHICAL UNION (IGU)

Created in 1922 as an international group with the aim of promoting the study of geographical issues. It maintains that aim through regular congresses, commissions and study groups. Activities include the initiation and co-ordination of international geographical research, the collection and dissemination of geographical information and the participation of its members in international organizations.

INTERNATIONAL GEOSPHERE-BIOSPHERE PROGRAM (IGBP)

An interdisciplinary research programme initiated by the **ICSU** in 1986, initially intended to continue for ten years. Its objective is to describe and understand the various processes – physical, chemical, biological – which together regulate the whole earth/atmosphere system. The IGBP is also concerned with the changes that are taking place in the system and the ways in which these changes are being initiated and influenced by human activities. The programme integrates other activities of the ICSU, for example, through its **Scientific Committee on Problems of the Environment** (SCOPE), and provides a framework for other interdisciplinary programmes which contribute to the study of global change. These include the **International Global Atmospheric Chemistry Project** (IGACP), Biological Aspects of the Hydrologic Cycle (BAHC) and the Joint Global Ocean Flux Study (JGOFS).

INTERNATIONAL GLOBAL ATMOSPHERIC CHEMISTRY PROJECT (IGACP)

A core project of the **IGBP** concerned with atmospheric chemistry and global **pollution**. It is particularly concerned with biogeochemical interactions in the **biosphere** and **atmosphere** and their impact on **climate**. The research activities involved in the project will ultimately provide a better understanding of the processes that regulate the climatologically active constituents of the atmosphere.

INTERNATIONAL HYDROLOGICAL PROGRAM (IHP)

An extension of the International Hydrologic Decade which ended in 1974 after ten years of international scientific co-operation on **water** problems. Under the auspices of **UNESCO**, the mandate of the IHP is to develop the scientific and technical capacity to permit the rational management of the world's freshwater resources, in terms of both quality and quantity.

INTERNATIONAL INSTITUTE FOR APPLIED SYSTEMS ANALYSIS (IIASA)

A non-governmental interdisciplinary research institute formed in 1972 as a result of co-operation between the United States and the former USSR, and sponsored in part by the **World Bank**. It is concerned mainly with the assessment of the interactions between human activities and the **environment**, with its research following three main themes: global environmental change, global economic and technological transitions, and methods for the analysis of global issues. The IIASA's research activities have depended heavily on the development of computer models, such as the **Regional Acidification Information and Simulation** Model (RAINS), which is used for teaching, research and policy development on **acid rain** in more than seventeen countries. Computer modelling has also been developed for the study of such issues as global **climate change**, world agricultural potential, demographic change and **energy** resources.

Further reading

Flohn, H. (1980) *Possible Climatic Consequences of a Man-Made Global Warming RR–80–30,* Laxenburg, Austria: International Institute for Applied Systems Analysis.

INTERNATIONAL JOINT COMMISSION (IJC)

A bi-national commission created in 1909 by Canada and the United States to oversee the provisions of the Boundary Water Treaty. The original purpose was to prevent disputes over use of the waters shared by the two countries, and to provide advice to both federal governments. Initial concerns involved changes to the natural flows and levels of boundary waters, but there was also provision for the monitoring and prevention of **pollution**. Despite the latter, in the first half-century of the Commission's existence, **water** pollution became a major problem particularly in the lower Great Lakes. This resulted in part from limited public concern over pollution, but it also reflected the fact that the Commission is an advisory body whose recommendations are not binding on the governments involved. More recently the advice of the Commission has been heeded. **Water quality** is now a major concern, particularly in the Great Lakes, and to that end the IJC has been instrumental in setting up Remedial Action Plans (RAPs) and Lakewide Management Plans (LMPs) under the Great Lakes Water Quality Agreement.

Further reading

Spencer, R., Kirton, J.J. and Nossal, K.M. (1981) *The International Joint Commission Seventy Years On,* Toronto: Centre for International Studies, University of Toronto.

INTERNATIONAL METEOROLOGICAL CONGRESS

Held in Vienna in 1873, the Congress created the International Meteorological Organization (IMO). The IMO promoted the standardization of meteorological instruments and observations, thus allowing global co-operation in **meteorology** and **climatology**. It was ultimately superseded by the **World Meteorologial Organization** (WMO).

INTERNATIONAL MONETARY FUND (IMF)

Founded in 1945 at the same time as the **International Bank for Reconstruction and Development (World Bank)**, the IMF was created to reform and stabilize international currencies following the disruption of the Second World War. It is involved in policy-based lending in which loans are provided to individual nations only if IMF-approved economic reforms are put in place.

Further reading
Driscoll, D. (1988) *What is the International Monetary Fund?*, Washington, DC: International Monetary Fund.

INTERNATIONAL NICKEL COMPANY (INCO)

A multinational **nickel** (Ni) mining and **smelting** company based in Canada. Its Sudbury, Ontario smelter was identified as one of the major sources of **sulphur dioxide** (SO_2) emissions in North America in the 1960s. These **acid gases** caused extensive damage to vegetation in the Sudbury area, and in an attempt to reduce local ground-level **pollution**, the company built a 400 m 'tall-stack' that released the pollutants higher into the **atmosphere**. Although this achieved a reduction in local pollution, it allowed the acid gases to be transported considerable distances downwind, to cause **acid rain** in eastern Ontario and Quebec. Since then the problem of acid gas emissions has been tackled by the introduction of **flue gas desulphurization**.

See also
Scrubbers, Tall-stacks policy.

INTERSTADIAL

A short period of **climate** amelioration during a major glacial. Ice sheets may cease to advance or may even retreat somewhat, but the amelior-ation is insufficient to produce the environ-mental improvements experienced during an **interglacial**. The milder conditions associated with the Bølling and Allerød interstadials, for example, which occurred separately between c.11,000 and 13,000 BP, are recognized in the European lateglacial sequence.

Further reading
Lowe, J.J. and Gray, M.J. (1980) 'The stratigraphic subdivision of the Lateglacial of NW Europe: a discussion', in J.J. Lowe, M.J. Gray and J.E. Robinson (eds) *Studies in the Lateglacial of North-west Europe*, Oxford: Pergamon.

INTERTROPICAL CONVERGENCE ZONE (ITCZ)

A thermal low-pressure belt that circles the earth in equatorial latitudes, between the tropical **Hadley Cells** positioned north and south of the equator. It owes its origin to convective uplift caused by strong surface heating, augmented by converging airflows from the north-east and south-east Trade Winds. The ITCZ moves north and south with the seasons, bringing the rains which relieve **seasonal drought** in areas such as the **Sahel**, India and northern Australia. This movement is not completely reliable, however. The ITCZ migrates at different rates and over different distances from one year to the next, and this inherent variability contributes to the problem of **drought**.

Figure I-4 The seasonal change in location of the ITCZ in Africa

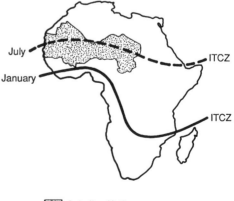

Sahelian Nations

See also
Monsoons.

Further reading
Hamilton, R.A. and Archbold, J.W. (1945) 'Meteorology over Nigeria and adjacent territory', *Quarterly Journal of the Royal Meteorological Society* 71: 231–62.
Musk, L. (1983) 'Outlook – changeable', *Geographical Magazine* 55: 532–3.
Riehl, H. (1979) *Climate and Weather in the Tropics*, New York: Academic Press.

INVERSION

See **temperature inversion**.

INVISIBLE DROUGHT

One of the four forms of drought identified by **C.W. Thornthwaite**. The condition is not obvious and can often be identified only by sophisicated instrumentation and statistical techniques. There may be no obvious lack of **precipitation**, but moisture requirements are not being met, the crops are not growing at their optimum rate and the potential yield is therefore reduced. Invisible drought is easily dealt with by **irrigation**.

See also
Contingent drought, Permanent drought, Seasonal drought.

Further reading
Thornthwaite, C.W. (1947) 'Climate and moisture conservation', *Annals of the Association of American Geographers* 37: 87–100.

IODINE (I)

A grey-black crystalline solid – one of the **halogen** group of **elements** – which is highly **volatile** and readily gives off a violet vapour. It is present in seaweed and occurs as an impurity in crude saltpetre (sodium nitrate $NaNO_3$). Dissolved in **alcohol** it produces tincture of iodine, widely used as an antiseptic. Iodine is essential for the proper functioning of the thyroid gland, which helps to regulate **metabolism** in humans, and to help to maintain appropriate levels of iodine in the body it may be consumed in the form of potassium iodide (KI) added to table salt –

sodium chloride. It is also used in chemical analysis and photography, and the **radioisotope** [131]iodine is used in the diagnosis and treatment of malfunctions of the thyroid gland.

ION

An **atom** or group of atoms that has become electrically charged by picking up or losing **electrons**. Since electrons are negatively charged, the addition of an electron creates a negative charge, whereas the removal of an electron creats a positive charge. Ions formed from **metals** are generally positively charged (**cations**); those from non-metals are negatively charged (**anions**). Many **compounds** undergo **dissociation** into their component ions when in **solution**, and **gases** can be ionized by passing an electrical charge through them.

Further reading
Boeker, E. and Van Grondelle, R. (1995) *Environmental Physics*, Chichester/New York: Wiley.

ION EXCHANGE

The removal or replacement of **ions** in **solution**, accomplished by using an exchange medium or filter that has the ability to capture the **cations** or **anions** in the solution. In **water** softeners, for example, the water is softened by removing the **calcium** ions and replacing them with **sodium** ions. When the capacity of the filter bed is full, it is flushed with brine which reverses the process and recharges the system. Ion exchange is also used in water purification and **sewage treatment**.

See also
Base exchange.

Further reading
Slater, M.J. (1991) *Principles of Ion Exchange Technology*, Oxford/Boston: Butterworth/Heinemann.

IONIZATION

The process by which **atoms** obtain a positive or negative charge and become **ions**.

IONIZING RADIATION

Radiation that is capable of causing **ionization**, for example, by dislodging **electrons** from **atoms**. Fast-moving particles such as electrons, alpha- and beta-particles are particularly effective in causing ionization, as is electromagnetic radiation in the form of **X-rays** and **gamma rays**. Ionizing radiation creates highly reactive atoms which have significant biological impacts, including the alteration of **DNA** and the initiation of **cancer**. Natural sources of ionizing radiation include cosmic rays, but currently the main concern is with the radiation sources released into the **environment** by human activities such as the improper disposal of radioactive materials or by nuclear accidents.

See also
Irradiation.

Further reading
Turner, J.G. (1995) *Atoms, Radiation and Radiation Protection* (2nd edition), New York: Wiley.

IONOSPHERE

That part of the upper **atmosphere** above *c.* 80 km in which free **electrons** and **ions** occur as a result of the **ionization** of atmospheric **gases** by **ultraviolet radiation** and by **X-rays** from the sun. The ionosphere is in a constant state of flux. Ionization peaks during the day, for example, when incoming **solar radiation** is at a maximum, but at night the free electrons recombine with the ions. Since it reflects radio waves and allows them to be transmitted around the curved surface of the earth, the ionosphere contributes to intercontinental radio communications.

See also
Electromagnetic spectrum.

Further reading
Ratcliffe, J.A. (1972) *An Introduction to the Ionosphere and Magnetosphere*, Cambridge: Cambridge University Press.

IPCC SUPPLEMENTARY REPORTS

See **Intergovernmental Panel on Climate Change.**

IRON (FE)

A silvery-grey magnetic **metal**. One of the most common **elements** in the earth's **crust**, iron is sufficiently reactive that it is rarely found in its natural form – it is readily oxidized, for example. Iron occurs as various **ores**, including magnetite and haematite, both iron oxides, from which the metal is extracted by **smelting** in a blast furnace. The resulting product is pig iron which is converted into other forms such as wrought iron or steel, by altering the proportion of **carbon** (C) it contains. Iron is also alloyed with other metals when specific properties are required. Stainless steel contains chromium (Cr), for example. The combined development of the iron and **coal** industries provided the impetus for the **Industrial Revolution**, and in its various forms iron remains an important industrial metal. It is also a micronutrient essential for human health, being present in haemoglobin, which facilitates the transfer of **oxygen** (O) around the body.

See also
Alloy, Coke.

Further reading
Trendall, A.F. and Morris, R.C. (eds) (1983) *Iron Formation: Facts and Problems,* Amsterdam/New York: Elsevier.
Lauffer, R.B. (ed.) (1992) *Iron and Human Disease,* Boca Raton, FL: CRC Press.

IRRADIATION

Exposure to any form of **radiation**, but frequently used to refer to exposure to **ionizing radiation**. Intense irradiation can alter the physical and chemical properties of materials. In **nuclear reactors**, for example, transuranic elements such as **plutonium** (Pu) are created through the irradiation of **uranium** (U). Irradiation at much smaller doses has potential uses in the food and agricultural industries. The sterilization of insects is possible by exposing them to low doses of ionizing radiation. The subsequent release of sterilized males into the environment would allow the **populations** of some insects to be controlled without the use of **pesticides**. Experiments with food have shown that the storage life of a variety of products, including fish and fruit, could be extended by irradiation. Widespread acceptance of such techniques would require strong evidence that the products could be safely consumed after irradiation.

IRRIGATION

The provision of **water** for crops in areas where the natural **precipitation** is inadequate for crop growth. The water may be obtained from natural or artificial surface storage systems (such as lakes or **reservoirs**) or from the **groundwater** system. Irrigation takes many forms, from the total flooding associated with paddy-rice production to sprinkler systems that attempt to emulate precipitation. Although irrigation is identified mainly with arid or semi-arid areas, it is used increasingly in more humid areas where **invisible drought** can lead to **soil moisture deficits**. Although irrigation can literally 'make the desert bloom', it is not without its economic and environmental consequences. Where the irrigation water is stored on the surface, it brings with it all the environmental

problems associated with the disruption of the **hydrological cycle** and the creation of large reservoirs. Modern pumping systems are so efficient that withdrawal of groundwater from **aquifers** easily exceeds recharge rates. The aquifers become depleted, leading to the need for deeper wells and increased pumping time, with a consequent rise in costs. The major environmental problem associated with irrigation is **salinization**, particularly in arid and semi-arid areas where **evaporation** rates are high. The **salts** that are left behind in the upper layers of the **soil** may be sufficient to render it unsuitable for cultivation.

See also
Aral Sea.

Further reading
Kirkby, A.V. (1971) 'Primitive irrigation', in R.J. Chorley (ed.) *Introduction to Geographical Hydrology*, London: Methuen.
Owen, O.S. and Chiras, D.D. (1995) *Natural Resource Conservation*, Englewood Cliffs, NJ: Prentice-Hall.
Reisner, M. (1986) *Cadillac Desert: The American West and its Disappearing Water,* New York: Penguin.

ISOMERS

Compounds that share the same chemical formula, but since the arrangement of **atoms** within the **molecules** is different they have

Figure I-5 The characteristics of selected irrigation systems

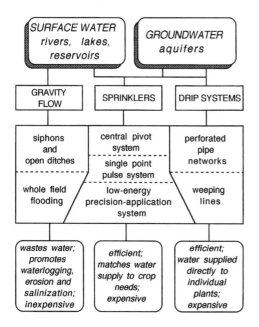

Figure I-6 The chemical structure of octane and its isomer iso-octane

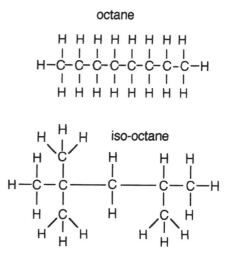

different structural formulae. Isomerism is common among the **hydrocarbons**. For example, iso-octane, a component of **gasoline,** is an isomer of octane. The two have the same formula (C_8H_{18}), but in iso-octane the atoms are arranged in a branched chain rather than the straight chain of octane. They also differ in their properties, with iso-octane less likely than normal octane to produce engine knock (uncontrolled explosive **combustion**) in gasoline engines.

Further reading
Zumdahl, S.S. (1993) *Chemistry* (3rd edition), Lexington, MA: D.C. Heath.

ISOPLETH

A line on a map connecting points of equal value. Perhaps the most common isopleths are the contour lines joining points of equal altitude on a topographic map, but there are many variations depending upon the type and purpose of the map. Examples of other commonly used isopleths and the **elements** they represent are provided in Table I–2.

Table I-2 Types of isopleths

TYPE OF ISOPLETH	ELEMENTS REPRESENTED
Isobar	Barometric pressure
Isobase	Uplift or subsidence during a specific time period
Isobath	Depth of water
Isohyet	Rainfall amount
Isotach	Velocity of wind or sound
Isotherm	Temperature

ISOSTASY

The concept of balance in the earth's **crust.** Lighter, rigid blocks of crustal material are seen to float on the denser, more plastic rock of the underlying mantle. **Erosion** or deposition of the crustal blocks disturbs the balance, causing them to float higher or

deeper in the underlying material until equilibrium is regained. The relatively slow rates of continental erosion and deposition make it difficult to recognize and measure isostatic movements, but the situation is well illustrated by the response of the earth's crust to glaciation. During the **ice ages** the build-up of land **ice** led to local depression of the crust. Once the ice melted the crust rebounded again, and even after 10,000 years that rebound continues in areas such as the Hudson Bay Lowlands in Canada and parts of Northern Europe where the ice was thickest during the last glaciation. Some of the landforms of formerly glaciated areas owe their existence to isostatic rebound. When the ice melted, for example, **sea-level** rose more rapidly than the depressed land surface could recover and many coastal areas were flooded. Once the sea-level had stabilized, however, isostatic recovery took over, and the coastal areas rose out of the sea to create the raised shorelines common in mid- to high latitudes in the northern hemisphere.

Further reading
Ritter, D.F., Kochel, R.C. and Miller, J.R. (1995) *Process Geomorphology,* Dubuque, IA: Wm C. Brown.
Sissons, J.B. (1983) 'Shorelines and isostasy in Scotland', in D.E. Smith and A.G. Dawson (eds) *Shorelines and Isostasy*, London: Academic Press.

ISOTHERMAL LAYER

An atmospheric layer in which the **lapse rate** is neutral; that is, the **temperature** remains constant with increasing altitude, in comparison with the normal situation in which temperature declines with altitude. The lower part of the **stratosphere** is isothermal.

See also
Environmental lapse rate.

ISOTOPES

Atoms that have the same **atomic number,** but different **mass numbers.** All the isotopes of a given **element** will have the same number of **protons** in their nuclei, but will differ in the number of **neutrons. Carbon** (C), for example, has an atomic number of 6 (i.e. six protons), but its three isotopes have mass numbers of 12, 13 and 14, reflecting the differences in the number of neutrons. Most elements consist of a mixture of different isotopes which are identical in chemical properties. Sometimes they differ in physical characteristics, however. The two isotopes of **oxygen** (^{16}O and ^{18}O), for example, respond slightly differently to **temperature** changes. This can lead to variations in the isotopic composition of **water** and the $^{16}O/^{18}O$ ratio in glacier **ice** has been used to provide information on past temperatures.

Further reading
Choppin, G.R. and Rydberg, J. (1980) *Nuclear Chemistry: Theory and Application*, Oxford/New York: Pergamon Press.
Johnson, S.J., Dansgaard, W., Clausen, H.B. and Langway, C.C. (1972) 'Oxygen isotope profiles through the Antarctic and Greenland ice sheets', *Nature* 235: 429–34.

ITAI-ITAI

A disease characterized by bone deterioration, caused by **cadmium** poisoning. It was first noted in Japan.

J

JET PROPULSION

Movement or locomotion produced when a fluid under pressure is released in a high-speed stream from a controlled nozzle. The reaction to the resulting thrust propels the object releasing the fluid, in a direction opposite to the discharge. Although present in nature – for example, used by squid for propulsion through **water** – its most common current use is in jet engines. Hot **gases** produced by the **combustion** of **kerosene** in **air** are exhausted at a sufficient rate to drive jet aircraft through the **atmosphere** at more than twice the speed of sound. In so doing, they release **combustion** products such as water, **oxides of nitrogen** (NOx) and unburned **hydrocarbons** to contribute to **atmospheric turbidity**, and provide a potential threat to the **ozone** layer.

See also
Supersonic transports.

JET STREAM

A fast-flowing stream of **air** in the upper **atmosphere** at about the level of the **tropopause**, associated with zones in which steep **temperature** gradients exist and where, in consequence, the **pressure** gradients are also steep. Well-defined jet streams are located in sub-tropical latitudes (the Subtropical jet) and in mid- to high latitudes (the Polar Front jet). Both flow from west to east in a sinuous path around the earth, changing position and path with the seasons. Speeds change with the season also, being greatest in the winter when maximum speeds in excess of 300 km/hr have been recorded. The Subtropical jet is much more persistent than the Polar Front jet, which is associated with migratory **weather** systems and tends to be seasonally irregular and discontinuous. An Easterly Tropical jet stream flowing from east to west has been identified in equatorial regions. It is a summer

Figure J-1 The location of jet streams in the atmosphere

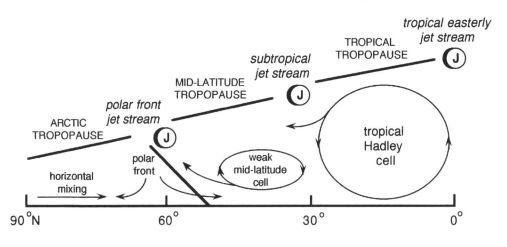

phenomenon produced by the regional reversal of the latitudinal temperature gradient over Africa and India. The effect of the jet streams extends into the lower atmosphere through their influence on the movement of the systems that produce the surface weather conditions, but their importance from an environmental point of view lies in their ability to transport pollutants over great distances in the upper atmosphere. **Smoke**, volcanic debris and acid particles are all spread by such transportation.

See also
Rossby waves.

Further reading
Eagleman, J.R. (1985) *Meteorology: The Atmos-phere in Action*, Belmont, CA: Wadsworth.
Harmon, J.R. (1967) *Tropospheric Waves, Jet Streams and United States Weather Patterns*, Washington, DC: Association of American Geographers.

JOULE (J)

The fundamental **SI unit** of work and **energy**. It represents the work done when a force of 1 newton acts through 1 metre. In the consideration of electrical **energy**, it is also the work done per second by a current of 1 ampere flowing through a resistance of 1 ohm.

See also
Watt.

K

K-STRATEGISTS

Organisms that are usually large, have relatively long lives and produce only a limited number of offspring. Most common in stable **environments**, they invest considerable time and **energy** providing for the survival of these offspring, so that they in turn can reproduce and ensure the continuation of the **species**. Larger animals such as deer, lions, bears and elephants are all K-strategists, as are humans. Although most obviously applied to animals, large, long-living plants such as trees also fit the concept.

See also
Carrying capacity, Demography, *r*-strategists.

Further reading
Pianka, E.R. (1970) 'On *r*- and *K*-selection', *American Naturalist* 104: 592–7.

KATABOLISM

See **catabolism**.

KELVIN

The **SI unit** of thermodynamic **temperature**. The kelvin is identical to the **Celsius** degree, but the Kelvin scale is based on the theoretically lowest temperature possible – **absolute zero** – which is equivalent to –273.15°C. Thus, the range between the freezing and boiling points of **water** on the Kelvin scale is between 273 K and 373 K. The relationship between the Kelvin and Celsius scales is represented by the following formula:

$$T_K = T_C + 273$$

KEROSENE

Also called paraffin oil, kerosene is a mixture of **hydrocarbons** in the form of medium, light oils obtained through the fractional distillation of **petroleum**. It is used as a **fuel** for lighting and space heating and is also used for cooking in many **Third World** countries. Large volumes of kerosene are consumed as jet fuel in airliners and military jets. Like all hydrocarbons, kerosene emits pollutants into the **atmosphere** when burned. Exhaust **gases** from jet engines are emitted at all levels in the atmosphere, contributing to **atmospheric turbidity** and **ozone depletion**. The **combustion** of kerosene in unvented space heaters also contributes to indoor air **pollution**. Even when functioning efficiently, these heaters emit **carbon monoxide** (CO), **nitrogen dioxide** (NO_2) and some **sulphur dioxide** (SO_2), which in an enclosed space can lead to health problems such as headaches and respiratory ailments.

See also
Jet propulsion.

KILOWATT (KW)

A unit of power equivalent to 1000 **watts**. Used mainly to measure electrical power, the unit for electrical **energy** (kilowatt hour – kWhr) is derived from it as follows:

power × time = energy
(kilowatt) (hour) (kilowatt hour)

KINETIC ENERGY

The **energy** an object possesses as a result of

its motion. It depends upon the **mass** and **velocity** of the moving object as indicated in the formula KE=1/2mv² with the result expressed in **joules**. KE is often produced by the conversion of other forms of energy. For example, the **thermal energy** obtained by burning **coal** can be used to produce steam that in turn provides kinetic energy through a mechanical device such as a piston engine or turbine. The kinetic energy in a moving hammer can be traced back to the chemical energy consumed in the form of food by the person wielding the hammer. Moving fluids such as **wind** and **water** also possess kinetic energy that can be used to drive windmills or hydroelectric generators.

Further reading
Reynolds, W.C. (1974) *Energy: From Nature to Man*, New York: McGraw-Hill.

KRAKATOA

A volcanic island in the Sunda Strait, Indonesia, which erupted explosively in 1883, sending at least 6 cubic km (and perhaps as much as 18 cubic km) of volcanic debris as high as 50 km into the **atmosphere**. The volcanic **dust** circled the earth, remaining in the upper atmosphere for several years. The resulting increase in **atmospheric turbidity** caused spectacular sunsets and lowered world temperatures – 1884 was the coolest year between 1880 and the mid-1990s. **H.H. Lamb** used the eruption of Krakatoa as the bench-mark against which other eruptions were measured when he developed his **dust veil index** (DVI).

Further reading
Austin, J. (1983) 'Krakatoa sunsets', *Weather* 38: 226–31.
Lamb, H.H. (1970) 'Volcanic dust in the atmosphere; with a chronology and assessment of its meteorological significance', *Philosophical Transactions of the Royal Society, A* 266: 435–533.

KUWAIT OIL FIRES

Major fires produced in the final stages of the Gulf War in 1991, when between 500 and 600 **oil** wells were set alight by the retreating Iraqi army. These wells continued to burn for

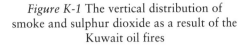

Figure K-1 The vertical distribution of smoke and sulphur dioxide as a result of the Kuwait oil fires

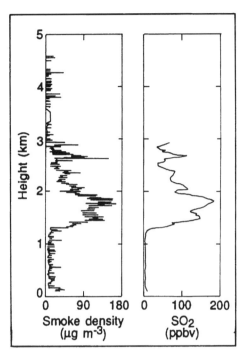

Source: After Jenkins, G.J., Johnson, D.W., McKenna, D.S. and Saunders, R.W. (1992) 'Aircraft measurements of the Gulf smoke plume', *Weather* 47 (6): 212–19

several months, kept alight by oil and **natural gas** brought to the surface under pressure from the underlying oilfields. During that time they introduced into the **atmosphere** probably the greatest amount of anthropogenically generated **aerosols** ever produced by a single event. Massive amounts of **smoke**, **sulphur dioxide** (SO_2), **carbon dioxide** (CO_2), unburned **hydrocarbons** and **oxides of nitrogen** (NO_x) were also emitted, mainly into the lower half of the **troposphere**. At the height of the fires, it was estimated that sulphur dioxide (SO_2) was being added to the atmosphere at an equivalent rate of 6.1 million tonnes per year and **soot** at 6.4 million tonnes per year. Most of the pollutants were retained close to the surface and the resulting reduction in incoming **solar radiation** was spectacular. Beneath the centre of the plume, **short-wave radiation** flux was measured at zero, leading to daytime **temperature** reductions of as much as 5.5°C. Mean monthly temperatures between March and September were reduced by 0.8 to 2.4°C and record low monthly temperatures were established in July and August. The low altitude of the initial plume, plus the rapidity with which the fires were extinguished, prevented them from having major impacts beyond the immediate region, although **acid rain** which fell in Iran, black **snow** observed in the mountains of Pakistan and unexpectedly high levels of **carbon** soot in the upper troposphere above Japan were all identified as products of the oil fires.

Further reading
Johnson, D.W., Kilsby, C.G., McKenna, D.S., Saunders, R.W., Jenkins, G.J., Smit, F.B. and Foot, J.S. (1991) 'Airborne observations of the physical and chemical characteristics of the Kuwait oil smoke plume', *Nature* 353: 617–21.
Seager, J. (1991) 'Operation Desert Disaster', *Ecodecision* September: 42–6.
Shaw, W.S. (1992) 'Smoke at Bahrain during the Kuwaiti oil fires', *Weather* 43: 208–10.

KWASHIORKOR

A form of **malnutrition** common in developing countries. It is the result of a low **protein**/high **carbohydrate** diet, and is particularly prevalent among young children between the ages of 1 and 3. Kwashiorkor is associated with poverty. It generally begins when the children are weaned and the protein-rich mother's milk is replaced by a diet of starchy foods such as plantains or cassava. The children suffer from anaemia, swelling of the abdomen, muscle wasting and exhibit general apathy. Kwashiorkor is often fatal, but even those who survive may suffer continued mental and physical disorders.

Further reading
Cheraskin, E., Ringsdorf, W.M. and Clark, J.W. (1968) *Diet and Disease*, Emmaus, PA: Rodale Books.
United Nations World Food Commission (1989) *The Global State of Hunger and Malnutrition*, New York: United Nations.

L

LA NIÑA

An intermittent cold current flowing from east to west across the equatorial Pacific Ocean, in those years when **El Niño** is absent. It is caused by strong equatorial easterly winds that push cold water, upwelling off the South American coast, far out into the ocean. Like El Niño, its influence appears to extend beyond the Pacific, being associated with increased **precipitation** in the **Sahel** and India, and below normal **temperatures** in central Canada.

See also
ENSO.

Further reading
Pearce, F. (1991) 'A sea change in the Sahel', *New Scientist* 130 (1757): 31–2.
Philander, G. (1989) *El Niño, La Niña and the Southern Oscillation*, Orlando, FL: Academic Press.

LAG TIME

The time between the occurrence of a phenomenon and its resulting effect. For example, peak stream discharge in a river basin may occur several hours after the peak storm rainfall which produced it. On a larger scale, many global **teleconnections** are time-lagged. There is a lag time of several months, for example, between the occurrence

Figure L-1 The lag time concept applied to precipitation and stream discharge

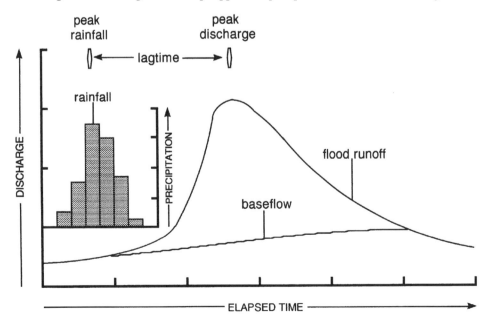

of an **ENSO** event and **drought** in Australia or India. Whenever cause and effect are separated by a period of lag time, there is the potential for the prediction of the effect. Knowing that a major storm has taken place in a river basin, the timing of the subsequent high water levels can be predicted and appropriate precautions taken. Modern developments aimed at predicting the timing and extent of droughts involve time-lagged teleconnections that would provide time to prepare a planned response to drought and therefore reduce its consequences.

LAMARCK, J.B. de (1744–1829)

A French biologist who published the first theory of **evolution** or 'transformism', based, in part, on the inheritance of acquired characteristics and habits. According to Lamarck's theory, modern giraffes, for example, have long necks because their ancestors stretched to reach the leaves on trees. The longer necks they acquired in this way were passed on to their offspring, creating a cumulative effect over many generations until current proportions were attained. Such modifications during the lifetime of an organism cannot be inherited. Lamarck's theories were superseded by those of **Charles Darwin** in the second half of the nineteenth century.

Further reading
Grant, V. (1985) *The Evolutionary Process, a Critical Review of Evolutionary Theory*, New York: Columbia University Press.
Stiling, P. (1992) *Introductory Ecology*, Englewood Cliffs, NJ: Prentice Hall.

LAMB, H.H.

British climatologist who had a major role in developing the investigation of **climate change**, particulary from the 1950s onwards. His interests are wide-ranging and include the physical aspects of change as well as the human impacts. Lamb developed the **dust veil index** (DVI), as a means of estimating the impact of increased **atmospheric turbidity** on **climate**, and was one of the first climatologists to appreciate the importance of primary historical sources such as the Domesday Book. In addition to undertaking his own investigations, Lamb has also assessed the work of other researchers, and through his publications has made them available to a wider audience.

Further reading
Lamb, H.H. (1970) 'Volcanic dust in the atmosphere with a chronology and assessment of its meteorological significance', *Philosophical Transactions of the Royal Society* A266: 435–533.
Lamb, H.H. (1972) *Climate: Present, Past and Future; Volume I, Fundamentals and Climate Now*, London: Methuen.
Lamb, H.H. (1977) *Climate: Present, Past and Future; Volume 2, Climatic History and the Future*, London: Methuen.
Lamb, H.H. (1996) *Climate, History and the Modern World* (2nd edition), London: Routledge.

LAMINAR FLOW

Non-turbulent, steady flow in a fluid. The flow takes the form of parallel layers that closely follow the shape of the underlying surface, and is best developed where the surface is smooth and flat or streamlined. If the surface is rough, however, the flow becomes turbulent or irregular. The nature of the flow is also related to its **viscosity**. Where the viscosity is low, as with **water** and **air**, laminar flow is uncommon.

See also
Turbulent flow.

Further reading
Rogers, D.F. (1992) *Laminar Flow Analysis*, New York: Cambridge University Press.

LAND CAPABILITY

A measure of the potential productive capacity of land for agricultural purposes. It is based mainly on the physical characteristics of the **soil** – for example, as revealed in the **soil profile** – and may include morphological **elements** such as gradient, susceptibility to **erosion** or drainage problems. Most land capability classifications range from a top category, that includes land with few limitations on its agricultural use, to the lowest, in which the land is best left in its

Figure L-2 A map showing the capability of land to support agriculture

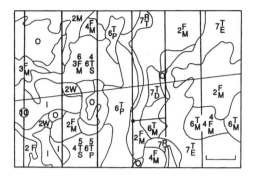

In this map the capability class is indicated by the first number, the subclass by letter and the proportion of the classes by the numbers located above the letters

e.g. 3F6T
     M S

means that 60% of the area is class 3 because of low fertility and lack of moisture and 40% is class 6 because of topography and soil conditions such as poor structure, low fertility and lack of moisture

Source: After Hoffman, D.W. (1976) 'Soil capability analysis and land resource development in Canada', in G.R. McBoyle and E. Sommerville (eds), *Canada's Natural Environment*, Toronto: Methuen

natural state or has been so altered that rehabilitation to make it suitable for agriculture is not feasible. Although land capability classifications were developed for agricultural purposes, they also have environmental implications. Land developed beyond its capability to support specific agricultural activities might suffer from **soil erosion**, for example. In theory, attention to the elements that determine land capability should prevent such problems, but in practice classifications may be too general for detailed decision making, or social and economic factors may override the suggested restrictions.

See also
Land use planning.

Further reading
Cutter, S.L., Renwick, H.L. and Renwick, W.H. (1991) *Exploitation, Conservation, Preservation: A Geographic Perspective on Natural Resource Use*, New York: Wiley.

LANDFILL

See **sanitary landfill.**

LANDSCAPE ARCHITECTURE

The conscious modification of an area to provide a more aesthetically pleasing, useful or enjoyable landscape. It has a long tradition stretching from the symbolic tranquility of China and Japan to the geometric formality of Renaissance Europe and the natural, picturesque parklands of eighteenth-century England. Modern landscaping is an integral part of the design and development of residential properties, public buildings, recreation areas and highways, although the results are not always environmentally appropriate. For example, the provision of lawns and golf courses, requiring large amounts of **irrigation** water in arid areas, creates a completely artificial landscape that is out of balance with the local **environment.**

LANDSCAPE ECOLOGY

The study of the patterns of **ecosystems** in the landscape and the ways in which the relationships among adjacent landscapes influences the functioning of individual units and of the landscape as a whole. **Runoff** and **erosion** in a mountainous landscape will impact on the landscapes in adjacent low ground, for example. The concept includes cultural as well as natural landscapes, and in many areas, such human activities as agriculture, forestry or urban development have a major impact on landscape ecology.

Further reading
Naveh, Z. and Lieberman, A.S. (1984) *Landscape Ecology: Theory and Applications*, New York: Springer-Verlag.
Vink, A.P.A. (1983) *Landscape Ecology and Land Use*, London: Longman.

LANDSCAPE MOSAIC

The combination of individual **ecosystems** across the landscape. Including human as well as natural landscapes, the study of these mosaics and the interactions among their

Figure L-3 A formal landscaped garden designed originally in the seventeenth century and restored in the 1950s

Photograph: The author

individual units is central to the concept of **landscape ecology**.

LAND USE PLANNING

The practice of assessing the best use for a piece of land, taking into account the widest range of **elements** possible. These may range from socioeconomic factors to environmental issues, and although all should receive equal treatment, the assessment on which the final land use is based is commonly the result of compromise with one or two elements receiving greater consideration than the others. The concept can be applied at different scales and in a variety of **environments**. In urban areas, land use planning, through zoning by-laws, allows an appropriate combination of residential, commercial and industrial uses. On a larger scale, in the natural environment, land use planning may require the consideration of land for **wilderness**, recreation or forestry, with the final decisions involving the assess-

ment of the best use for thousands of square kilometres of land.

See also
Land capability.

Further reading
Lounsbury, J.F., Sommers, L.M. and Fernald, E.A. (eds) (1981) *Land Use: A Spatial Approach,* Dubuque, IA: Kendall/Hunt.

LAPSE RATE

See **environmental lapse rate, adiabatic processes**.

LATENT HEAT

The quantity of heat absorbed or released by a substance during a change of state. Latent heat of fusion is absorbed from the **environment** in the transformation of a **solid** into a **liquid** and is retained in the liquid until it is transformed into a solid again, at which time

Figure L-4 Energy transfer during the change in state of water

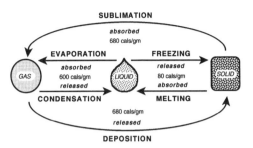

it is released. Similarly, latent heat of vaporization is absorbed when a liquid is converted into a **gas**, and retained in the gas until reconversion into a liquid. The latent heat redistribution during the change in the state of **water** has important implications for the earth's **energy budget**. For example, **energy** absorbed during the conversion of liquid water to **water vapour** is retained by the latter in the form of latent heat until the process is reversed. The water vapour may travel over great distances in the **atmosphere** in the period between the absorption and re-release of the energy, and in this way energy absorbed in one location is transported elsewhere in the earth/atmosphere system. The release of latent heat during **condensation** in rising air contributes to the continued buoyancy of that air, allowing **cloud** development at higher levels in the atmosphere.

Further reading
Lock, G.S.H. (1994) *Latent Heat Transfer: An Introduction to Fundamentals*, Oxford/New York: Oxford University Press.

LATERITE

A surface accumulation, consisting mainly of **iron** and **aluminum** oxides and **hydroxides**, common in humid tropical regions. The high **temperatures** and abundant **precipitation** of these regions contributes to very effective chemical **weathering** and the rapid **leaching** of **silica** (desilication) and various alkaline minerals from the upper layers of the **soil**, leaving the iron (Fe) and aluminum (Al) behind. The thickness of the accumulation varies from a few centimetres to tens of

metres. If sufficiently thick and with high levels of aluminum, it can be mined as bauxite, the **ore** of aluminum. Laterite usually develops as a hard, **impermeable** layer, or duricrust, with few **nutrients,** and the latosols that form on it are generally of low quality. The removal of surface vegetation in tropical regions exposes the surface to heat and heavy precipitation which encourages laterization and leads to reduced soil fertility.

Further reading
Bourman, R.P. (1993) 'Perennial problems in the study of laterite; a review', *Australian Journal of Earth Sciences* 40 (4): 387–401.
MacFarlane, M. (1983) 'Laterites', in A.S. Goudie and K. Pye (eds) *Chemical Sediments and Geomorphology,* London: Academic Press.

LAVA

Molten rock material or **magma** extruded from **volcanoes** or fissures in the earth's surface. Lava consists mainly of silicates, but also contains other minerals, such as **quartz**, and a variety of **gases**. Acid lavas have a higher proportion of **silica** than basic lavas. The former tend to be more viscous and solidify more readily than basic lavas. As a result, they may block the volcanic vents before an eruption is complete, leading to a build-up of pressure in the **crust** and a subsequent explosion. Basic lavas tend to flow more freely and cover hundreds of square kilometres before solidifying. The basaltic rocks of the Columbia–Snake Plateau of the Pacific North-west in the United States and parts of the Deccan Plateau in India represent flows of basic lava that occurred millions of years ago.

See also
Igneous rocks.

Further reading
Hyndman, D.W. (1985) *Petrology of Igneous and Metamorphic Rocks* (2nd edition), New York: McGraw-Hill.

LEACHATE

Liquid containing dissolved **solids** produced by the process of **leaching**. Although leachate is produced by natural processes, the term is commonly used to refer to polluted liquids released into the **environment** as a result of human activities. Leachate seeping from **waste disposal** sites, for example, is characterized by a high **BOD** and high acidity and may contain toxic chemicals capable of contaminating local surface and **groundwater** supplies.

See also
Sanitary landfill.

Further reading
McArdle, J.L., Arozarena, M.M. and Gallagher, W.E. (1988) *Treatment of Hazardous Waste Leachate*, Park Ridge, NJ: Noyes Data Corporation.
Qasim, S.R. and Chiang, W. (1994) *Sanitary Landfill Leachate: Generation, Control and Treatment*, Lancaster, PA: Technomic Publishing.

LEACHING

The process by which soluble **solids** are removed from **soils** or **waste disposal** sites by percolating **water**. In the **environment**, leaching is a common process in most soils, with the minerals removed from the upper layers of the soil either being redeposited lower down in the profile or removed completely. Human activities that remove organic matter from the soil or expose the soil to greater rates of infiltration (for example, by removing vegetation) encourage more effective leaching and lead to reduced soil fertility. Chemical **fertilizer** applied to the soil at rates that exceed the soil's absorption capacity will also be leached out. This has created a **pollution** problem in some agricultural areas where the over-application of fertilizers has led to the contamination of water supplies with **nitrates**. Accelerated leaching, particularly of **calcium ions**, also takes place in areas subject to **acid rain**. Modern **sanitary landfill** systems include an **impermeable** barrier such as a layer of **clay**, rubber or **plastic** to prevent contaminants from being carried into the natural environment.

See also
Base exchange.

Further reading
Brady, N.C. (1990) *The Nature and Properties of Soils* (10th edition), New York/London: Macmillan/Collier Macmillan.
Rowe, R.K., Quigley, R.M. and Booker, J.R. (1995) *Clayey Barrier Systems for Waste Disposal Facilities*, London: Spon.

LEAD (Pb)

A soft, grey, **heavy metal** produced by roasting the **ore** galena (lead sulphide (PbS)). Lead has four natural **isotopes**, three of which are the end products of the **radioactive** decay of **uranium** (U) and thorium (Th). It has the highest **density** of all **metals** in common use and, because of this, it provides a very effective shield against **X-rays** and **gamma rays**. Lead is therefore widely used in protective shielding for X-ray machines and **nuclear reactors**. Being malleable and therefore easily worked, lead has a long tradition of use in areas such as plumbing, printing and ceramics. Alloyed with tin (Sn), it produces a solder which is resistant to **corrosion** and stronger than the lead alone. It has also been used in lead-acid batteries, in paint and as an anti-knock additive in **gasoline**. Since lead compounds are poisonous, some, such as lead arsenate, have been used as **pesticides**. However, as the toxicity of lead becomes increasingly apparent, many of its traditional uses have been banned or severely restricted.

See also
Lead poisoning.

Further reading
Blaskett, D.R. and Boxall, D. (1990) *Lead and its Alloys*, New York: Ellis Horwood.

LEAD POISONING

Lead (Pb) and lead compounds are highly toxic when consumed and the symptoms of lead poisoning have long been recognized in miners and workers employed in industries using lead or lead-based products. It is only within the last several decades, however, that the true extent of lead poisoning has become apparent. According to the Centers for Disease Control and Prevention, it is the number one environmental health threat to children in the United States. The ingestion of small amounts of lead causes loss of appetite, headaches, tiredness and a variety of behavioural changes. The effects are usually much greater in children than in adults, and pregnant women and their unborn children are also particularly vulnerable. However, because the symptoms of low-level lead poisoning are shared with other ailments, the problem can remain undiagnosed. Since lead accumulates in the body, the continued consumption of even small amounts may ultimately lead to much greater problems such as brain damage, paralysis and eventually death. The main sources of lead in the environment include, or have included, emissions from industries producing lead or lead-based products, emissions from automobile engines burning leaded **gasoline**, emissions from **incinerators**, lead piping used in water supply, lead-based paints and glazes and **leachate** from landfill sites. In the developed nations of North America and Europe, legislation has been introduced to deal with these sources, by controlling emissions, banning the use of leaded gasoline and lead-based paints or prohibiting the use of lead piping, but large quantities of the lead **compounds** produced in the past remain in the **environment** and thousands of houses still have lead piping in their plumbing systems and lead-based paints on their walls and even furniture. Thus, lead poisoning will remain a recognizable environmental health problem even in those countries where attempts are being made to deal with it, and is likely to grow in the less developed nations where regulations are not yet in place or are inadequately enforced.

Further reading

Centers for Disease Control and Prevention (1991) *Preventing Lead Poisoning in Young Children,* Atlanta: Centers for Disease Control and Prevention.
Wallace, B. and Cooper, K. (1986) *The Citizen's Guide to Lead: Uncovering a Hidden Health Hazard,* Toronto: NC Press.

LEGUME

See **leguminous plants**.

LEGUMINOUS PLANTS

A large group of pod (or legume) bearing plants (Leguminosae) that includes trees (acacia, carob), shrubs (whin or gorse) and a variety of commercially important **species**, such as peas, beans, alfalfa, clover, lentils and soybeans. Nodules on the roots of leguminous plants contain **nitrogen**-fixing **bacteria** (*Rhizobium*) which have an important role in the earth/atmosphere **nitrogen cycle**. They absorb atmospheric nitrogen (N) and convert it into **amino acids**, a form in which it can be used by the plants. Leguminous fodder crops, such as clover and alfalfa, are often grown as green manures, being ploughed back into the **soil** to improve its nitrogen level and organic content.

Further reading

Dilworth, M.J. and Glenn, A.R. (eds) (1991) *Biology and Biochemistry of Nitrogen Fixation,* New York: Elsevier.
Nutman, P.S. (ed.) (1976) *Symbiotic Nitrogen Fixation in Plants,* Cambridge/New York: Cambridge University Press.

LEOPOLD, A. (1887–1962)

Aldo Leopold trained as a forester and worked for the United States Forest Service, before becoming a professor of game management at the University of Wisconsin.. His interests became focused on the **ecology** of large game **populations** and in the United States he is regarded as the father of scientific wildlife management. He appreciated the interrelationships among the various components of the **environment** and saw the concept of the **ecosystem** as the essential element in the management of nature. In his work, he also examined the ethical factors involved in the human relationship with nature. Practising what he preached, Leopold supported the establishment of **wilderness** reserves and was a founding member of the **Wilderness Society**. Although now more than half-a-century old, his writings are still considered as basic references for students of wildlife ecology.

Further reading

Leopold, A. (1933) *Game Management,* New York: Scribner & Sons.
Leopold, A. (1949) *A Sand County Almanac,* New York: Oxford University Press.

LEVÉES

Natural ridges of **alluvium** that build up along the banks of a river as a result of the deposition of sediments when the river overflows its banks during a **flood**. As a result of this deposition natural levées are higher than the adjacent **floodplain**, and under conditions of normal flow the river remains within them. Along many rivers, particularly where the floodplain has been developed for human use, the levées are raised artificially to protect against flooding.

Further reading
Pavel, P. (1982) *Canal and River Levées*, Amsterdam/New York: Elsevier.

LICHENS

Compound organisms based on the association of **algae** and **fungi** in a symbiotic relationship. They grow on a variety of surfaces from trees to bare rock, obtaining their **nutrients** from these surfaces. Lichens are usually the primary colonizers of newly exposed surfaces, and begin the series of processes that break down the rock surface and ultimately lead to the formation of **soil**. They dominate Arctic and high mountain **environments**, where conditions are too harsh for plants to grow. Being primary colonizers and growing progressively larger with time, lichens can be used in a dating technique called lichenometry. Measurements of the largest lichens present in an area can provide an estimate of the date at which a rock surface was exposed or a deposit stabilized. Lichens are also sensitive to **pollution** and can be used as biological indicators of pollution levels in an area.

See also
Symbiosis.

Further reading
Hale, M.E. (1983) *The Biology of Lichens* (3rd edition), London: Edward Arnold.

LIFE CYCLE ANALYSIS

A development of the economic concept of life cycle cost, which considers not only the initial cost of a commodity, but also its lifetime operating costs. Applied to environmental issues, it involves an assessment of the environmental impacts expected from a product and the activities or processes associated with its manufacture, use and final disposal. The concept takes into account the environmental costs of raw material extraction, **energy** costs at all stages in the life of the product and the costs which for some products remain even after disposal. In the nuclear industry, for example, where products retain their potential to cause environmental damage for thousands of years, costs may continue to accrue long after the product has been scrapped. In theory, life cycle analysis leads to more efficient use of **resources** with less environmental impact.

LIGHT

See **visible light**.

LIGHTNING

Luminous electric discharges usually associated with thunderstorms, but also occurring in volcanic clouds and snowstorms. The rapid expansion of the **gases** in the **atmosphere** caused by the lightning produces the sound of thunder. For lightning to occur, charge separation must take place, either within a **cloud**, between clouds or between a cloud and the earth's surface. The origin of the charge separation is complex and not fully understood, but it appears to be associated with rapid vertical movement and the formation of **precipitation** within a mature cumulonimbus cloud. With time, the upper part of the cloud becomes positively charged and the base negatively charged. The negative charge at the cloud base in turn induces a positive charge at the earth's surface. Since air is a poor **conductor** of electricity, the potential gradients produced may exceed 1 million **volts**, and it is the rapid flow of **electrons** required to resolve these differences in the charges that produces the lightning flash. Lightning causes major economic losses and several hundred deaths around the world every year. In **grassland** and forest areas, it has an important role as an ecological agent because of the fires it causes.

Further reading
Golde, R.H. (ed.) (1977) *Lightning,* London/New York: Academic Press.
Lutgens, F.K. and Tarbuck, E.J. (1989) *The Atmosphere,* Englewood Cliffs, NJ: Prentice-Hall.

LIGNIN

A complex **polymer** which provides strength and rigidity to the **cell** walls in woody plants. It constitutes as much as 40 per cent of the wood in some trees and must be separated chemically from the **cellulose** required to produce pulp for the paper and rayon industries.

LIKENS, G.

A pioneer in the study of **acid rain** in North America, Gene Likens along with his colleague Herbert Bormann set up a multidisciplinary study of a small New Hampshire watershed in 1963. The chemistry of the local rainwater, which was part of the study, indicated the presence of highly acidic **precipitation** in the area despite its remoteness from sources of acid **gas** emissions. When the results were reported in the journal *Science* in 1974, they led to increased scientific interest in the problem and, when followed up by the national media, initiated public awarness of the presence of the problem in the United States.

Further reading
Likens, G.E. (1976) 'Acid precipitation', *Chemical and Engineering News* 54: 29–37.
Likens, G.E. and Bormann, F.H. (1974) 'Acid rain – a serious regional environmental problem', *Science* 184: 1176–9.
Park, C.C. (1987) *Acid Rain: Rhetoric and Reality,* London: Methuen.

LIME INJECTION MULTI-STAGE BURNING (LIMB)

A technique developed to reduce acid **gas** emissions from **coal**-burning furnaces. Fine lime is injected into the **combustion** chamber where it fixes the **sulphur** (S) released from the burning coal to produce a sulphate-rich lime ash. LIMB can reduce **sulphur dioxide** (SO_2) emissions by 35–50 per cent and oxides of **nitrogen** (NO_x) emissions by 30 per cent.

See also
Fluidized bed combustion.

Further reading
Burdett, N.A., Cooper, J.R.P., Dearnley, S., Kyte, W.S. and Turnicliffe, M.F. (1985) 'The application of direct lime injection to U.K. power stations', *Journal of the Institute of Engineering,* 58: 64–9.

LIMESTONE

A **sedimentary rock** consisting mainly of **calcium carbonate** ($CaCO_3$). It is formed by **evaporation** from carbonate-rich **solutions**, by the accumulation of the skeletons of dead marine organisms, by the accumulation of existing limestone fragments or some combination of two or more of these processes. Being porous, limestones are

Figure L-5 The limestone cliffs of the Cheddar Gorge in south-west England

Photograph: Courtesy of Heather Kemp

important **aquifers** and can act as **reservoirs** for **petroleum hydrocarbons**. Limestones have been widely used as building stones and can be combined with **clay** to produce cement. When limestone is heated, **carbon dioxide** (CO_2) is driven off and calcium oxide (CaO) or lime is left. Lime has been used for centuries to sweeten acid **soils**, and as a **base** it is widely used to neutralize the acid **gas** emissions responsible for **acid rain**.

Further reading
Selley, R.C. (1996) *Ancient Sedimentary Environments and their Sub-Surface Diagnosis* (4th edition), London: Chapman and Hall.

LIMITING FACTORS

Environmental factors that restrict the growth and distribution of an organism or group of organisms. They may be physical in nature including, for example, **temperature** and **precipitation**, or chemical such as the absence of specific **nutrients** in the **soil**. They help to determine the nature of a particular **ecosystem**, but in addition, successful agricultural development often depends upon the identification of these limits and the adoption of techniques to accommodate them.

Further reading
Cox, B.C. and Moore, P.D. (1993) *Biogeography: An Ecological and Evolutionary Approach*, Oxford: Blackwell.
Park, C.C. (1980) *Ecology and Environmental Management*, Folkestone: Dawson.

LIMITS TO GROWTH

A report commissioned by the **Club of Rome** as part of its project to examine the *Predicament of Mankind*. Published in 1972, it contained the results of the computer analysis of world **population** growth, **resources**, food supply, **pollution** and industrial output, carried out by a group of scientists at the Massachusetts Institute of Technology (MIT). It indicated that the human population of the earth would exceed the **carrying capacity** of the planet within a century if the growth rates current in the 1960s and 1970s continued. Proposed solutions included **zero population growth**,

Figure L-6 Potential trends in population, pollution and world resource use identified in the 'Club of Rome' project on the predicament of mankind. This standard model run assumes no major changes in the relationships that have historically governed the development of the world system

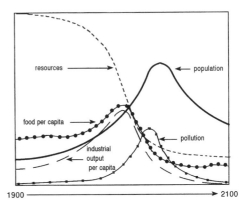

Source: After Meadows, D.H., Meadows, D.L. Randers, J. and Behrens, W.W. (1974) *The Limits to Growth*, London: Pan Books

pollution control, **recycling** programmes and appropriate **soil** and resource management, all of which are now accepted in some form or other as integral to the concept of **sustainable development**.

See also
Club of Rome.

Further reading
Cole, H.S.D. (1973) *Thinking About the Future: A Critique of 'The Limits to Growth'*, London: Chatto & Windus/Sussex University Press.
Meadows, D.H., Meadows, D.L., Randers, J. and Behrens, W.W. (1972)*The Limits to Growth*, New York: Universe Books.

LIMNOLOGY

The study of the physical, chemical and biological components of the **environment** of freshwater ponds, lakes and **reservoirs**.

LIQUEFIED NATURAL GAS (LNG)

Natural gas, mainly **methane** (CH_4), which has been liquefied by cooling below 150K.

Under most circumstances, natural gas is transported via pipeline (*c.* 75 per cent), but if that option is not available it is most easily transported in its **liquid** state, in insulated or refrigerated containers. LNG tanker traffic is dominated by Algeria and Indonesia, and Japan is the main consumer of LNG with its supplies coming mainly from south-east Asia and Australia.

LIQUEFIED PETROLEUM GAS (LPG)

A mixture of **petroleum gases** – butane (C_4H_{10}), **propane** (C_3H_8) and pentane (C_5H_{12}) – liquefied under pressure and stored in metal containers. It can be used as an engine **fuel** and, when available in portable pressurized containers, it is widely used for cooking, heating and lighting. It is a clean-burning fuel producing little or no atmospheric **pollution**.

LIQUID

Intermediate between a **solid** and a **gas**, it is a substance in which the **molecules** are relatively free to move, but within limits imposed by cohesive forces between them. Liquids maintain a fixed volume, but offer no resistance to a change of shape, and therefore assume the shape of the vessel in which they are contained. Liquids are only slightly compressible, which makes them suitable for use in hydraulic control systems.

LIQUID PHASE REACTION

The conversion of acid **gases** into **liquid** acids, the reactions taking place in **solution**. The process is very efficient, with the conversion of **sulphur dioxide** (SO_2) into **sulphuric acid** (H_2SO_4) over Britain being measured at rates as high as 100 per cent per hour in the summer and 20 per cent per hour in the winter.

See also
Acid precipitation, Gas phase reaction.

Further reading
Mason, B.J. (1990) 'Acid rain – cause and consequence', *Weather* 45: 70–9.

LITHOSPHERE

The outermost layer of the solid earth. It consists of the **crust** and the upper, rigid part of the **mantle**. The lithosphere is thinnest under the ocean basins, being only a few kilometres thick in the vicinity of the mid-oceanic ridges, and thickest under the continents, where it may reach thicknesses of 300 km or more.

See also
Plate tectonics.

LITTLE (CLIMATIC) OPTIMUM

A phase in early medieval times from about AD 750 to 1200, sometimes called the

Figure L-7 The Little Optimum (Medieval Warm Period) and the Little Ice Age in the long-term temperature record

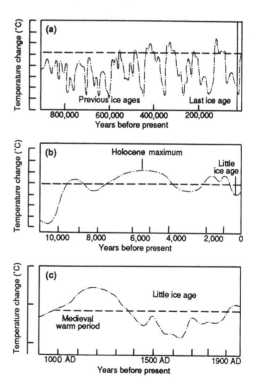

Source: IPCC (1990) *Climate Change: The IPCC Scientific Assessment*, Cambridge: Cambridge University Press

medieval warm spell, when climatic conditions ameliorated in Europe and North America and probably elsewhere. The improvement was sufficient to reduce the amount of sea ice in the North Atlantic allowing the colonization of Iceland and Greenland by settlers from the Nordic countries. The Greenland settlements were subsequently abandoned during the Little Ice Age. During the warm spell, however, crop yields improved in marginal areas such as Scotland, and vines were cultivated in Britain several hundred kilometres beyond the current northern limit of the vine.

Further reading
Grove, J.M. and Switsur, R. (1994) 'Glacial geological evidence for the Medieval Warm Period', *Climatic Change* 26: 143–69.
Lamb, H.H. (1996) *Climate, History and the Modern World* (2nd edition), London: Routledge.
Zhang, D. (1994) 'Evidence for the existence of a Medieval Warm Period in China', *Climatic Change* 26: 289–97.

LITTLE ICE AGE

Also referred to as the neoglacial and the Fernau glaciation, the Little Ice Age was a period of global cooling lasting for about 300 years from the mid-sixteenth to the mid-nineteenth century. There is evidence, however, that the cooling started as early as the mid-fourteenth century in some areas. It included prolonged periods of particularly cool and wet conditions – as in the 1690s, for example – but it was also characterized by considerable variability with some periods that were warm and dry – for example, the 1730s. Most of the information available is from the northern hemisphere, but the Little Ice Age was a global event, although specific elements were not always completely synchronous. The shorter and less intense growing seasons, along with the cool summers and longer winters, had a significant effect on socioeconomic conditions in North-west Europe, Iceland, Greenland and eastern North America. Although the causes are not known, some researchers have suggested that greater **atmospheric turbidity** following increased volcanic activity was a contributory factor. There may be links with individual

volcanic events and particularly cool years, such as the eruption of **Mount Tambora** in 1815 and the 'year without a summer' in 1816, but in general the degree of contemporaneity between increased volcanic activity and climatic deterioration is limited.

Further reading
Grove, J.M. (1988) *The Little Ice Age*, London: Routledge.
Lamb, H.H. (1996) *Climate, History and the Modern World* (2nd edition), London: Routledge.
Mannion, A.M. (1991) *Global Environmental Change: A Natural and Cultural Environmental History*, London: Longman.
Moore, P.D. (1995) 'Back to the Little Ice Age', *Nature*: 684–5.

LITTORAL ZONE

The zone along the landward margins of lakes and seas. On lakes, it extends from the shoreline to the limit of rooted vegetation. Along the edges of the **oceans**, it is equivalent to the intertidal zone. The littoral environment frequently suffers major disruption as a result of **pollution** following oil spills.

LIVE AID

An organization set up in 1985 to help the victims of **drought** and **famine** in Ethiopia. It raised money through two major concerts in England and the United States at which the leading popular entertainers of the day performed. Extensive television coverage raised worldwide public concern to new heights, and helped to make the event a most effective fund-raiser for famine relief.

LIVESTOCK WASTE

Sewage produced by farm animals. Under traditional agricultural operations, the **waste** products were collected and **recycled** as organic **fertilizers**. Modern agribusiness techniques in which animals, particularly cattle, are prepared for market in feedlots that may cover several square kilometres, involve the concentration of large quantities of livestock waste. This presents serious disposal problems, which if handled inappropriately can lead to **groundwater** and surface water **pollution**.

LOESS

A sedimentary deposit formed towards the end of **ice ages**, when fine, silt-sized rock flour produced by glacial **erosion** is carried away by strong **winds** to be deposited sometimes as far as several hundred kilometres from the source. Some loess may have been derived from desert sources. Loess deposits are located mainly in the northern hemisphere, in North America, Europe and Asia. **Soils** formed on loess are very fertile because they are fine-grained and contain a mixture of all of the minerals derived from the bedrock eroded by the glaciers. The thickest deposits, averaging 75 m and up to 180 m in places are found in northern China, where agricultural production from the fertile soils supports a dense **population**, living in dwellings tunnelled out of the loess.

Further reading
Pye, K. (1987) *Aeolian Dust and Dust Deposits*, London: Academic Press.
Smalley, I.J. (ed.) (1975) *Loess: Lithology and Genesis*, Stroudsberg, PA: Dowden, Hutchinson and Ross.

LONDON MINISTERIAL CONFERENCE ON OZONE (1990)

A follow-up to the **Montreal Protocol**, the London Ministerial Conference was successful in strengthening the restrictions on the production of **CFCs** first proposed in Montreal. It also set up a fund of US$240 million as a means of helping **Third World** countries to obtain the technology necessary to develop alternatives to CFCs.

See also
Ozone depletion.

LONDON SMOG (1952)

A major **pollution** event in London, England caused by a combination of meteorological conditions (low **temperatures**, high **atmospheric pressure**, poor ventilation) and **energy** use. The burning of high-sulphur **coal** released **sulphur dioxide** (SO$_2$) into the urban **atmosphere** where it was converted into

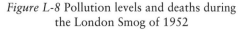

Figure L-8 Pollution levels and deaths during the London Smog of 1952

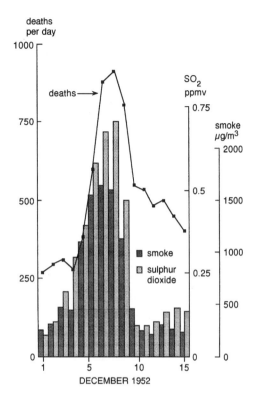

Source: Derived from data in Bates, D.W. (1972) *A Citizen's Guide to Pollution*, Montreal: McGill-Queen's University Press

droplets of **sulphuric acid** (H$_2$SO$_4$). In combination with **smoke** and **soot** particles, this reduced visibility, and created a health risk. An estimated 4000 extra deaths were attributed to the smog, and the event was credited with encouraging the passage of the **clean air legislation** that followed in the late 1950s. A similar episode occurred in 1962, with a death-toll of some 700, mainly elderly people or those suffering from chronic respiratory ailments.

See also
Acid precipitation, Atmospheric turbidity.

Further reading
Brimblecombe, P. (1987) *The Big Smoke: A History of Air Pollution in London since Medieval Times*, London/New York: Methuen.

LONG RANGE TRANSPORTATION OF AIR POLLUTION (LRTAP)

The transportation of air **pollution** over great distances – usually in excess of 500 km – by the prevailing **winds** in the **atmosphere**. Pollutants released into the upper atmosphere, where the **winds** tend to be more persistent, are particularly prone to long range transportation. The introduction of the **tall stacks policy**, which increased the height of emissions, encouraged LRTAP and contributed to the spread of **acid rain** damage. Acid **gas** emissions from British industry, for example, were carried eastwards, being converted into **sulphuric acid** (H_2SO_4) during the journey, to fall as acid rain in Scandinavia.

See also
Acid precipitation, INCO.

Further reading
Ottar, B. (1976) 'Monitoring long range transport of air pollution: the OECD study', *Ambio* 5: 203–6.

LONG-WAVE RADIATION

Relatively **low-energy radiation** from the infrared sector of the **electromagnetic spectrum**. **Terrestrial radiation** is long wave.

See also
Infrared radiation.

LOVE CANAL

A name now synonymous in North America with the dangers associated with the uncontrolled dumping of toxic and **hazardous wastes**. In the 1940s and early 1950s, near Niagara Falls, New York, **waste** materials from a **pesticides** and **plastics** operation were dumped in an area excavated for a canal but never developed. Some 20,000 tonnes of hazardous chemicals, sealed in drums, were dumped in the canal and when full the dump was backfilled with **clay** and covered with **topsoil**. Some time later a school was built on the property and the adjacent area was developed for residential use. By the 1970s it became clear to the residents that there were problems with the site. The area smelled of chemicals, toxic waste began to migrate into basements, storm sewers, gardens and the school playground. Children playing in the area suffered chemical burns, and adverse health effects such as eye and nose irritation, rashes and headaches became commonplace. There was also some indication of an increase in more serious disorders, such as various forms of **cancer**, nerve and kidney disease and birth defects. After twenty years underground, the barrels containing the waste chemicals had corroded, allowing the contents to escape into the local **environment**. There was little immediate response to the problems, but persistent lobbying by the residents plus intense and unfavourable coverage by the media eventually forced the state government of New York to take action. Environmental tests indicated the presence of more than eighty different chemicals, and health surveys revealed that the incidence of birth defects, miscarriages, respiratory disease, nasal and sinus infections and urinary tract disorders was well above the national average. Some doubts have been raised about the statistical significance of these tests, but as a result of such evidence the site was declared a disaster area by the federal government in 1980, the school was closed and the residents evacuated. After twelve years and some US$250 million spent on clean-up and rehabilitation of the site, about two-thirds of the area originally contaminated was declared fit for habitation and by 1993 houses were again being built and sold in the area. Following the Love Canal incident, it became clear that the dump was only one of many. At a minimum, tens of thousands of similar sites exist in the United States alone and a comparable number is likely for Europe, where the situation in the former Eastern Bloc nations and the republics that were once part of the USSR is particularly serious. To escape the restraints of modern environmental controls, some companies have contracted to dump waste products in **Third World** countries, creating the potential for future problems in these areas.

Further reading
Chiras, D.D. (1994) *Environmental Science: Action for a Sustainable Future*, Redwood City, CA:

Benjamin/Cummings.
Gibbs, L. (1982) *The Love Canal: My Story*, Albany: State University of New York Press.
Kolata, G.B. (1980) 'Love Canal: false alarm caused by botched study', *Science* 208: 1239–42.

LOVELOCK, J.E.

British scientist and inventor. His detection of CFCs in the upper **atmosphere** helped to pave the way towards an understanding of **ozone depletion**. James Lovelock is probably most widely recognized, however, for his development of the **Gaia hypothesis**, which viewed the earth as a single integrated organism.

Further reading
Lovelock, J.E. (1972) 'Gaia as seen through the atmosphere', *Atmospheric Environment* 6: 579–80.
Lovelock, J.E. (1986) 'Gaia: the world as living organism', *New Scientist* 112 (1539): 25–8.
Lovelock, J.E. (1988) *The Ages of Gaia*, New York: Norton.
Lovelock, J.E. (1995) *Gaia: A New Look at Life on Earth* (2nd edition), Oxford: Oxford University Press.

LOW-INPUT SUSTAINABLE AGRICULTURE (LISA)

An acronym for low-input **sustainable agriculture**, a programme developed by the US Department of Agriculture in the mid-1980s to encourage natural farming with less synthetic inputs. Farmers practising LISA continue to use synthetic **fertilizers** and **pesticides**, but in limited amounts and where possible only as a last resort. It is estimated that some 30–40 per cent of farmers in the US have taken some steps to reduce dependency on agricultural chemicals. The programme also encourages the movement away from monoculture, and has the overall aim of all sustainable agricultural practices of maintaining **soil** fertility and productivity.

Further reading
Kaufman, D.G. and Franz, C.M. (1993) *Biosphere 2000: Protecting Our Global Environment*, New York: HarperCollins.

LULU

Locally unwanted land use.

LYSIMETER

A device for measuring **evapotranspiration** from a vegetated **soil** column. A column of soil with the vegetation growing on its surface is isolated from its surroundings in a container. By comparing the input of **precipitation** with the drainage through the base of the system, and by weighing the column to estimate **soil moisture storage**, it is possible to calculate a **water budget** for the whole column. Since the **environment** is controlled, the input of moisture measured in a rain gauge minus the sum of the drainage and storage elements provides an estimate of the moisture lost from the column by **evaporation** and **transpiration**.

Further reading
Forsgate, J.A., Hosegood, P.H. and McCulloch, J.S.G. (1965) 'Design and installation of semi-enclosed hydraulic lysimeters', *Agricultural Meteorology* 2: 43–52.

M

MAGMA

Molten rock material consisting mainly of silicates with included **gases** and some **solids**. It originates in the lower part of the earth's **crust** or **mantle**, where it reaches **temperatures** as high as 1000°C. When it solidifies, it forms **igneous rock**. Magma may extrude at the earth's surface through volcanic vents or fissures in the form of **lava** or it may remain intrusive, solidifying within or between existing rocks.

Further reading
Hyndman, D.W. (1985) *Petrology of Igneous and Metamorphic Rocks* (2nd edition), New York: McGraw-Hill.

MAGNESIUM (Mg)

A light, silver-white **metal** which tarnishes on exposure to **air** and burns intensely to form magnesium oxide (MgO). In the natural **environment**, it is a constituent of a number of carbonate minerals such as magnesite ($MgCO_3$) and dolomite ($MgCO_3.CaCO_3$) and is present in **chlorophyll**. Magnesium **compounds** also cause **hard water**. It is used in industry in the production of lightweight **alloys**, in photography and in medicine.

MAGNETOHYDRODYNAMIC GENERATOR (MHD)

A direct **energy**-conversion **system** which produces electricity by passing a stream of very high **temperature** ionized **gas** through a magnetic field. The gas is produced by burning finely crushed **coal**, seeded with **potassium** (K) and caesium (Cs) to encourage **ionization**, at very high temperature. The electricity is drawn off by means of **electrodes** inserted into the gas flow. Since it involves high **combustion** temperatures and no moving steam turbines or generators, an MHD system has a conversion efficiency of about 60 per cent compared to the 30–40 per cent of a conventional thermal generating system. The MHD system also produces less acid gases than the conventional system. Once the gas has passed through the system, it is still very hot and can be used to generate steam in a normal generator coupled to the downstream end of the MHD system.

See also
Cogeneration.

Further reading
Kleinbach, M.H. and Salvagin, C.E. (1986) *Energy Technologies and Conversion Systems*, Englewood Cliffs, NJ: Prentice-Hall.

MAGNETOSPHERE

The space around the earth and other planets in which there is an external magnetic field. Unlike the field around a common magnet, the magnetosphere is not symmetrically distributed around the earth. As a result of interactions with the **solar wind**, it is distorted into the shape of a comet with the tail stretching away from the sun. On the side of the earth closest to the sun the magnetosphere extends some 60,000 km out into space, but on the opposite side the tail carries the magnetic field much further out, perhaps as far as 6 million km. The entrapment of **radiation** particles within the magnetosphere is responsible for such phenomena as the aurora.

Further reading
Ratcliffe, J.A. (1972) *An Introduction to the Ionosphere and Magnetosphere*, Cambridge: Cambridge University Press.

MALARIA

A parasitic disease endemic to many parts of the tropics in Africa, south-east Asia and Latin America. It is spread when the Anopheles mosquito injects the parasite into the blood of its victims during feeding. Symptoms include general malaise, fever and sweating, that recur with some regularity often over many years as the parasite passes through its life cycle. Malarial infection of the brain can lead to coma and death. Malaria has traditionally been treated with quinine-based drugs, but in a number of areas the parasite has developed an immunity to them and even to some of the new drugs designed to replace them. The best method of attacking the disease is to treat the causes rather than the symptoms. Mosquito breeding areas such as swamps and pools can be drained, or treated with **pesticides**, for example.

See also
DDT, Insecticides.

Further reading
Harrison, G.A. (1978) *Mosquitoes, Malaria and Man*, New York: Dutton.
National Academy of Sciences (1991) *Malaria; Obstacles and Opportunities*, Washington, DC: National Academy Press.

MALNUTRITION

A result of the consumption of essential **nutrients** at levels inadequate to maintain good health. Malnutrition can occur even when total **calorie** requirements are being met. People forced to live on a low **protein**, high **starch** diet, for example, may suffer from malnutrition as a result of deficiencies in proteins or other **nutrients**. Individuals suffering from malnutrition are usually prone to disease, too weak to work effectively and too tired to think clearly. Children may have their physical and mental development retarded by malnourishment in their growing years. Many die prematurely as a direct result of the malnutrition or because of diseases which they cannot resist in their malnourished state. In the past, malnutrition was common in marginal areas such as the **Sahel**, where **drought** and harvest failure regularly led to **famine**. Despite modern worldwide

Figure M-1 The global distribution of malaria

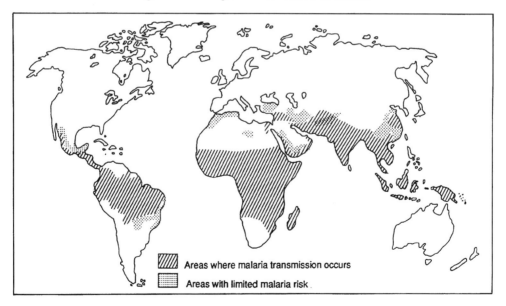

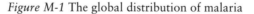
Areas where malaria transmission occurs
Areas with limited malaria risk

Source: Various

food distribution systems, that situation still prevails to some extent, but with the growth of world **population** the problem has become more dispersed. Life-threatening malnutrition is now endemic among the **Third World**'s landless rural labourers and urban poor. More than a billion people around the world suffer malnutrition and some 40 million die prematurely every year as a direct or indirect result.

See also
Kwashiorkor.

Further reading
Brown, L.R. (1994) 'Facing food insecurity', in L.R. Brown (ed.) *State of the World 1994*, New York: W.W. Norton.

MALTHUS, T.R. (1766–1834)

Thomas Robert Malthus, an English clergyman and economist, was one of the first to address the problems associated with a rapidly growing **population**. In 1789 he published a short book entitled *An Essay on the Principle of Population as it Affects the Future Improvement of Society*, followed by a second edition in 1803, in which he set out his concerns and his conclusions. He recognized that whereas population growth tended to follow a geometric progression (1, 2, 4, 8, 16, 32, 64 . . .), the growth in food production, on which the population depended, was arithmetic (1, 2, 3, 4, 5, 6, 7 . . .) and concluded that population growth would ultimately exceed the available food supply. At that stage, **famine**, disease and warfare would reduce the population until it was once again within the bounds of the available subsistence. As an alternative, Malthus suggested that the practice of 'moral restraint' – continence and delayed marriage – could reduce population growth to manageable rates. Although this seemed preferable to alternatives such as poverty, vice and misery, Malthus seemed to have little confidence that the population would exercise the necessary restraint and foresaw that the more drastic events would provide the main checks on population growth. The opening up of new lands for European migration and improved agricultural technology that boosted food production later in the nineteenth century confounded his predictions, and the development of birth control and other family planning techniques showed that 'moral restraint' was possible. Continuing high productivity levels in agriculture into the twentieth century, coupled with a significant reduction in fertility in developed nations seemed to show that Malthus's fears were unfounded. In the latter half of the century, however, rapidly expanding populations in **Third World** regions, unable to produce enough food for their own survival, and too numerous to be supplied with surpluses from elsewhere, are again experiencing the starvation, disease and war foreseen by Malthus as the main natural checks on population growth.

See also
Carrying capacity, Darwin, C.R., Exponential growth.

Further reading
Ehrlich, P.R. and Ehrlich, A.H. (1990) *The Population Explosion*, New York: Simon & Schuster.
Lloyd, W.F. (1977) 'On the checks to population', in G. Hardin and J. Baden (eds) *Managing the Commons,* San Francisco, CA: Freeman.

MAN AND BIOSPHERE PROGRAM (MAB UNESCO)

An international scientific programme, launched in 1971 under the sponsorship of the **United Nations Educational, Scientific and Cultural Organization** (UNESCO), to study the extent and nature of the human impact on the **biosphere**. Involving 110 national committees, its main achievement in the first two decades of its existence was the creation of a network of biosphere reserves to be examined and researched by scientists from a wide range of disciplines in an attempt to establish the scientific basis for **sustainable development**. From the preliminary work on the reserves, a number of priority areas were recognized as having potential for future development. These included continuing work on approaches to sustainable development, the conservation and sustainable use of **biodiversity**, the effective communication of information on environmental and development issues and the setting up of institutional functions and capacities to deal with the emerging problems.

MAN'S ROLE IN CHANGING THE FACE OF THE EARTH (MRCFE)

The first major international conference to address the impact of human activities on the **environment**, held in Chicago in 1956 and attended by a variety of academics including geographers, biologists, economists and historians. Although in large part a review of the historical development of environmental issues to that time, it also drew attention to such problems as **air** and **water pollution**, **resource** depletion, **waste disposal** and **soil erosion**, which were to become of major concern in the decades that followed.

Further reading
Thomas, W.L. (1956) *Man's Role in Changing the Face of the Earth*, Chicago, IL: University of Chicago Press.

MANKIND AT THE TURNING POINT

The second report sponsored by the **Club of Rome**, published in 1975. Like the earlier report **Limits to Growth**, it was based on computer analysis designed to provide

Figure M-2 Projected population growth and protein requirements in south-east Asia according to The Second Report of the Club of Rome

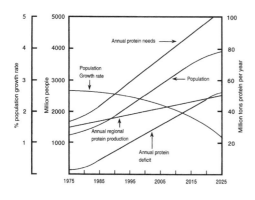

Source: After Mesarovic, M. and Pestel, E. (1975) *Mankind at the Turning Point*, London: Hutchinson

information on world **population** growth and the availability of **resources**. Unlike Limits to Growth, however, it did not present a doomsday prophecy, but suggested a number of scenarios, obtained from a computerized world development model, that the authors claim can avert potential catastrophe.

Further reading
Mesarovic, M. and Pestel, E. (1975) *Mankind at the Turning Point*, London: Hutchinson.

MANTLE

That part of the interior of the earth lying between the **crust** and the core. It extends between 50–70 km and 2900 km below the surface. Part of the upper mantle is rigid and with the crust forms the **lithosphere**, whereas the asthenosphere immediately below it is at least partially molten. Convective activity in the lower mantle in combination with active subduction is thought to be responsible for the movement of crustal plates. **Magma** from the mantle may escape to the surface during volcanic activity.

See also
Plate tectonics.

Further reading
Poirier, J.P. (1991) *Introduction to the Physics of the Earth's Interior*, Cambridge/New York: Cambridge University Press.

MARITIME TROPICAL AIR MASS (mT)

An **air mass** originating over the **oceans** in tropical latitudes, and therefore hot and moist. Maritime tropical air is also inherently unstable and capable of producing large amounts of **precipitation**. The rainy season in **monsoon** climates, for example, is associated with the arrival of maritime tropical air.

Further reading
Nieuwolt, S. (1977) *Tropical Climatology: An Introduction to the Climates of Low Latitudes*, London/Toronto: Wiley.

MARSH

See **wetlands**.

MARSH, G.P. (1801–1882)

One of a group of nineteenth-century environmental pioneers, including **Emerson** and **Thoreau**, who drew attention to the dynamic character of nature and through their writings stimulated public interest in the natural **environment**. Marsh's main contribution was his appreciation of the human impact on nature and natural **resources**, and his ideas, as expressed in his book *Man and Nature*, were carried through into the mid-twentieth century through such events as the 1956 Chicago Conference on **Man's Role in Changing the Face of the Earth.**

Further reading
Marsh, G.P. (1864) *Man and Nature*, New York: Scribner.
Thomas, W.L. (1956) *Man's Role in Changing the Face of the Earth*, Chicago: University of Chicago Press.

MASS

The quantity of matter in a body. The mass of a body is defined in terms of its inertia or resistance to a change in motion. An object with a large mass, for example, will be more difficult to set in motion than one with a small mass. Similarly, once in motion, a body with a large mass will be more difficult to stop.

MASS BURN INCINERATOR

An **incinerator** that burns mixed **garbage** or unsorted municipal **waste**. The presence of non-combustible material in such waste may reduce the **combustion** efficiency of the incinerator and raise its **air pollution** potential. Municipal waste often includes products that contain hazardous material – for example, **mercury** batteries – which is released into the **atmosphere** as a result of incineration.

MASS MOVEMENT

The downslope movement of material under the force of **gravity**. Movement may be slow – for example, **soil** creep – and not readily apparent, or it may be very rapid, as in the case of a mud flow. The speed of the flow generally depends upon such factors as the nature of the material involved, the steepness of the slope and the amount of **water** available to provide lubrication. All the processes involved are natural features of the **environment**, but human activities can increase the frequency with which they occur and the impact that they have, sometimes with disastrous consequences.

Further reading
Turner, A.K. and Schuster, R.L. (eds) (1996) *Landslides: Investigation and Mitigation*, Washington, DC: National Academy Press.

MASS NUMBER

The number of **protons** and **neutrons** in the nucleus of an **atom**.

MAUNDER MINIMUM

A prolonged period of low **sunspot** activity between 1645 and 1715, first identified by the astronomer E.W. Maunder. That same time period included some of the worst years of the **Little Ice Age**. The cold, wet **weather** of the 1690s, for example, which caused havoc in Northern Europe, is well documented. A similar period of low sunspot activity, named the 'Spörer Minimum', occurred between 1460 and 1550.

Further reading
Eddy, J.A. (1976) 'The Maunder Minimum', *Science* 192: 1189–202.
Parry, M.L. (1978) *Climatic Change, Agriculture and Settlement*, Folkestone: Dawson.

MAXIMUM ALLOWABLE CONCENTRATION (MAC) OR MAXIMUM PERMISSIBLE CONCENTRATION (MPC)

The upper limit for the concentration of pollutants in the workplace. Exposure to noxious or toxic pollutants up to the level of the MAC/MPC should be harmless to healthy adults.

Further reading
Wilson, C. (1993) *Chemical Exposure and Human Health*, Jefferson, NC: McFarland & Co.

MAXIMUM PERMISSIBLE DOSE

The maximum amount of **ionizing radiation** that an individual should be allowed to absorb over a given period of time. The dose is sufficiently low that no injury will occur during the lifetime of the individual as a result of the exposure and no genetic damage will be carried over to the individual's offspring. In most countries, maximum permissible doses for occupational exposures are set by law, and are usually based on recommendations from the International Commission on Radiological Protection (ICRP). Maximum permissible dose limits are calculated according to the best available knowledge and are subject to change as research continues. To allow for unknown **elements**, the ICRP emphasizes the need to keep all doses as low as readily achievable.

See also
Rem, Sievert.

Further reading
Turner, J.E. (1995) *Atoms, Radiation and Radiation Protection*, New York: Wiley.

Table M-1 Maximum permissible doses for occupational exposure in Canada

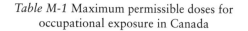

Gonads and red bone marrow (whole body in the case of uniform irradiation)	5 rems in a year
Skin; thyroid; bone	30 rems in a year
Hands and forearms; feet and ankles	75 rems in a year
All other organs	15 rems in a year

Source: Atomic Energy of Canada Limited

MAXIMUM SUSTAINABLE YIELD

See **sustainable yield.**

MELANIN

A dark-brown pigment, produced in certain skin **cells** called melanocytes, which contributes to the colour of skin and hair in many

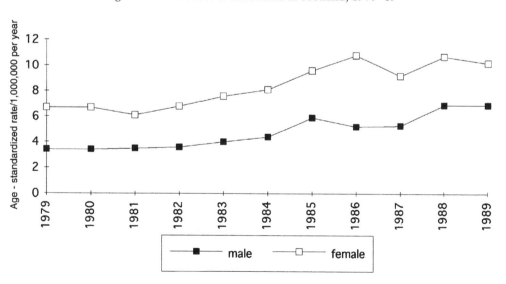

Figure M-3 Incidence of melanoma in Scotland, 1979–89

Source: After Mackie, R., Hunter, J.A.A., Aitchison, T.C., Hole, D., McLaren, K., Rankin, R., Blessing, K., Evans, A.T., Hutcheon, A.W., Jones, D.H., Soutar, D.S., Watson, A.C.H., Cornbleet, M.A. and Smith, J.F. (1992) 'Cutaneous malignant melanoma, Scotland 1979–89', *The Lancet* 339: 971–5

animals including human beings. Its main function is to protect the skin from burning by absorbing **solar radiation**, particularly that from the **ultraviolet** end of the spectrum. Initial exposure to solar radiation stimulates the melanocytes to produce melanin, which provides protection against subsequent exposure. The number of melanocytes, and therefore the ability to produce melanin, is genetically determined and, as a result, some individuals have more protection than others. Dermatological studies indicate that few people can produce sufficient melanin to provide complete protection from the sun's rays and increasingly artificial protection in the form of sun screen lotion is being recommended.

See also
Melanoma, Ozone depletion, Ultraviolet radiation.

MELANOMA

A malignant, normally fatal form of skin **cancer**, associated with over-exposure to ultraviolet-B radiation (UVB). The increased occurrence of melanoma has been linked to the thinning of the **ozone layer** and the subsequent increase in **ultraviolet radiation** reaching the earth's surface, but it may also be associated with lifestyle factors, such as increased sunbathing.

Further reading
Concar, D. (1992) 'The resistible rise of skin cancer', *New Scientist* 134 (1821): 23–8.
Schwartz, R.A. (ed.) (1988) *Skin Cancer: Recognition and Management*, New York: Springer-Verlag.

MELTDOWN

The melting of the **fuel elements** in a **nuclear reactor** as a result of the overheating of the reactor core. Overheating may be caused by a loss of **coolant** or a malfunction in the mechanical systems, such as the control rod mechanisms, necessary for the safe and efficient operation of the facility. Partial meltdowns have occurred, such as that at **Three Mile Island** in the United States in 1979, but the sequence that would occur in the event of a complete meltdown remains

speculative. In theory the molten mass, containing the fuel and incorporating parts of the reactor, could melt its way through the concrete floor of the containment structure into the ground beneath. This gave rise to the idea of the **China Syndrome** that saw the molten material working its way through the earth to appear through the **crust** on the other side. A more likely scenario is a steam explosion caused when the hot fuel encounters **groundwater** beneath the structure. In practice, the process is probably self-limiting. As the fuel becomes dispersed, the efficiency of the **fission** process declines and the **temperature** of the mass falls. However, before that stage is reached, a complete meltdown in a nuclear reactor has the potential to cause major environmental damage. The odds against such an event are often quoted as several million to one, but as the existing reactors age, the odds will decline and whatever the calculated odds, even one minor human error can set the meltdown sequence in motion.

See also
Chernobyl.

Further reading
Flavin, C. (1987) *Reassessing Nuclear Power: The Fallout from Chernobyl*, Washington, DC: Worldwatch Institute.

MERCAPTANS

A group of **organic compounds** that contain **sulphur** (S) and are characterized by an offensive odour, sometimes likened to rotting cabbage. Produced naturally by decaying organic matter, they are also by-products of various industrial activities. The residents of **petroleum** refining centres or towns which support a **pulp and paper industry** are well aware of the obnoxious **odours** associated with the emission of mercaptans. Although they can be removed by installing **scrubbers** in the industrial plants, they are recognizable at very low concentrations. The most common mercaptans – methyl mercaptan (CH_3SH) and ethyl mercaptan (C_2H_5SH) – have odour thresholds of only a few parts per billion. Thus, only small amounts need to escape before they become a **nuisance**.

Further reading
Patai, S. (1974) *The Chemistry of the Thiol Group*, London/New York: Wiley.

MERCURY (Hg)

A **heavy metal** that exists as a silvery-grey **liquid** (quicksilver) at normal **temperature** and **pressure**. It is extracted from its **ore**, cinnabar (HgS), by roasting in air. As well as its traditional uses in scientific instruments such as **thermometers** and **barometers**, mercury is widely used as a **catalyst** in the chemical and **plastics** industry. Mercury **alloys** (amalgams) remain the most commonly used form of dental fillings, although their safety is being questioned. Mercury compounds are poisonous, but some have a use in medicine as antiseptics and **fungicides**.

See also
Mercury poisoning.

Further reading
Hammond, A.L. (1971) 'Mercury in the environment: natural and human factors', *Science* 171 (3973): 788–9.

MERCURY POISONING

Poisoning caused by the ingestion, inhalation or **absorption** of **mercury** or its **compounds**. Organic mercury compounds, such as methyl mercury, are particularly toxic. Mercury poisoning affects the brain, kidneys and bowel, producing symptoms that include amnesia, insomnia, fatigue, blindness and emotional or mental disorientation. The direct effects of mercury on humans have been known since at least the nineteenth century through observations of those who mined the **ore** and those who used mercury in manufacturing. (The Mad Hatter in *Alice in Wonderland* is considered a good example of an individual who is mentally deranged by exposure to mercury used in making felt hats.) However, the impact of the wider distribution of mercury in the **environment** has only received attention much more recently. Mercury released into the environment, whether by natural or human activities, may ultimately become part of the **food chain**, accumulating in the body tissues of organisms to levels sufficiently high to cause the symptoms of mercury poisoning. Predators at the upper levels of the chains suffer most as a result of this **bioaccumulation**. Humans are particularly vulnerable, but there is evidence that fish and birds also exhibit signs of mercury poisoning. A series of events in the 1960s and 1970s which included the identification of **Minamata disease**, the discovery of extraordinarily high levels of mercury in fish in locations as far apart as the Great Lakes and the Irish Sea, and in species ranging from lake trout to tuna and swordfish, plus several accidents involving the consumption of seed grain treated with mercury-based **fungicides**, led to the reconsideration of the role of mercury in the environment. Permissible levels of mercury in foodstuffs were established by national health bodies and by the **World Health Organization** (WHO), restrictions were placed on the consumption of fish, particularly in North America, the use of mercury compounds for agricultural purposes was re-examined, and strict controls were instituted for the disposal of wastes containing mercury. Despite this, mercury levels in the environment remain high, particularly in the **aquatic environment**, and with several thousand tonnes being produced annually, there is a need for continued vigilance.

Further reading
Borg, K., Wanntrop, H., Erne, K. and Hanko, E. (1966) 'Mercury poisoning in Swedish wildlife', *Journal of Applied Ecology* 3: 171.
Hutchinson, T.C. (ed.) (1987) *Lead, Mercury, Cadmium and Arsenic in the Environment*, New York: Wiley.
Nriagu, J.O. (1990) 'Global metal pollution: poisoning the biosphere', *Environment* 32: 7–32.

MERIDIONAL CIRCULATION

The movement of **air** in the **atmosphere** in a north–south or south–north direction, i.e. parallel to the meridians or lines of longitude.

See also
Zonal circulation, Zonal index.

MESOPAUSE

The boundary between the **mesosphere** and

thermosphere lying some 80 km above the earth's surface.

MESOSPHERE

The layer of the **atmosphere** lying above the **stratosphere**. Temperature declines with increasing altitude, from close to 0°C at the **stratopause** to –80°C at the **mesopause**, the upper limit of the mesosphere.

MESOTROPHIC (LAKES)

Lakes that are moderately productive in terms of their output of organic matter. They contain levels of **nutrients** that place them somewhere between nutrient-poor **oligotrophic** lakes and nutrient-rich **eutrophic** lakes.

METABOLISM

The sum of the chemical processes that take place in a living organism. Including both **anabolism** and **catabolism**, it provides not only the raw materials for maintenance and growth, but also the **energy** needed for growth, development and other activities. The various reactions involved in metabolic processes are generally controlled by **enzymes**.

Further reading
Pandian, T.J. and Vernberg, F.J. (eds) (1987) *Animal Energetics*, San Diego: Academic Press.

METAL

An element which is malleable, ductile, has a high **relative density** (**specific gravity**) and is a good conductor of heat and electricity. Metals are also characterized by their ability to take on a shine, the so-called 'metallic lustre', and most have a high melting point which makes them **solid** at normal **temperatures**. Metals have been important to society for thousands of years, as reflected in terms such as the 'Bronze Age' and the 'Iron Age', and modern society could not function as it does without metals. Their malleability allows them to be hammered or pressed into a variety of forms; their ductility allows them to be drawn out into fine and flexible wires;

their conductivity makes them indispensable in the **energy** and communications industries; their ability to form **alloys** with each other allows specific combinations of metals to be developed to meet special needs. The many advantages of metals come with an environmental cost, however. Modern extraction methods are usually large scale, creating major landscape change through **strip mining** or the excavation of pits and causing disruption to the **ecology** and **hydrology** of the area. Through such **elements** as **acid mine drainage** and the escape of **dust** particles into the **atmosphere**, the impact of the activity is carried beyond the boundaries of the mine site itself. Few metals occur in their free or elemental state in nature. They exist as **ores** in combination with other elements, and to be used they must be released using some form of **smelting** process. The unwanted products are released to contaminate the **environment**. In the smelting of non-ferrous ores, for example, the **sulphur** (S) driven off contributes to **acid precipitation**, while more direct health effects are associated with poorly controlled **mercury** and **lead** smelting.

Since the proportion of a given metal in any ore is usually small, the smelting process produces substantial amounts of **waste** which require disposal. In the extraction and refining of metals such as **uranium**, the problems are complicated by the presence of **radioactivity**. On the positive side, most metals can be recycled, without losing any of their properties. Precious metals such as **gold** have been recycled for thousands of years, and there is a regular scrap metal trade in metals such as **iron**, **copper** and **aluminum**. Most metal **recycling** is cost driven, however, and because of fluctuating metal prices, the environmental benefits of recycling are not always fully realized. The continued high demand for metals has led to the exploitation of lower quality ores. This in turn creates greater stress on the environment. More ore has to be extracted and smelted to produce the same amount of metal, more energy is used in the process and a larger volume of waste remains for disposal. The dependence of modern society on metals is likely to continue and, despite environmental legislation, the threat to the environment from extraction

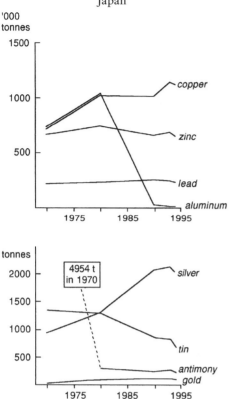

Figure M-4 Changing metal production in Japan

Source: Derived from data in *The Japan Almanac 1995*, Tokyo: Asahi Shimbun

to final disposal will remain. The situation is particularly serious in many less developed nations where environmental concerns in mineral production are forced to take second place to economic issues.

See also
Beneficiation, Heavy metals, INCO, Industrial Revolution, Lead poisoning, Mercury poisoning, Open-pit mining.

Further reading
McLaren, D.J. and Skinner, B.J. (eds) (1987) *Resources and World Development*, New York: Wiley.
Tilton, J.E. (ed.) (1990) *World Metal Demand*, Washington, DC: Resources for the Future.

METAMORPHIC ROCK

Rock formed when the composition,

structure and texture of existing **igneous** or **sedimentary rock** is altered through the action of heat and/or **pressure**. Local metamorphism is caused by heat alone – for example, the alteration of rocks adjacent to an igneous intrusion – or by pressure alone – for example, the alteration of rocks along a fault or thrust line. Heat and pressure combined are responsible for the large-scale development of metamorphic rocks produced in mountain belts during **orogenesis**. Sedimentary rocks such as **limestone** or mudstone, subjected to heat and pressure, may be transformed into marble and slate respectively, while igneous rock such as granite may be metamorphosed into gneiss, a coarse-grained, banded or foliated rock.

See also
Lava, Magma.

Further reading
Turner, F.J. (1981) *Metamorphic Petrology: Mineralogical, Field and Tectonic Aspects*, Washington, DC/New York: Hemisphere Publishing/McGraw-Hill.

METEOR/METEOROID

A **solid** extraterrestrial body. Meteors or meteoroids regularly enter the earth's **atmosphere**, where most burn up as a result of the heat generated by **friction**, and appear in the sky as 'shooting stars'. Larger meteors that reach the surface are called meteorites. They are composed mainly of **iron** (Fe) and **nickel** (Ni) or silicate minerals, but some have been found to contain complex organic **molecules** such as **hydrocarbons** and **amino acids**. Since most meteorites appear to be between four and five billion years old, they have been studied as a source of information on the origins of the solar system which is considered to be of a similar age. Few large meteorites have struck the earth in recent times, but there is speculation that a massive meteorite which struck the earth's surface some 65 million years ago at the end of the Cretaceous period was responsible for the mass extinction of dinosaurs. The great volume of **dust** and debris which was injected into the atmosphere by the collision reduced incoming **solar radiation** sufficiently to slow or even halt **photosynthesis**, and the resulting loss of their plant food supply led to the demise of the dinosaurs.

See also
Extinction, Silica.

Further reading
Alvarez, L.W., Alvarez, W., Asaro, F. and Michel, H.V. (1980) 'Extraterrestrial cause for the Cretaceous-Tertiary extinction', *Science* 221: 1256–64.

METEOROLOGY

The study of the **atmosphere** including its structure, composition and the physical processes that produce the **weather**. In theory, the term applies to all atmospheres, but usually refers to the earth's atmosphere.

See also
Agrometeorology, Climate.

Further reading
Ahrens, C.D. (1994) *Meteorology Today* (5th edition), St Paul, MN: West Publishing.

METHANE

A simple **hydrocarbon gas** (CH_4) produced during the decomposition of organic material under **anaerobic** conditions. It is the main constituent of **natural gas** and therefore an important **fuel**. A powerful **greenhouse gas**, it is about twenty-one times more effective than **carbon dioxide** (CO_2), **molecule** for molecule, and its atmospheric concentration is growing at a rate of about 0.9 per cent per year, almost twice the rate of carbon dioxide. The main causes of increasing methane levels in the **atmosphere** are to be found in agricultural development, including land clearing, rising **populations** of domestic cattle, sheep and pigs, and increases in rice cultivation. Some 18 per cent of the world's methane production is contributed by the animals alone. The escape of methane during natural gas extraction, transportation and use may account for as much as 15 per cent of global methane emissions. Methane has an atmospheric lifetime of about ten years. It is removed from the atmosphere by the **hydroxyl radical** (OH)

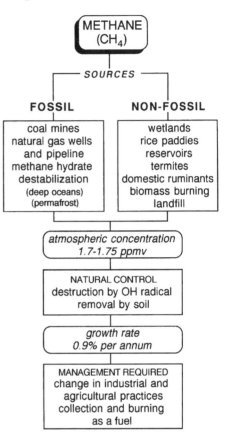

Figure M-5 Sources, control and management of methane in the environment

and as a result any reduction in OH will contribute to the potential rise in methane levels.

Further reading
Blake, D.R. and Rowland, F.S. (1988) 'Continuing worldwide increase in tropospheric methane', *Science* 239: 1129–31.
Nisbet, E.G. (1991) *Leaving Eden: To Protect and Manage the Earth*, Cambridge: Cambridge University Press.
Smith, A.T. (1995) 'Environmental factors affecting the global atmospheric methane concentrations', *Progress in Physical Geography* 19: 322–35.

METHANE SULPHONIC ACID (MSA)

A **gas** produced by the oxidation of **dimethyl sulphide** (DMS) released by **phytoplankton** during its seasonal bloom. MSA is ultimately converted into sulphate in the **atmosphere**, therefore adding to natural atmospheric acidity.

METHANOL

Methyl **alcohol** (CH_3OH). Formerly produced by the **destructive distillation** of wood, hence its popular name wood alcohol or wood spirit, it is now commonly obtained by the catalytic **oxidation** of **methane** (CH_4). It is used as a **denaturing** agent and as a **solvent** and is important in the chemical industry as a raw material for the production of synthetic resins. It has been actively promoted as an engine **fuel**, either on its own or mixed with **gasoline**. This would reduce the emission of **carbon monoxide** (CO) from **internal combustion engines**, and if used on a large scale it would help to conserve **oil** supplies. Methanol has only about half the **energy** content of gasoline, however, and unless there is a major supply crisis or radical pricing change, methanol is unlikely to replace existing motor fuels.

Further reading
Heath, M. (1989) *Towards a Commercial Future: Ethanol and Methanol as Alternative Transportation Fuels*, Calgary: University of Calgary Press.
Supp, E. (1990) *How to Produce Methanol from Coal*, Berlin: Springer-Verlag.

METHYL BROMIDE

A fumigant used as a **pesticide** since the 1960s in the fruit and vegetable industry. Its role in destroying **ozone** has only been appreciated since 1991, but with an **ozone depletion potential** (ODP) similar to most **CFCs**, it is now realized that it may be responsible for as much as 10 per cent of existing ozone depletion. At a meeting of the world's environment ministers in Copenhagen in 1992, it was added to the list of banned, ozone-destroying **gases**, with emissions to be frozen at the 1991 level by 1995. When the situation was reviewed again in 1995 in Vienna, it was agreed that the year 2010 would be the target date for the complete phase-out of the production and use of methyl bromide.

Further reading
MacKenzie, D. (1992) 'Agreement reduces damage to ozone layer', *New Scientist* 136 (1850): 10.

METHYL CHLOROFORM

See **chloroform**.

MICROBES

Microscopic organisms. See **bacteria**.

MID-LATITUDE FRONTAL MODEL

A model of mid-latitude cyclonic circulation that identifies the interactions between the **air masses** in a **cyclone** and explains the resulting **weather** patterns. Developed between 1915 and 1920 by Norwegian **weather** forecasters, it remained an important forecasting tool in mid-latitudes for at least fifty years and the **elements** of the **model** – such as cold front, warm front, warm sector – remain very much part of the current weather forecasting vocabulary. See Figure M-6 overleaf.

Further reading
Carlson, T.N. (1991) *Mid-latitude Weather Systems*, London: Routledge.

MILANKOVITCH HYPOTHESIS

A model developed by a Yugoslavian mathematician and physicist Milutin Milankovitch in the 1920s. Based in part on the work of James Croll in the nineteenth century, it used changing astronomical relationships between the earth and the sun in an attempt to explain the **ice ages**. The hypothesis is based on the observation that the earth's orbital path around the sun and its attitude in relation to the sun are not constant, but change with measurable periodicity. This in turn produces changes in the amount and distribution of **solar radiation** received at the earth's surface. Three cyclical phenomena are involved: changes in the eccentricity of the orbit – ranging between a circle and an extended ellipse (periodicity of 96,000 years); changes in the obliquity of the ecliptic – or tilt of the earth's axis (periodicity of 40,000 years); the

Figure M-6 The development of a mid-latitude cyclone, according to the mid-latitude frontal model

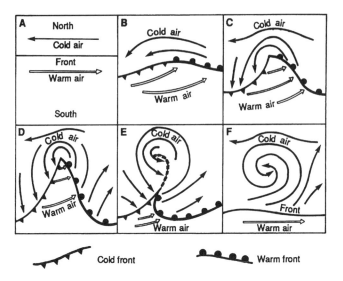

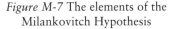

precession of the equinoxes – the time of year when the earth is closest to the sun (periodicity of 21,000 years). The changing eccentricity of the earth's orbit affects the total solar radiation received by the earth whereas the other two produce seasonal variations in **insolation**. For example, the total solar radiation receipt will be greatest when the orbit is completely circular, and will be reduced as the orbit becomes an extended ellipse. The earth is closest to the sun at perihelion which is currently in January. In 10,500 years, however, perihelion will occur in June, placing it in the northern summer. The changing tilt of the earth's axis – between 21.8° and 24.4° – will also modify the seasonal summer/winter contrast in both hemispheres. Reconciling the three different rhythms is not easy, and when originally put forward the variability involved appeared to be insufficient to cause changes as great as the ice ages. As a result, the hypothesis was generally dismissed. More recent investigations of deep ocean cores suggest some degree of correlation between the timing of heating and cooling during the **Pleistocene** and the periodicity of the Milankovitch cycles. The most important of the cycles was the longest, associated with the changing eccentricity of

Figure M-7 The elements of the Milankovitch Hypothesis

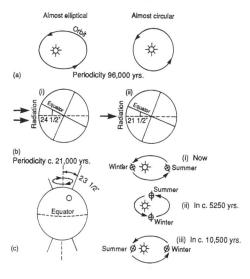

Source: Moore, P.D., Chaloner, B. and Stott, P. (1996) *Global Environmental Change*, Oxford: Blackwell Science

the orbit. Despite this, the general consensus is that in themselves the changes in insolation are insufficient to explain the climatic events in the Pleistocene. They may have initiated change or contributed to it, but the magnitude of the changes required contributions from other sources.

Further reading
Calder, N. (1974) *The Weather Machine and the Threat of Ice*, London: BBC Publications.
Dawson, A.G. (1992) *Ice Age Earth: Late Quaternary Geology and Climate*, London: Routledge.
Graves, J. and Reavey, D. (1996) *Global Environmental Change*, London: Longman.
Imbrie, J. and Imbrie, K.P. (1979) *Ice Ages: Solving the Mystery*, London: Macmillan.

MINAMATA DISEASE

A disease of the central nervous system, caused by **mercury poisoning** and named after the town of Minamata on the west coast of Kyushu in Japan, where it was identified in the 1950s. The poisoning was caused by the consumption of fish and shellfish contaminated with methyl **mercury** contained in effluent released into Minamata Bay by a **plastics** manufacturing plant. The symptoms of the disease included numbness of the hands and lips, slurred speech and deafness, followed in severe cases by seizures, paralysis and death. The old and very young suffered most and even unborn children were poisoned when their mothers consumed contaminated fish. Between 1953 and 1960, forty-three people died from the disease and more than sixty suffered permanent damage. Total deaths may have been as high as 800 and, although the real figure may never be known, it seems likely that several thousand people were affected. These high figures reflect the importance of fish and shellfish in the diet of the inhabitants of Minamata. At about the same time, in north-western Ontario in Canada, the consumption of fish contaminated with mercury from pulp mill effluent caused similar problems on two Indian reserves, where fish was also an important part of the diet. In both cases, it took more than twenty years before blame was assessed, responsibility for the problem acknowledged and compensation arranged.

Further reading
Hutchison, G. and Wallace, D. (1977) *Grassy Narrows*, Scarborough, Ont: Van Nostrand Reinhold.
Kurland, L.T., Faro, S.N. and Seidler, H. (1960) 'Minamata disease: the outbreak of a neurological disorder in Minamata, Japan and its relationship to the ingestion of seafood contaminated by mercury', *World Neurology* 1 (5): 370–91.

MINIMUM ACCEPTABLE FLOW

A concept developed in the 1962 UK Water Resources Act that required local river authorities to work towards determining the flow that would allow the normal uses of a river to be maintained. It included consideration of such factors as public health requirements, industrial, agricultural and domestic **water** use, in addition to the needs of navigation, fisheries and drainage. A minimum acceptable flow will vary from river to river and perhaps from year to year or season to season on any one river, because of changing local conditions. For example, physical factors such as **temperature** and **precipitation** introduce variability into the concept, as do human factors such as **population** growth and changing land use.

MINIMUM VIABLE POPULATION (MVP)

The critical **population** size below which survival as a **species** may not be possible. The actual numbers involved vary from species to species. Some may survive with only a few breeding pairs, whereas others require large numbers to remain viable. The classic example is the passenger pigeon, which was on the path to ultimate extinction when its numbers fell below 2000. Large colonies were needed to provide the appropriate interaction necessary to propagate the species. For many **endangered** species the minimum viable population size is not known. Without that knowledge, even the most strenuous efforts to save a species might be unsuccessful.

See also
Extinction.

Further reading
Chiras, D.D. (1994) *Environmental Science*, Redwood City, CA: Benjamin/Cummings.

MINOR GASES

The gases that occupy the 1 per cent of the volume of the **atmosphere** not occupied by **oxygen** (O) and **nitrogen** (N). The most common is argon (Ar), but being **inert** it plays little part in earth/atmosphere processes. In contrast, the **greenhouse gases**, which are all minor gases, have a role out of all proportion to their low volume.

MODEL EVALUATION CONSORTIUM FOR CLIMATE ASSESSMENT (MECCA)

An international consortium involving industrial, governmental and academic groups from the US, Japan, France and Italy, created to model **greenhouse gas**-induced climate change. It is based at the **National Center for Atmospheric Research** (NCAR) in the United States, using its computing facilities to quantify the uncertainty associated with many **models**, and thus making it easier for policy makers to co-ordinate policy with the findings of the scientific community.

MODELS

Idealized, and usually simplified, mathematical or physical representations of complex phenomena, models are used to describe and explain as well as forecast the effects of change. Models have been used extensively in attempts to unravel the complexity of the earth/atmosphere system.

See also
Carbon cycle models, Climate models, Coupled models, General circulation models.

Further reading
Schneider, S.H. (1987) 'Climatic modelling', *Scientific American* 256: 72–89.
Washington, W.M. and Parkinson, C.L. (1986) *An Introduction to Three Dimensional Modelling*, Mill Valley, CA: University Science Books.

MOISTURE DEFICIT

In theory, a moisture deficit exists in a region when **evapotranspiration** exceeds **precipitation**.

However, all **soils** have the ability to store moisture that will offset the effects of any deficit. As a result, a true deficit may not exist until the soil water storage has been used up, and for most practical purposes – **drought** evaluation, for example – the focus is on **soil moisture deficit** (SMD) rather than the simple relationship between **precipitation** and **evapotranspiration**.

See also
Potential evapotranspiration.

MOISTURE INDEX

A representation of moisture availability in an area, often used in **drought** and **aridity** studies. Most indices incorporate a combination of meteorological elements including **precipitation, evapotranspiration, solar radiation, temperature** and **wind** speed. The concept of a moisture index, based on the comparison of incoming moisture (precipitation) and outgoing moisture (**potential evapotranspiration**) can be used in climate classification.

Further reading
Thornthwaite, C.W. (1948) 'An approach towards a rational classification of climate', *Geographical Review* 38: 55–94.

MOLECULE

The smallest part of an **element** or **compound** which retains the composition and chemical properties of the element or compound. Molecules may consist of single **atoms** or as many as 10,000 bonded together chemically into a single unit.

MONSOON

Derived from the Arabic word *mausin* meaning season, the term monsoon has come to refer to the seasonal reversal of **winds** and associated **air masses**. Originally it was applied to the situation in the Indian Ocean between East Africa and the Indian subcontinent, where in the winter months the winds blow from the north-east and in the summer from the south-west. Traditionally in

India, the summer monsoon brought the rains and the winter monsoon the dry season. When extended over eastern Asia, however, the direction of the winds and timing of the **precipitation** varies. The summer monsoon over China, for example, may blow from the south-east or south, and in areas like Indochina, the east coast of Vietnam receives its main precipitation from the winter, north-east monsoon. The monsoon concept is also applied to areas in West Africa and northern Australia where wind reversals are an integral part of the **climate** of the area. Most monsoon areas have distinct wet and dry seasons, causing **seasonal drought**. The failure of the rain-bearing monsoon winds can extend the dry season and cause serious **water** supply problems for the inhabitants of the region.

See also
Atmospheric circulation, Drought, ITCZ, Sahel.

Further reading
Chang, C.P. (1987) *Monsoon Meteorology*, Oxford: Oxford University Press.
Das, K.P. (1972) *The Monsoons*, London: Edward Arnold.

MONTREAL PROTOCOL

An agreement reached in Montreal, Canada in 1987 aimed at reducing the destruction of the **ozone layer**. It was the culmination of a series of events which had been initiated two years previously at the Vienna Convention for the Protection of the Ozone Layer. Signatories from thirty-one countries agreed to a 50 per cent cut in the production of **chlorofluorocarbons** (CFCs) by the end of the century. That figure is deceptive, however, since **Third World** countries were to be allowed to increase their use of CFCs for a decade to allow technological improvements in such areas as refrigeration. The net result turned out to be only a 35 per cent reduction in total CFC production by the end of the century, based on 1986 totals. To deal with that problem, the Protocol included a provision for an Interim Multilateral Fund to assist developing countries in reducing their dependence on ozone-depleting chemicals. Subsequent meetings in Helsinki (1989),

London (1990) and Copenhagen (1992) produced agreement among the signatories – now numbering 112 – to bring about the complete elimination of the production and use of CFCs by the year 2000. In most cases – **CFCs, carbon tetrachloride** and methyl **chloroform** – production bans were brought forward from 2000 to 1996, but in the case of **halons** the ban was implemented in 1994. **Hydrochlorofluorocarbons** (HCFCs), widely used as substitutes for CFCs, have been banned progressively over a 35-year period ending in 2030. **Methyl bromide** was added to the list of banned substances, with emissions frozen at the 1991 level in 1995.

Further reading
Environment Canada (1993) *A Primer on Ozone Depletion*, Ottawa: Atmospheric Environment Service.
Kemp, D.D. (1994) *Global Environmental Issues: A Climatological Approach* (2nd edition), London/ New York: Routledge.
MacKenzie, D. (1992) 'Agreement reduces damage to ozone layer', *New Scientist* 136 (1850): 10.
United Nations Environment Program (1991) *Handbook for the Montreal Protocol on Substances that Deplete the Ozone Layer*, Nairobi: UNEP.

MOUNT AGUNG

An Indonesian **volcano** that erupted in 1963, sending an estimated 10 million tonnes of **particulate matter** and **gases** to an altitude of at least 20 km and perhaps as high as 50 km. One of the first major eruptions to be intensely monitored, over a period of several months debris from the eruption spread around the earth, ultimately reducing net **radiation** by 6 per cent. Despite this, its impact on global **temperatures** was limited – possibly because of low sulphate emissions – being measured at only a few tenths of a degree Celsius for a year or two, and therefore well within the normal range of annual temperature variation.

See also
Atmospheric turbidity, Dust veil index, Krakatoa, Mount Pinatubo, Mount St Helens, Mount Tambora.

Further reading
Dyer, A.J. and Hicks, B.B. (1965) 'Stratospheric

transport of volcanic dust inferred from solar radiation measurements', *Nature* 94: 545–54.

Lamb, H.H. (1970) 'Volcanic dust in the atmosphere; with a chronology and assessment of its meteorological significance', *Philosophical Transactions of the Royal Society A* 266: 435–533.

MOUNT PINATUBO

A **volcano** in the Philippines that erupted spectacularly in June 1991. It released an estimated 30 million tonnes of **particulate matter**, most of it in the form of **sulphur dioxide** (SO_2), which ultimately produced sulphuric acid **aerosols**. A large proportion of the debris was pushed up into the **stratosphere**, to heights as great as 25–30 km, and over a period of several months spread polewards to create a veil over the entire earth. As a result, net **radiation** in the ten months following the eruption declined by an average of 2.7 per cent, with individual monthly values reaching as much as 5 per cent. The eruption of Mount Pinatubo was blamed for the cool summer of 1992 in eastern North America and, by September of that year, it was linked with a global temperature decline of 0.5°C. **Heterogeneous chemical reactions** on the surface of the **sulphate particles** produced by Mount Pinatubo were implicated in the major thinning of the **ozone layer** which took place in 1992. Thinning was particularly well marked over the Antarctic, but global **ozone** (O_3) levels were also more than 4 per cent below normal that year.

See also
Atmospheric turbidity, Dust veil index, Krakatoa, Mount Agung, Mount St Helens, Mount Tambora.

Further reading
Brasseur, G. and Granier, C. (1992) 'Mount Pinatubo aerosols, chlorofluorocarbons and ozone depletion', *Science* 257: 1239–42.

Gobbi, G.P., Congeduti, F. and Adriani, A. (1992) 'Early stratospheric effects of the Pinatubo eruption', *Geophysical Research Letters* 19: 997–1000.

Hansen, J., Lacis, A., Ruedy, R. and Sato, M. (1992) 'Potential climatic impact of Mount Pinatubo eruption', *Geophysical Research Letters* 15: 323–6.

Kiernan, V. (1993) 'Atmospheric ozone hits a new low', *New Scientist* 138 (1871): 8.

Figure M-8 Spread of the ash cloud from Mount Pinatubo

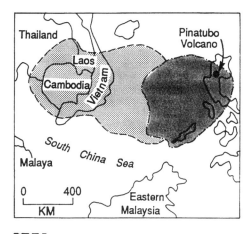

 Dense, high altitude ash cloud

 Thinner, high altitude ash cloud

Source: After Park, C.C. (1997) *The Environment: Principles and Applications*, London: Routledge

MOUNT ST HELENS

A **volcano** that erupted explosively in the western United States in May 1980, after a series of minor eruptions, releasing about 5 million tonnes of debris into the **stratosphere**. Since the debris was mainly in the form of ash and **dust** particles rather than **sulphate particles**, it returned to the surface relatively quickly and, in comparison to other major eruptions, Mount St Helens was relatively insignificant in climatological terms. It did, however, cause major landscape changes, brought about by landslides, mudflows and **floods**, which devastated a large area in the state of Washington, in the US north-west.

See also
Atmospheric turbidity, Dust veil index, Krakatoa, Mount Agung, Mount Pinatubo, Mount Tambora.

Further reading
Burroughs, W.J. (1981) 'Mount St. Helens: a review', *Weather* 36: 238–40.

Findley, R. (1981) 'The day the sky fell', *National Geographic* 159: 50–65.

MOUNT TAMBORA

When Mount Tambora in Indonesia erupted in 1815, it produced what was probably the most violent volcanic eruption on historical record. It sent an estimated 80 cubic km of ejecta (equivalent to 150 million tonnes) into the **atmosphere**, producing a DVI of 3000. The net result was a major disruption of the earth's **energy budget**, that in 1816 produced the 'year without a summer'. In Europe and North America, the year 1816 is remembered for its summer snowstorms and unseasonable frosts. Its net effect was a reduction of the mean annual global **temperature** by 0.7°C, but the impact in mid-latitudes in the northern hemisphere was greater. There, the mean annual temperature was reduced by 1°C and the average summer temperatures in parts of England were 2–3°C below normal. The eruption of Tambora was the worst of several between 1811 and 1818, and together they may have contributed to the cooling associated with the later stages of the **Little Ice Age**.

See also
Atmospheric turbidity, Krakatoa, Mount Agung, Mount St Helens.

Further reading
Lamb, H.H. (1970) 'Volcanic dust in the atmosphere; with a chronology and assessment of its meteorological significance', *Philosophical Transactions of the Royal Society*, A 266: 435–533.

MUIR, J. (1838–1914)

Born in Scotland, John Muir spent most of his life in the United States, where his thoughts on nature and **wilderness** were influenced by the writings of **Emerson** and **Thoreau**. As a naturalist, he was particularly concerned about the damage being done to the mountain and forest **environments** of the American West, and towards the end of the nineteenth century turned to writing to promote their **conservation**. He advocated strong government participation – for example, through the National Parks and Monuments system – but was also a popular activist whose efforts helped to create the **Sierra Club**.

Further reading
Fleck, R.F. (1985) *Henry Thoreau and John Muir among the Indians*, Hamden, CT: Archon Books.
Muir, J. (1894) *The Mountains of California*, New York: Century.
Muir, J. (1901) *Our National Parks*, Boston, MA: Houghton Mifflin.

MULTIPLE CHEMICAL SENSITIVITY (MCS)

An illness characterized by a variety of symptoms associated with the central nervous system, the gastrointestinal tract and the respiratory system, and brought on by exposure to chemical sensitizers, such as resins, **mercury compounds**, **pesticides** and **solvents**. The specific sensitizers will vary from individual to individual, but those who suffer from MCS find it impossible to tolerate even very low levels of the chemicals to which they are sensitive.

MUNICIPAL SOLID WASTE

See **garbage**.

MUTAGEN

A substance capable of causing genetic change or **mutation** in an organism, that may be passed on to succeeding generations. Many mutagens are also **carcinogens**.

See also
Cancer, DNA, Genes.

MUTATION

A change in the genetic make-up of an organism resulting from a change in its **DNA**. Changes are usually restricted to individual **genes**, but they may also involve entire **chromosomes**. In humans, mutation is often associated with the action of **mutagens**, that lead to **cancer** and birth defects, but mutation is common in the natural **environment**. Mutation is usually harmful to a **species**, but if it produces a change that is beneficial it may improve the reproductive capability of a species and therefore its chance of survival. This is a form of adaptation which provides an organism with an advantage over its competitors.

Further reading
Klekowski, E.J. (1988) *Mutation, Developmental Selection and Plant Evolution*, New York: Columbia University Press.

MYCENAEAN CIVILIZATION

A civilization which flourished in southern Greece up to some 3000 years ago. At that time it went into irreversible decline. The rapidity and extent of the decline and archaeological evidence suggested that it was the result of invasion by Greeks from regions to the north. Reassessment of the evidence has raised the possibility that **drought**, followed by **famine**, social unrest and migration, led to the downfall of the Mycenaeans.

See also
Harappan Civilization.

Further reading
Bryson, R.A., Lamb, H.H. and Donley, D.L. (1974) 'Drought and the decline of the Mycenae', *Antiquity* 48: 46–50.
Carpenter, R. (1968) *Discontinuity in Greek Civilization*, New York: W.W. Norton.
Parry, M.L. (1978) *Climate Change, Agriculture and Settlement*, Folkstone: Dawson.

MYXOMATOSIS

A viral disease of rabbits endemic to South America, that was deliberately introduced into Australia in the early 1950s in an attempt to deal with the overpopulation of rabbits on the continent. It also appeared in Britain shortly after its introduction to Australia, perhaps accidentally but probably deliberately. Almost all the rabbits in many parts of Britain were killed off, causing obvious changes in vegetation and forcing predators to look for alternative prey. In both Britain and Australia, the rabbit **population** has rebounded gradually, and there is evidence that some communities are building up an immunity to the **virus**, since recent outbreaks have been much less damaging than earlier ones.

See also
Ecological introductions.

Further reading
Fenner, F. and Ratcliffe, F.N. (1965) *Myxomatosis*, Cambridge: Cambridge University Press.

N

NAIROBI DECLARATION ON CLIMATE CHANGE

A statement released at the conclusion of an International Conference on Global Warming and Climate Change held in Nairobi, Kenya in 1990. It contained a 'Call for Action' to deal with the major problems that the projected **global warming** would bring to Africa. Although most responses to **climate change** are reactive, the Declaration called for a new approach which would emphasize anticipation and prevention. It recognized the reluctance of governments to respond to global warming in the absence of a clear timeframe and convincing evidence of its potential impact, but pointed out that, even if the projected warming did not occur, the proposed preventive measures would still be required to allow the development of the African economies along sustainable lines.

NATIONAL ACID PRECIPITATION ASSESSMENT PROGRAM (NAPAP)

A programme set up by the **Environmental Protection Agency** (EPA) in the 1980s in which US$600 million was spent to assess the extent of the **acid precipitation** problem in the United States. It involved the establishment of essential factors such as emission densities as well as studies of the impact of acid precipitation in different **environments**. The programme included a surface **water** survey, for example, which assessed the level of acidification in several thousand lakes and streams. As a result of this survey, it became apparent that acid deposition had caused the acidification of a significant number of lakes and streams in the north-eastern part of the United States.

NATIONAL AMBIENT AIR QUALITY STANDARDS (NAAQS)

Maximum allowable levels of pollutants in ambient **air** measured over a specific time period. Established under the US **Clean Air Act (1970)**, they are based on the levels that can be permitted without compromising public health and welfare. See Table N-1.

See also
Maximum allowable concentration or Maximum permissible concentration.

NATIONAL AUDUBON SOCIETY

A society founded in the United States in 1905 to protect the nation's bird **species**. It was named after the naturalist and artist John James Audubon (1785–1851), who, through his paintings, recorded the bird **population** of the United States as it was in the early part of the nineteenth century. The society is still primarily concerned with the protection of birds, but is also involved in the much wider issues of wildlife **conservation**.

NATIONAL CENTER FOR ATMOSPHERIC RESEARCH (NCAR)

Located in Boulder, Colorado, NCAR was established in 1961 as the major United States atmospheric research facility. Since then it has gained widespread international recognition

Table N-1 National Ambient Air Quality Standards

POLLUTANT	STANDARD VALUE		STANDARD TYPE
Carbon monoxide			
8-hour average	9 ppm	10mg/m³	Primary
1-hour average	35 ppm	40 mg/m³	Primary
Nitrogen dioxide			
Annual arithmetic mean	0.053 ppm	100 μg/m³	Primary and Secondary
Ozone			
1-hour average	0.12 ppm	235 μg/m³	Primary and Secondary
Lead			
Quarterly average		1.5 μg/m³	Primary and Secondary
Particulates < 10 μm			
Annual arithmetic mean		50 μg/m³	Primary and Secondary
24-hour average		150 μg/m³	Primary and Secondary
Sulphur dioxide			
Annual arithmetic mean	0.03 ppm	80 μg/m³	Primary
24-hour average	0.14 ppm	365 μg/m³	Primary
3-hour average	0.50 ppm	1300 μg/m³	Secondary

Source: U.S. Environmental Protection Agency Web Page
Note: Primary standards apply to humans, Secondary standards to plants and animals.

for the quality of the work carried out by its scientists. NCAR scientists are involved in both the traditional and interdisciplinary aspects of atmospheric research. Of particular importance is their development and use of computerized modelling techniques that have been applied to a wide range of topics from **ice age** climates through **nuclear winter** to **global warming** and other aspects of global change.

See also
General circulation models (GCMs).

Further reading
Levenson, T. (1989) *Ice Time: Climate, Science and Life on Earth*, New York: Harper & Row.

NATIONAL ENVIRONMENTAL POLICY ACT (NEPA)

An act passed by the US Congress in 1969 and considered at the time to be one of the most significant environmental laws in the country. It recognized the importance of the relationships that existed between society and **environment** and had the overall aim of restoring and maintaining these relationships. To deal with that situation NEPA included the requirement that all actions (including those involving federal agencies) with environmental implications needed an environmental impact statement (EIS) and established the Council on Environmental Quality (CEQ) which developed the regulations governing the EIS process. The CEQ also had the important role of advising the President on environmental issues. Implementation of the provisions of the act was not always easy. As an early attempt at comprehensive environmental legislation, for example, its language was not always precise, and the courts had to determine the real meaning of such phrases as 'fullest extent possible' and 'significant environmental impact'. In addition, since it included no provision for enforcement, the courts had to provide the mechanisms required to ensure that the agencies involved took appropriate action. In practice, NEPA has been used in a wide range of situations involving both the natural and built environments.

See also
Environmental impact assessment.

Further reading
Brown, R.D., Ouellette, R.P. and Cheremisinoff, P. (1983) *National Environmental Policies and Research Programs*, Lancaster, PA: Technomic Publishing.
Smith, Z. A. (1992) *The Environmental Policy Paradox*, Englewood Cliffs, NJ: Prentice-Hall.

NATIONAL WILDLIFE REFUGE SYSTEM

A **system** set up for the protection of migratory birds. Its primary purpose is to protect breeding, resting or wintering areas used by waterfowl and other migratory birds, but the refuges may also be safe areas for other animals. Such areas cover some 13 million hectares in the US alone and a comparable system exists in Canada. Protection is not complete, however. Some refuges are used for timber cutting, grazing, farming, **oil** and **natural gas** development and mining, which has brought major opposition from environmentalists. Since 1985, oil and **gas** companies have been lobbying to have the Arctic National Wildlife Refuge in Alaska opened for **petroleum** exploration. Although the refuge contains nearly one-fifth of all the land in the US wildlife refuge system, it seems likely that only strong and continued opposition from environmental groups has prevented it being opened for oil and gas development.

See also
Flyways.

Further reading
National Wildlife Federation (1987) *The Arctic National Wildlife Refuge Coastal Plain: A Perspective for the Future*, Washington, DC: National Wildlife Federation.

NATURAL ENVIRONMENT

See **environment**.

NATURAL GAS

A mixture of **hydrocarbons** in gaseous form, found in pockets beneath the earth's surface, usually in association with liquid **petroleum** products. It consists largely of **methane** (CH_4) (*c.* 85 per cent), but contains other hydrocarbons such as **ethane** (C_2H_6) and **propane** (C_3H_8), and was formed as the result of the **anaerobic** decay of organic matter. As a high **energy**, clean burning product it is much in demand as a **fuel**, and is also used as a feedstock in the chemical industry. In the past, natural gas was flared off in some oilfields, particularly in the Middle East, because of distance from markets and low prices. With the general rise in energy prices since the 1970s, however, and the development of technology that allows it to be transported in its **liquid** form, the gas is no longer wasted.

See also
Coal gas, Liquefied natural gas.

Further reading
Hay, N.E. (ed.) (1992) *Guide to Natural Gas Cogeneration*, Lilburn, GA/Englewood Cliffs, NJ: Fairmont Press/Prentice-Hall.
Selley, R.C. (1985) *Elements of Petroleum Geology*, New York: W.H. Freeman.

NATURAL HAZARD

An element or circumstance in the natural **environment** which has the potential to cause harm to persons or property. Natural hazards may be extreme physical events such as **earthquakes** and **tornadoes** or they may involve less violent biological phenomena such as insect infestations or viral infections. Such events are the result of natural biophysical processes in the earth/atmosphere system. They only become hazardous when humans are involved. For example, in some areas landslides are geomorphological agents with an important role in shaping the landscape. Elsewhere, similar events are

Figure N-1 Natural hazards: origin and type

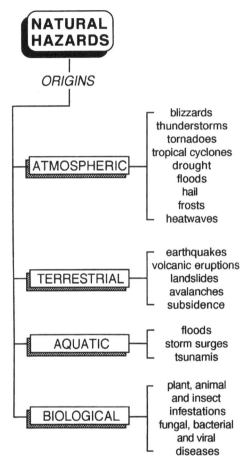

considered natural hazards because they cut highways or overwhelm settlements. Because of this human element, the impact of a natural hazard is normally measured in terms of the property damage or loss of life it causes, which is ultimately a function of the nature of the hazard and the human use of the area involved. Features such as the nature of the hazardous event, its magnitude, frequency and rate of development combine with such elements of the human component as **population** density, cultural and technological characteristics, and the degree of adaptation to the hazard, to determine the final impact of the event. As a result of the interplay of these variables, a similar event may have completely different consequences depending upon its location. Natural hazards

in developing nations often produce major loss of life and widespread property damage, whereas in developed nations a similar event might be characterized by little loss of life but involve property damage far more costly than that in the developing nations. Some success has been achieved in controlling natural hazards such as **floods** and avalanches, but more extreme events such as earthquakes and **hurricanes** continue to cause disaster with some regularity. There is some evidence that the impact of natural disasters produced by hazardous events is increasing, but that may be a reflection of growing population pressure necessitating the use of hazard lands. Work continues to mitigate the effects of natural hazards. Prevention is not always possible, but appropriate planning and forecasting can allow precautions to be taken and perhaps reduce loss of life and damage to property.

Further reading
Blaikie, P., Cannon, T., Davis, I. and Wisner, B. (1994) *At Risk: Natural Hazards, People's Vulnerability and Disasters*, London: Routledge.
Burton, I., Kates, R.W. and White, G.F. (1993) *The Environment as Hazard*, New York: Guilford Press.
National Academy of Sciences (1991) *A Safer Future: Reducing the Impacts of Natural Disasters*, Washington, DC: National Academy Press.
Smith, K. (1992) *Environmental Hazards*, London: Routledge.

NATURAL RESOURCES DEFENSE COUNCIL (NRDC)

A non-governmental environmental organization established in the United States in response to declining public concern about the **environment** in the 1970s. It monitors **resource** use and **pollution** problems, and over the years has been involved in a range of issues from **water** pollution and **pesticide** use to **ozone depletion**. The NRDC is particularly concerned with the regulatory environment and legislation as it is applied to resource issues.

Further reading
Smith, Z. A. (1992) *The Environmental Policy Paradox*, Englewood Cliffs, NJ: Prentice-Hall.

NATURAL SELECTION

See **Darwin, C.R.**

NATURAL VEGETATION

The combination of plants produced in an area as a result of the interactions of climatic and **edaphic factors**. When the various environmental elements are in equilibrium a **climax community** will exist. In popular usage, the term is applied to a vegetation cover that has not been modified by human activity.

See also
Biome, Ecosystem.

Further reading
Scott, G.A.J. (1995) *Canada's Vegetation: A World Perspective*, Montreal: McGill-Queen's University Press.
Woodward, F. (1987) *Climate and Plant Distribution*, Cambridge/New York: Cambridge University Press.

NATURE CONSERVANCY (THE) (TNC)

An environmental advocacy group in the United States, involved in the **conservation** and restoration of natural areas that are in danger from development. These include areas in the Tall Grass prairies of the **Great Plains** and in the Florida Everglades, where land has been purchased and set aside to protect wildlife and plant **species**. Founded in 1951, The Nature Conservancy has had an international programme since 1980, in which it is working towards the building of inventories of natural diversity in individual countries and the setting up of conservation areas to protect that diversity.

NATURE CONSERVANCY COUNCIL (NCC)

An organization established by Act of Parliament in Britain in 1973 to deal with the **conservation** of all aspects of the natural **environment**, including **flora, fauna**, geological and physiographical features. It does

this through the establishment of nature reserves at both the local and national level and through the designation of Sites of Special Scientific Interest. The latter are established in co-operation with landowners, occupiers and local planning authorities and have statutory protection. The NCC is also active in education and research. In the early 1990s it was reorganized into the regional units, English Nature, Scottish Natural Heritage and the National Countryside Council for Wales.

NECROSIS

Death of a **cell** or piece of tissue that is still attached to a living organism. An example is the disintegration of leaf tissue caused by the degeneration of cells in direct contact with **acid precipitation**.

NEGATIVE FEEDBACK

See **feedback**.

NEPHANALYSIS

The analysis of **clouds** usually from data provided by **weather** satellites, although in pre-satellite days synoptic charts were used. Both visible and infrared imagery may be used. Cloud patterns, colour and thickness can supply information on the characteristics of weather systems, including flow patterns, **temperatures** and **precipitation** intensity and distribution.

NERITIC ZONE

That part of the ocean that lies over the continental shelf, in which the waters are usually less than 200 m deep. Because **light** penetrates most of this zone and adjacent rivers supply **nutrients**, it supports an abundance of plant and animal life. Environmental problems arise because its productivity, relatively shallow water and proximity to the coast attract a variety of users. **Overfishing** can become a problem, but conflicts between the fisheries and other users are also common. **Oil** exploration and production is common in the neritic zone and the same rivers that bring down nutrients also bring in pollutants.

NET PRIMARY PRODUCTIVITY (NPP)

The net gain in plant material or **biomass** per unit of area per unit of time, either in terms of **energy** (kcal m^{-2} year^{-1}) or more commonly in terms of **mass** (g m^{-2} year^{-1}). It may also be considered the amount of stored **cell** energy produced by **photosynthesis**, and is represented by the amount of atmospheric **carbon** (C) sequestered by green plants. NPP is always less than the theoretical productivity of an area since some energy is lost through plant **respiration** during **metabolism**. The relationship can be expressed as follows:

$$NPP = \text{Gross primary productivity} - \text{Respiration}$$

On land, NPP ranges from < 250 g m^{-2} year^{-1} in **desert** and **tundra** to as much as 5000 g m^{-2} year^{-1} in the **equatorial rainforest**, whereas in the **oceans** the greatest NPP values seldom reach 1000 g m^{-2} year^{-1} in the open ocean and fall as low as 2 g m^{-2} year^{-1} beneath the Arctic ice cap. Only in special locations, such as coral reefs or mangrove swamps, do the marine NPP figures approach those of the rainforest. Maximum values for cultivated crops using modern techniques range from 750 – 1,500 g m^{-2} year^{-1}.

Further reading
Leith, H. (ed.) (1978) *Patterns of Primary Production in the Biosphere*, Stroudsburg, PA/New York: Dowden, Hutchinson and Ross/Academic Press.
Kershaw, H.M., Jeglum, J.K. and Morris, D.M. (1996) *Long-term Productivity of Boreal Forest Ecosystems*, Sault Ste Marie, Ont: Great Lakes Forestry Centre.

NET RADIATION

Also referred to as the radiation balance, net radiation is the difference between **solar radiation** and **terrestrial radiation** passing through the **atmosphere**. The flow is considered positive when the incoming solar radiation exceeds the outgoing terrestrial radiation and negative when the terrestrial radiation is in excess. Thus, the balance is usually positive during the day and negative at night.

NEUTRON

An elementary or subatomic particle that carries no electrical charge. Neutrons are constituents of the nuclei of all **atoms** except **hydrogen** (H). They have a **mass** only slightly greater than the **protons** with which they share an atomic **nucleus**.

Further reading
Das, A. and Ferbel, T. (1994) *Introduction to Nuclear and Particle Physics*, New York: Wiley.

NICHE

The position of an **organism** within a **habitat**, as defined by its role in that habitat. It includes not only physical location, but also the functional role of the organism in the **community** as determined by its needs and its interrelationships with other components of the **environment**. In occupying a specific niche an organism is making use of the set of conditions that are best suited to its survival. If something happens to change these conditions then the survival of the organism may be threatened.

Further reading
Giller, P.S. (1984) *Community Structure and the Niche*, London/New York: Chapman and Hall.
Odum, E.P. (1971) *Fundamentals of Ecology* (3rd edition), Philadelphia, PA: Saunders.
Whittaker, R.H. and Levin, S.A. (eds) (1975) *Niche: Theory and Application*, Stroudsberg, PA: Dowden, Hutchinson and Ross.

NICKEL (Ni)

A silvery-white corrosion-resistant **metal**, resembling **iron** (Fe) and, like iron, magnetic. It is used in nickel plating – for example, for coins – is alloyed with other metals such as steel and **silver** (Ag) and can be used as a **catalyst**. Ores of nickel usually contain **sulphur** (S) and **arsenic** (As), and during the initial **smelting** process, which involves roasting the **ore**, these are released into the atmosphere. In the past, the resulting increase in environmental acidity caused major damage to the adjacent vegetation and increased the acidity of local lakes and waterways. The **tall stacks policy** was introduced to deal with that problem, but while it reduced local acidity, it caused an increase in **acid precipitation** downwind from the smelter. Nickel processing has been linked to the development of asthma as well as lung and sino-nasal **cancers**.

See also
Alloy, INCO, LRTAP.

Further reading
Environment Canada/Health Canada (1994) *Nickel and its Compounds*, Ottawa: Environment Canada.

NIMBY

An acronym for Not In My Back Yard, a phrase which indicates opposition to a proposed development likely to lower the quality of life or reduce property values in an area. There may also be the underlying implication that such a development in a back yard belonging to someone else would be less of a problem. The opposition is not necessarily raised as a matter of principle, but rather is rooted in the personal impact of the development. For example, it reflects the desire of people to have the products of an industry, but not the **wastes** generated in their manufacture.

Further reading
Portnoy, K.E. (1992) *Siting Hazardous Waste Treatment Facilities: The NIMBY Syndrome*, New York: Auban House.

NITRATE

A **salt** or **ester** of nitric acid (HNO_3). Nitrates, formed naturally in the **soil** by microorganisms, are the chief source of **nitrogen** (N) available to plants. Nitrogen-based **fertilizers** are also a major source of nitrates in cultivated soil, but excessive or improper use has led to nitrate **pollution**. Nitrates are highly **soluble** in water and, as a result, can be easily leached out of the soil to pollute **groundwater** and surface

water supplies. Most adults can cope with the presence of nitrates in food or drinking-water, but infants are less tolerant and may contract methaemoglobinaemia by consuming even small amounts of nitrate – for example, >45–50 ppm in drinking-water. Methaemoglobinaemia disrupts the transfer and distribution of **oxygen** (O) by the circulatory system, causing asphyxiation in the most serious cases. Most jurisdictions now restrict the level of nitrates in drinking-water, the European Union having a maximum permitted level of 50 ppm, for example.

See also
Eutrophication, Leaching, Nitrification.

Further reading
Keleti, C. (1985) *Nitric Acid and Fertilizer Nitrates*, New York: M. Dekker.

NITRIC OXIDE (NO)

One of the **oxides of nitrogen** (NO_x), in which each molecule consists of one **atom** of **nitrogen** (N) and one of **oxygen** (O). It is an important natural destroyer of **ozone** (O_3), as a participant in a long **catalytic chain reaction**, and responsible for perhaps as much as 50 to 70 per cent of the natural destruction of stratospheric ozone.

NITRIFICATION

The conversion of organic **nitrogen** (N) **compounds** into **nitrates** by nitrifying **bacteria** in the **soil**. The nitrates are then available to provide nitrogen to growing plants. Nitrogenous **waste** products are broken down into various components including ammonia (NH_3), and the nitrates are produced from that ammonia. The process involves two stages, with the initial production of **nitrites** being followed by conversion of the nitrites into nitrates.

Oxidation of ammonia: (performed by the bacterium **Nitrosomonas**)

$$NH_3 \quad + \quad O_2 \quad \rightarrow \quad NO_2^- \quad + \quad H_2O$$
ammonia   oxygen   nitrite ion   water

Nitrification: (performed by the bacterium **Nitrobacter**)

$$NO_2- + O_2 \rightarrow O_3^-$$
nitrite ion   oxygen   nitrate ion

See also
Denitrification, Nitrogen cycle.

Further reading
Haynes, R.J. (1986) *Mineral Nitrogen in the Plant-Soil System*, Orlando, FL: Academic Press.

NITRIFYING BACTERIA

See **nitrification**.

NITRITE

A **salt** or **ester** of nitrous acid (HNO_2).

NITROGEN (N)

A colourless, odourless **gas** that makes up about 78 per cent of the volume of the **atmosphere**. It is an essential **element** for all living organisms, being present in **proteins** and **nucleic acids**. Molecular nitrogen is **inert** and may be considered as a dilutant for the **oxygen** (O) with which it shares most of the atmosphere. It seldom becomes directly involved in atmospheric chemical or biological processes, except under extraordinary circumstances. During thunderstorms, for example, the enormous **energy** flow in a **lightning** strike may cause it to combine with oxygen (O) to form **oxides of nitrogen** (NO_x), a group of gases involved in a number of current environmental problems. On a less spectacular but ultimately more important level, certain **soil bacteria** are able to fix the atmospheric nitrogen (N) necessary for the synthesis of the complex nitrogen compounds found in all forms of life on earth.

See also
Nitric oxide, Nitrifying bacteria, Nitrogen dioxide, Nitrous oxide.

Further reading
Bailey, P.D. and Morgan, K.M. (1996) *Organo–nitrogen Chemistry*, Oxford/New York: Oxford University Press.

Bacon, P.E. (ed.) (1995) *Nitrogen Fertilization in the Environment*, New York: M. Dekker.

NITROGEN CYCLE

The circulation of **nitrogen** (N) through the earth/atmosphere system. Since gaseous nitrogen is relatively inactive, the bulk of the nitrogen in the cycle is that in organic and inorganic nitrogen **compounds** present in the **soil** and in living organisms. Inorganic nitrogen compounds such as **nitrates** are absorbed by plants and converted into more complex compounds such as **amino acids** or **proteins**. When the plants die, **bacteria** convert the organic compounds in the vegetable matter back into nitrates that can enter the loop again. Some of the nitrates, however, may be subject to **denitrification**, resulting in the release of nitrogen **gas** or **nitrous oxide** (N_2O) into the **atmosphere**. When plants are eaten by animals, the nitrogen compounds are absorbed into the bodies of the animals and remain there until the animals die. They are then released into the **environment** again to be converted into inorganic products such as nitrates and **nitrites** or perhaps returned to the atmosphere as nitrogen. Nitrogenous **waste**, containing ammonia (NH_3) and urea ($CO(NH_2)_2$) excreted by animals, is also involved in the cycle. Gaseous nitrogen is

Figure N-2 The nitrogen cycle

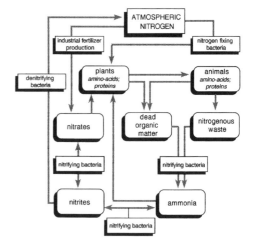

reintroduced into the main part of the cycle through the activities of nitrogen fixing bacteria. Thus, the nitrogen cycle is not a simple cycle but rather a complex group of integrated loops. Human activity has disrupted the volumes involved in the cycle through agricultural and industrial practices. Nitrogen **fertilizers**, for example, add to the quantities of nitrate available in the soil; the growing number of people and animals increases the volume of nitrogenous waste; industrial activities cause the formation of **oxides of nitrogen** (NO_x), that bring additional atmospheric nitrogen into the system. This has resulted in the creation or augmentation of a range of environmental problems – from **eutrophication** to **ozone depletion** – involving nitrogen compounds.

See also
Nitrogen fixation.

Further reading
Bolin, B. and Cook, R.B. (1983) *The Major Biogeochemical Cycles and their Interactions*, New York: Wiley.
Delwiche, C.C. (1970) 'The nitrogen cycle', *Scientific American* 223 (3): 148–58.
Sprent, J.I. (1987) *The Ecology of the Nitrogen Cycle*, Cambridge/New York: Cambridge University Press.

NITROGEN DIOXIDE (NO$_2$)

A red-brown toxic **gas**, in which each **molecule** consists of one **atom** of **nitrogen** (N) and two of **oxygen** (O). It is a major constituent of automobile exhaust gases and a common component of urban **photochemical smog**.

NITROGEN FIXATION

The formation of **nitrogen** (N) **compounds** from the free nitrogen in the **air**. The nitrogen is converted first into ammonia (NH_3), then subsequently into more complex compounds. The process is made possible by the activities of **bacteria** which live in the **soil** (for example, *Azotobacter* and *Clostridium*) and have the ability to fix atmospheric nitrogen. Other **bacteria** (for example, *Rhizobium*) living in the root nodules of **leguminous plants** also have this ability.

Further reading
Dilworth, M.J. and Glenn, A.R. (eds) (1991) *Biology and Biochemistry of Nitrogen Fixation*, New York: Elsevier.
Sprent, J.I. and Sprent, P. (1990) *Nitrogen Fixing Organisms: Pure and Applied Aspects* (2nd edition), London/New York: Chapman and Hall.

NITROGEN OXIDES

See **oxides of nitrogen**.

NITROUS OXIDE (N$_2$O)

One of the **oxides of nitrogen** (NO_x) in which each **molecule** consists of two **atoms** of **nitrogen** (N) and one of **oxygen** (O). Nitrous oxide is a naturally produced **greenhouse gas**, but owes its current growth to the increased use of agricultural **fertilizers** and the burning of **fossil fuels**.

Further reading
Bouwman, A.F., Van der Hoek, K.W. and Olivier, J.G. (1995) 'Uncertainties in the global source distribution of nitrous oxide', *Journal of Geophysical Research* 100: 2785–800.

NOBLE GASES

The **gases** helium, neon, argon, krypton, xenon and radon, once thought to be chemically **inert**. **Compounds** of krypton and xenon have been synthesized, however, although they are generally unstable. As a result, the term 'inert gases' is no longer commonly used for this group.

NOISE

As an environmental issue, noise is sometimes defined as unwanted **sound**, sound without value or sound that causes sufficient disturbance and annoyance that it has social and medical implications. The intensity of sound is expressed in decibels (dB) obtained by comparing the power of a specific sound with a reference level. On the decibel (A) scale (dB(A)), most commonly used when dealing with human hearing, the reference level is based on the threshold of hearing. The scale is logarithmic, so that a sound with a dB rating of 20 is ten times as intense as one with a dB

Table N-2 Noise pollution: sources and impact

SOUND LEVEL (dB)	SOURCE	PERCEIVED LOUDNESS	HEARING DAMAGE
150	Jet plane take-off	Painful	Injurious range
130	Rock music	Painful/uncomfortably loud	Danger zone/progressive hearing loss
90	Diesel truck at 80 km/hr	Very loud	Damage after long exposure
50	Light traffic 30 m away	Moderately loud	Little chance of damage
40	Living room/Bedroom	Quiet	No damage
20	Broadcasting studio	Very quiet	No damage
0	Threshold of hearing		

rating of 10 and 100 times the intensity of a sound with a dB rating of 0. The impact of noise on people depends not only on the intensity of the sound produced, but also on such factors as its frequency and duration. The most obvious effect of noise is its impact on hearing. This is usually progressive, with individuals exposed to loud noises in the workplace, for example, gradually losing their hearing, but it can be catastrophic when, for example, an explosion causes an ear-drum to rupture. Other effects of noise include fatigue, headaches, nausea and general irritability. The main sources of noise in urban areas are industrial activities plus road, rail and air traffic. In rural areas, agricultural activities were once the main sources of noise, but in some areas at least noise levels have increased significantly with the advent of all terrain vehicles (ATVs), dirt bikes and snow machines.

See also
Noise abatement, Noise exposure forecast.

Further reading
Kryter, K.L. (1985) *The Effects of Noise* (2nd edition), New York: Academic Press.
Lipscomb, D.M. (1974) *Noise: The Unwanted Sounds*, Chicago: Nelson-Hall.

NOISE ABATEMENT

The reduction or control of **noise** levels. This can be achieved in three ways: reduction at source, transmission control, receiver control. Reduction at source may include modifying or changing a specific piece of equipment or it may deal with the location or timing of the use of unaltered machinery or equipment. Jack hammers or pneumatic drills can be provided with some degree of silencing, for example, and exhaust systems on cars or trucks are designed to muffle sound. Similarly, the replacement of conventional turbo-jet aircraft engines with turbo-fan bypass jets helps reduce noise production. On the larger scale most airports have restrictions on times of operation and those in urban areas may require noise abatement procedures that include power reduction or routing changes following take-off. Once the sound has been generated, its effects can be reduced by interrupting its transmission path. In buildings this can be accomplished by using acoustical materials to absorb the sound. In the **natural environment** it is more difficult, but artificial ridges or berms, sometimes in combination with vegetation can absorb or deflect noise. Engineering techniques such as the routing of urban motorways through cuttings or along embankments are used to restrict the transmission of noise or to allow it to disperse easily. If the noise cannot be reduced or diverted, then individuals exposed to high noise levels can take personal precautions. Operators of industrial machinery, for example, can protect their hearing by wearing ear-plugs, ear muffs or a combination of the two. Such protection is essential for chain-saw operators or aircraft service personnel, but it can also benefit individuals in gun clubs or even the homeowner using a powered lawn-mower.

See also
Noise and number index, Noise exposure forecast.

Further reading
Bell. L.H. and Bell, D.H. (1994) *Industrial Noise Control: Fundamentals and Applications* (2nd edition), New York: M. Dekker.
Foreman, J.E.K. (1990) *Sound Analysis and Noise Control*, New York: Van Nostrand Reinhold.

NOISE AND NUMBER INDEX (NNI)

An index which combines perceived **noise** levels at an airport with the number of aircraft using the airport to provide a value for daily air traffic noise levels. It was developed at Heathrow Airport in London, based on a survey of human disturbance by aircraft noise in the vicinity of the airport.

See also
Noise exposure forecast.

NOISE EXPOSURE FORECAST (NEF)

Similar to the **NNI** in that it reflects certain aspects of human sensitivity to airport **noise**, the noise exposure forecast system provides an index of the nuisance value of noise to human beings living or working in the vicinity of an airport. It includes information on perceived noise levels and the number of flights, but also considers additional information on the timing of flights, overhead flight profiles and the frequency of specific **sound** tones. The results of NEF calculations for a particular airport are usually repre-

sented in the form of a noise footprint or isoline map in which the lines join points of equal NEF. For single runway airports the noise footprint is relatively simple, being elongated in the same direction as the runway, but at large international airports the footprint becomes very complex. The NEF index has been developed as a land use planning tool. Since different activities have different tolerances for noise, the knowledge of the distribution of noise around an airport as indicated by its noise footprint allows specific land use to be encouraged in one area and discouraged elsewhere. Residential development, for example, which is particularly sensitive to noise, would not normally be allowed within the 35 dB(A) contour in the footprint, whereas railyards, **water** treatment plants, **sanitary landfill** sites and warehouses are commonly found in such noisy areas. In many cases, land use patterns had already been established by the time the NEF system came into common use. It can still be used, however, in property re-zoning plans and for providing information on areas that include buildings that require renovation to meet noise standards. Around many older airports, for example, residential buildings have been retrofitted with double glazing or acoustical **insulation** to levels based on the local NEF index.

Figure N-3 A noise footprint at an airport with a single runway and taxiway

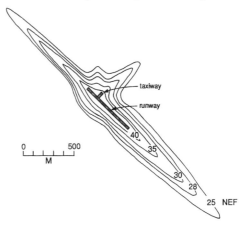

Source: After Environment Canada (1983) *Stress on Land in Canada*, Ottawa: Ministry of Supply and Services

Further reading
Rujigrok, G.J.J. (1993) *Elements of Aviation Acoustics*, Delft: Delft University Press.
Stevenson, G.M. (1972) *The Politics of Airport Noise*, Belmont, CA: Duxbury Press.

NOMADISM

A way of life in which groups of people move from place to place in search of food for themselves or their animals. Before the development of agriculture nomadic hunting groups were the norm, and with the domestication of animals pastoral farmers migrated in accordance with the availability of **forage** for their flocks and herds. The inhabitants of the **Sahel**, for example, followed a typical nomadic existence, travelling north and south seasonally so that their herds could graze on the grasses produced during the rainy season. Transhumance, the seasonal movement of animals in the mountain areas of Europe to the high pastures in the spring and back to the valley in the autumn, is another form of nomadism.

Further reading
Johnson. D.L. (1969) *The Nature of Nomadism*, Chicago: Department of Geography, University of Chicago.

NON-CONSUMPTIVE USE

An activity that does not result in the consumption or depletion of **resources**. The use of a forested area for hiking, camping or bird-watching is a non-consumptive use, for example, compared to most forestry practices involving the removal of timber. The use of **water** to produce **hydroelectricity** or the use of lakes and rivers for navigation are also non-consumptive uses. In most cases, the environmental impact of the non-consumptive use of resources is low, although well-travelled navigable waterways are often polluted and the heavy use of hiking trails can lead to **erosion**.

NON-RENEWABLE RESOURCE

A natural **resource** that cannot be replaced after it has been consumed. It applies particularly to the **fossil fuels**, which can only be used once, but it also describes other mineral resources that are present in only fixed quantities in the earth's **crust**, although **metals** can be reused through **recycling**. Central to the concept is the human time frame. **Oil** and **natural gas** are being formed beneath the earth's surface at present and new mineral **ores** are also being created. However, replacement may take millions of years, and society can consume them much more rapidly than they can be replaced. Thus in human terms they are effectively non-renewable.

See also
Renewable resources.

Further reading
Rees, J. (1990) *Natural Resources: Allocation, Economics and Policy* (2nd edition), London: Routledge.

NOORDWIJK DECLARATION

The closing declaration from a conference on atmospheric **pollution** and **climate change** held at Noordwijk in the Netherlands in 1989. Recognizing that many uncertainties remained in the scientific knowledge of global change, the participants identified certain areas that needed immediate attention and urged individual nations and international organizations to increase their activities in climate change research, paying particular attention to the role of **greenhouse gases** other than **carbon dioxide** (CO_2).

NORMALS (CLIMATE)

Data representing average climatic conditions, usually calculated over a thirty-year period. The period is adjusted every decade (for example, 1950–1979; 1960–1989).

NORTH AMERICAN WATER AND POWER ALLIANCE (NAWPA)

A scheme, first proposed in 1964, to divert Canadian rivers, flowing into the Arctic Ocean, southwards into the United States, by means of a continental-scale network of **reservoirs**, canals, aqueducts and pumping stations. The diverted **water** would help to flush **pollution** out of the Great Lakes-St Lawrence system and the Mississippi River, but the main purpose of the scheme was the provision of water for **irrigation** and municipal supply in the arid west and south-west of the United States.

See also

Interbasin transfer.

Further reading

Bryan, R. (1973) *Much Is Taken, Much Remains*, North Scituate, MA: Duxbury Press.
Schindler, D.W. and Bailey, S.E. (1990) 'Fresh waters in cycle', in C. Mungall and D.J. McLaren (eds) *Planet under Stress*, Toronto: Oxford University Press.

NUCLEAR AUTUMN (FALL)

See **nuclear winter.**

NUCLEAR ENERGY

The **energy** released during **nuclear fission**. If released in an uncontrolled manner it can cause an explosion, but under the controlled conditions of a **nuclear reactor** can be used to generate electricity. Energy is also produced by nuclear fusion, although, as yet, it cannot be made available in a commercial form. In the early 1990s, some 17 per cent of the world's electricity was being produced from about 420 nuclear reactors in twenty-five countries. France leads with 73 per cent of its electricity generated by nuclear reactors, followed by Belgium at 66 per cent. In Britain the figure is 20 per cent, and for Europe as a whole 30 per cent. In Canada and the United States about 20 per cent of the electricity is produced in nuclear stations, and in parts of Asia the amounts are about double that figure, with 49 per cent in Korea and 41 per cent in Taiwan. The nuclear energy industry has been stagnant since the 1980s, in part because of costs, in part because of reduced demand and in part because of safety concerns. Although tonne for tonne, nuclear **fuels** contain much more energy than conventional fuels, when the construction, operating and decommissioning costs of nuclear plants are taken into account, nuclear fission is arguably the least cost-effective method of producing electricity.

Further reading

Ahearne, J.F. (1993) 'The future of nuclear power', *American Scientist* 81: 24–35.
Burton, B. (1990) *Nuclear Power, Pollution and Politics*, London: Routledge.
National Academy of Sciences (1991) *Nuclear Power: Technical and Institutional Options for the Future*, Washington, DC: National Academy Press.

NUCLEAR FISSION

A reaction in which the **nucleus** of an **atom** of a **heavy metal** such as **uranium** (U) splits into two relatively equal parts, emitting **neutrons** as it does so and releasing large amounts of **energy**. The fission process may be spontaneous or it may be initiated by bombarding the atomic nucleus with a neutron. The release of additional neutrons sets up a **chain reaction** which is accompanied by the continued generation of thermal energy. The development of this process allowed the creation of the **atomic bomb** in which an uncontrolled chain reaction releases energy so rapidly that it causes an explosion equivalent to thousands of tonnes of **TNT**. When controlled within a **nuclear reactor**, the fission process provides the basis for the commercial production of **nuclear energy**. The energy available in one tonne of ^{235}uranium is equivalent to that in about 3,500 tonnes of **coal**.

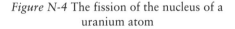

Figure N-4 The fission of the nucleus of a uranium atom

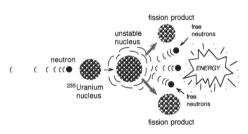

Further reading

Murray, R.L. (1988) *Nuclear Energy: An Introduction to the Concepts, Systems and Applications of Nuclear Processes* (3rd edition), Oxford/New York: Pergamon Press.
Vandenbosch, R. and Huizenga, J.R. (1973) *Nuclear Fission*, New York: Academic Press.

NUCLEAR FUSION

A nuclear reaction in which the nuclei of light **atoms** are fused to form a heavier **nucleus**. In the process, large amounts of **energy** are released. Such a system produces the tremendous destructive power of a **thermonuclear device** or hydrogen bomb, and supplies the energy of the sun and other stars. In the sun, for example, the fusion of the nuclei of **hydrogen** (H) atoms creates helium (He) atoms, producing in the process enough energy to maintain the surface **temperature** of the sun at c. 6000°C. Nuclear fusion has been seen by some scientists as the solution to all society's energy problems. The hydrogen nuclei used in the fusion process are available in virtually unlimited amounts in sea **water**; the quantities of energy produced are enormous – a 600 MW power station would require a net daily **fuel** input of only 15 tonnes of ordinary water – and the environmental problems seem likely to be less than with energy produced by **nuclear fission**. One major and possibly insurmountable problem remains, however. Fusion requires very high temperatures – as much as 50,000,000°C – and **pressures** sustained for long periods, and as yet there is no material capable of withstanding the extreme conditions involved. Experimental reactors have been built using magnetic confinement as a form of non-material container and these have met with only limited success. Until such engineering problems can be solved, the commercial production of energy from fusion is not feasible. Cold fusion, which would remove the problems created by high temperatures, is theoretically possible, but claims that a cold fusion reaction has been created in the laboratory have not been confirmed scientifically.

Further reading
Herman, R. (1990) *Fusion: The Search for Endless Energy*, New York: Cambridge University Press.

NUCLEAR REACTOR

A device in which the **nuclear fission** process is initiated and allowed to proceed in a controlled manner for the production of energy. Nuclear reactors differ in their detailed engineering but all designs have essential elements in common. The nuclear **fuel** is the core of the reactor system. It is usually ^{235}**uranium** in natural or enriched uranium oxide (UO_2), housed in a shielded reactor vessel. A moderator such as light water, **heavy water** or graphite, is used to slow the **neutrons** and increase the efficiency of the fission process, and control rods are used to manage the rate of the reaction. Some form of **coolant** is required to prevent the system from overheating. In light or heavy water moderated reactors, the moderator may also act as a coolant, but **carbon dioxide** (CO_2), helium (He) and liquid **sodium** (Na) are also used. The heat released during the reaction is transported via a heat exchanger to a boiler where steam is produced. From that point in the process, the nuclear system is no different from that in a conventional thermal power plant where high-pressure steam-powered turbines are linked to generators that produce electricity, the most common end-product of nuclear reactors. All reactors incorporate systems for refuelling the core and containing or disposing of the **nuclear waste** products created in the fission process. Most reactors are burner reactors that consume fuel, but some are breeder reactors that produce additional fissionable products during the fission process. Loaded with ^{235}uranium and ^{238}uranium, an unmoderated reactor will ultimately produce ^{239}**plutonium** (Pu). Plutonium was once in demand for use in nuclear weapons, but in terms of energy output it is similar to uranium and can also be used as a fuel. In the breeder reactor, more fissionable material is produced than consumed, which suggests that breeder reactors could be an important source of fuel for burner reactors. Costs are high in time and money, however, and there are safety concerns with breeder technology. As a result, the development of breeder reactors has effectively ceased. Commercial nuclear reactors are usually described in terms of variables such as the moderator or coolant system they use. Some examples of common types of fission reactors are indicated in Figure N-5 (a–d).

Figure N-5 The form and characteristics of selected nuclear reactors currently in use
(a) A gas-cooled reactor (b) A boiling water reactor (c) A light water reactor
(d) A breeder reactor

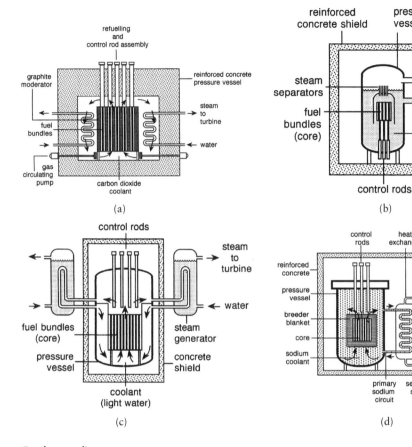

(a)

(b)

(c)

(d)

Further reading

Cole, H.A. (1988) *Understanding Nuclear Power: A Technical Guide to Industry and its Processes*, Aldershot/Brookfield, VT: Gower Technical Press/ Gower Publishing.

Judd, A.M. (1981) *Fast Breeder Reactors: An Engineering Introduction*, Oxford/New York: Pergamon Press.

Kleinbach, M.H. and Salvagin, C.E. (1986) *Energy Technologies and Conversion Systems*, Englewood Cliffs, NJ: Prentice-Hall.

Ramage, J. (1983) *Energy: A Guidebook*, Oxford: Oxford University Press.

NUCLEAR WASTE

Waste products produced by the nuclear industry and characterized by their **radio-activity**. Radioactivity creates special problems in the disposal of nuclear waste. Waste is produced during the mining and milling processes required to produce the **fuel** for **nuclear reactors**. The products are usually considered as low-level wastes, but rock waste and the **liquids** used in processing **uranium** (U) must be stored until their radioactivity is reduced to acceptable levels. Other low-level waste includes contaminated clothing, filters used to remove radioactive particles from the **air** and from liquids, piping and contaminated fluids. In the past these low-level wastes were routinely disposed of in landfill sites or into the sea, but in most cases they are now stored in sealed drums until they lose their radioactivity. The problem wastes are those that emit high levels of radioactivity. When the nuclear fuel cycle is complete, the spent fuel is removed from the reactor. This so-called spent

Figure N-6 An underground disposal facility for high level nuclear waste

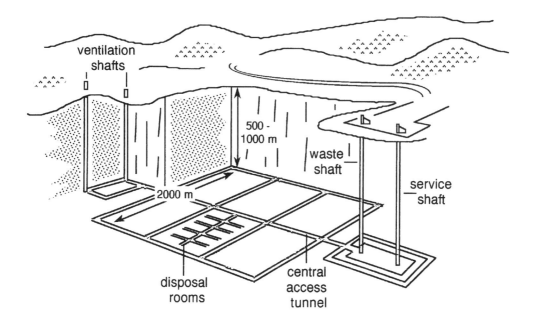

fuel still contains fissionable uranium, but it also contains other products that prevent the nuclear reactions from taking place efficiently. These include radioactive fission products such as strontium (Sr), caesium (Cs) and **iodine** (I), as well as **actinides** such as **plutonium** (Pu), neptunium (Np) and curium (Cm), all of which are fissionable. The spent fuel may be reprocessed to obtain the remaining uranium and other fissionable products, but even after that radioactive waste will remain. Once the used fuel is removed from the reactor it must be cooled, contained and shielded. Preliminary storage in **water** bays can deal with all of these. The heat is dissipated in the water, and a 1-metre thick concrete bay containing 3 metres of water will provide both shielding and containment. All nuclear plants have these bays, and once the wastes have cooled and lost an appropriate amount of their radioactivity, they can be transferred to above-ground sealed concrete storage bins. Some of the waste products have very long **half-lives**, however, and require permanent storage. As yet, there has been little progress on this aspect of nuclear waste disposal, in part because the industry has not

been growing in recent years and has therefore produced less waste than expected. Thus, the urgency to find suitable disposal systems has declined. The most promising approach appears to be deep burial in stable geological formations. Atomic Energy of Canada has been exploring the possibility of burying nuclear waste in the ancient rocks of the Canadian Shield, and the British government has considered nuclear waste burial beneath the Irish Sea. Such disposal methods are not universally accepted, however, because they still involve some potential for the escape of radioactivity into the **environment**. Since the sources of the waste and the preferred disposal sites are often well separated, the potential for contamination during the transportation of the waste must also be considered. Work in North America and Europe suggests that safe transportation is possible using specially designed containers. A final source of waste in the nuclear industry is the **decommissioning wastes** that result when a plant is closed down.

Further reading
Krausekopf, K.B. (1988) *Radioactive Waste Disposal and Geology*, London: Chapman and Hall.

Lenssen, N. (1991) *Nuclear Waste: The Problem That Won't Go Away*, Washington, DC: Worldwatch Institute.
Pasqualetti, M.J. (1990) *Nuclear Decommissioning and Society: Public Links to a Technical Task*, New York: Routledge & Kegan Paul.

NUCLEAR WINTER

The result of a major nuclear war, according to results from a theoretical **model** that simulated the consequences of multiple nuclear explosions. Calculations indicated that the explosions and the fires that followed would throw sufficient debris and **smoke** into the **atmosphere** to block incoming **solar radiation** and cause **temperatures** to fall rapidly to winter levels even in midsummer. Such an event would have major environmental and socioeconomic consequences. Named the **TTAPS scenario** after the scientists who developed and ran the first model in 1983, the hypothesis was widely criticized and modified by other researchers using improved models. The results of one of the modifications suggested that cooling would be less than first expected, with temperatures declining to values more common in autumn or fall than in winter – hence the term **nuclear autumn (fall)**. With the end of the so-called 'cold war', interest in nuclear winter waned, but in 1990 the original TTAPS team re-examined the theory using new information from laboratory studies, field experiments and numerical modelling. While uncertainties remain, the new research affirms the basic findings of the original work, including the cooling estimates upon which nuclear winter was based.

See also
Atmospheric turbidity, ENUWAR.

Further reading
Maddox, J. (1984) 'Nuclear winter not yet established', *Nature* 308: 11.
Sagan, C. and Turco, R.P. (1990) *A Path Where No Man Thought: Nuclear Winter and the End of the Arms Race*, New York: Random House.
Thompson, S.L. and Schneider, S.H. (1986) 'Nuclear winter reappraised', *Foreign Affairs* 64: 981–1005.
Turco, R.P., Toon, O.B., Ackerman, T.P., Pollack,

Figure N-7 The development of nuclear winter: (a) the conflict; (b) the post-conflict fires; (c) nuclear winter; (d) the after effects

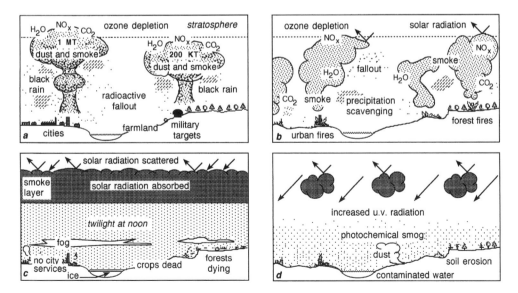

Source: Kemp, D.D. (1994) *Global Environmental Issues: A Climatological Approach*, London/New York: Routledge

J.B. and Sagan, C. (1983) 'Nuclear winter: global consequences of multiple nuclear explosions', *Science* 222: 1283–92.

Turco, R.P., Toon, O.B., Ackerman, T.P., Pollack, J.B. and Sagan, C. (1990) 'Climate and smoke: an appraisal of nuclear winter', *Science* 247: 166–76.

NUCLEIC ACIDS

Polymers of nucleotides – **compounds** consisting of a nitrogenous base plus **sugar** and a **phosphate** group – and probably the most important compounds found in living organisms. **DNA** and RNA, for example, are nucleic acids that contain the genetic codes necessary for the development and functioning of living organisms.

Further reading
Hecht, S.M. (ed.) (1996) *Bioorganic Chemistry: Nucleic Acids*, New York: Oxford University Press.

NUCLEUS

Used in general terms to refer to the core of an object. An atomic nucleus, for example, is the core of an **atom**. It consists of **protons** and **neutrons**, and although occupying only a very small fraction of the volume of the atom it accounts for almost all its **mass**. Living **cells** also have a nucleus in the form of a distinct unit contained within a membrane. The **DNA** molecules which control the characteristics of the cell are located in the nucleus. In the **atmosphere**, small particles that attract moisture are called **condensation nuclei**.

NUCLIDE

The **nucleus** of an **atom**, characterized by its **atomic number** and **mass number**.

See also
Isotope, Radionuclide.

NUISANCE

A cause of action in common law that has been used with some success in environmental cases. In its simplest terms, a nuisance is any unreasonable or unnecessary interference with the use or enjoyment of a property or piece of land by its owner. It can be used to deal with **air pollution**, **noise**, vibration and **odours**, for example, which have caused damage to the property or loss of enjoyment of the property. Action against a private nuisance may be initiated by an individual, but if the problem is more widespread, it is considered a public nuisance and action must be initiated by a government representative. The aim of the plaintiff in such cases is to obtain an injunction that will bring about the abatement of the nuisance and receive compensation for the damages caused.

Further reading
Estrin, D., Swaigen, J. and Carswell, M. A. (1978) *Environment on Trial*, Toronto: Canadian Environmental Law Research Foundation.

NUTRIENT

A raw material required by plants and animals for growth and development. In their basic form nutrients may be simple chemicals, such as **calcium**, (Ca) **magnesium** (Mg) or **iron** (Fe), available in the **soil** and absorbed by growing plants. Such minerals are also important for animals, but the nutrients consumed by animals are usually presented in a more complex form. In addition to minerals they include **carbohydrates**, fats, **proteins**, **vitamins**, inorganic **salts** and **water**. They are consumed in forms and amounts that vary according to the needs of the organisms involved, and an excess or deficit of specific nutrients can have an impact on growth and development. A distinction is usually made between macronutrients – such as **nitrogen** (N), **potassium** (K) and **phosphorus** (P) required in relatively large amounts – and micronutrients – such as **copper** (Cu), **zinc** (Zn) or cobalt (Co) required in only small amounts. The concept of nutrient status or nutrient availability can be applied both to soils and to lakes. Nutrients in soils are provided by rock **weathering** or the **recycling** of organic material and removed by plants or lost to **leaching**. If nutrients are being used more rapidly than they are being replaced, then the fertility of the soil will decline. Nutrient levels in lakes are related to age – old lakes generally having a higher nutrient status than younger lakes – but human activities can cause nutrient levels to rise rapidly, leading to premature **eutrophication**.

See also
Eutrophic lakes, Malnutrition, Metabolism, Oligotrophic (lakes).

Further reading
Paoletti, M.G., Foissner, W. and Coleman, D.C. (eds) (1993) *Soil Biota, Nutrient Cycling and Farming Systems*, Boca Raton, FL: Lewis.

O

OASIS

A location in an arid region in which there is sufficient **water** to support a plant and animal **community**. It is usually the result of the emergence of the local **water table** at the surface. Even major oases, such as those at Kufra and Jalo in Libya, can never support large permanent populations, but historically they were important centres on the desert trade routes. The drilling of **boreholes** can create artificial oasis-like conditions by bringing **groundwater** to the surface.

See also
Sahel.

OCCLUSION

The final stage in the development of a mid-latitude frontal **depression**, caused when the warm sector **air** is pinched out between the cold air ahead of the warm front and the advancing cold air behind the cold front, and forced to rise. The specific form of the occlusion will vary according to the **temperature** difference between the two cold **air masses**. The uplift of the warm air is accompanied by cooling and **condensation**, causing the **weather** beneath an occlusion to be cloudy and wet. Depending upon the relative temperatures of the warm and cold air, **precipitation** may take the form of **rain**, freezing rain or **snow**.

See also
Mid-latitude frontal model.

Further reading
Carlson, T. (1991) *Mid-latitude Weather Systems*, London/New York: Routledge.

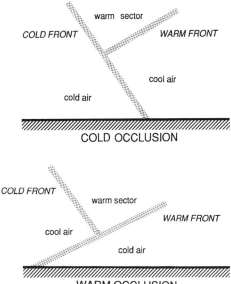

Figure O-1 Vertical cross section through two types of occlusion

OCCULT DEPOSITION

Acid precipitation composed of fine **fog** and **cloud** droplets. These droplets condense around sulphate and nitrate **aerosols** and as a result contain concentrations of sulphuric and nitric acid that may be as much as twenty times that of normal acid rainfall. The process is most often experienced in urban areas where **smog** is common, in areas prone to coastal fog and in hilly or mountainous areas, where the summits are frequently in cloud. Because the droplets are small enough to remain in suspension, their contribution to total **precipitation** is not always adequately represented in **rain** gauge totals. Occult deposition may be underestimated by up to 20 per cent, particularly in forested areas where trees are very effective in capturing the fine droplets.

Further reading
Park, C.C. (1987) *Acid Rain: Rhetoric and Reality*, London: Methuen.

OCEANIC CIRCULATION

The organized movement of **water** in the earth's ocean basins. The oceanic circulation is intimately linked to the circulation of the **atmosphere**. The prevailing **winds** in the atmosphere, for example, drive water across the ocean surface at speeds of up to 5 km per hour, in the form of broad, relatively shallow drifts. The **Coriolis effect** and the shape of the ocean basins also help to determine the flow of these currents. In some cases, they carry warm water polewards while in others they carry cooler water into lower latitudes. In addition to these surface flows, density differences, in part thermally induced, cause horizontal and vertical movements within the **oceans**. For example, cold, dense water, sinking in the Antarctic, flows along the bottom of the ocean basins and spreads northwards into the Southern Pacific, Atlantic and Indian oceans. There is also some evidence that the difference between the saltier, denser water of the Atlantic Ocean and the more dilute, less dense waters of the Pacific contributes to an overall deep water flow from Atlantic to Pacific and a surface flow in the opposite direction. This has been referred to as the 'Great oceanic conveyer belt'. All of these processes help to offset the imbalance of **energy** that develops between equatorial regions and the poles. This is illustrated particularly well in the North Atlantic, where the warm waters of the Gulf Stream Drift ensure that areas as far north as the Arctic Circle are anomalously mild during the winter months. One ocean current which does not appear as part of the well-

Figure O-2 The surface circulation of the oceans

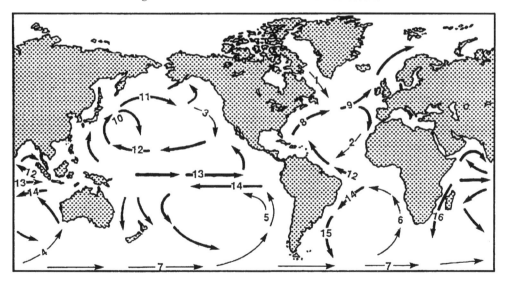

COLD CURRENTS		WARM CURRENTS	
1 Labrador	5 Humboldt	8 Gulf Stream	13 Equatorial Counter
2 Canary	6 Benguela	9 N. Atlantic Drift	14 S. Equatorial
3 California	7 West Wind Drift	10 Kuroshio	15 Brazil
4 West Australia		11 N. Pacific Drift	16 Agulhas
		12 N. Equatorial	

Figure O-3 The great oceanic conveyer belt

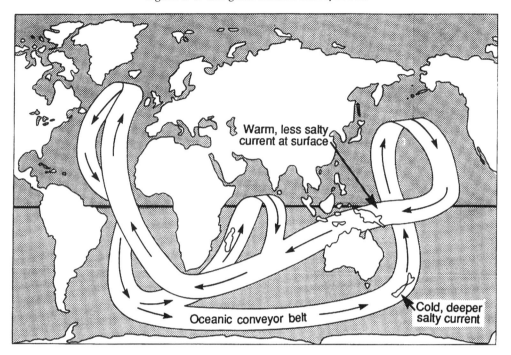

Source: After Moore, P.D., Chaloner, B. and Stott, P. (1996) *Global Environmental Change*, Oxford: Blackwell Science

established pattern, but which makes a major contribution to energy transfer is **El Niño**. This is a flow of warm surface water which appears with some frequency in the equatorial regions of the eastern Pacific. On average, oceanic transport accounts for 40 per cent of the total poleward transfer of energy in the earth/atmosphere system, with the **atmospheric circulation** accounting for the remaining 60 per cent.

See also
ENSO, Gyre.

Further reading
Broecker, W.S. (1991) 'The great ocean conveyer', *Oceanography* 4 (2): 79–89.
Strahler, A.H. and Strahler, A.N. (1992) *Modern Physical Geography* (4th edition), New York: Wiley.

OCEAN MODELS

See **coupled ocean/atmosphere climate models, general circulation models.**

OCEANOGRAPHY

The study of all aspects of the world's **oceans**, including the physical nature and structure of the ocean basins, the physics and chemistry of the ocean waters and the biology of the marine organisms that inhabit the oceans.

Further reading
Pickard, G.L. and Emery, W.J. (1990) *Descriptive Physical Oceanography: An Introduction* (5th edition), Oxford: Pergamon.

OCEANS

The large bodies of saltwater that cover about 71 per cent of the earth's surface and contain 97 per cent of the world's total **water** supply. Parts of the ocean that are deeper than 200 m belong to the oceanic zone, whereas those that are less than 200 m deep – for example, in coastal areas and above the continental shelf – are in the **neritic zone**. The neritic zone is much more

Figure O-4 The morphology of the ocean basins

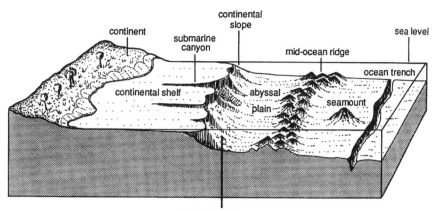

variable in terms of **temperature, salinity** and **sedimentation** than the ocean deeps, and because it is penetrated by **solar radiation**, it also has a much richer plant and animal **community**. Beneath the ocean surface, the landscape of the ocean floor includes a complex combination of morphological features from extensive plains to deep troughs and high mountain ridges.

See also
Abyssal zone.

Further reading
deBlij, H.J. and Muller, P.O. (1993) *Physical Geography of the Global Environment*, New York: Wiley.
King, C.A.M. (1962) *Oceanography for Geographers*, London: Edward Arnold.

ODD HYDROGENS

See **hydrogen oxides.**

ODOURS

In environmental terms, the smells produced by the release of (organic) **vapours** into the atmosphere and therefore a form of gaseous **air pollution**. The main sources of odours include farming operations, meat-processing plants, **pulp and paper mills, sewage** works and the chemical industry. The impact of this type of pollution varies with the odour and with the individual. For some, the odours may be merely disagreeable, but for others they may bring on nausea and insomnia. In most cases, the human nose can detect odours long before the vapours reach life-threatening concentrations, providing an early warning of a potentially more serious situation.

See also
Mercaptans.

Further reading
Harper, R., Bate-Smith, E.G. and Land, D.G. (1968) *Odour Description and Odour Classification: A Multidisciplinary Examination*, New York: Elsevier.

OIL

See **petroleum.**

OIL POLLUTION

In a modern industrial society dependent upon **petroleum** as its main source of **energy**, it is no surprise that the contamination of the **environment** by petroleum products is widespread and common. Major sources of oil pollution on land include motor vehicle operations and maintenance, refineries, pipelines, petrochemical plants and other industrial operations. Most oil spills on land are small – generally less than

1000 gallons – but occasionally they reach disastrous proportions. In late 1994, for example, a major leak in a pipeline carrying oil from the Arctic to central Russia allowed some 200,000 tonnes of crude to spill on to the Siberian **tundra** (Pain and Kleiner 1994). Such a massive spill compares with the largest **oil tanker** spills. However, nothing as yet is comparable to the volumes of oil released into the environment as a result of the Gulf War in 1991, when the defeated Iraqi forces sabotaged some 800 oil wells in Kuwait. Some they set alight, others were allowed to spew oil over the surrounding **desert**. The Kuwait oil fires burned for several months; at their peak, the spills amounted to more than 7 m tonnes of oil, and formed lakes that covered some 49 km^2 of Kuwait. As late as 1995, despite enormous amounts of time and money spent on clean-up, an estimated 0.5 m tonnes remained (Pearce 1995a).

In the **oceans**, pollutants are provided by shipping activities – both tanker and non-tanker – and by offshore petroleum exploration and production. Large spills from oil tanker accidents are the major sources of sudden, large-volume pollution episodes, but offshore oil operations also make major contributions. In 1977, for example, a blow-out from the *Ekofisk Bravo* platform released 14,000 tonnes of crude oil into the North Sea (Jenkins 1980), and the largest oil spill yet recorded – nearly 500,000 tonnes – was the result of a blow-out at the *Ixtoc-1* well in the Bay of Campeche, off the coast of Mexico (Cutter *et al.* 1991). Even regular activities in the offshore environment can contribute significant amounts of oil pollution. More than 16,000 tonnes of oil were added to the Norwegian sector of the North Sea between 1984 and 1990 in the form of oil-based drilling muds. Over the same period perhaps as much as eight times that amount was discharged into the British sector (Pearce 1995b). In addition to ongoing shipping and offshore oil production activities, there are the unexpected events, such as the release of more than 800,000 tonnes of oil into the Persian Gulf during the Gulf War in 1991, which may be few in number but are no less serious.

The classification into land-based and ocean-based spills is not perfect, since petroleum products released on land can be carried into the oceans in **runoff**, and the oil slicks created by spills at sea can be washed up onshore. Oil pollution receives most attention following major spills, both on land and in the oceans, but the regular ongoing small-scale contamination from leaking underground storage tanks or from the bilgewater discharges of ocean shipping, for example, commonly exceeds the irregular contributions of major pipeline ruptures or oil tanker accidents.

The spills that normally receive most attention are those which occur in coastal areas, where they foul great stretches of sandy beach or rocky shoreline. Reports in the media carry pictures of oil-soaked seabirds and mammals, many destined to die because the oil has destroyed their natural **insulation** and buoyancy or because they have ingested the toxic chemicals in the oil. The effects of a spill reach far beyond these more obvious impacts, however. Organisms living in the intertidal zone are poisoned and bottom-dwelling shellfish become contaminated as the effects of the spill descend through the water. Often the **food chain** is broken, either directly when an entire group of organisms is wiped out by the initial spill, or indirectly when predators begin to die after ingesting a contaminated food supply. Thus the entire **ecosystem** suffers. Similarly, on land, the oil lakes that flooded large areas of the desert in Kuwait were only the more obvious impacts of the spills. Vegetation is destroyed and life in the upper layers of the **soil** is no longer possible. Oil spills on land often migrate into local water bodies, through which contamination may be carried far beyond the original source, and pollution may remain in the area for decades if the oil seeps into the slow moving **groundwater** system. The terrestrial food chain is disrupted and toxic chemicals are spread through the system in much the same way as they are in the aquatic ecosystem.

Many of these ecological impacts have

economic implications. The aesthetic and physical degradation that follows coastal spills, for example, may bring major economic losses through the disruption of the tourist industry. Commercial shellfish beds may have to be closed because oysters and clams have been tainted by the oily water they have ingested. Similarly, local fisheries may be decimated if spills occur near fish-spawning beds or on migration routes. On land, soil that has been contaminated is no longer suitable for agriculture, and contaminated pasture or natural grazing will no longer support animals. The size of the particular oil spill is not always a good indication of its ultimate impact. The blow-out of 500,000 tonnes at the *Ixtoc-1* well off Mexico in 1979 was the largest offshore oil spill on record. However, a combination of **winds** and currents prevented the main oil slicks from coming ashore, and most of the oil dispersed in the open ocean. Damage to the oceanic environment must have taken place, but it was much less than would have occurred in the more complex and productive coastal environment. In contrast, the release of a much smaller amount of oil – 38,000 tonnes – by the *Exxon Valdez* in Alaska killed thousands of seabirds and mammals and caused incalculable environmental and economic damage.

Although the pollutants are mainly **hydrocarbons**, and therefore subject to **biodegradation** by **bacteria** and other organisms in the **natural environment**, the process is slow and the impact of the pollutants may be felt for many years after the initial contamination. Test sites set up in Alaska in 1976, for example, showed little improvement in the level of contamination when re-examimed in 1991 (Pain and Kleiner 1994). As a result, the area damaged by the oil spill requires some form of direct clean-up or rehabilitation. On land, free oil in pools can be pumped into tanks, but the contaminated soil remains. In Kuwait, after the liquid oil had been removed from the surface of the desert, the soil and **sand** beneath remained contaminated to a depth of 1.5 m (Pearce 1995a). Even if it is possible to remove the polluted soil, the local environ-

ment is destroyed and the problem of disposing of the polluted soil remains. Oil spills at sea are attacked using floating booms, that contain the oil until it can be removed by pumps or skimmers; by spraying the slick with chemical dispersants that break up the oil; and by burning the oil on the surface or in the vessel from which it is escaping. None of these systems is ideal. Booms do not work well when seas are rough. Chemical dispersants were once popular because they caused the oil to sink and the slicks to disappear quite quickly. Evidence from the areas contaminated by the Torrey Canyon and Amoco Cadiz spills where large volumes of dispersants were used suggests that the contamination of bottom-dwelling organisms by oil and the chemical dispersants persisted some ten to fifteen years after the events (Pearce 1993). Once the oil comes ashore, clean-up becomes difficult, particularly on rocky shores. Absorbents such as straw or **peat** can be used to soak up the oil, contaminated sand removed and steam used to clean coated rocks. Nothing is completely successful, however. Absorbents cannot remove all of the oil and the removal of shoreline material may have significant ecological and morphological consequences. Steam cleaning was used extensively following the **Exxon Valdez** accident, and although it effectively removed the oil, it also tended to scald and kill the organisms that had escaped the initial effects of the spill.

Despite ongoing attempts to improve clean-up technology, the success rate is not impressive. Scientists estimate that even under ideal conditions, with state-of-the-art technology provided rapidly by well-trained technicians, no more than 10–15 per cent of the oil from a major spill can be recovered (Miller 1994). The residue remains in some form or other in the environment. As with all pollution problems, prevention rather than response after the event would seem to be the most appropriate approach.

References and further reading
Cutter, S., Renwick, H.L. and Renwick, W.H. (1991) *Exploitation, Conservation, Preservation,*

New York: Wiley.
Jenkins, R.H. (1980) 'Oily water discharges from offshore oil developments', in C.R. Upton (ed.) *Ninth Environmenatal Workshop on Offshore Hydrocarbon Development*, Calgary: Arctic Institute of North America.
Miller, G.T. (1994) *Living in the Environment* (8th edition), Belmont, CA: Wadsworth.
Pain, S. and Kleiner, K. (1994) 'Frustrated West watches as Arctic oil spill grows', *New Scientist*

144 (1950): 8–9.
Pearce, F. (1993) 'What turns an oil spill into a disaster?', *New Scientist* 137 (1858): 11–13.
Pearce, F. (1995a) 'Devastation in the desert', *New Scientist* 146 (1971): 40–3.
Pearce, F. (1995b) 'Dirty rigs choke North Sea to death', *New Scientist* 146 (1976): 4.
Revelle, C. and Revelle, P. (1974) *Sourcebook on the Environment: The Scientific Perspective*, Boston: Houghton Mifflin.

Figure O-5 The distribution of major oceanic oil spills, 1962–89. Each dot represents a spill of more than 5000 tons

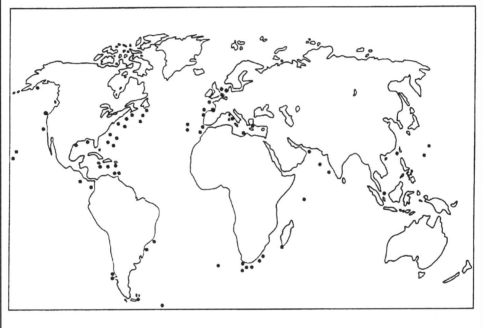

Source: Cutter, S.L., Renwick, H.L. and Renwick, W.H. (1991) *Exploitation, Conservation, Preservation: A Geographic Perspective on Natural Resource Use*, New York: Wiley

OIL SHALE

A fine-grained **sedimentary rock** containing kerogen, a **hydrocarbon** of botanical origin, from which gaseous and liquid **petroleum** can be extracted by **destructive distillation**. Oil shale was an important source of oil in Europe in the nineteenth century with the industry centred in Estonia and central Scotland. The Scottish operations produced a variety of products including **gasoline**, **kerosene**, lubricating oils, paraffin wax and tar, with the **fertilizer** sulphate of ammonia as

an important by-product. Production ceased in the early twentieth century, when conventional petroleum became more readily available. The largest existing deposits are in the Green River formation of Colorado, Wyoming and Utah in the United States where there are reserves of as much as 1800 billion barrels of **oil**. Significant deposits also exist in the Baltic region of Europe and in Brazil. Oil shale received much attention in the 1970s following the **OPEC**-generated oil crisis, when it was seen as an attractive alternative to the increasingly expensive and politically sensitive

Figure O-6 The products of oil shale processing from a nineteenth-century operation in central Scotland

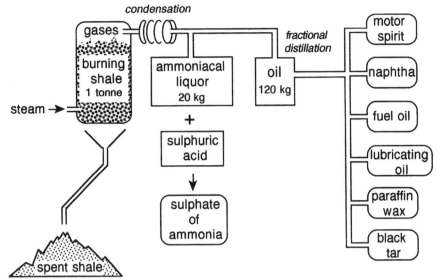

OPEC crude. However, on average nearly 1.3 tonnes of shale must be processed to produce one barrel of oil, requiring the excavation of some 500 million tonnes of shale per year to make the operation economically feasible. In addition, the process is accompanied by serious environmental concerns. When cooked to release the oil, the shale expands to several times the volume of the original rock, creating **waste disposal** problems. **Leachate** contamination of the adjacent **water** systems and **air pollution** from the refining process are also potential hazards. To overcome some of these environmental problems several oil companies explored the in situ conversion of oil shale into liquid oil by heating the shale underground and recovering the oil by way of wells. Although technically possible, it was not taken beyond the experimental stage because it was expensive and not without environmental problems. Oil shale remains a potential source of **energy**, but it is unlikely to be developed as long as relatively inexpensive conventional crude oil remains available.

Further reading
Kleinbach, M.H. and Salvagin, C.E. (1986) *Energy Technologies and Conversion Systems*, Englewood Cliffs, NJ: Prentice-Hall.

OIL SPILL

See **oil pollution**.

OIL TANKERS

Shipping activities are responsible for about 50 per cent of the **oil** entering the world **oceans**, largely as a result of tank flushing and bilge pumping, and perhaps as little as 2 per cent on average is released in tanker accidents. However, since the 1960s, tankers have become increasingly larger, with the so-called very large crude carriers (VLCCs) capable of transporting between 250,000 and 500,000 tonnes of crude oil. When such ships run aground or collide with another vessel, the result is normally catastrophic for the **environment**, because of the concentration of large amounts of oil in a limited area. The causes of tanker accidents fall into three main groups: human error, equipment failure and structural failure. Although the human factor can never be eliminated completely, it could be reduced by better training and the development of improved navigation techniques, involving, for example, global positioning systems. Major equipment failure has resulted

Figure O-7 A design for a safer tanker

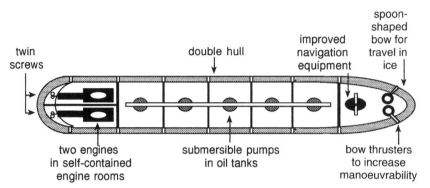

in a number of tanker accidents. Most tankers have only one engine driving a single screw. In the event of a breakdown, the vessel is at the mercy of the **elements**, and may well drift or be driven ashore. Twin-engined tankers with engines in separate engine rooms would be able to cope with the loss of an engine, and would be much more manoeuvr-able under normal circumstances. The failure of other components such as navigational aids or pumping systems could be dealt with by introducing the redundant system approach used in aircraft. Structurally, oil tankers are not particularly strong. They are subject to flexing in heavy seas, which may cause the hull to leak or, in the worst cases, the ship to

Table O-1 Major oil spills from tanker accidents

YEAR	SHIP	LOCATION	TONNES SPILLED
1967	Torrey Canyon	Scilly Islands, UK	120,000
1972	Sea Star	Persian Gulf	115,000
1976	Showa Maru	off Malaysia	70,000
1976	Argo Merchant	North Atlantic	28,000
1976	Urquiola	NW Spain	100,000
1976	Haven	Mediterranean Sea	40,000
1978	Amoco Cadiz	Brittany, France	220,000
1978	Andros Patria	NW Spain	20,000
1979	Atlantic Express	South America	300,000
1979	Burmah Agate	Gulf of Mexico	42,000
1989	Kharg-5	S Spain	82,000
1989	Exxon Valdez	Alaska, USA	38,000
1992	Aegean Sea	NW Spain	70,000
1993	Braer	Shetland Islands, UK	84,000

Source: Various, including Pearce (1993)

break up. The relatively thin hull plates are unable to withstand much more than a minor collision with another vessel or with rocks without leaking oil. Double-hulled tankers, in which the oil is carried in the inner hull separated from the outer hull by air-space or ballast **water** tanks, would prevent spills following many minor accidents, but both hulls could still be punctured in a major grounding or collision. All these improvements are technically possible but would be costly, ultimately driving up the price of oil. Without them the world's oceans and sea coasts will continue to be subject to major oil pollution episodes, and the environmental toll will be greater than it need be.

See also
Exxon Valdez, Oil pollution.

Further reading
Pearce, F. (1993) 'What turns an oil spill into a disaster?', *New Scientist* 137 (1858): 11–13.

OKIES

A somewhat disparaging term for farmers from Oklahoma who migrated westwards to California to escape the ravages of the **Dustbowl** on the **Great Plains** of North America in the 1930s. Their plight was described in John Steinbeck's fictional account of the Dustbowl, *The Grapes of Wrath*.

OLIGOTROPHIC (LAKES)

Water bodies that have a low concentration of **nutrients** and are therefore low in organic productivity. Oligotrophic lakes tend to be clear because of the limited amounts of organic matter they contain. This is characteristic of young lakes which have had insufficient time to develop an adequate nutrient supply and **recycling** system. Recently filled **reservoirs**, for example, are generally oligotrophic.

See also
Eutrophic, Eutrophication.

OPACITY

See **opaque**.

OPAQUE

Not **transparent** to **light**. The term may also be applied to the ability of a substance to block **sound** waves or **X-rays**. In the latter case, the substance would be 'radio-opaque'. Opacity may not be absolute. Some materials, for example, may be opaque to certain wavelengths in the **electromagnetic spectrum**, but transparent to others.

OPEC

See **Organization of Petroleum Exporting Countries**.

OPEC OIL EMBARGO

In October 1973, the Arab-dominated **Organization of Petroleum Exporting Countries** (OPEC) supported Egypt and Syria in the Yom Kippur War against Israel. Its support took the form of an oil embargo against Western nations seen to favour Israel. **Oil** exports were reduced to some nations, and banned completely to the United States and the Netherlands. The embargo, which lasted until March 1974, was accompanied by a series of rapid unilateral price increases which doubled the price of crude oil almost overnight and led to 1973 being referred to as 'the last year of cheap oil'. By 1973, the economies of the world's industrial nations had become seriously dependent upon cheap imported oil, mainly from the **OPEC** nations who at that time supplied more than 80 per cent of all imported oil. As a result, the effects of the combined embargo and price increases were immediate and almost universally devastating to the non-exporting nations. The global economy rapidly went into a recession, characterized by double-digit inflation, rising interest rates and spiralling debts incurred by the oil importing nations. To combat this situation, energy **conservation** was encouraged and an effort was made to substitute other **energy** sources for oil. In addition, the higher cost of oil encouraged increased exploration for new sources of non-OPEC oil, and in the second half of the 1970s production increased in Canada,

Figure O-8 Oil price trends following the 1973 OPEC oil embargo

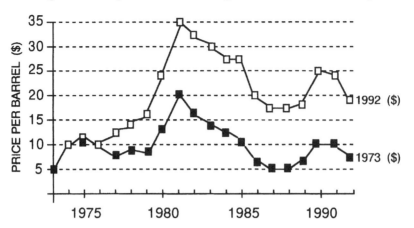

Source: After Miller, G.T. (1994) *Living in the Environment* (8th edition), Belmont, CA: Wadsworth

Mexico and the North Sea. This improved the supply and reduced the demand for oil, so that by 1978 the price had stabilized at about US$12 per barrel – only slightly above what it had been following the initial rapid price rises in 1974–1975. Further price increases followed, partly brought on by OPEC's manipulation of the supply and partly as a result of the shut-down of Iranian oil production during that country's Islamic revolution in 1979. By 1981, the price of a barrel of oil ranged from US$35–40, helping to produce a repeat of the recessionary conditions of the 1970s, but had declined to the US$18–20 range by the mid-1990s. In real terms, with adjustments for inflation, the price of oil in the mid-1990s is about the same as it was in the mid-1970s. The OPEC embargo was a major political and economic event which showed the power of energy-rich nations in a world dependent upon **petroleum** products. The embargo, and the responses of the non-OPEC nations to it, brought about radical changes to the geography of oil production and consumption, and hence to the global economy.

Further reading
Hanink, D.M. (1994) *The International Economy: A Geographical Perspective*, New York: John Wiley.
Khouja, M.W. (ed.) (1981) *The Challenge of Energy: Policies in the Making*, London/New York: Longman.

OPEN-CAST MINING

See **open-pit mining**.

OPEN-PIT MINING

A form of mining used to extract minerals that are present close to the earth's surface. Also referred to as open-cast mining, it involves the removal of the overlying **soil** and rock – the overburden – to expose the **ore** beneath. Open-pit mining is generally less costly than underground mining, which means that it can be used to extract lower grade **ores** profitably. In most cases, however, it causes greater damage to the **environment** than underground mining. The largest open pits are between 1 and 4 km in diameter and may range from 0.3 to 0.8 km in depth. The Bingham Canyon copper mine at Bingham, Utah is the largest open pit in the world. It is 4 km in diameter and 0.8 km deep. Excavating such a pit changes the landscape permanently, disrupts the local **hydrology** and creates large amounts of **waste**. With low grade ore, concentration or **beneficiation** is normally carried out on-site, to reduce the amount of waste transported to the smelter, and this produces **tailings** which may contain toxic chemicals. Although rehabilitation of the site is desirable, and commonly required by law, the landscape can never be returned to its original state, and full restoration of the pre-operational environment is impossible, particularly with larger pits.

Further reading
Griggs, G.B. and Gilchrist, J.A. (1983) *Geological Hazards, Resources and Environmental Planning* (2nd edition), Belmont, CA: Wadsworth.

OPEN SYSTEM

See **systems**.

ORE

A naturally occurring mineral deposit that contains an **element** at concentrations well above the normal crustal average for that element.

The term can include **fuels**, such as **coal** and **oil**, and chemical deposits, such as **salt** (NaCl) or **sulphur** (S), but it is most commonly applied to metallic elements. The desired element is obtained by the extraction of the ore and the separation of the **metal** from the surrounding unwanted minerals – gangue – by **smelting** or chemical extraction. The

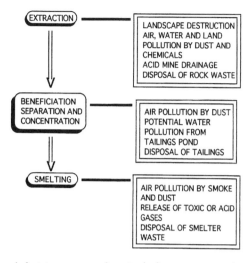

Figure O-9 The processing of metallic ores and related environmental problems

definition may also include an economic factor, in that a particular deposit might not be considered an ore body unless the element it contains can be extracted at a profit. The level at which profitability becomes possible may be represented by the minimum ore body concentration, but profitability will also

Table O-2 A comparison of crustal concentration and ore body concentrations for selected minerals

ELEMENT	NATURAL CRUSTAL CONCENTRATION (%)	MINIMUM ORE BODY CONCENTRATION (%)
Aluminum	8.0	24–32
Iron	5.0	20–30
Copper	0.0058	0.5–0.8
Nickel	0.0072	1.08
Zinc	0.0082	2.46
Titanium	0.5	5
Chromium	0.02	20
Lead	0.0001	0.2
Platinum	0.0000005	0.003
Silver	0.000007	0.01
Gold	0.0000002	0.0008

depend upon such geological factors as the grade, size, shape and depth of the ore body, and non-geological factors such as prices, geographical accessibility, labour costs and government policies. Since the non-geological factors can be quite changeable, they exert considerable influence on the potential profitability of an ore body and therefore on the timing and rate of its development. Modern technological societies depend heavily on metals and development is closely tied to the avaliability of a wide variety of types, from the relatively mundane and abundant, such as **iron** (Fe) and steel, to the more esoteric and scarcer, such as **gold** (Au) and germanium (Ge). Such a dependence has come with environmental costs, however, at all stages in the process from the original extraction of the ore to the manufacture of the final metal product.

Further reading

Barnes, J.W. (1988) *Ores and Minerals: Introducing Economic Geology*, Milton Keynes/ Philadelphia, PA: Open University Press.
Evans, A.M. (1993) *Ore Geology and Industrial Minerals: An Introduction* (3rd edition), Oxford/ Boston: Blackwell Scientific.
Ripley, E.A., Redman, R.E. and Crowder, A.A. (1995) *Environmental Effects of Mining*, New York: St Lucie Press.

ORGANIC COMPOUNDS

Chemical compounds which contain **carbon** (C) combined with **hydrogen** (H) and often with a variety of other **elements** including **oxygen** (O) and **nitrogen** (N). Organic compounds are molecularly very complex, commonly consisting of large numbers of **atoms** arranged in chains or rings.

See also

Alcohols, Carbohydrates, Organochlorides.

ORGANIC FARMING

Farming without the use of artificial (chemically manufactured) **fertilizers** or **pesticides**. Fertilization is provided by natural products such as animal manure, which contains organic **nitrogen** (N), stimulates soil **bacteria** and maintains the **soil structure** through its organic content. Green manure,

Figure O-10 The chemical structure of a simple organic compound – urea and a complex organic compound – mimosine

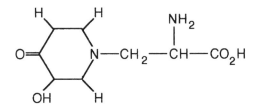

provided by ploughing in growing green plants and **compost**, also help to fertilize and condition the soil. Pests are controlled by cultivation practices which restrict the **habitat** necessary for the pests, by the **genetic engineering** of plant species to make them resistant to pests, by encouraging the natural predators of the pests and by using biopesticides such as the pyrethroids obtained from plants of the chrysanthemum family. Such an approach maintains the quaility of the soil and reduces the build-up of synthetic chemicals in the **environment**, although, in the short term at least, crop yields and cash profits may be less than those from a modern chemically based agricultural system.

Further reading

Little, C.E. (1987) *Green Fields Forever: The Conservation Tillage Revolution in America*, Washington, DC: Island Press.
Kaufmann, D.G. and Franz C.M. (1993) *Biosphere 2000: Protecting our Global Environment*, New York: HarperCollins.

ORGANIZATION FOR ECONOMIC CO-OPERATION AND DEVELOPMENT (OECD)

An intergovernmental institution, consisting of the world's twenty-four most developed nations, set up in 1962 to co-ordinate economic policies and encourage economic growth and world trade to the mutual benefit of its members. As the apparent link between economic growth and a variety of environmental issues increases, the OECD has become involved in the assessment of the socioeconomic impacts of specific issues. It is also concerned with the environmental aspects of foreign aid, and its impact on **sustainable development**.

Further reading
OECD (1991) *Climate Change: Evaluating the Socio-Economic Impacts*, Paris: Organization for Economic Cooperation and Development.
Starke, L. (1990) *Signs of Hope: Working Towards our Common Future*, Oxford: Oxford University Press.

ORGANIZATION OF PETROLEUM EXPORTING COUNTRIES (OPEC)

A group of Middle Eastern, Asian, African and Latin American nations that includes the world's major **petroleum** producers and exporters. They came together in 1960, recognizing the importance of **oil** as a source of future development, and with the intention of using their petroleum **resources** to advance their economic interests. By controlling prices and production, for example, they were able to increase their oil revenues sufficiently to allow major investment in their social and economic infrastructures. OPEC operated very successfully in the 1970s and early 1980s, but since then, internal disputes and significant changes in the nature of the world's petroleum economy have reduced its importance. However, since the thirteen nations (Algeria, Ecuador, Gabon, Indonesia, Iran, Iraq, Kuwait, Libya, Nigeria, Qatar, Saudi Arabia, United Arab Emirates, Venezuela) that make up the organization control the bulk of the world's petroleum **reserves** (*c.* 67 per cent), their potential to influence international economics and politics remains high. They are generally unwilling, for example, to support environmental initiatives such as restrictions on **carbon dioxide** (CO_2) emissions which would have an impact on the use of their petroleum products.

Further reading
Danielsen, A.L. (1982) *The Evolution of OPEC*,
New York: Harcourt Brace Jovanovich.
Flavin, C. (1985) *World Oil: Coping with the Dangers of Success*, Worldwatch Paper 66, Washington, DC: Worldwatch.
Hallwood, P. and Sinclair, S.W. (1981) *Oil, Debt and Development: OPEC in the Third World*, London/Boston: Allen & Unwin.

ORGANOCHLORIDES

A group of **organic compounds** that contain **chlorine** (Cl). They have a variety of forms and uses including **aerosol** propellants, plasticizers, transformer **coolants** (**PCBs**), food packaging (**PVCs**), electrical **insulation** and construction materials (PVCs), but probably their greatest use was as **pesticides** in the form of **DDT**, Aldrine and Lindane. Strongly biocidal, they act through the central nervous system, and initially when they were introduced in the 1940s they were very effective against insect pests such as the malarial mosquito. With time, however, many pests have developed an immunity to them. It has also become clear that the characteristics that made them good pesticides – persistence, mobility and high biological activity – also posed dangers for the **environment**. Organochlorides accumulate in the fatty tissue of animals, and through biomagnification in the **food chain** may reach toxic levels in predators. Because of side-effects such as sterilty, birth defects, **cancer** and damage to the nervous system, they have been banned or had their use severely restricted in most parts of the world.

Further reading
Brooks, G.T. (1974) *Chlorinated Insecticides*, Cleveland: CRC Press.
Hargrave, B. (1989) *Distribution of Chlorinated Hydrocarbons and PCBs in the Arctic Ocean*, Dartmouth, NS: Department of Fisheries and Oceans: Canada.
Simonich, S.L. and Hites, R.A. (1995) 'Global distribution of persistent organochloride compounds', *Science* 269: 1851–4.

ORGANOPHOSPHORUS COMPOUNDS

A group of **pesticides** that work by blocking the central nervous systems of the organisms exposed to them. Malathion and diazonon are the most commonly used organophosphates. They are highly effective against insects, but break down rapidly in the **environment** and do not bioaccumulate. For these reasons, they are preferred over organochloride pesticides. Although generally considered safer than the **organochlorides**, they are highly toxic to humans and other mammals and may be carcinogenic.

Further reading
Baarschers, W.H. (1996) *Eco-facts and Eco-fiction*, London/New York: Routledge.
Toy, A.D.F. and Walsh, E.N. (1987) *Phosphorous Chemistry in Everyday Living* (2nd edition), Washington, DC: American Chemical Society.

OROGENESIS

The process of mountain building. It is an integral part of the theory of **plate tectonics**, that views the earth's **crust** as a series of rigid but mobile plates. As the plates move, they come into contact with each other, and it is these contacts that cause orogenesis. The European Alps, for example, were formed when the African and European plates collided, folding, deforming and uplifting the materials that had been deposited in the sedimentary basin between them. In some areas, contact is followed by one of the plates sliding beneath the other – a process called subduction – creating a trench into which materials eroded from the plates are deposited. Ultimately these sediments undergo orogenesis. The Western Cordillera of North America appear to have

originated in this way in the subduction zone formed where the North American plate plunges beneath the Pacific plate. The crustal stresses and strains created by such activities encourage **earthquakes** and volcanic activity, both of which are characteristic of orogenesis. Three major orogenic periods have occurred in the past 500–600 million years. There is geological evidence of even earlier episodes in **Precambrian** times, and the continued movement of the earth's crustal plates suggests that orogenesis is an ongoing process.

Further reading
Duff, P.M.D. (1993) *Holmes' Principles of Physical Geology* (4th edition), London/New York: Chapman and Hall.
Tarbuck, E.J. and Lutgens, F.K. (1993) *The Earth: An Introduction to Physical Geology*, New York: Macmillan.

OSMOSIS

The diffusion of a **solvent** such as **water** through a semi-permeable membrane, from a **solution** of low concentration to one of a higher concentration. The semi-permeable membrane permits the movement of the solvent but prevents the dissolved substances from passing through. Ultimately, as a result of the solvent flow, the concentrations of the solutions on either side of the membrane will tend to become equal. Osmosis, through the semi-permeable membranes which surround individual **cells**, helps to control the flow of water through living organisms. Reverse

Figure O-11 The formation of mountains during orogenesis

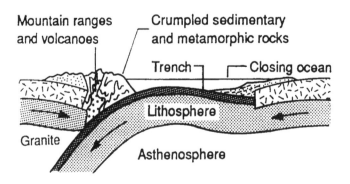

Source: After Goudie, A. (1989) *The Nature of the Environment* (2nd edition), Oxford: Blackwell

osmosis is used in the **desalination** and purification of water. Saltwater or contaminated fresh water is pumped under pressure through a semi-permeable membrane. The water molecules are allowed to pass through the membrane, whereas the **salt** and other impurities are retained in the original solution. The process is quite effective, but costly.

Further reading
Amjad, Z. (ed.) (1993) *Reverse Osmosis: Membrane Technology, Water Chemistry and Industrial Applications*, New York: Van Nostrand Reinhold.
Zumdahl, S.S. (1993) *Chemistry* (3rd edition), Lexington, MA/Toronto: D.C. Heath.

OUR COMMON FUTURE

The final report of the UN sponsored **World Commission on Environment and Development**, published in 1987.

See also
Brundtland Commission.

Further reading
Starke, L. (1990) *Signs of Hope: Working Towards our Common Future*, Oxford: Oxford University Press.

OVERFISHING

The taking of a fish **species** at a rate which exceeds its reproductive capacity. With prolonged overfishing, the breeding stock becomes smaller and smaller, a **sustainable yield** cannot be maintained and ultimately, when too few fish remain to make catching them profitable, the species faces commercial extinction. At that stage, fishing fleets begin to exploit other species, reducing the pressure on the original fish stock and therefore allowing it to recover. Recovery is not automatic, however. The overfished species may find it difficult to regain its original **population** because of competition from another species, or because the **habitat** it requires for recovery has been altered by **climate change** or **pollution**. The species might also be reduced to below its **minimum viable population** (MVP) size, but the minimum population that allows profitable fishing is usually reached before that point, and pressure on the species is reduced. Global fish harvests peaked in 1989 and have declined since then. Overfishing may not be the only factor involved, but it is estimated that the annual global fish harvest would be 20–24 per cent

Figure O-12 The decline of the Pacific herring stock

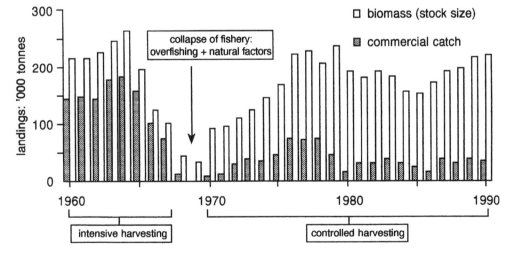

Source: After Ministry of Environment, Lands and Parks, British Columbia and Environment Canada (1993) *State of the Environment Report for British Columbia*, Victoria, BC: Ministry of Environment, Lands and Parks/Environment Canada

higher if no overfishing took place. International agreements such as the 1982 UN Convention on the Law of the Sea have attempted to control overfishing, but they are difficult to enforce, and it is estimated that more than forty species have been overfished since then. These include major commercial species such as the Newfoundland cod, North Sea herring, Pacific and Atlantic salmon and Peruvian anchovy.

Further reading
Miller, G.T. (1994) *Living in the Environment* (8th edition), Belmont, CA: Wadsworth.
Worldwatch Institute (1994) *State of the World*, New York: W.W. Norton.

OVERPOPULATION

A situation in which the **population** exceeds the **carrying capacity** of the **environment**. If allowed to continue, it will cause the population to crash, as the **resources** upon which it depends are used up. For example, a rapidly growing population often outstrips its food supply, leading to **famine**, starvation and an increase in the death rate, until a new balance is reached. In human terms, the increasing levels of environmental degradation that accompany the high rates of resource use associated with overpopulation are an important element in the relationship between the population and the carrying capacity of the environment.

See also
Malthus, T.R.

Further reading
Brown, L.R. and Kane, H. (1995) *Full House: Reassessing the Earth's Population Carrying Capacity*, London: Earthscan.

OVERTURN

The vertical circulation of lake **water** in spring and autumn, associated with seasonal changes in water **temperature** and **density**. Overturn involves the sinking of **oxygen**-rich surface water and the upwelling of **nutrient**-rich bottom water. Together, these have an important role in improving the productivity of the lake, particularly in the spring.

See also
Epilimnion, Hypolimnion, Thermocline.

OXIDATION

The combination of **oxygen** (O) with another **element** or **compound** to form an oxide. It may also be defined as a process in which **hydrogen** (H) is removed from a compound. An oxidation reaction involving a specific element produces an end-product that is a combination of that element and oxygen. With **magnesium**, for example, the end product is magnesium oxide.

$$2Mg + O_2 \rightarrow 2MgO$$

With organic compounds, complete oxidation will always give **carbon dioxide** (CO_2) and **water** (H_2O), whatever the original compound.

$$C_6H_{12}O_6 + 6O_2 \rightarrow 6CO_2 + 2H_2O$$
glucose

or

$$CH_4 + 2O_2 \rightarrow CO_2 + 2H_2O$$
methane

See also
Combustion, Corrosion, Metabolism, Reduction.

Further reading
Hudicky, M. (1990) *Oxidation in Organic Chemistry*, Washington, DC: American Chemical Society.

OXIDE

See **oxidation**.

OXIDES OF NITROGEN (NO_x)

A group of **gases** formed by the combination of **oxygen** (O) and **nitrogen** (N), often under high **energy** conditions, such as the bombardment of the upper **atmosphere** by **cosmic radiation**, during **lightning** storms, in high **temperature** furnaces and in **internal combustion engines**. They include **nitric oxide**, (N_2O) **nitrous oxide** (NO) and **nitrogen dioxide** (NO_2), commonly referred to as a group by the designation NO_x. Nitrogen dioxide may exist as dinitrogen tetroxide (N_2O_4), formed when two molecules of nitrogen dioxide combine. All the gases have an important role in the **nitrogen cycle**, but are also involved in environmental issues such as **photochemical smog** and **acid rain**. The presence of NO_x in the upper atmosphere also contributes to the depletion of the **ozone layer**.

Further reading

Lee, S.D. (1980) *Nitrogen Oxides and their Effects on Health*, Ann Arbor: Ann Arbor Science.
Seinfeld, J.H. (1986) *Atmospheric Chemistry and Physics of Air Pollution*, New York: Wiley.

OXYGEN (O)

A colourless, odourless **gas** that, as a gas and in combination with other substances, is the most abundant **element** in the earth/atmosphere system, occurring in rocks, **water**, **air** and a variety of organic materials. Oxygen makes up 21 per cent of the volume of the **atmosphere**, that amount being kept relatively constant through the process of **photosynthesis**. It occurs in several forms, such as atomic oxygen (O) and the triatomic **allotrope, ozone** (O_3), but most commonly as diatomic oxygen (O_2). It is a highly reactive chemical, combining readily with other elements to form **oxides**. Oxygen is essential for life on earth, being absorbed by animals during **respiration** and used to release **energy** in reactions with other chemicals.

See also

Combustion, Oxidation.

Further reading

Sawyer, D.T. (1991) *Oxygen Chemistry*, New York: Oxford University Press.

OXYGEN SAG CURVE

A graph of dissolved **oxygen** (O) levels against distance downstream from a known source of biodegradable effluent. When organic **effluent** is added to a stream, the demand for oxygen from **bacteria** and other organisms which will digest it, and from chemical **oxidation** processes, is met by the oxygen dissolved in the **water**. Thus, immediately downstream from the effluent source, the dissolved oxygen content of the stream falls. As the organic material is gradually

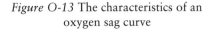

Figure O-13 The characteristics of an oxygen sag curve

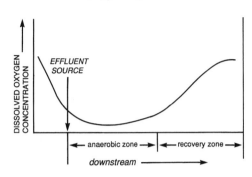

decomposed and converted into **carbon dioxide** (CO_2) and water, the demand for oxygen declines. At the same time, reaeration processes, associated with such factors as the turbulence of the stream, return oxygen to the water. Ultimately, the addition of oxygen exceeds its use and dissolved oxygen levels in the stream rise again. The characteristic dip in the curve as the dissolved oxygen level declines and then recovers is termed the oxygen 'sag'. The depth and extent of the sag depend upon such factors as the initial dissolved oxygen content of the stream, the **biochemical oxygen demand** (BOD) of the effluent, the rate at which the effluent is added and the stream's capacity for reaeration.

See also

Biodegradation.

Further reading

Chiras, D.D. (1994) *Environmental Science: Action for a Sustainable Future*, Redwood City, CA: Benjamin/Cummings.
Nemerow, N.L. (1974) *Scientific Stream Pollution Analysis*, Washington, DC: Scripta Book Company.

OZONE (O_3)

A blue **gas** with a pungent **odour**, ozone is an **allotrope** of **oxygen** (O) in which each **molecule** contains three **atoms** rather than the two of normal atmospheric oxygen. It is a very powerful oxidizing agent. Ozone is present in both the **troposphere** and the **stratosphere**, with tropospheric ozone accounting for about 10 per cent of the total ozone column. In the troposphere, where it is a constituent of

Figure 0-14 Schematic representation of the formation of stratospheric ozone

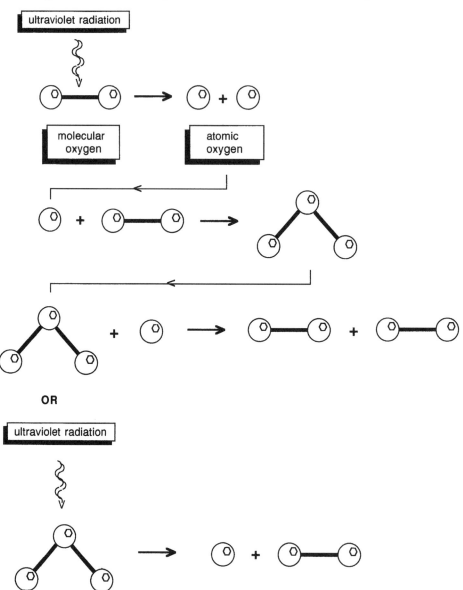

Source: Kemp, D.D. (1994) *Global Environmental Issues: A Climatological Approach,* London/New York: Routledge

photochemical smog, it is normally considered to be a pollutant, irritating to eyes and respiratory tissues, and harmful to plants. In contrast, stratospheric ozone is an essential component of the earth/atmosphere system, because of its ability to protect the **biosphere** from excess **ultraviolet radiation.**

See also
Ozone depletion, Ozone layer.

Further reading
Fishman, J. and Kalish, R. (1990) *Global Alert: The Ozone Pollution Crisis,* New York: Plenum Press.
Sawyer, D.T. (1991) *Oxygen Chemistry,* New York: Oxford University Press.

OZONE DEPLETION

The vulnerability of stratospheric **ozone** (O_3) to human interference was discovered almost by accident in the mid-1970s, but in the two decades that followed, depletion of the **ozone layer** became a major environmental issue, with a very high public profile.

The amount of ozone in the upper **atmosphere** is not fixed; it may fluctuate by as much as 30 per cent from day to day, and by 10 per cent over several years (Hammond and Maugh 1974). Since the formation of ozone is the result of a series of dynamic physical and chemical processes such fluctuations are not unexpected, although they are normally limited by checks and balances built into the system. By the early 1970s, however, there were indications that these controls were unable to prevent a gradual decline in ozone levels. Inadvertent human interference in the chemistry of the ozone layer was identified as the cause of the decline, and there was growing concern over the potentially disastrous consequences of such a development.

Ozone owes its existence to the impact of **ultraviolet radiation** on **oxygen** (O_2) **molecules** in the upper atmosphere, with the main production taking place in the tropical **stratosphere**, where radiation levels are high (Rodriguez 1993). Oxygen molecules normally consist of two **atoms**, and in the lower atmosphere they retain that configuration. At the high **energy** levels associated with ultraviolet radiation in the upper atmosphere, however, these molecules split apart to produce atomic oxygen (see Figure O-14). Before long, these free atoms combine with the available molecular oxygen to create triatomic oxygen or ozone. That reaction is reversible. The ozone molecule may break down again into its original components – molecular oxygen and atomic oxygen – as a result of further absorption of ultraviolet radiation, or it may combine with atomic oxygen to be reconverted to the molecular form (Crutzen 1974). Ozone molecules are also broken down in chemical reactions with various **compounds** containing **nitrogen**, (N) **hydrogen** (H) and **chlorine** (Cl), all of which occur naturally in the atmosphere. Although present in only small amounts, they become efficient ozone destroyers as a result of their ability to initiate **catalytic chain reactions**. The total amount of ozone in the stratosphere at any given time represents a balance between the rate at which the **gas** is being produced and the rate at which it is being destroyed. Thus, although the ozone layer is in a continuing state of flux, the total concentration should remain relatively constant.

Data on stratospheric ozone are available only as far back as the 1950s, but they indicate that ozone levels remained relatively stable, fluctuating only within the normal limits, until the mid- to late 1970s. By that time, it had become clear that human activities had the potential to bring about sufficient degradation of the ozone layer that it might never recover. The threat was seen to come from four main sources, associated with modern technological developments in warfare, aviation, agriculture and lifestyle, and involving a variety of complex chemical compounds, both old and new.

Nuclear war, **supersonic transports** (SSTs), cruising in the stratosphere and agricultural techniques increasingly dependent upon nitrogen-based **fertilizers** were seen as potential sources of increasing amounts of **oxides of nitrogen** (NO_x), a group of highly potent destroyers of ozone. Computer simulations suggested that following a major nuclear conflict 50–70 per cent of the ozone layer might be destroyed mainly as a result of the synthesis of oxides of nitrogen from atmospheric oxygen and nitrogen at the high **temperatures** produced by the thermonuclear explosions (Dotto and Schiff 1978). With the end of the **cold war** and agreements to reduce the numbers of nuclear weapons, the contribution of nuclear war to ozone depletion is no longer given serious consideration. Similarly, the hue and cry that accompanied the development of the original supersonic transports in the 1970s has died away. Estimates of the production of oxides of nitrogen and other ozone-destroying pollutants such as **hydrogen oxides** (HO$_x$) were based on fleets of several hundred SSTs. Fewer than ten SSTs remained in operation by the mid-1990s, and their effects on the ozone layer are generally considered to be negligible. The role of the oxides of nitrogen released during agricultural activities is also questionable. Although increasing amounts of the chemicals have been released into the **environment** as a result of the use of nitrogen-based fertilizers, there is, as yet, no proof that they have contributed to ozone depletion.

If there is some doubt about the impact of SST exhaust emissions on the ozone layer, or the contribution of nitrogen fertilizers to ozone depletion, the effects of certain other chemicals are now well established. **Chlorofluorocarbons** (CFCs) and related **bromofluorocarbons** or **halons** have made a major contribution to ozone depletion and continue to pose a significant threat to the ozone layer, despite successful attempts at controlling their output and use. They were developed in the 1930s, and after 1960 their production increased rapidly. CFCs were used in refrigeration and air-conditioning systems, and as propellants in **aerosol** spray cans dispensing a wide range of products from deodorants to paints and insect repellants. They were also used as foaming agents in the production of insulating foams, **polymer** foams for upholstery and foam containers for the fast food industry. Halons were found to be ideal for use in fire extinguishers and fire protection systems for aircraft, computer centres and industrial control rooms where conventional fire extinguishing materials such as **water** or foam would cause damage to delicate instruments.

The popularity of CFCs and halons stemmed from their stability and low toxicity which meant, for example, that they could be used as propellants in the inhalers required by those suffering respiratory problems without changing the efficacy of the medication or causing harm to the user. Being **inert** they were also ideal for cleaning delicate electronic components such as computer chips. However, these very properties that made CFCs and halons so useful ultimately allowed them to become the major contributors to ozone depletion. Their stability allowed them to accumulate in the environment relatively unchanged. With time they gradually diffused into the upper atmosphere, where they encountered conditions under which they were no longer inert, and broke down to release by-products with a great capacity to deplete stratospheric ozone.

The process was first explained in 1974 by Mario Molina and Sherwood Rowland, atmospheric chemists working at the University of California, Irvine. They recognized that the high levels of ultraviolet radiation in the upper atmosphere caused the photochemical degradation of the normally inert CFCs, and the release of chlorine (Cl) into the ozone layer. Catalytic chain reactions initiated by the free chlorine then began the process of depleting the ozone layer. Ironically, the survival of the CFCs in the lower atmosphere was possible only because the ozone layer protected them from the excess ultraviolet radiation which would have caused their destruction. Conclusions similar to those of Molina and Rowland were reached independently at about the same time by other researchers (Cicerone *et al.* 1974; Crutzen 1974; Wofsy *et al.* 1975), and with the knowledge that the use of CFCs had been growing since the late 1950s, the stage seemed set for an increasingly rapid thinning of the earth's ozone shield.

The research also indicated that the stability of the CFCs and halons, which allowed them to remain in the atmosphere for periods of 40 to 150 years, would exacerbate their overall effect on ozone depletion. It was estimated that even after a complete ban on the production of CFCs, the effects on the ozone layer might continue to be felt for a further 20 to 30 years and, under certain circumstances, for as long as 200 years after production ceased (Crutzen 1974; Wofsy *et al.* 1975).

The net effect of all of this on long-term ozone depletion proved difficult to predict. Molina and Rowland's original estimate of a 7–13 per cent reduction in steady-state ozone depletion was increased to as much as 20 per cent by some studies but reduced to as little as 5–7 per cent by others (Molina and Rowland 1994). Although all the researchers acknowledged that the results were at best preliminary because of inadequate knowledge of the photochemistry of the stratosphere, the concept was widely accepted and led to much speculation on the effects of the increasing levels of ultraviolet radiation that would follow ozone depletion of that magnitude.

One of the characteristics of the 1970s debate on ozone depletion was the way in which aerosol spray cans took much of the blame for the thinning of the ozone layer. Although CFCs were being employed as refrigerants and used in the production of various foams, the problem was usually presented as one in which the convenience of the aerosol spray can was being bought at the expense of the global environment. In 1975 there was some justification for this, since at that time two-thirds of CFCs were used as propellants in aerosol spray cans

(Molina and Rowland 1994). When the campaign against that product grew rapidly the multi-million dollar aerosol industry reacted strongly. Through advertising and participation in US National Academy of Sciences (NAS) hearings, they emphasized the speculative nature of the Molina-Rowland hypothesis, and the lack of hard scientific facts to support it. The level of concern was high, however, and the anti-aerosol forces met with considerable success. By 1978 a ban on the use of CFCs in hair and deodorant sprays was in place in the US and Canada followed with similar legislation in 1980. This proved to be the peak of the CFC controversy and by the late 1970s ozone depletion had already ceased to make headlines. Academic studies in stratospheric chemistry continued, but the level of public concern fell in the early 1980s, until quite unexpectedly, in 1985, scientists working at the Halley Bay base of the British Antarctic Survey announced that they had discovered a 'hole' in the ozone layer (Farman *et al.* 1985). All the fears that had been raised during the aerosol spray can debate suddenly returned.

Seasonal fluctuations in the ozone layer above Antarctica are part of the normal

Figure O-15 Changes in total ozone: British Antarctic Survey Halley Station: 26 August 1996 to 16 January 1997

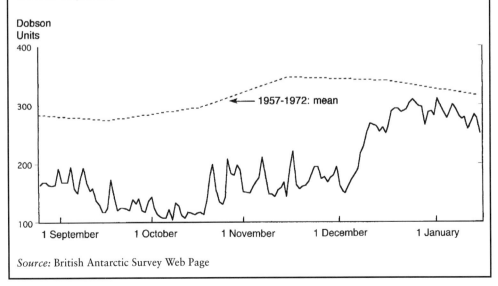

Source: British Antarctic Survey Web Page

variability of the atmosphere in that area. Thinning in the southern spring would become evident as early as late August and continue until mid-October when ozone levels would average between 250 and 300 **Dobson units** (DU). The ozone would begin to thicken again in November to reach an average of 350 to 360 DU by midsummer (December). In the 1980s, however, the hole began to develop earlier, grow more extensive and persist for longer. At the same time, total ozone values were declining, with reductions of as much as 40 to 65 per cent from 1960s levels not uncommon. By

Figure O-16 The breakdown of a chlorofluorocarbon molecule and its effect on ozone

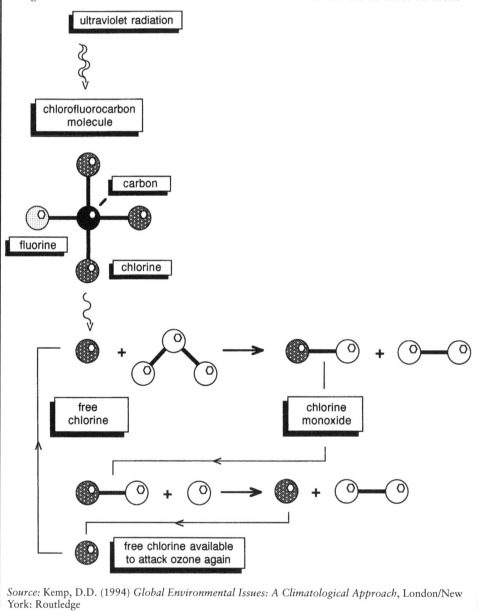

Source: Kemp, D.D. (1994) *Global Environmental Issues: A Climatological Approach*, London/New York: Routledge

the mid-1990s daily values within the hole were often below 130 DU (see Figure O-15) and it had grown to cover as much as 10 per cent of the Southern Hemisphere for at least one month every spring. At its most extensive its edges reached as far as Australia, New Zealand and the southern tip of South America.

When the hole was first discovered, blame fell almost immediately on CFCs and measurements of these chemicals in the stratosphere at the South Pole tended to support that. However, even with a continuous increase in CFCs of 5 per cent per year there appeared to be insufficient chlorine released through normal **gas phase reactions** to cause such an intense thinning of the ozone layer. Further investigation of the problem indicated that the process involved **heterogeneous chemical reactions** associated with the formation of **polar stratospheric clouds**. Although the polar stratosphere is very dry it becomes saturated at the very low temperatures (for example, 195 K) reached during the winter months, and **clouds** form. These consist of **ice particles**, on the surface of which nitric and hydrochloric acid particles enter into a complex series of reactions which ultimately lead to the release of chlorine. By way of a chlorine – **chlorine monoxide (ClO)** – cycle, it then attacks the ozone (Shine 1988). The **evaporation** of the polar stratospheric clouds in the spring, as temperatures rise, brings an end to the reactions and allows the ozone layer to recover. Similar heterogeneous chemical reactions take place on the surface of **sulphate particles** released into the stratosphere during major volcanic eruptions (Brasseur and Granier 1992). The eruptions of **Mount Pinatubo** and Mount Hudson in 1991, for example, were followed by a rapid destruction of the ozone layer at levels between 9 and 13 km (Deshler *et al.* 1992). In the following year, record low ozone levels were reported over the South Pole and over southern Chile and Argentina, while global ozone levels were 4 per cent below normal (Kiernan 1993). Although the Antarctic ozone hole

provides the most spectacular example of ozone depletion, thinning has also been identified elsewhere. Ozone depletion associated with the eruption of Mount Pinatubo was observed in Canada in 1993 when values averaged 14 per cent below normal between January and March (Kerr *et al.* 1993), and decreases of between 10 and 20 per cent were recorded over the Arctic in early 1992 (Concar 1992). Similar situations occurred over Japan, the United States, Russia and parts of Europe.

Growing concern over the consequences of ozone depletion led to the **Montreal Protocol**, a landmark international environmental agreement on the protection of the ozone layer, which was signed in 1987. Together with subsequent amendments agreed upon in Helsinki (1989), London (1990) and Copenhagen (1992), it provided a timetable for a reduction in the production and use of ozone-destroying chemicals, and their replacement by ozone-friendly products.

Global atmospheric concentrations of indicator chemicals such as CFC-11 and CFC-12 continue to increase, but since 1989 their growth rates have decreased significantly, presumably as a result of the Montreal

Table O-3 Phase-out dates for some common anthropogenically produced ozone destroying chemicals

CHEMICAL COMPOUND	PHASE-OUT DATE
CFC-11	1996
CFC-12	1996
CFC-113	1996
CFC-115	1996
Halon 1211	1994
Halon 1301	1994
Methyl chloroform	1996
Carbon tetrachloride	1996

Source: Kemp (1994)

Protocol and its amendments (Environment Canada 1994). However, the stability of CFCs and halons ensures that ozone depletion will continue even after production ceases, and it is likely that stratospheric ozone depletion has not yet reached its maximum.

References and further reading

Brasseur, G. and Granier, C. (1992) 'Mount Pinatubo aerosols, chlorofluorocarbons and ozone depletion', *Science* 257: 1239–42.

Cicerone, R.J., Stolarski, R.S. and Walters, S. (1974) 'Stratospheric ozone destruction by man-made chlorofluoromethanes', *Science* 185: 1165–7.

Concar, D. (1992) 'The resistable rise of skin cancer', *New Scientist* 134 (1821): 23–8.

Crutzen, P.J. (1974) 'Estimates of possible variations in total ozone due to natural causes and human activities', *Ambio* 3: 201–10.

Deshler, T., Adrians, A., Gobbi, G.P., Hofmann, D.J., Di Donfrancesco, G. and Johnson, B.J. (1992) 'Volcanic aerosol and ozone depletion within the Antarctic polar vortex during the austral spring of 1991', *Geophysical Research Letters* 19: 1819–22.

Dotto, L. and Schiff, H. (1978) *The Ozone War*, Garden City, NY: Doubleday.

Environment Canada (1993) *A Primer on Ozone Depletion*, Ottawa: Environment Canada.

Environment Canada (1994) *Stratospheric Ozone Depletion*, SOE Bulletin No. 94–6, Ottawa: Environment Canada.

Farman, J.C., Gardiner, B.G. and Shanklin, J.D. (1985) 'Large losses of total ozone in Antarctica reveal seasonal ClO_x/NO_x interaction', *Nature* 315: 207–10.

Hammond, A.L. and Maugh, T.H. (1974) 'Stratospheric pollution: Multiple threats to earth's ozone', *Science* 186: 335–8.

Kemp, D.D. (1994) *Global Environmental Issues: A Climatological Approach* (2nd edition), London/New York: Routledge.

Kerr, J.B., Wardle, D.I. and Tarasick, D.W. (1993) 'Record low ozone levels over Canada in early 1993', *Geophysical Research Letters* 20: 1979–82.

Kiernan, V. (1993) 'Atmospheric ozone hits a new low', *New Scientist* 138 (1871): 8.

Molina, M.J. and Rowland, F.S. (1974) 'Stratospheric sink for chlorofluoromethanes: chlorine atom-catalysed destruction of ozone', *Nature* 249: 810–12.

Molina, M.J. and Rowland, F.S. (1994) 'Ozone depletion: 20 years after the alarm', *Chemical and Engineering News* 72 (33): 8–13.

Rodriguez, J.M. (1993) 'Probing atmospheric ozone', *Science* 261: 1128–9.

Shine, K. (1988) 'Antarctic ozone – an extended meeting report', *Weather* 43: 208–10.

Wofsy, S.C., McElroy, M.B. and Sze, N.D. (1975) 'Freon consumption implications for atmospheric ozone', *Science* 187: 535–7.

OZONE DEPLETION POTENTIAL (ODP)

A measure of the capacity of a specific chemical to destroy **ozone** (O_3). CFC-11, with an ODP of 1.0 is the standard against which all other chemicals are measured. The ability of **Halon 1301** (ODP 10) to destroy ozone, for example, is ten times greater than that of CFC-11.

See also

Chlorofluorocarbons, Halons.

Further reading

MacKenzie, D. (1992) 'Agreement reduces damage to ozone layer', *New Scientist* 136 (1850): 10.

Tickell, O. (1992) 'Fire-fighters find gas that's easy on ozone', *New Scientist* 134 (1818):19

Table O-4 The ozone depletion potential of selected chemicals

CHEMICAL COMPOUND	ODP
CFC-11	1.0
CFC-12	1.0
CFC-113	0.8
CFC-115	0.6
Halon 1211	3.0
Halon 1301	10.0
Methyl chloroform	0.1
Carbon tetrachloride	1.1
Methyl bromide	0.3-0.9
HCFC-22	0.05
HCFC-123	0.02
HFC-134a	0.0

Sources: MacKenzie (1992); Tickell (1992)

OZONE LAYER

A diffuse layer of **ozone** (O_3) at heights of 20–50 km in the **stratosphere** which protects the earth's surface from the effects of excess **ultraviolet radiation**. At any given time, the total amount of ozone is small – if brought to normal **atmospheric pressure** at **sea level**, for example, it would form a band no more than 3 mm thick. However, through a dynamic, reversible process in which the **gas** is continually broken down and reformed, this relatively minor amount of ozone retains the ability to absorb ultraviolet radiation and prevent it from reaching the earth's surface. The growth in the volume of ozone-destroying chemicals, such as **chlorofluorocarbons** (CFCs), has disrupted the process, allowing greater amounts of ultraviolet radiation to pass through to the surface, raising fears of the increased occurrence of skin **cancer**, eye damage and genetic **mutation** in terrestrial organisms. Concern over such developments led to the signing of the **Montreal Protocol** which banned the use of ozone-destroying chemicals.

See also
Ozone depletion.

Further reading
Shea, C.P. (1999) *Protecting Life on Earth: Steps to Save the Ozone Layer*, Washington, DC: Worldwatch Institute.

OZONE PROTECTION ACT

Legislation passed by the Australian government in 1989 aimed at eliminating **chlorofluorocarbon** and **halon** use by 1994. It is an example of the type of legislation required at the national level to make international agreements, such as the **Montreal Protocol,** work.

P

PALAEOCLIMATOLOGY

(The study of) **climate** in the period prior to the development of the instrumental record. It includes the climates of the geological past, but most palaeoclimate research has involved the investigation of the climates of the **Quaternary** period from the **ice ages** to the beginning of the instrumental record in the nineteenth century. The investigation of palaeoclimates is based on the use of **proxy data** obtained from **elements** containing a climatic component. These include **ice cores**, ocean sediments, terrestrial deposits (e.g. glacial sediments, **loess**, lacustrine sediments), biological elements (e.g. **tree rings**, **pollen**, plant and insect **fossils**), historical records (e.g. private journals and diaries, company and government papers). Such sources vary in quantity and quality, but the choice generally improves with time. For example, the glacial fluctuations which provide evidence of climatic conditions during the last **glaciation** are identified mainly from ice cores and the stratigraphy of glacial sediments. Such sources are also available to provide evidence of postglacial changes, but for the later **Holocene**, additional physical sources are available including tree rings and pollen sequences. Proxy data based on human elements such as archaeological and agricultural records were subsequently added to the list (see Figure C-13). The resulting increase in the range of sources allows easier comparison and cross-checking of evidence, and the overall reliability of the palaeoclimate record is better for later time periods.

See also
Dendrochronology, Dendroclimatology, Palynology.

Further reading
Bradley, R.S. (1985) *Quaternary Palaeoclimatology*, London: Chapman and Hall.
Crowley, T.J. and North, G.R. (1991) *Palaeoclimatology*, New York: Oxford University Press.

PALAEOECOLOGY

(The study of) relationships between living organisms and their **environment** in the past. It normally involves the more recent (late- and postglacial) past and uses plant and animal **fossil** remains to establish former biogeographical environments. By examining the distribution of fossils such as **pollen** grains and insect remains in time and place, it may be possible to establish the history of environmental change in a particular area, which in turn may provide **proxy data** for the study of **climate change**.

See also
Palaeoclimatology, Palynology.

Further reading
Berglund, B.E. (ed.) (1986) *Handbook of Holocene Palaeohydrology and Palaeoecology*, Chichester: Wiley.
Lowe, J.T. and Walker, M.J.C. (1996) *Reconstructing Quaternary Environments* (2nd edition), London: Longman.

PALAEOGEOGRAPHY

The geography of a particular time and/or place in the past. It will include consideration of landforms, **climate** and other environmental factors, and may or may not include a human element. Descriptions of the nature and distribution of glacial landforms, for example, are essentially palaeogeographical in nature as are the travelogues produced by explorers in the past, particularly when the European influence began to spread to other parts of the world. In geological literature the term is used to refer to the geographical distribution of the continents in the past.

See also

Palaeoclimatology, Palaeoecology.

Further reading

McGowran, B. (1990) 'Fifty million years ago', *Scientific American* 78 (1): 30–9.
Sissons, J.B. (1967) *The Evolution of Scotland's Scenery*, Edinburgh: Oliver and Boyd.
Stamp, L.D. (1962) *Britain's Structure and Scenery*, Edinburgh: Collins.
Thwaites, R.G. (ed.) (1905) *Original Journals of the Lewis and Clark Expedition, 1804-1806*, New York: Dodd, Mead & Co.

PALAEONTOLOGY

A branch of geology which involves the study of the remains of plants and animals preserved in rocks in the form of **fossils**. By examining the nature, form and distribution of fossil types, it is possible to infer the environmental conditions under which they originally lived, and the ways in which different organisms have evolved over time. Although palaeontology is popularly seen as concerned with fossils present in solid rock, fossils preserved in soft deposits such as **peat** or **clay** are more useful for the study of recent environmental change. The study of fossil **pollen** and insects, for example, has made a major contribution to the understanding of changing environmental conditions during the **Pleistocene** and **Holocene**.

See also

Darwin, C.R., Evolution, Palaeoclimatology, Palaeoecology, Palynology.

Further reading

Coope, G.R. (1986) 'Coleoptera analysis', in B.E.

Berglund (ed.) *Handbook of Holocene Palaeoecology and Palaeohydrology*, Chichester: Wiley.
Paul, C.R.C. (1980) *The Natural History of Fossils*, London: Weidenfeld & Nicolson.

PALMER DROUGHT SEVERITY INDEX (PDSI)

A **drought** index developed in the 1960s by W.C. Palmer in the United States and widely used there to evaluate **soil moisture** conditions. It differs from many earlier indexes in that it does not attempt to tie drought to arbitrary amounts of **precipitation** or soil moisture levels. Instead, Palmer's Index takes an applied approach based on the **water** requirements of the normal level of economic activities in a region. Drought only exists when these water requirements cannot be met and the activities are adversely affected. Index values range from +4.0 (very much wetter than usual) to –4.0 (extreme drought), derived from a formula which includes consideration of antecedent precipitation, **potential evapotranspiration**, moisture storage in two soil layers and soil moisture recharge. The resulting budget is revised monthly to provide an ongoing evaluation of the severity of the drought.

See also

Soil moisture deficit (SMD), Soil moisture storage, Thornthwaite, C.W., Water balance.

Further reading

Briffa, K.R., Jones, P.D. and Hulme, M. (1994) 'Summer moisture variability across Europe, 1892-1991: an analysis based on the Palmer Drought Severity Index', *International Journal of Climatology* 14: 474–506.
Palmer, W.C. (1965) *Meteorological Drought*, Research Paper 45, Washington, DC: US Weather Bureau.
Mather, J.R. (1974) *Climatology: Fundamentals and Applications*, New York: McGraw-Hill.

PALYNOLOGY

Pollen analysis – the study of **fossil** spores and pollen grains. Spores and pollen are microscopic particles (commonly 5 to 200μm in diameter), produced as part of the reproductive process in certain organisms, that have provided significant information on the nature of palaeoenvironments as far back as the Palaeozoic

Figure P-1 A sample pollen diagram from south-east Scotland

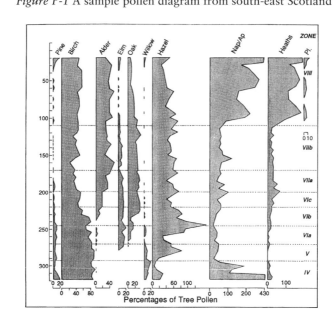

Source: After Sissons, J.B. (1967) *The Evolution of Scotland's Scenery*, Edinburgh: Oliver and Boyd

era. Most current palynological studies, however, are concerned with the analysis of more recent deposits, particularly those of the **Quaternary**. The abundance of pollen produced by flowering plants, trees, shrubs and grasses, particularly those which depend upon the **wind** for distribution, plus the durability of the grains, ensures that sufficent pollen survives for analysis. The grains are most often preserved in lakes, **peat** bogs and **soil**. Samples removed by hand boring from these sources are treated chemically to separate the pollen from the matrix, allowing individual grains to be identified and counted. Chemical processing of the samples is possible because of the durability of the exoskeleton of the pollen grains, that under normal conditions protects the living **cell** chemicals and a cellulose-based internal skeleton.

Following the burial of the pollen grains, the living cell and the internal skeleton normally decay, but the exoskeleton, or exine, survives. It is composed of sporopollenin, an inert, durable, organic **compound** that is able to withstand, for example, the highly acidic environment of a peat bog. The

distinctive shape, size, surface morphology and distribution of pores of individual pollen grains and spores allows them to be used to identify the plants that produced them, sometimes to the level of particular **species**. Microscopic inspection of the grains allows the development of a pollen spectrum that indicates the proportion of the various plant genera at a specific stratigraphic horizon and reflects the plant assemblage that existed at the time the pollen or spores were originally produced. Subsequent sampling of adjacent horizons extends the record of the plant **communities** that occupied the site. Pollen grains were first identified in peat in the late nineteenth century, and Swedish scientists had begun the first systematic use of pollen analysis by 1916. The main development, however, began in the 1950s, and since then, palynology has been used extensively and successfully to reconstruct palaeoenvironments during the Quaternary. In Northern Europe, for example, the poleward migration of vegetation and associated environmental changes as the Wurm/Weischel **glaciers** retreated was established using pollen analysis. Changing climatic conditions were then inferred from the results and a time frame was created using ^{14}C dating.

Reconstructions covering similar time periods have been produced for North America and parts of Asia. Pollen analysis is one of the most common and perhaps most successful techniques used in the study of environmental change. In addition, palynology has an important role in archaeology – for example, establishing the plants grown by a former agricultural community – and although the amount of pollen surviving from pre-Quaternary times is limited, the presence of fossil pollen in **sedimentary rocks** may help with the correlation of strata or the identification of strata likely to contain **coal** or **oil**.

See also
Palaeoclimatology, Palaeoecology, Proxy data.

Further reading
Bernabo, J.C. and Webb, T (1977) 'Changing patterns in the Holocene pollen record of Northeastern North America: a mapped summary', *Quaternary Research* VIII: 70–1.
Faegri, K. and Iversen, J. (1950) *Text-book of Modern Pollen Analysis*, Copenhagen: Munksgard. (4th edition revised by K. Faegri, P.E. Kalan and K. Kryswinski (1990) Chichester: Wiley.)
Moore, P.D., Webb, J.A. and Collinson, M.E. (1991) *Pollen Analysis*, Oxford: Blackwell Scientific.
Traverse, A. (1988) *Palaeopalynology*, Boston: Allen & Unwin.
Tzedakis, P.C., Bennett, K.D. and Magri, D. (1994) 'Climate and the pollen record', *Nature* 370: 513.

PAN

See **Peroxyacetylnitrate**.

PARAFFIN

See **kerosene**.

PARAMETERIZATION

The method by which regional scale processes are included in global climate **models**. Parameterization involves the establishment of statistical relationships between small-scale or regional processes and variables which can be measured at the grid scale of the model. Since the latter can be calculated by the model, the values of the small-scale processes can then be estimated. For example, cloudiness, which is very much a local factor, can be represented using **temperature** and **humidity** values calculated at the model grid points. **Radiation** and **evaporation** can also be estimated in a similar fashion.

See also
General circulation models.

Further reading
Hengeveld, H.G. (1991) *Understanding Atmospheric Change*, SOE Report 91–2, Ottawa: Environment Canada.
Houghton, J.T. (ed.) (1984) *The Global Climate*, Cambridge: Cambridge University Press.
Schneider, S.H. (1987) 'Climatic modelling', *Scientific American* 256: 72–89.

PARASITE

An organism that lives in or on another organism and uses that relationship to obtain food **energy**. Parasites are usually smaller than their hosts, but can range in size, from micro-organisms such as **viruses** and **bacteria** to intestinal tapeworms several metres long found in many large mammals, including humans. Many parasites are host specific, but some may infest an intermediate host at some stage in their life cycle. Some tapeworms, for example, undergo initial development in snails or mice, only reaching maturity when consumed by larger grazing animals or predators. In theory, it is not in the parasite's best interest to kill the host, but some do. Others may make the host sufficiently weak that it succumbs to some form of disease. Parasites are integral parts of all **ecosystems**, but modern society spends considerable time and money ensuring that parasites are kept under control. This can be done by maintaining a clean **water** supply, for example, by the strict enforcement of food inspection or by the use of **pesticides**. In **Third World** nations, however, overcrowding, generally insanitary living conditions and the absence of an efficient health care infrastructure permits the easy spread of parasites, causing sickness, blindness and death, and generally reducing the ability of the society to reach its full potential.

See also
Commensalism, Symbiosis.

Further reading
Crawley, M.J. (ed.) (1992) *Natural Enemies: The Population Biology of Predators, Parasites and Diseases*, Oxford: Blackwell Scientific.

PARTICLE COAGULATION

An atmospheric process in which individual particles floating in the **atmosphere** combine together to form larger particles. Individual **soot** particles, for example, are ~0.1μm in diameter, but readily link together in branching chains or loose aggregates which may reach ten times that size. These larger particles may be too heavy to remain suspended and, under the effect of **gravity**, fall back to the surface, helping to cleanse the atmosphere in the process.

See also
Particulate matter, Particulates.

PARTICULATE MATTER/ PARTICULATES

A collective name for fine **solid** or **liquid** particles added to the **atmosphere** by processes at the earth's surface. Particulate matter includes **dust, smoke, soot, pollen** and **soil** particles. Globally, 85–90 per cent of it is of natural origin, with human activities accounting for the other 10–15 per cent. Locally, the proportion of anthropogenically produced particulate matter can be higher – for example, in urban/industrial areas – and from time to time events such as fires (**Kuwait oil fires**) or industrial explosions (**Chernobyl**) also increase the human contribution. Once in the atmosphere, particulates are redistributed by way of the **wind** and **pressure** patterns, remaining in suspension for periods ranging from several hours to several years depending upon particle size and altitude attained. Particulates exert an influence on **climate** by disrupting the flow of **radiation** within the earth/atmosphere system. By increasing **atmospheric turbidity**, they cause the **attenuation** of **solar radiation**, the net effect being local or even global cooling. Cooling following such events as the eruption of **Mount Pinatubo** appears to support this, but there is also some evidence that the

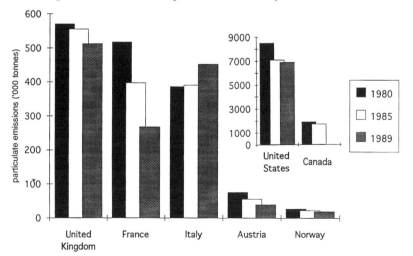

Figure P-2: Emissions of particulate matter by selected nations

Source: Based on data in World Resources Institute (1992) *World Resources 1992–93, A Guide to the Global Environment*, New York: Oxford University Press

presence of particulate matter in the atmosphere might produce a slight warming.

See also
Aerosols, Arctic haze, Dust veil index, Fall-out, Nuclear winter, Smog, Sulphate particles.

Further reading
Fennelly, P.F. (1981) 'The origin and influence of airborne particulates', in B.J. Skinner (ed.) *Climates, Past and Present*, Los Altos, CA: Kauffmann.
Groisman, P.Y. (1992) 'Possible regional climate consequences of the Pinatubo eruption: an empirical approach', *Geophysical Research Letters* 19: 1603–6.

PASTORAL AGRICULTURE

A form of agriculture, based on the herding of grazing animals such as cattle, sheep, goats and camels, common in semi-arid tropical and temperate natural **grasslands**. It often involves **nomadism**, necessary to allow grazed vegetation to regenerate or to accommodate seasonal changes in **weather** conditions. Permanent pastoralism is made possible by the cultivation of **forage**, such as hay, using **fertilizers** and irrigating where necessary.

See also
Arable agriculture, Desertification, Sahel, Soil erosion.

PATHOGEN

A living organism which causes a transmissible or communicable disease. Common pathogens are **bacteria**, **viruses**, **protozoa** and parasitic worms, spread in **air**, **water**, food, body **fluids** and insects. They are responsible for diseases such as **cholera**, typhoid fever, **malaria** and dysentry, which cause the deaths of millions of people every year, particularly in less developed countries, where poor nutrition, overcrowding, poor sanitation and unsafe drinking-water supplies encourage and increase the spread of pathogens.

Further reading
Mottet, N.K. (ed.) (1985) *Environmental Pathology*, New York: Oxford University Press.

PCBs

See **polychlorinated biphenyls**.

PEAT/PEATLANDS

Peat is a soft, fibrous, organic substance, brown or black in colour and consisting of partially decomposed plant material, found in waterlogged or poorly drained areas. Waterlogging allows only **anaerobic** decomposition, which is a relatively slow process, and as a result the organic material can accumulate to depths of as much as 10 m. Because the process of decomposition is so slow, parts of plants remain recognizable even after thousands of years. Peat deposits have therefore become prime locations for investigating palaeoenvironments. The examination of such plant **fossils** as leaves, stems, seeds and **pollen**, for example, can provide a record of changing environmental conditions at a particular site. Peat will develop in any location where the drainage is poor and continues to be poor for some time. It is part of the sedimentary succession in the evolution of lakes, for example, growing on the waterlogged sediments present near the end of the infill process. Ideal conditions for peat growth are also provided by estuarine mudflats and in flat-bottomed valleys with incompetent streams. Many of the peatlands now in existence owe their formation to the disruption of drainage during the last glacial. The hollows in hummocky glacial landscapes, for example, often contain peat, and the flat floors of glaciated valleys provide the waterlogged conditions required for peat growth. Climatologically, high amounts of **precipitation** coupled with low **evapotranspiration** rates and poor **runoff** also encourage peat growth, often in the form of a blanket bog, that covers large areas of both upland and lowland. Such conditions exist in higher latitudes in the northern hemisphere, and, as a result, thousands of hectares of peatlands are located in a zone stretching from Russia, through Northern Europe to Canada and Alaska. Peatlands may be subdivided according to their **nutrient** status into fens (**eutrophic**) and bogs (**oligotrophic**).

The former are generally more rich in nutrients and usually occur where the inflow of **water** brings with it minerals from the surrounding area. If the local rocks are rich in **bases**, the **pH** of such fens may reach as much as 7 or 8. Fens are characterized by grasses and sedges and in places may support moisture-tolerant trees such as willow, alder and even birch. Peat bogs have low nutrient levels in part because they receive water draining from base-poor rocks such as granite or deposits such as **sand** and gravel that are low in available minerals. Peatlands which are maintained by an excess of precipitation, that provides no minerals, are also nutrient poor. Bogs are characterized by an abundance of *Sphagnum* mosses in combination with heather (*Calluna*) and heath (*Erica*) species, which can survive the relatively high acidity of the **environment**. For centuries, much effort has gone into clearing and draining fens and lowland bogs to provide additional land for agriculture. Peat was long the traditional fuel in areas such as Ireland and Scotland, where coal was not readily available. In more recent times, the large-scale burning of peat to produce electricity has been developed in Finland and Ireland, and explored in North America. Given the low **energy** content of peat, however, it is only feasible where the peat is easily accessible and cheap to extract. Dried peat also has a variety of applications in horticulture. Once seen as unproductive and uninteresting, peatlands are now recognized as being significant in the **hydrology** of many areas through their ability to retain water, and having a role in the **greenhouse effect** through their capacity to store **carbon** (C). In addition, the release of **methane** (CH_4), one of the products of anaerobic decomposition, from peat bogs may contribute to **global warming**. Increasingly environmentalists consider them as representing a distinct environment, often unaltered by human activity and therefore worth preserving.

See also
Palynology, Wetlands.

Further reading
Crum, H.A. and Planisek, S. (1988) *A Focus on* *Peatlands and Peat Mosses*, Ann Arbor: University of Michigan Press.
Godwin, H. (1981) *The Archives of the Peat Bogs*, Cambridge: Cambridge University Press.
Moore, P.D. and Bellamy, D.J. (1974) *Peatlands*, London: Elek.
Pearce, F. (1994) 'Peat bogs hold bulk of Britain's carbon', *New Scientist* 144: 6.

PEDOLOGY

The study of the morphology and distribution of **soils** including consideration of their internal properties and the processes involved in their development.

See also
Soil classification, Soil structure, Soil texture.

Further reading
Foth, H.D. (1990) *Fundamentals of Soil Science* (8th edition), New York: Wiley.

PERCOLATION

Vertical movement of **water** downwards through **soil** or rock in the unsaturated zone immediately beneath the surface.

PERIGLACIAL

Periglacial **environments** occur in cold regions which experience near glacial conditions and where frost action is the predominant geomorphological process. Periglacial conditions exist adjacent to **ice** sheets and **glaciers**, but also extend into areas in high latitudes and at high altitudes where average **temperatures** are low, winters are long and the ground is permanently frozen (**permafrost**). It is estimated that 20–25 per cent of the earth's surface experiences periglacial conditions, mostly in the polar and subpolar regions of the northern hemisphere and in the mountainous areas of Europe, Asia and North America. With much less landmass in high latitudes, the southern hemisphere has only small areas exhibiting periglacial conditions, in Antarctica, Tierra del Fuego and the southern Andes mountains. The land surface in periglacial regions may include **snow** and ice, bare rock and **tundra** vegetation. In some periglacial areas the summer temperatures may be sufficiently high to permit tree growth. Morphological features in periglacial zones are associated with frost action and **mass movement**. The freezing and expansion of water in cracks and joints aids the destruction of rock, for example, while on a larger scale thermal contraction cracks, ice wedges and ice-cored conical hills or pingoes are all characteristic of periglacial areas. Mass movement in the form of frost creep and **solifluction** also contributes to periglacial landscapes, particularly those formed in unconsolidated sediments. Where partial thawing of the permafrost has taken place, a hummocky landscape with water-filled enclosed hollows – thermokarst – is formed. Elsewhere, patterned ground in the form of polygons created by interlocking ice wedges, stone stripes or stone circles is common. Many of these features are present in relict or fossil form in areas that are now no longer experiencing periglacial conditions. They are found, for example, in middle latitudes across central and eastern North America, northern and north-western Europe and eastern Asia. They usually represent features which were active at the time of the **ice ages**, when in the northern hemisphere

the periglacial zone moved south ahead of the advancing ice. With **global warming**, existing periglacial features in northern latitudes might also become inactive. Human use of areas in the periglacial zone is currently quite limited, but it is clear from such activities as **oil** and other mineral development that the periglacial environment is easily disturbed and possesses its own peculiar engineering, construction, maintenance and **pollution** control problems.

Further reading

French, H.M. and Slaymaker, O. (eds) (1993) *Canada's Cold Environments*, Montreal: McGill-Queen's University Press.
Washburn, A.L. (1979) *Geocryology: A Survey of Periglacial Processes and Environments,* New York: Wiley.
Williams, P.J. and Smith, M.W. (1989) *The Frozen Earth: Fundamentals of Geocryology*, London/New York: Cambridge University Press.

PERIODIC TABLE

A classification of the **elements** in tabular form, originally developed in 1869 by the Russian chemist D.I. Mendeleev, and based on his periodic law which stated that the properties of

Figure P-3 The periodic table of the elements

Group numbers 1 - 18 represent the system recommended by the International Union of Pure and Applied Chemistry

elements were related to their atomic weights. Although a number of elements were as yet unidentified when Mendeleev produced his table, its organization was sufficiently systematic that he was able to predict the existence and properties of missing elements and left spaces in his table to accommodate them. Modern versions of the periodic table arrange the elements in order of their **atomic numbers**. Each horizontal row or period includes elements with distinctly different properties in sequence. When the sequence is repeated a series of horizontal groups is produced in which the elements have similar properties. Descriptive names have been applied to some of the groups to reflect their characteristics. Group II is the alkaline earth **metals**, for example, Group VII contains the **halogens** and Group VIII the **noble gases**. The constituents of the lower numbered groups tend to be metals, whereas those in the higher numbered groups are mainly non-metallic. Located between Groups II and III are the transition elements, that share some of the characteristics of the adjacent groups but differ in terms of such elements as **electron** distribution and **valence**. All the transition elements are metallic and include most of the commercially important metals – for example, **iron, copper, silver, gold**. Although all of the information contained in the periodic table is available in a variety of other sources, its main advantage is the provision of easy correlation of the wide range of physical and chemical properties of the elements.

Further reading

Faughn, J.S., Turk, J. and Turk, A. (1991) *Physical Science*, Philadelphia, PA: Saunders.
Zumdahl, S.S. (1993) *Chemistry* (3rd edition), Lexington, MA/Toronto: D.C. Heath.

PERMAFROST

A contraction of 'permanently frozen ground'. Popularly, permafrost is characterized by the presence of perennial **ice** beneath the surface of the earth. Technically, it refers to a condition in which subsurface **temperatures** remain below 0°C for at least two consecutive winters and during the intervening summer. **Water** may or may not be present. If it is present it may not be frozen – for example, because of the mineralization of **groundwater**. Even under the most severe conditions, the ground may not always be frozen completely to the surface. In most areas experiencing permafrost conditions there is a surface, active layer which is subject to seasonal freezing and thawing. It varies between *c*.15 cm and several metres in thickness, depending upon such factors as seasonal temperature variation, **snow** cover, vegetation cover and the nature of the **soil** and rock in the area. Beneath that, where temperatures remain below freezing, is the true permafrost, its upper boundary being referred to as the permafrost table. Permafrost underlies as much as 25 per cent of the earth's surface mainly in high latitudes in the northern hemisphere. It is extensive in Alaska, northern Canada, Siberia and China, being thickest in polar regions and generally thinning towards the south, but with local differences because of variations in **climate** and terrain. Thicknesses range from a few metres to as much as 500 m in parts of northern Canada with reported thicknesses as great as 1200 m in far north-eastern Siberia. Along the southern margins of the permafrost zone, the presence of frozen ground is discontinuous, and its distribution fluctuates as a result of even short-term **climate** variations. Discontinuous permafrost is found almost as far south as the 50th parallel in Canada and China. If the predicted **global warming** occurs, the limits of permafrost will retreat polewards and the area underlain by permanently frozen ground will be much restricted. The presence of permafrost creates problems for the human use of northern areas, mainly through the disruption of **energy** flow in the active layer. Under stable conditions, the thickness of the active layer and the location of the permafrost table is in balance with the flow of heat into and out of the surface layer. That balance is disturbed during such activities as the erection of buildings, road

Figure P-4 Vertical cross-sections across (a) Asia and (b) North America, showing the thickness and changing nature of the permafrost layer

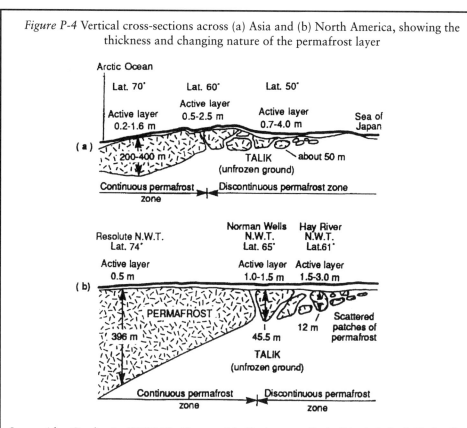

Source: After Goudie, A. (1989) *The Nature of the Environment* (2nd edition), Oxford: Blackwell

and pipeline construction and mineral extraction. The removal or destruction of **tundra** vegetation, for example, allows an increased inflow of heat into the ground, because the insulating value of the vegetation is lost. This in turn increases the thickness of the active layer, makes additional water available in the surface layer, and since that water is prevented from percolating downwards because of the presence of the permafrost, it collects in pools or runs across the surface causing erosion. Similarly, any heated structure built on permafrost will also disrupt the heat flow. The resulting localized melting of the permafrost will cause the building to sink or tilt on its foundations, with resulting structural damage. The construction of pipelines in the north represents a serious threat to permafrost. During the construction phase, the building of access roads, excavation for the line and construction of pumphouses and other facilities all disrupt the energy balance of the permafrost. Once the **oil** begins to flow, the loss of heat from the pipeline itself has the potential to degrade the permafrost. Stresses and strains placed on the structure by uneven melting might then lead to the rupture of the pipe with a consequent escape of oil into the **environment**. Modern engineering and construction techniques have allowed some of these problems to be mitigated. For example, where possible, construction work on the tundra is carried out in the winter months when the active layer is frozen and snow protects the vegetation from damage; structures are placed on gravel pads and insulated to prevent the escape of heat into the permafrost; pipelines may be raised above the surface, rather than buried, in areas where the permafrost is particularly unstable, to allow the heat from the system to be dissipated into the **atmosphere**. Despite such developments, permafrost environments remain delicately balanced and any advance of human activities into such areas must be approached with caution and a clear understanding of the environmental limitations involved.

See also
Alaska Pipeline, Periglacial, Solifluction.

Further reading
Andersland, O.B. and Ladanyi, B. (1994) *An Introduction to Frozen Ground Engineering*, New York: Chapman and Hall.
Anisimov, O.A. and Nelson, F.E. (1996) 'Permafrost distribution in the Northern Hemisphere under scenarios of climatic change', *Global and Planetary Change* 14: 59–72.
Harris, S.A. (1986) *The Permafrost Environment*, Totowa, NJ: Rowman and Littlefield.
Williams, P.J. and Smith, M.W. (1989) *The Frozen Earth: Fundamentals of Geocryology*, London/New York: Cambridge University Press.

PERMANENT DROUGHT

One of the forms of **drought** identified by **C.W. Thornthwaite**. Agriculture is not normally possible, since there is insufficient moisture for anything but the **xerophytic plants** which have adapted to the arid environment. The provision of moisture through **irrigation** may allow crops to be produced, but costs are high.

See also
Contingent drought, Invisible drought, Seasonal drought.

Further reading
Thornthwaite, C.W. (1947) 'Climate and moisture conservation', *Annals of the Association of American Geographers* 37: 87–100.

PERMEABILITY

A measure of the ability of rock or **soil** to allow the tranmission of **water** or other fluids. Permeability depends not only upon the amount of pore space in the rock, but also upon the way in which the pore spaces are interconnected. For example, a rock structure may have large pore spaces but if they are sealed off from each other the permeability of the rock will be low. **Sand** and gravel deposits with large interconnected pore spaces are commonly much more permeable than **clays** that have small pores poorly connected with each other. For most purposes, in fact, clays may be considered to be **impermeable**. The permeability of an **aquifer** will determine the rate at which it can be pumped and also the rate at which it will be replenished.

See also
Porosity, Semi-permeable membrane.

Further reading
Todd, D.K. (1980) *Groundwater Hydrology* (2nd edition), New York/Chichester: Wiley.
Walton, W.C. (1991) *Principles of Groundwater Engineering*, Chelsea, MI: Lewis Publishers.

PEROXYACETYL NITRATE (PAN)

One of a group of highly potent oxidants, which is present in **photochemical smog**. Although details of its synthesis are not well established, the raw materials required for the production of PAN include **hydrocarbons, oxides of nitrogen** (NO and NO_2) and **ozone** (O_3), all of which are common in an **atmosphere** polluted by automobile emissions. **Solar radiation** provides the **energy** to initiate hundreds of chemical reactions involving these pollutants and one of the by-products is PAN. It is a highly reactive, relatively unstable **compound**, that is constantly being broken down and reformed in the dynamic chemical environment of urban photochemical smog. Like other components of photochemical smog, PAN is responsible for irritation to the eyes and respiratory system, and is the major source of **smog** damage to plants. Since the formation of PAN depends upon the presence of hydrocarbons in the atmosphere, any attempt to reduce PAN levels must initially involve a reduction in hydrocarbon emissions.

Further reading
Aubrecht, G.J. (1989) *Energy*, Columbus, OH: Merrill.
McCormick, J. (1991) *Urban Air Pollution*,

Nairobi: UNEP.
Seinfeld, J.H. (1986) *Atmospheric Chemistry and Physics of Air Pollution*, New York: Wiley.

PERVIOUS

Allowing the transmission of fluids. Commonly considered synonymous with permeable, but with transmission being possible because of the presence of cracks, joints and fissures rather than because of the original texture of the rock.

See also
Permeability

PESTICIDES

Chemical products designed to kill or restrict the development of pests – organisms deemed undesirable by society. They include **fungicides**, **herbicides** and **insecticides**. Pesticides range from relatively simple **elements** such as

sulphur (S) to complex chemical compounds such as **chlorinated hydrocarbons** and may be broad-spectrum or narrow-spectrum agents. A broad-spectrum herbicide, for example, could be used to clear all the vegetation along a pipeline right of way, whereas a narrow-spectrum herbicide might be used to clear broad-leaved weeds from a lawn without damaging the grass. Pesticides also vary in their persistence in the **environment**, and in general, the longer they remain chemically stable the greater is their potential for environmental damage. The use of pesticides has undoubtedly benefited society, by preventing disease, improving food supply and contributing to economic development. At the same time, ignorance of the environmental impact of pesticides, the indiscriminate use of certain products and inadequate control of the production and use of pesticides has created problems for wildlife and natural vegetation and has threatened human health.

Table P-1 Some advantages and disadvantages of pesticide use

ADVANTAGES	DISADVANTAGES
• Used in the appropriate manner, pesticides save lives, by killing pathogens	• Non-threatening organisms may also be destroyed, particularly if broad-spectrum pesticides are used
• Pesticides kill pests that eat or destroy food, and therefore improve the quality and quantity of food available	• If used excessively, or at the wrong time in the growing cycle, pesticide residues may remain on the food to cause health problems for consumers
• By saving lives, by protecting food supplies, pesticides provide a benefit that outweighs their economic, environmental and health cost	• Immediate or short-term effects of pesticide use have to be considered against the impact of the longer term effects of exposure of people, wildlife and the environment in general to low levels of toxic agents
• Health risks of pesticide use are becoming better understood, and can be offset by responsible handling or the development of safer products	• Unsafe manufacturing techniques and improper handling of the product can cause illness and death for those working with pesticides
• Pesticides are cheaper to produce and are faster and more effective than many alternatives	• The development of genetic immunity among organisms reduces the effectiveness of many pesticides and the necessity to develop stronger chemicals or to use larger amounts reduces their economic benefit considerably

See also
Bhopal, DDT, Integrated pest management, Organochlorides, Organophosphates.

Further reading
Baarschers, W. (1996) *Eco-facts and Eco-fiction*, London/New York: Routledge.
Briggs, S.A. (1992) *Basic Guide to Pesticides, Their Characteristics and Hazards*, Washington, DC: Hemisphere Publishing.
Miller, G.T. (1994) *Living in the Environment: Principles, Connections and Solutions*, Belmont, CA: Wadsworth.
Ware, G.W. (1991) *Fundamentals of Pesticides: A Self Instruction Guide* (3rd edition), Fresno, CA: Thomson Publishing.

PETROCHEMICALS

Chemicals derived from **oil** and **natural gas** – for example, ethylene, propylene, toluene –
which act as feedstocks for the manufacture of products such as **plastics, pesticides, fertilizers**, antiseptics and pharmaceuticals. Petrochemicals play a very important role in modern society, but they also create **pollution** problems. Plastics are a major component of solid **waste**, for example, fertilizers contribute to **eutrophication** of lakes and rivers and pesticide residues in food and **water** present health problems.

Further reading
Burdick, D.L. and Leffler, W.L. (1990) *Petrochemicals in Non-Technical Language* (2nd edition), Tulsa, OK: PennWell Publishing Co.

PETROL

See **gasoline.**

PETROLEUM

A mixture of naturally occurring **hydrocarbons**, that may exist in a **solid** (e.g. bitumen), **liquid** (e.g. crude **oil**) or gaseous state (e.g. **natural gas**). It commonly contains variable amounts of other chemicals such as **sulphur** (S) and **nitrogen** (N). Petroleum is the end-product of the partial decay of living organisms which once inhabited the world's **oceans**. As they died they sank to the bottom of the oceans, where the **anaerobic** conditions allowed them to be preserved. A combination of heat, **sedimentation** and **pressure** converted the organic material into its different forms. In the case of oil and natural gas, migration through the pores and cracks in the rocks took place allowing them to accumulate in underground pools or **reservoirs** or, in some cases, leak through to the surface. Crude oil is the most complex of the various forms of petroleum, with crude containing higher proportions of lighter fluids such as **gasoline** (petrol) and **kerosene** being most valuable. The different components of the oil are separated from each other by the process of fractional **distillation**, allowing the individual constituents of the crude mixture to be collected. Petroleum is the predominant source of **energy** in the

transportation industry (gasoline, kerosene) and is widely used for space heating (fuel

Figure P-5 Examples of petroleum reservoirs

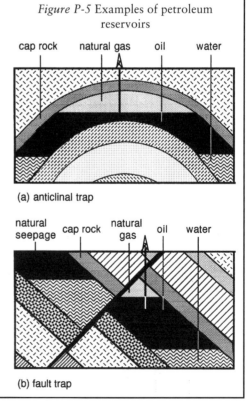

(a) anticlinal trap

(b) fault trap

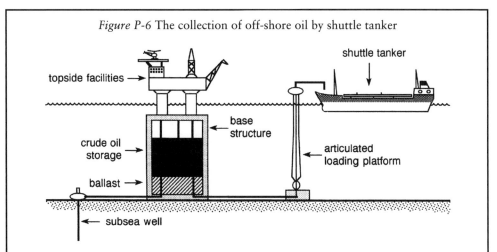

Figure P-6 The collection of off-shore oil by shuttle tanker

oil, natural gas), with the heavier components (heavy oils, waxes, bitumen) used for lubrication, waterproofing and road construction.

The various forms of petroleum also provide the feedstock for the **petrochemicals** industry. The first commercial production of petroleum from conventional sources took place in North America in the mid-nineteenth century, and the North American control of the industry continued until the 1960s, despite a shift in the main source of oil from the United States to the Middle East. Multinational oil companies based in the United States dominated all aspects of the industry outside the communist bloc. With the growth of **OPEC** in the 1960s, power shifted to the major producing nations and the role of the multinationals declined dramatically, particularly on the production side. The development of non-OPEC sources of petroleum has since reduced the influence of OPEC in the world oil market, although the Middle Eastern OPEC members remain the world's major suppliers of oil. The United States remains a major oil producer, although production has stabilized and most fields are past their peak. Prior to the breakup of the Union, the USSR was the world's major petroleum producer outside the Middle East. Since then, output from the former USSR has stabilized, although the potential for increased production appears more promising than in the United

States. Exploration in offshore and frontier locations since the 1960s has brought oil and gas fields in the North Sea, Alaska and Siberia into production, and further development in these areas as well as locations in the Canadian Arctic and on sections of the continental shelf as far apart as the east coast of Canada, the Falkland Islands, Western Europe and southern China is already taking place. Such areas offer promising additions to the world's petroleum **reserves**, although the Persian Gulf states continue to hold some 55 per cent of the total. Once such conventional sources of petroleum have been used up, oil will still be available in large quantities in such non-conventional sources as **oil shale** and **tar sands**.

The dependence of society on petroleum has had a major impact on the geography of the world as a whole, but more particularly in certain areas, such as major cities, for example, where the urban geography is very much a reflection of the dominant role of the automobile. The geography of nations such as Scotland and Norway has changed significantly as a result of the discovery of oil beneath the North Sea while the development of a modern infrastructure in Saudi Arabia and other nations around the Persian Gulf would not have been possible without the riches provided by the oil beneath the **desert**. The complex political geography of the region is also in large part a result of the presence of oil. Along with the economic benefits associated with the use of petroleum as a source of energy, there are also some distinct drawbacks. Spills during the production and transportation of oil have been responsible for some of the world's major environmental disasters, and such events are likely to increase in number as exploration and production expand into the more fragile **environments** of the frontier zones and the ocean margins. Even when petroleum products are used as intended the effects are not benign. Emissions from automobiles, residential heating and industrial processes add to atmospheric **pollution,** and contribute to such global environmental issues as **atmospheric turbidity, acid rain** and the enhancement of **greenhouse gas** levels. Attempts at dealing with these problems have met with only limited success, and as long as the production and use of petroleum remains cheaper than the alternatives, that situation is likely to continue.

See also
Alaska Pipeline, Oil pollution, Oil tanker accidents, OPEC, OPEC oil embargo

Further reading
Chapman, J.D. (1989) *Geography and Energy: Commercial Energy Systems and National Policy,* Harlow/New York: Longman/Wiley.
Chapman, K. (1976) *North Sea Oil and Gas: A Geographical Perspective,* Newton Abbot: David & Charles.
Livingston, J. (1981) *Arctic Oil,* Toronto: CBC Publications.
Mushrush, G.W. and Speight, J.G. (1995) *Petroleum Products: Unstability and Incompatibility,* Washington, DC: Taylor & Francis.
Shwandran, B. (1977) *Middle East Oil Issues and Problems,* Cambridge, MA: Schenkman Publishing.

pH (POTENTIAL HYDROGEN)

The representation of the acidity or akalinity of a substance on a logarithmic scale of 0 to 14 based on **hydrogen ion concentration.** Acid substances have pH values between 0 and 7; alkaline substances range between 7 and 14 and a pH of 7 is considered neutral.

Figure P-7 The pH scale: showing the pH level of common substances

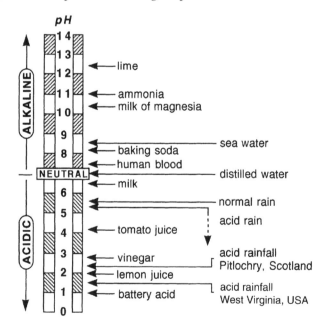

PHENOLOGY

The study of periodic biological phenomena and their relationship to **weather** and **climate**. It includes consideration of such elements as the growth of plants and the migratory activities of animals. The stages of development of certain plants **species** – for example, flowering, fruiting, ripening – are called phenophases. Phenology has been used extensively in the study of **climate change** to provide **proxy data** from which past climate conditions can be established. In Europe, the study of phenological data from the cultivation of grapes has proved a fruitful source of information on climate change in France and Germany.

See also
Palaeoclimatology.

Further reading
Le Roy Ladurie, E. (1980) 'Grape harvests from the fifteenth through the nineteenth centuries', *The Journal of Interdisciplinary History* 10 (4): 839–50.

PHOSPHATES

Salts of phosphoric acid (H_3PO_4), phosphates are a source of the **phosphorus** (P) essential for plant growth. They may be supplied naturally through the **weathering** of rocks or may be added to the **soil** through the use of phosphate rock prepared as a **fertilizer**. Phosphate rock is treated with **sulphuric acid** (H_2SO_4) to produce superphosphates, a form in which the phosphorus is more readily available to plants. The **leaching** of excess phosphates from the soil or its addition to waterways in **sewage** can lead to **eutrophication**. The development of phosphate-free **detergents** has done much to alleviate that problem.

See also
Nutrient.

Further reading
Toy, A.D.F. and Walsh, E.N. (1987) *Phosphorus Chemistry in Everyday Living* (2nd edition), Washington, DC: American Chemical Society.

PHOSPHORUS (P)

An **element** that occurs in several allotropic forms, white and red being the most common. The former is highly flammable and poisonous whereas the latter has a low flammability and is non-poisonous. Phosphorus does not occur free in the **environment**, but usually in combination in the form of a **phosphate** – for example, calcium phosphate – $Ca_3(PO_4)_2$. It is an important macronutrient which as a component of **nucleic acids** and **adenosine triphosphate** is essential to life. Phosphorus is recycled through the environment, mainly by living organisms that absorb the phosphorus as they grow and return it to the **soil** following death and decomposition. New phosphorus is added to the system through the **weathering** of **phosphates** in rocks. Phosphorus **compounds** are used as **fertilizers** and **detergents**.

See also
Allotropy.

Further reading
Jahnke, R.A. (1992) 'The phosphorus cycle', in S.S. Butcher, R.J. Carlson, G.H. Orians and G.V. Wolfe (eds) *Global Biogeochemical Cycles*, London: Academic Press.
Toy, A.D.F. and Walsh, E.N. (1987) *Phosphorus Chemistry in Everyday Living*, Washington, DC: American Chemical Society.

PHOTOCHEMICAL PROCESSES

Chemical processes induced by the presence of sunlight which provides the **energy** required for the reactions involved. Photochemical processes have an integral role in **photosynthesis**. The production of **melanin** in the human body is initiated by photochemical processes. They also cause the **photochemical smog** common in many urban areas and contribute to global environmental issues such as **ozone depletion** and **acid rain**.

PHOTOCHEMICAL SMOG

Smog which is produced by the action of **photochemical processes** on primary **combustion** products, particularly those such as

Figure P-8 The formation of photochemical smog by the action of sunlight on pollutants

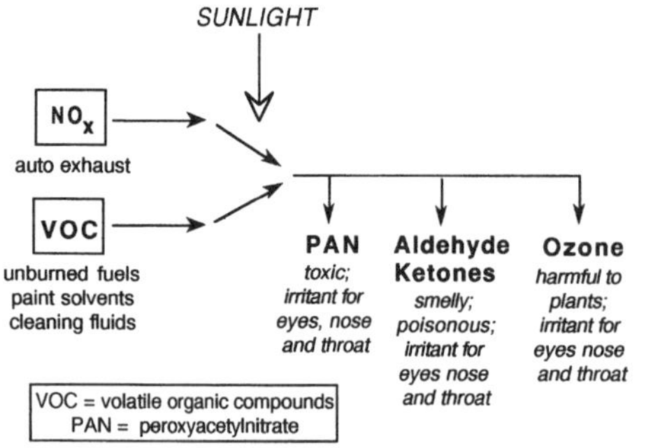

National Academy of Sciences (1988) *Air Pollution, the Automobile and Human Health*, Washington, DC: National Academy Press.

PHOTON

A discrete bundle of electromagnetic **energy** emitted by a **radiation** or light source. A light beam consists of a stream of photons, each of which carries a specific amount of energy related to the frequency of radiation. Thus, the photons of higher frequency radiation at the violet end of the **electromagnetic spectrum** carry more energy than photons of red light at the other end of the spectrum.

PHOTOSYNTHESIS

A biochemical process in which green plants absorb **solar radiation** and convert it into chemical **energy**. It is made possible by **chlorophyll**, the green pigment that gives plants their colour, which initially converts the radiant energy into chemical energy in the form of **adenosine triphosphate** (ATP). That energy is subsequently used in a series of reactions which bring about the **reduction** of carbon dioxide (CO_2) to simple **sugars**. Although complex in detail, photosynthesis can be summarized as a process in which carbon dioxide and **water** are consumed, **carbohydrates** are produced and stored, and **oxygen** (O) is released.

$$6CO_2 + 6H_2O \xrightarrow[chlorophyll]{light\ energy} C_6H_{12}O_6 + 6O_2$$

In green plants, the **hydrogen** (H) for the formation of the carbohydrate is supplied by water, but certain photosynthetic **bacteria** obtain their hydrogen from **hydrogen sulphide** (H_2S) and release **sulphur** (S) as a by-product. Animals depend upon the ability of plants to convert radiant energy into chemical energy in the form of carbohydrates, since they lack the ability to synthesize their own food in that way. Thus photosynthesis is the base on which the earth's food supply is built. Photosynthesis also helps to maintain the oxygen/carbon dioxide balance in the

hydrocarbons and **oxides of nitrogen** (NO_x), produced by the **internal combustion engine**. Los Angeles in California is commonly considered the classic example of a **community** which suffers the effects of photochemical smog, but the problem is also common in other large cities – from Sao Paulo in Brazil to Bangkok in Thailand – which have heavy automobile traffic and abundant sunshine. Under such circumstances, complex photochemical reactions produce a variety of toxic chemicals such as **ozone** (O_3), **aldehydes** and **peroxyacetyl nitrate** (PAN), creating an **atmosphere** that causes eye and lung irritation as well as plant damage. Since the processes involved depend upon sunlight, the greatest development of photochemical smog is usually in the early afternoon, when sunlight intensity is highest. Clear, stable atmospheric conditions – such as under high **atmospheric pressure** – and physical barriers to atmospheric mixing – such as strategically located high ground – also contribute to the intensification of photochemical smog. Despite some forty years of legislation designed to curb emissions of hydrocarbons and oxides of nitrogen (NO_x), photochemical pollution remains endemic in many of the world's largest cities, with severe smog episodes leading to increases in respiratory disease and death rates.

Further reading
Degobert, P. (1995) *Automobiles and Pollution*, Warrendale, PA: Society of Automotive Engineers.
McCormick, J. (1991) *Urban Air Pollution*, Nairobi: UNEP.

atmosphere, but agricultural activities and **deforestation** have reduced the amount of photosynthesis taking place, allowing levels of atmospheric carbon dioxide to rise and contribute to **global warming.**

Further reading
Foyer, C.H. (1984) *Photosynthesis*, New York: Wiley.

PHYTOPLANKTON

Microscopic plants that live in the upper layers of fresh and salt **water environments,** moved around by **wind,** waves and currents. They consist mainly of unicellular **algae,** such as **diatoms.** Because of their light requirements, they are found mainly in the upper 5 to 100 m of the water – the photic zone – into which light can penetrate. Containing **chlorophyll,** phytoplankton are capable of **photosynthesis,** and as primary producers they provide a base for aquatic **food chains.** Their **nutrient** requirements are met by the natural flow of nutrients into the ocean from the land, or by the upwelling of nutrient-rich waters from deeper parts of the ocean – for example, off the coast of Peru. A rapid increase in the availability of nutrients, particularly **nitrates** and **phosphates,** can lead to massive **algal blooms,** that can cause the release of toxins into the water and create serious environmental problems. During their seasonal blooms, phytoplankton also release **dimethyl sulphide** into the **atmosphere** in quantities sufficiently large to contribute to **acid precipitation.** Phytoplankton are also involved in the **carbon cycle,** through their consumption of **carbon dioxide** (CO_2) during photosynthesis, thus helping to maintain the atmosphere's oxygen/carbon dioxide balance.

See also
Plankton, Zooplankton.

Further reading
Boney, A.D. (1975) *Phytoplankton*, London: Edward Arnold.

PLANKTON

Free-floating plants and animals, usually microscopic, that live in the upper layers of both fresh and salt **water** bodies where they are moved around by **wind,** waves and currents. **Populations** vary as a result of changes in **nutrient** availability and **climate.** The microscopic plants are **phytoplankton** and the animals are **zooplankton.** Phytoplankton are the primary producers in the system, grazed on by **zooplankton** at the first stage in the aquatic **food chain.** The zooplankton are in turn consumed by other zooplankton through a series of perhaps three or four **trophic levels** before larger fish enter the chain. Since there is a 90 per cent loss of **energy** at each trophic level, the higher elements in the planktonic food chains – sea mammals, humans – receive only a very small amount of the original energy provided by the phytoplankton. Direct harvesting of phytoplankton has been suggested as a means of countering that loss and providing food from the sea more efficiently by short-circuiting the chain. However, the cost of such a system would be prohibitive, and might have serious ecological consequences. Farming shellfish, such as oysters and mussels, is a potentially workable alternative. They obtain their food by naturally filtering out plankton and since they occupy the second trophic level the total energy loss is reduced. Because they form the base of all aquatic food chains, any damage to the plankton population will have far-reaching repercussions, extending beyond marine organisms to human communities which depend on fish for their food supply. Destruction of phytoplankton also has the potential to impact on **global warming** through a reduction in the **recycling** of **carbon dioxide** (CO_2) that takes place during **photosynthesis.**

Further reading
Moore, P.D. (1994) 'Does plankton hold the key to carbon budgets?', *Science Watch* 5 (6): 7–8.
Platt, T. and Li, W.K.W. (eds) (1986) *Photosynthetic Picoplankton*, Ottawa: Department of Oceans and Fisheries.

PLANT COMMUNITY

The **population** of all the plant **species** living within a particular **habitat.**

Further reading
Bazzaz, F.A. (1996) *Plants in Changing Environments: Linking Physiological, Population and Community Ecology*, Cambridge: Cambridge University Press.

PLASTICS

Organic materials, commonly **petroleum** based, which can be made sufficiently flexible that they can be shaped or moulded under heat, **pressure** or both to produce a product that is stable under normal conditions. Most plastics are **polymers**, formed through the combination of **molecules** into long chains. They are usually classified into thermoplastics and thermosetting plastics. The former can be heated or melted and re-formed without losing any of their properties. **Polythene**, **polyvinyl chloride** (PVC), **polystyrene** and high impact polystyrene (HIPS) all belong to this group. Thermosetting plastics which include resins cannot be reused because of the permanent chemical changes that take place during their formation. Because of such properties as strength, lightness, flexibility, durability,

insulating value and resistance to **corrosion**, plastics are used in a wide range of products from household utensils to construction materials and have replaced wood and **metal** for many purposes. Along with such utility comes an environmental price. The creation of toxic **wastes** during the production process and the extended survival rates of plastics in the **environment** are serious problems in some areas.

See also
Plastics disposal, Plastics recycling.

Further reading
Elias, H-G. (1993) *An Introduction to Plastics*, Weinheim, NY: VCH.
Wolf, N. and Feldman, E. (1990) *Plastics: America's Packaging Dilemma*, Covelo, CA: Island Press.

PLASTICS DISPOSAL

The strength and durability of plastics, while a major advantage when they are in use, creates serious disposal problems. Many plastics will last for several hundred years

Figure P-9 Types of plastic and their use

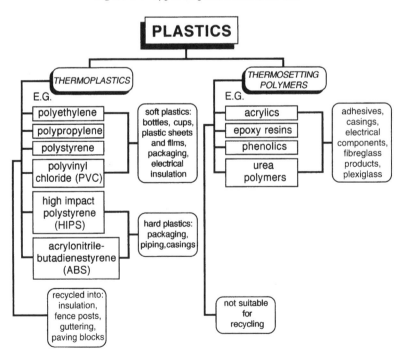

once discarded. Old fishing nets, beer can holders, coffee cups or shopping bags all create aesthetic pollution, but also pose threats to wildlife, which may swallow or get caught in them. Plastics now account for between 7 and 10 per cent of **solid waste** by weight and between 20 and 25 per cent by volume. Until recently, most were disposed of in **landfill** sites or incinerated, but because of the pressure on land and concern over the release of toxic **gases** during **incineration**, other methods of disposal are being considered and practised. Degradable plastics have been developed. Some contain non-plastic material such as cornstarch which breaks down under microbial action and destroys the integrity of the plastic. Others are photodegradable, with **molecules** that degrade on continued exposure to light. Despite the attractiveness of the technology, little is known about the long-term environmental impact of the degradation products, and many environmentalists consider the **recycling** of plastics to be more appropriate.

Further reading

Environmental Protection Agency (1990) *Methods to Manage and Control Plastic Wastes*, Washington, DC: Government Printing Office.
Kupchella, C.E. and Hyland, M.C. (1993) *Environmental Science: Living within the System of Nature*, Englewood Cliffs, NJ: Prentice-Hall.

PLASTICS RECYCLING

As with all **recycling**, the recycling of **plastics** is seen as a means of conserving **resources**, saving money and reducing the flow of **waste** products to the **environment**. In theory, all **thermoplastics** are recyclable, but because of contaminants – for example, colouring pigments – and products containing a combination of plastics or plastics and other materials, it is not possible to return all thermoplastics to their original state. The most commonly recycled plastics are PET (polyethylene terephthalate) used in soft drink bottles, HDPE (high density polyethylene) used in larger containers, particularly those for corrosive **liquids,** and **polystyrene** used in fast food containers, cups and plates. Recycling of these products is encouraged at the consumer level by community-based recycling programmes and the establishment of coding systems to allow consumers to recognize different types of plastic. This allows the sorting of these recyclable plastics from other solid **waste** and helps to make the process viable. The imposition of cash deposits on non-refillable liquid containers also encourages consumers to return the plastic to a collection centre for recycling. Plastics are not usually re-formed into their original products. The PET of plastic drink bottles may be reused as plastic tiles, automotive parts, carpet backing or containers for non-food products, while recycled HDPE may appear in garbage cans, flowerpots or so-called 'plastic lumber' used for fencing or house siding. Polystyrene, cleaned and pelletized, is most commonly used for **insulation** and packaging. Where plastics are not separated for recycling, they are shredded and converted into products in which the properties of a specific type of plastic are not required. Despite the availability of technology to recycle plastics, it is not generally viable without selective collection or some form of automatic sorting. Separation from domestic **garbage** is not economical and as a result only about 20 per cent of the easily recyled plastics are recycled. The remainder continue to be disposed of in landfill sites or incinerated. The EC has developed waste management guidelines which should increase recycling to at least 50 per cent, and in the United States and Canada, governments, particularly at state and provincial levels, are responding to strong demands from environmental groups to increase the level of plastics recycling.

Further reading

Powelson, D.R. and Powelson, M.A. (1992) *The Recycler's Manual for Business, Governments and the Environmental Community*, New York: Van Nostrand Reinhold.

PLATE TECTONICS

The theory that the **lithosphere** consists of a number of rigid plates that are capable of movement over the asthenosphere, the most easily deformed part of the underlying **mantle**. As they move, they cause major

Figure P-10 The location of the earth's main crustal plates

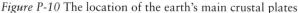
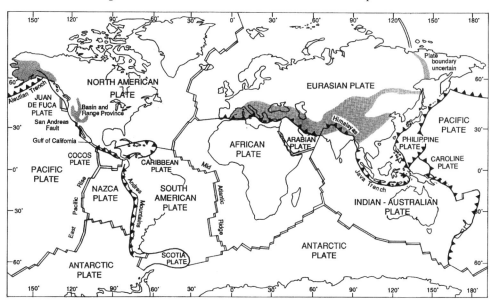

Source: After Goudie, A. (1989) *The Nature of the Environment* (2nd edition), Oxford: Blackwell

tectonic activity, which may be rapid and potentially catastrophic as with **earthquakes** and volcanic eruptions, or slower and less obvious as with **orogenesis** and sea-floor spreading. Plates moving away from each other (sea-floor spreading) have created the mid-oceanic ridges beneath the Atlantic and Pacific oceans. Colliding plates have produced major mountain systems such as the Himalayas and Alps (orogenesis). Where one of the plates slides beneath the other on contact, subduction zones are created. These areas, where parts of the lithosphere are reabsorbed into the mantle, are characterized by frequent seismic activity and volcanic eruptions. Subduction zones are often marked by deep oceanic trenches, but they are also sites where orogenesis can occur. Where the plates slide past each other, as is the case along much of the west coast of North America, earthquakes are common. The movement of the plates is thought to be associated with **convection** currents in the mantle, but this may not be the only factor involved.

Further reading
Condie, K.C. (1989) *Plate Tectonics and Crustal Evolution* (3rd edition), Oxford/New York: Pergamon Press.
Kearey, P. and Vine, F.G. (1990) *Global Tectonics*, Oxford: Blackwell Scientific.
Summerfield, M.A. (1991) *Global Geomorphology*, London/New York: Longman Scientific/ Wiley.

PLEISTOCENE

The first epoch of the **Quaternary** period, beginning between 1.6 and 2 million years ago at the end of the Pliocene and continuing until some 10,000 years ago when the **Holocene** began. However, because of different approaches towards dating – for example, stratigraphical, biostratigraphical or archaeological – the Pliocene/Pleistocene boundary has not been firmly set, the Pleistocene/Holocene boundary is to some extent arbitrary and the chronology of events within the epoch remains controversial. The Pleistocene was dominated by **ice**. Great ice sheets covered the continental landmasses in the northern hemisphere, extending from the pole south beyond the 50th parallel and flowing down from the mountain ranges of western North America, Scandinavia and Central Europe on to the adjacent lowlands.

Figure P-11 Glacials and interglacials of the Pleistocene

	ALPINE EUROPE	NORTH/NORTHWEST EUROPE	NORTH AMERICA	YRS. X 10³
IG (?)		Holocene	Holocene	10
GL	*WURM*	*WEICHSEL*	*WISCONSIN*	122
IG	Mondsee and Somberg	Eemian	Sangamonian	130
GL	*RISS*	*SAALE (Saalian complex)*	*ILLINOIAN*	360
IG	M/R	Holstein	Yarmouthian	430
GL	*MINDEL*	*ELSTER*	*KANSAN*	550
IG	G/M	Cromer	Aftonian	600
GL	*GUNZ*	?	*NEBRASKAN*	?
IG	D/G	?	?	?
GL	*DONAU*	?	*PRE-NEBRASKAN*	

In the southern hemisphere, the absence of large landmasses in high latitudes limited glaciation to Antarctica and higher altitudes in South America and New Zealand. At least six major Pleistocene glaciations have been identified and there are indications that as many as sixteen or twenty may have taken place. The **interglacials** and **interstadials** that separate the **ice ages** represent periods of climatic amelioration. As the ice advanced and retreated across mid- to high latitudes in the northern hemisphere, it repeatedly disrupted the distribution patterns of plants and animals in these areas and significantly altered the landscape. Plants and animals have responded to new post-Pleistocene environmental conditions, but the cirques, glaciated valleys, ice-moulded landscapes, moraines, fluvioglacial features and old lake beds, found in Alaska and Canada, through Greenland and Iceland to North-western Europe and Siberia are evidence of the enduring imprint of the Pleistocene on the landscape. Although extensive land glaciation is considered to be the main characteristic of the Pleistocene, significant geographical changes also took place beyond the ice margins. **Tundra** vegetation now typical of the Arctic covered much of Europe and North America south of the ice front, and the other vegetation belts were also displaced. Changing amounts and distribution of **precipitation** caused the desert margins to fluctuate, and

lake levels in inland drainage basins such as Salt Lake in the United States, Lake Chad in Africa and Lake Eyre in Australia varied considerably. **Winds** blowing off the ice carried large amounts of fine **dust** which settled out to produce the **loess** deposits of Northern China. The most widespread landscape changes outside the glaciated areas, however, were brought about by the eustatic changes produced as the ice advanced and retreated. Relic marine features such as raised and drowned beaches and valleys or **deltas** graded to a lower **sea-level** and now submerged provide evidence of these changes. In areas adjacent to the ice sheets in the northern hemisphere subject to isostatic change, complex combinations of raised, submerged and tilted strandlines reflect the interplay of eustatic and isostatic processes. In addition to the changes that took place in the physical landscape and the vegetation that covered it, significant changes in the nature and distribution of animal **species** – including the human animal – were a feature of the Pleistocene. In addition to modern species, the **fauna** of the Pleistocene included such large mammals as the elephant-like mammoth, the cave bear and the sabre-toothed tiger, some of which survived into postglacial times before becoming extinct. There is some indication that **extinction** might have been accomplished finally by hunting pressure from human groups. The **evolution** of the human species from the upright hominid (*Homo erectus*) to modern *Homo sapiens* took place during the Pleistocene, and developing technological skills may well have allowed hunting groups to finish off these already declining giant species – perhaps an early indication of the potential of the human species to bring about environmental change. By the end of the **ice ages**, *Homo sapiens* was sufficiently well established to accommodate and even take advantage of the steadily ameliorating conditions that followed the decay of the ice sheets, and led into the Holocene.

See also
Eustasy, Glacier, Isostasy.

Further reading
Deynoux, M. (ed.) (1994) *Earth's Glacial Record*, Cambridge/New York: Cambridge University Press.
Flint, R.F. (1971) *Glacial and Quaternary Geology*, New York: Wiley.
Jones, R. and Keen, D. (1992) *Pleistocene Environments in the British Isles,* London: Chapman and Hall.
Rice, S.K. and Giles, L. (1994) 'Climate in the Pleistocene', *Nature* 371: 111.

PLUME

Emissions of **gases** and **particulate matter** from a chimney that are not immediately dispersed, but retain a distinct identity as they move away from the source. **Water** droplets, **smoke** particles and **dust** make the plume visible, with a sharp boundary between the emissions and surrounding non-polluted air, but in some cases gaseous emissions form plumes but remain invisible. The size, shape and persistence of a plume will depend upon such factors as the **temperature** and composition of the discharge, its emission rate and the stability of the **atmosphere** at the time of emission. These factors also govern the potential of the plume to cause **pollution**. Although most commonly used in the **atmospheric environment**, the plume concept can also be applied to the release of effluent into water bodies.

See also
Atmospheric stability/instability.

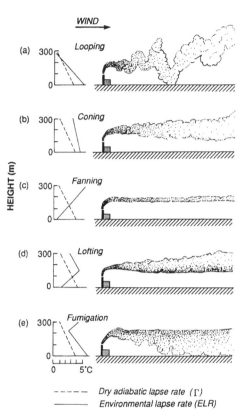

Figure P-12 Types of plume and the environmental conditions that produce them

- - - - - Dry adiabatic lapse rate (Γ)
———— Environmental lapse rate (ELR)

Source: After Oke, T. (1978) *Boundary Layer Climates*, London: Methuen

Further reading
Oke, T. (1987) *Boundary Layer Climates* (2nd edition), London: Methuen.
Pasquill, F. (1974) *Atmospheric Diffusion: The Dispersion of Windborne Material from Industrial and Other Sources* (2nd edition), New York/Toronto: Halstead Press.

PLUTON

See **batholith**.

PLUTONIUM (Pu)

A silvery, dense **metal**. Plutonium is an **actinide**, produced as a result of nuclear reactions. For example, [239]plutonium, one of thirteen **isotopes** of the **element,** is produced through the bombardment of ^{238}**uranium** by **neutrons**. Since it is fissile it can be used as **fuel** in **nuclear reactors** or to provide the explosive power in nuclear weapons. One kilogram of plutonium contains the **energy** equivalent to 10^{14} joules, or more than 150,000 times that available in a barrel of **oil**. ^{239}Plutonium has a **half-life** of 24,400 years, and the high level of **radioactivity** it emits during that time makes it one of the most hazardous elements known.

See also
Nuclear fission.

Further reading
Seaberg, G.T. and Loveland, W.D. (1990) *The Elements Beyond Uranium*, New York: Wiley.

PNEUMONOCONIOSIS

A chronic lung disease caused when the inhalation of **dust**, usually over a prolonged period, leads to a significant reduction in lung function. It is most common among miners where it is referred to as silicosis, or 'black lung', where **coal** is involved.

PODZOL

A **soil** in which percolating **water** has leached **iron** and **aluminum oxides** and **hydroxides** as well as organic matter from the upper horizons of the **soil profile** and redeposited them in illuvial horizons deeper in the profile. As a result of the removal of **nutrients** from the upper layers, podzols have low fertility. They are common in cool, moist regions, where low **temperatures** inhibit the bacterial activity necessary to break down organic matter to replace the leached nutrients, and there is sufficient **precipitation** to maintain the rate of **leaching**. Podzols are naturally acidic and sensitive to additional acidity such as that provided by **acid rain**. They are located in areas such as the Canadian Shield in North America and Scandinavia and Russia in Europe. With low fertility and moderate to high acidity, they are capable of only limited agricultural development and are probably best left under their **natural vegetation** of coniferous, evergreen forest.

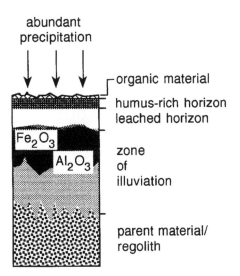

Figure P-13 The characteristics of a podzolic soil horizon

See also
Soil classification.

Further reading
Hassett, J.J. and Banwart, W.L. (1992) *Soils and their Environment*, Englewood Cliffs, NJ: Prentice-Hall.
Paton, T.R., Humphreys, G.S. and Mitchell, P.B. (1995) *Soils: A New Global View*, New Haven, CT: Yale University Press.

PODZOLIZATION

The processes by which **podzols** are formed.

POLAR FRONT

The transition zone separating **air masses** of polar origin and those of tropical or subtropical origin. Conceived at the time of the First World War by meteorologists in Norway, the concept remains an integral part of modern **meteorology** and **weather forecasting**. The front is best developed in the winter when the **temperature** contrast between the air masses is greatest. At that season it is also commonly located between 40° and 50° latitude, whereas in the summer it is found at higher latitudes and is often discontinuous. Extra tropical **cyclones**, or frontal **depressions**, which develop and move along the front make a major contribution to weather patterns in mid- to high latitudes. Associated with the surface front is the polar front **jet stream** located at the **tropopause**.

Further reading
Barry, R.G. and Chorley, R.J. (1992) *Atmosphere, Weather and Climate*, (6th edition), London: Routledge.

POLAR STRATOSPHERIC CLOUDS

Clouds of **ice** particles that form in the **stratosphere** above the poles during the winter. **Heterogeneous chemical reactions**, involving nitric and hydrochloric acid, on the surface of these particles may lead to the release of **chlorine** (Cl) which destroys **ozone** (O_3) molecules and leads to the thinning of the **ozone layer**. Such chemical changes made possible by the formation of polar stratospheric **clouds** have been implicated in the development of the **Antarctic ozone hole**. In the Arctic, the stratosphere is generally warmer than over southern polar regions. As a result, polar stratospheric clouds are less ready to form and ozone destruction is less efficient. The **evaporation** of the clouds as **temperatures** rise in the spring brings an end to the reactions, and allows the ozone layer to recover.

Further reading
Hofmann, D.J. and Deshler, T. (1991) 'Stratospheric cloud observations during formation of the Antarctic ozone hole in 1989', *Journal of Geophysical Research* 96: 2897–912.
Shine, K. (1988) 'Antarctic ozone – an extended meeting report', *Weather* 43: 208–10.

POLAR VORTEX

See **circumpolar vortex**.

POLLEN

Small grains containing the male reproductive **cells** of seed-producing plants. They are spread by **wind**, birds and insects.

POLLEN ANALYSIS

See **palynology**.

'POLLUTER PAYS' PRINCIPLE

The concept that those who cause **pollution** or are likely to cause pollution should pay for its clean-up or prevention. It applies at the level of both the producer and the consumer. A producer causing an **oil spill**, for example, would be expected to finance the clean-up, or an industry emitting toxic pollutants would be forced to take measures to prevent the emissions; a consumer might be assessed a tax on a product likely to cause pollution when used or discarded – for example, **gasoline**, disposable drinks containers or car tyres. In some cases, producers or manufacturers might be required to post bonds to ensure that money is available to deal with unexpected environmental damage. In practice, government subsidies or tax incentives may be used to encourage producers to reduce pollution, and the consumer normally ends up paying, either through higher taxation required to offset the tax relief for companies or by assuming costs easily passed on by the producer through an increase in the price of the commodity to the consumer.

Further reading
Goldfarb, T.D. (ed.) (1993) *Taking Sides: Clashing Views on Controversial Environmental Issues* (5th edition), Guildford, CT: Dushkin Publishing Group.

POLLUTION

See **environmental pollution**.

POLYCHLORINATED BIPHENYLS (PCBs)

A group of highly stable **chlorinated hydrocarbons** used as **liquid** insulators in the electricity distribution industry and as plasticizers and synthetic resins in the **plastics** industry. Being stable, they tend to accumulate in the **environment**, mainly by passage through **food chains**. As a result, they are found in locations far removed from their industrial sources. Seals, fish and humans in the Arctic, for example, have traces of PCBs in their bodies. Tests have indicated that

PCBs are carcinogenic and may impair the immune system, perhaps in part because of impurities, such as **dioxins**, that they often contain. Because of this, their production and use has been banned in most industrial countries since the late 1970s, but considerable quantities remain in use in **Third World** countries or in storage in developed nations. Storage has been necessary since they are too toxic to be disposed of in the usual manner – for example, in land-fill. Special high-**temperature** incinerators have been developed as an effective alternative. Even with no additional production or use, the stability of PCBs will ensure that they remain in the environment for at least several decades.

Further reading
Baarschers, W.H. (1996) *Eco-facts and Eco-fiction*, London/New York: Routledge.
Hutzinger, O., Safe, S. and Zitko, V. (1974) *The Chemistry of PCBs*, Cleveland: CRC Press.

POLYMER

A chemical produced by the combination of individual **molecules** of a **compound** to produce larger molecules, often in the form of long chains. Most **plastics** are polymers and natural polymers include **cellulose**, **proteins** and **nucleic acids**.

Further reading
Young, R.J. and Lovell, P.A. (1991) *Introduction to Polymers* (2nd edition), London/New York: Chapman and Hall.

POLYMER FOAMS

Synthetic foam produced by bubbling **chlorofluorocarbons** (CFCs) through liquid **plastic**. Foams produced in this way are used in **insulation** materials, upholstery cushioning and food containers. The escape of CFCs during production, and from the foam as it ages, has contributed to **ozone depletion**, and resulted in the banning of some foams or the search for replacements.

POLYSTYRENE

A thermoplastic **polymer** used as a thermal and electrical insulator. Easily moulded, it is a common packaging material. It is frequently foamed to improve its thermal insulating properties.

See also
Plastics, Polymer foams.

POLYTHENE

Polyethylene. A thermoplastic material that is tough, flexible and resistant to most chemicals. It is frequently used for liquid containers and formed into thin flexible sheets suitable for a variety of purposes from shopping bags to vapour or **moisture** barriers.

See also
Plastics.

POLYVINYL CHLORIDE (PVC)

Probably the most commonly used **plastic**. It has the same origin as **polythene** – the **hydrocarbon**, ethylene – but with the addition of a **chlorine** (Cl) **atom**. The result is a more versatile plastic that with the addition of plasticizers and pigments can be used for a variety of products from clothing and food packaging to coverings for electrical cables. It also uses less raw material and less **energy** to produce than most comparable plastics and recycles well. In its powdered form, however, it can cause **pneumoconiosis** and if incinerated it releases hydrochloric acid **gas** and **dioxins** into the **atmosphere**. The former can easily be removed by **scrubbers** during controlled **incineration**, but dioxins are much more difficult to remove. They are persistant chemicals which have been implicated in the development of **cancers**, neurological problems and birth defects in those exposed to them.

POPULATION – ENVIRONMENTAL IMPACTS

A population is a group of individuals, usually of the same **species** or related species (for example, birds or fish), occupying a specific area. The number of individuals in the group will depend upon the **carrying capacity** of the area. In human terms, the total population is theoretically governed by the carrying capacity of the whole earth, but in practice different parts of the earth are more able to support life than others – compare the Arctic with the more populous regions of North America and Europe to the south, for example. For thousands of years, the earth's total human population changed little until the **demographic transition** of the eighteenth and nineteenth centuries when it increased rapidly. In the last decade of the twentieth century it is approaching 6 billion. Whether this is close to the earth's maximum sustainable population is not easy to calculate, but in many areas numbers are sufficiently high that the available **resources** are inadequate to meet the needs of the population, and the inhabitants suffer from poverty, **malnutrition** and disease. In these areas also the **environment** suffers. Population pressures cause **soil** and **water pollution**, the destruction of vegetation and animal populations and **soil erosion**. Society's ability to cause such disruption is a relatively recent phenomenon, strongly influenced by **demography** and technological development. Primitive peoples, for example, being few in number, and operating at low **energy** levels with only basic tools, did very little to alter their environment. In truth, they were almost entirely dominated by it. When it was benign, survival was assured. When it was malevolent, survival was threatened.

Population totals changed little for thousands of years, but slowly, and in only a few areas at first, the dominance of the environment began to be challenged. Central to that challenge was the development of technology which allowed the more efficient use of energy. It was the ability to concentrate and then expend larger and larger amounts of energy that made the earth's human population uniquely able to alter the environment. The impact remained local or regional until the so-called **Industrial Revolution**, when major developments in technology and a significant increase in population made a global impact possible. Since then, energy consumption has increased six-fold and the world population is now at least five times greater than it was in 1800. The exact relationship between population growth and technology remains a matter of controversy, but there can be no denying that, in combination, these two elements were responsible for the increasingly rapid environmental change which began in the mid-eighteenth century. Because of technology, populations did not have to be large to have a significant impact on the environment. Western industrialized nations, for example, place much greater pressure on **resources** and have a much greater impact on the environment than **Third World** nations which have much larger and faster growing populations. However, the developed nations with reduced population pressures to contend with should be able to use technology to reverse the environmental problems they have created.

Further reading
Ehrlich, P.R. and Ehrlich, A.H. (1990) *The Population Explosion*, New York: Doubleday.
Hardin, G. (1993) *Living within Limits: Ecology, Economics and Population Taboos*, New York: Oxford University Press.
United Nations (1994) *Population, Environment and Development*, New York: United Nations.

POPULATION EXPLOSION

See **ecological explosion**.

POROSITY

A measure of the amount of pore space in a rock, usually expressed as a percentage.

See also
Permeability.

POSITIVE FEEDBACK

See **feedback.**

POTASSIUM (K)

A soft, silvery-white **metal** similar to **sodium** (Na) in its properties. It is highly reactive and is usually found in the form of **salts**. Potassium is an essential **nutrient** for the human body, and various potassium salts are used as **fertilizers**, oxidizing agents, disinfectants and antacids.

POTENTIAL ENERGY

See **energy.**

POTENTIAL EVAPOTRANSPIRATION (PE)

The amount of **evaporation** and **transpiration** that will take place if sufficient moisture is available to fill the **environment's** capacity for **evapotranspiration.** Measurable evaporation ceases when **water** is no longer available, but the environment may retain the ability to cause additional evapotranspiration through such **elements** as **temperature, radiation, humidity** and **wind.** Potential evapotranspiration is the theoretical value that represents that ability. The difference between actual (measurable) and potential (theoretical) evapotranspiration can be considered as a measure of **moisture deficit,** and farmers in agricultural areas experiencing moisture deficiency can use this approach to estimate the appropriate amount of moisture required to combat **drought** or to allow crops to grow at their full potential.

Further reading
Thornthwaite, C.W. (1948) 'An approach towards a rational classification of climate,' *Geographical Review* 38: 55–94.

POWER

A measure of work done or energy expended per unit of time:

$$P = \frac{E}{t} \text{ (where P = power; E = energy; t = time)}$$

The unit of power is the **watt.**

Figure P-14 A sample climatic water budget for a mid-latitude station in North America or Europe, based on the Thornthwaite model

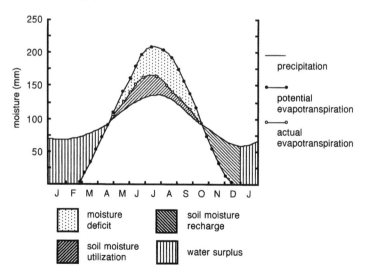

Source: Kemp, D.D. (1994) *Global Environmental Issues: A Climatological Approach*, London/New York: Routledge

PRECAMBRIAN SHIELD

An area of ancient rocks, mainly **igneous** and **metamorphic** in origin, formed perhaps as early as 4.5 billion years ago. Long exposure to **erosion** has worn them down to a subdued rounded landscape. Being composed mainly of acid-rich rock, the Precambrian Shield areas of the world are susceptible to **acid rain** damage. They also contain mineral-rich intrusions that are worked to provide the **ores** of **copper**, **nickle** and **zinc**. The Laurentian Shield in Canada and the Fenno-Scandian Shield in Sweden and Finland are Precambrian in origin.

See also
INCO.

PRECESSION OF THE EQUINOXES

The gradual movement of the equinoxes around the elliptic, brought about by the gyroscopic effect of a wobble in the earth's axis. The net result is to change the season of the year at which the earth is closest to the sun (perihelion). At present, perihelion occurs in January. In about 10,500 years it will occur in July, since the precession of the equinoxes has a periodicity of about 21,000 years. Such cyclical variations have implications for **climate change**, since they alter the amount and timing of the **solar radiation** reaching the earth. The precession of the equinoxes is a central **element** in the **Milankovitch hypothesis**, one of the **models** used to explain the climate fluctuations that produced the **ice ages** during the **Pleistocene**.

Further reading
Birchall, G. and McCutcheon, J. (1993) *Planet Earth: A Physical Geography*, Toronto: Wiley.
Pickering, K.T. and Owen, L.A. (1994) *An Introduction to Global Environmental Issues*, London/New York: Routledge.

PRECIPITATION (CHEMICAL)

The formation of an insoluble substance – a precipitate – as a result of a chemical reaction in a **solution**. Precipitation can be used in **waste** treatment to remove hazardous chemicals from a solution, or in refining to allow an economically important substance

to be separated out of a solution.

See also
Coagulation.

PRECIPITATION

Any **solid** or **liquid water** particles falling to the earth's surface from the **atmosphere**. It includes **rain**, **snow**, hail and sleet, but 'precipitation' and 'rain' are often treated as synonyms. The amount, distribution and nature of precipitation make an important contribution to variations in the physical and biological processes in the earth/atmosphere system – for example, compare tropical **rainforests** and tropical **deserts**. In human terms the availability of precipitation is never ideal, and considerable time, money and effort is expended in dealing with its uneven distribution. Too much will lead to the need for drainage systems or even **flood** control; too little may require the development of water transfer systems and **irrigation**.

PRECIPITATION SCAVENGING

One of the **atmosphere's** self-cleansing mechanisms. It is the process by which **rain** and **snow** wash **particulate matter** out of the atmosphere, thus helping to cleanse it. Rain coloured by **dust** or **smoke** during the process may be described as 'red rain' or 'black rain'. Precipitation scavenging is known to accompany forest fires, and black, sooty rain fell following the bombing of Hamburg and Dresden during the Second World War, helping to clear the dense smoke clouds from the burning cities. Black rain also fell at Hiroshima and Nagasaki following the nuclear bombing of these cities.

Further reading
Peczkic, J. (1988) 'Initial uncertainties in "Nuclear Winter": a proposed test based on the Dresden firestorm', *Climatic Change* 12: 198–208.
Tullett, M.T. (1984) 'Saharan dust-fall in Northern Ireland', *Weather* 39: 151–2.

PRESSURE

Pressure is force per unit area. Atmospheric pressure is created by the weight of the atmosphere on the earth's surface, for example. A column of **air**, 1 inch square, reaching from the earth's surface to the outer edge of the atmosphere, has an average weight of 14.7 lb. This represents a **sea-level** pressure of 14.7 lb in⁻². Atmospheric pressure can also be expressed as inches or mm of **mercury** (Hg), representing the height of the column of mercury that can be supported by the atmosphere (29.9 in or 760 mm on average), as millibars (1013.25 mb on average) or more commonly in **SI units** as kilopascals (101.3 kpa on average).

PRIMARY AEROSOLS

Large particles with diameters between 1 and 100 μm. They include **soil, dust** and a variety of industrial emissions formed by the break-up of material at the earth's surface.

See also
Secondary aerosols.

PROTECTION OF THE GLOBAL ATMOSPHERE (THE HAGUE 1989)

Declaration signed by twenty-four nations which recognized the global extent of such environmental issues as **global warming** and **ozone depletion** and the urgent need to deal with them before the threats to life that they represented became unmanageable. It called for improved decision making and enforcement at the international level, but also recognized that not all nations were equally able to implement these proposals, and therefore included provisions for assisting developing countries to introduce higher environmental standards.

PROTEINS

Complex nitrogenous compounds that are basic components of living organisms. Proteins are organic **polymers** consisting of **amino acids** linked into chains. A specific protein may include as many as twenty amino acids and it is the arrangement of these acids in the protein **molecule** that provides the characteristic properties of the protein.

See also
Enzymes.

Further reading
Light, A. (1974) *Proteins: Structure and Function*, Englewood Cliffs, NJ: Prentice-Hall.

PROTON

A stable elementary or subatomic particle that carries an electrical charge equal to but opposite in sign to that of an **electron**. Protons share the nuclei of **atoms** with **neutrons**. The number of protons in the **nucleus** of an atom is an indication of its **atomic number**.

PROTOPLASM

The living matter of which **cells** are made. It includes the **nucleus** of the cell plus the non-nucleic material or cyctoplasm contained within a thin membrane.

PROTOZOA

Microscopic unicellular organisms. The **cell** may be relatively large and complex, with a well-defined **nucleus** and some mechanism, such as cilia (short threads) or flagella (relatively long whip-like threads) that allow them to move. Protozoa occupy a great variety of **habitats**. Some are parasitic – for example, causing sleeping sickness – and many have an important role in the **environment** because of their ability to consume dead organic matter.

See also
Activated sewage sludge, Parasite.

PROXY DATA

Data used to study a situation, phenomenon or condition for which no direct information – such as instrumental measurements – is available. Proxy data are widely used in the study of **climate change** to extend the meteorological record back beyond the mid-nineteenth century when instrumental measurements began. Such indicators of past **climates** take many forms. They may, for example, be biological, stratigraphical, archaeological, agricultural, glaciological or historical in nature, but all reflect to a greater or lesser degree the climatic conditions that prevailed at the time they developed. They vary in quality; some, such as **tree rings** and fossil **pollen**, allow past conditions to be quantified with some precision, whereas others, such as some of the historical data, may provide only qualitative results. The calibration of the data is also variable. Tree rings and historical documents can provide specific dates for meteorological events, but other data may provide only relative dating or at best establish a range within which an event may have occurred. Gaps in the proxy record also create problems. Few long-term climate reconstructions depend entirely upon one source of proxy data. They incorporate information from a variety of sources, not only to fill gaps, but also to improve the reliability of the results. Using the available proxy data, scientists have reconstructed the climatic history of the last 10,000 years with some reliability, and in places there is sufficient evidence to extend the record back further into glacial times.

See also
Dendroclimatology, Palynology, Phenology.

Further reading
Ball, T. (1986) 'Historical evidence and climatic implications of a shift in the boreal forest–tundra transition in Central Canada', *Climatic Change* 8: 421–34.
Bradley, R.S. and Jones, P.D. (eds) (1992) *Climate since A.D. 1500*, London: Routledge.
Cook, E.R. (1995) 'Temperature histories from tree rings and corals', *Climate Dynamics* 11: 211–22.
Lorius, C., Jouzel, J., Ritz, C., Merlivat, I., Barkov, N.I., Korotkevich, Y.S. and Kotlyakov, V.M. (1985) 'A 150,000 year climatic record from Antarctic ice', *Nature* 316: 591–6.

PUDDLING

The compaction of **soil**, brought about by such factors as heavy, persistent **precipitation** on bare ground and trampling by livestock or humans. The net effect is to decrease the **permeability** of the **soil**. This reduces its ability to support vegetation and encourages **soil erosion** since **water** unable to percolate into the soil runs off the surface. It may occur in forest cut-overs where the soil is no longer protected from heavy **rain** by vegetation or in high traffic areas such as pasture gates or livestock feeding stations, and it has become a problem in popular recreation areas as a result of heavy use by walkers and hikers.

PUEBLO DROUGHT

A serious **drought** which struck southwestern North America – particularly what are now the states of Arizona and New Mexico – in the thirteenth century. Named after the Pueblo group of Indian tribes who inhabited the area at the time and who were strongly affected by the drought and the **famine** that accompanied it.

Further reading
Rosenberg, M.J. (ed.) (1978) *North American Droughts*, Boulder, CO: Westview Press.

PULP AND PAPER INDUSTRY

An industry that processes long-fibred, **cellulose**-rich material into pulp and paper. It is the economic mainstay of many **communities** in North America – particularly Canada – and Scandinavia. Trees are the main source of fibres, but esparto grass, hemp, bamboo and jute are also used. More than 90 per cent of the world's paper is produced from trees, although the proportion of tree and treeless paper production varies. In China, for example, 80 per cent of the paper pulp produced is from non-wood sources. Wood pulp is produced mechanically, by grinding down the wood to separate the fibres, or chemically, by dissolving the **lignin** in the wood to release the cellulose fibres. The environmental impact of the industry, from the harvesting of the pulpwood to the production of the paper, is significant, and

Table P-2 Pulp and paper mill effluents

EFFLUENT	PROBLEMS	SOLUTIONS
Fibre	Fibre mats on streambeds; suspended solids; increased BOD	Recovery and removal of fibre through secondary treatment; aeration of the effluent in ponds
Liquor (chemical by-products of the wood digestion process)	Toxic to aquatic organisms	In-plant recovery and recycling of chemicals
Slime inhibitors (e.g. organo-mercury compounds)	Bioaccumulation of mercury in fish; mercury poisoning	Replacement of organo-mercury compounds
Bleaching by-products	Bioaccumulation of chlorinated organic compounds	Replace chlorine bleach with chlorine dioxide or hydrogen peroxide
Gases (e.g. methyl sulphides; methyl mercaptans)	Obnoxious odours	Installation of scrubbers, but difficult to remove completely

after the paper has left the mill, its ultimate disposal has important environmental implications. **Clear cutting** or the complete removal of timber was once the norm in all pulpwood producing areas. This led to changes in local **temperature** regimes, changes to the **hydrological cycle**, changes in **soil** conditions and changes to the animal **populations** of the area. The exposure of the soil to direct **precipitation** caused soil compaction, reduced infiltration and allowed greater **runoff**, causing the potential for **soil erosion** to increase. Clear cutting is now banned in many areas and reforestation is common as the need to conserve **resources** has led to programmes which involve sustainable yields. However, the amount of pulpwood harvested continues to exceed the amount replaced and even where reforestation programmes are well established the extended life cycle of trees means that complete replacement takes decades. Environmental problems increase through the production process. Pulp and paper mills require large amounts of **water** and when that water is returned to the system it is usually polluted. Groundwood pulp produces large amounts of suspended **solids** and organic **waste** which raise the **biochemical oxygen demand** (BOD) of water bodies into which they are dumped. Chemical pulping adds toxic chemicals such as organic **mercury** (Hg), bleaching agents, sulphites and **dioxins** to the water, causing contamination of aquatic organisms and in some cases creating serious health problems for communities using the water or eating fish caught in it. Effluent control legislation has reduced many of these problems. Used liquor from the pulping process is recycled, mercury has been banned and the increased use of non-bleached pulp and paper has helped to reduce the amount of **chlorine** (Cl) released into the **environment** which has in turn caused the production of dioxins to decline. Air **pollution** from the industry is relatively easy to control, although **odours** caused by some processes, while not proven hazardous, continue to reduce environmental quality in areas adjacent to pulp mills. Once the paper is produced and used, the remaining environmental problem is its disposal. Burial in **landfill** sites or **incineration** were once the most common forms of disposal, but **recycling** is now seen as the preferred option. The recycled paper can be mixed in with new pulp to produce a product that can match one made completely from new pulp. With such attention to environmental impacts at all stages of production and use, as well as the consideration given to sustainable yields, the pulp and paper industry is responding to public pressure. Such responses must continue if the potential environmental impact of the rising world demand for paper products –

projected to double between 1995 and 2010 – is to be minimized.

Further reading
Ferguson, K. (ed.) (1991) *Environmental Solutions for the Pulp and Paper Industry*, San Francisco, CA: Miller Freeman.
Postel, S. (1994) 'Carrying capacity: earth's bottom line', in L. Brown (ed.) *State of the World 1994*, New York: Norton.
Servos, M.R. (ed.) (1996) *Environmental Fate and Effects of Pulp and Paper Mill Effluents*, Delray Beach, FL: St Lucie Press.

PUMPED STORAGE SYSTEM

A system used in the production of **hydro-electricity** involving two **reservoirs** with a pumping/generating station between the two. Under normal working conditions **water** released from the upper reservoir passes through the generator and produces electricity. At times when the demand is low – for example, overnight – the excess baseload electricity from the electrical grid is used to pump water from the lower reservoir to the upper. The energy expended in the pumping operation is stored as potential **energy** in the water in the upper reservoir. When the demand for electricity rises again, the water is released to flow down through the turbines of the generator. Advantages of such systems include the ability to provide electricity rapidly when needed and a reduced dependence upon **precipitation** since the same water can be used several times.

Further reading
Ramage, J. (1983) *Energy: A Guidebook*, Oxford/ New York: Oxford University Press.

PYRETHRUM

See **insecticides.**

PYROLYSIS

The **destructive distillation** of organic **wastes**. Domestic refuse or **garbage** contains considerable amounts of **hydrocarbon**-based material. When heated in an **oxygen**-free environment, one ton of such waste can produce the **energy** equivalent to one barrel of **oil** in the form of **gas** and **liquid**. The best results are obtained from wastes containing high proportions of rubber or **plastic**. Pyrolysis can be incorporated into integrated **waste disposal** facilities, which include the sorting and **recycling** of materials, with the energy produced contributing to the running of the system. The energy recovered from garbage through pyrolysis is less than that recovered by direct **combustion**, but being in the form of a gas or liquid, the pyrolysis product has advantages for storage and transportation. The residue remaining after pyrolysis – char – may be used as a low-calorie **fuel** or disposed of in **landfill** sites.

Further reading
Wampler, T.P. (1995) *Applied Pyrolysis Handbook*, New York: M. Dekker.

Q

QUASI-BIENNIAL OSCILLATION (QBO)

The reversal of easterly and westerly winds in the equatorial **stratosphere**. The entire oscillation occurs over a period of 26–30 months with an easterly flow dominating for 12–16 months, followed by a reversal which allows the westerly winds to prevail for 12–16 months. In conjunction with **sunspot** activity, the QBO appears to contribute to **climate change** although the mechanisms involved remain unclear. During periods of low solar activity, for example, when the QBO is westerly, winters in North America are colder than normal. Under the same QBO conditions, but with an active sun, winters are warmer.

See also
Quiet sun, Sunspots.

Further reading
Mannion, A.M. (1991) *Global Environmental Change*, London/New York: Longman/Wiley.

QUATERNARY

The second period of the Cainozoic (Cenozoic) era, following the Tertiary. It includes the **Pleistocene** and **Holocene** epochs. Delimiting the Quaternary has been a subject of some debate. The Pliocene/Pleistocene boundary, marking the end of the Tertiary and beginning of the Quaternary, was originally dated at some 3 million years BP based on the deterioration of **climate** and the expansion of **glaciers** and **ice** sheets into mid-latitudes. More recently, **floral** and **faunal** evidence suggests that the base of the Quaternary should be dated at some 1.6 to 1.8 million years ago. The Pleistocene was characterized by widespread glacial episodes, although the onset of glaciation had already begun in the Pliocene if the 1.6 to 1.8 million-year dating of the base of the Pleistocene is accepted. The Holocene, which began about 10,000 years BP, has been a period of general amelioration of climate (although not without periods of deterioration also) during which modern post-glacial environmental patterns developed and human beings evolved culturally and socially.

See also
Ice ages.

Further reading
Flint, R.F. (1971) *Glacial and Quaternary Geology*, New York: Wiley.
Bradley, R. (1985) *Quaternary Palaeontology*, London: Chapman and Hall.
Gordon, J.E. and Sutherland, D.G. (1993) *The Quaternary of Scotland*, London/New York: Chapman and Hall.
Jenkins, D.G. (1987) 'Was the Pliocene–Pleistocene boundary placed at the wrong level?', *Quaternary Science Reviews* 6: 41–2.

QUARTZ

Natural **silica** (SiO_2), a common rock-forming mineral. It is characteristic of acid **igneous rocks** such as granite, but because of its resistance to **weathering** it is also found in **sedimentary** and **metamorphic rocks,** such as sandstones and schists. Quartz occurs in crystalline form in igneous rocks, with pure quartz being clear, and coloured varieties such as amethyst (purple), citrine (yellow) and rose quartz (pink) being caused by chemical impurities in the crystal. Quartz is a raw material for glass, produced by the melting and cooling of pure **silica.**

Further reading
Hearney, P.J., Prewitt, C.T. and Gibbs, G.V. (1994) *Silica: Physical Behaviour, Geochemistry and Materials Applications*, Washington, DC: Minerological Society of America.

QUIET SUN

A situation in which the sun displays no sunspots or **solar flares.** This reduction in solar activity causes the **solar wind** to decline, reduces the frequency of the aurora and decreases the level of radio and magnetic interference on the earth. It may also have an impact on **weather** and **climate,** by reducing the flow of **energy** into the earth/atmosphere system.

See also
Magnetosphere.

R

r-STRATEGISTS

Organisms that are usually small and have relatively short life-spans during which they mature quickly and produce large numbers of offspring. *r* refers to the maximum **population** growth rate, and *r*-strategists do little to support their offspring, but depend on quantity rather than quality for the survival of the **species**. Numbers fluctuate wildly. Efficient reproduction allows the population to grow rapidly as long as the conditions suitable for growth are present. If these conditions change – for example, food supply is curtailed – the population will crash and remain low until favourable conditions return again. Many insects are *r*-strategists as well as some small mammals such as mice. Although the concept is most commonly applied to animals, dandelions and other plants classified as weeds may be considered as *r*-strategists.

See also
Carrying capacity. Demography, K-strategists.

Further reading
Enger, E.D. and Smith, B.F. (1995) *Environmental Science: A Study of Interrelationships*, Dubuque, IA: Wm C. Brown.

RAD

An acronym for radiation absorbed dose – a measure of the amount of **ionizing radiation** absorbed by living tissue. One rad is equivalent to the absorption of 0.01 **joules** per kilogram of the tissue being irradiated. The rad has been replaced by the **gray** which is equivalent to 100 rads.

RADIANT ENERGY

Energy transmitted in the form of **radiation** – i.e. as rays, waves or streams of particles. The main source of radiant energy in the earth/ atmosphere system is the sun.

RADIATION

The emission of electromagnetic **energy** from a source in the form of rays, waves or streams of particles. The term is also applied to the energy itself. The nature and intensity of the emission varies with the energy status of the source, and this is reflected in the range of energy levels in the **electromagnetic spectrum**. With a surface **temperature** close to 6000 K, the sun emits high intensity radiation mainly in the **ultraviolet** (9 per cent) and **visible light** (45 per cent) sectors of the spectrum, whereas the earth, with an average surface temperature of 285 K, emits low intensity radiation in the **infrared** range. Other high intensity radiation sources include certain radioactive **elements**, which emit, for example, **gamma rays**. These are capable of causing **ionization**, leading to **cell** damage and the initiation of various types of **cancer**. Lower energy radiation such as microwaves and radiowaves have important commercial applications and infrared radiation has a significant role in **remote sensing**.

See also
Ionizing radiation, Radiation absorption, Radioactivity, Solar radiation, Terrestrial radiation.

Further reading
Marion, J.B. and Heald, M.A. (1989) *Classical Electromagnetic Radiation* (2nd edition), New York/Toronto: Academic Press.

Figure R-1 Spectral distribution of solar and terrestrial radiation

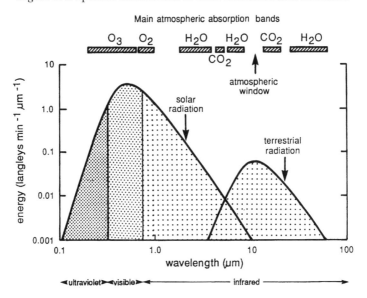

Source: Kemp, D.D. (1994) *Global Environmental Issues: A Climatological Approach*, London/New York: Routledge

RADIATION ABSORPTION

The intake of **radiant energy** by an object. Usually this causes the **temperature** of the object to rise, and allows it to become a radiating body in its own right. The earth's absorption of **solar radiation**, for example, allows it to emit **terrestrial radiation**. When radiation is intercepted by an object, several things can happen to it, depending upon the nature of the object and the intensity of the radiation. It may be transmitted if the object is **transparent**; it may be reflected or scattered if the object is bright or polished; it may be absorbed if the object is dark; it may be transmitted, reflected or absorbed in varying proportions depending upon the exact nature of the object. Only the radiation that is absorbed can cause the temperature of the object to change. The **greenhouse effect** depends upon the selective ability of the **atmosphere** to allow solar radiation to be transmitted but to cause terrestrial radiation to be absorbed. Thus it is the latter that causes the atmosphere to warm up. The absorption of high-energy radiation can cause damage to living tissue. The link between **ultraviolet radiation** and skin **cancer**

is well established, for example, and nuclear radiation can cause death, serious injury, **cancer** or genetic defects, depending upon the size and timing of the dose. The absorption of radiation by a body or object can be controlled through the use of **insulation** or shielding.

See also
Backscatter, Electromagnetic spectrum, Rad, Radioactivity, Rem.

Further reading
Faughn, J.S., Turk, J. and Turk, A. (1991) *Physical Science*, Philadelphia, PA: Saunders.
Seigel, R. and Howell, J.R. (1992) *Thermal Radiation Heat Transfer* (3rd edition), Washington, DC: Hemisphere Publishing.

RADIATION BLINDNESS

Loss of sight caused by damage to the eye from exposure to excess **solar radiation**. It usually takes the form of cataracts in which the normally clear lens of the eye becomes **opaque**, causing reduced light transmission and loss of visual perception. Although cataracts can have a variety of other causes,

Table R-1 Biological effects of exposure to radiation

DOSAGE	EFFECTS
0.005 rem	Maximum annual radiation at perimeter of nuclear generating station. No known effects.
0.220 rem	Level of normal background radiation at sea-level. Probability of cancer is 1 in 100,000 people exposed.
10 rem	No obvious illness following instantaneous exposure, but 1 in 1000 chance of delayed cancer among those exposed.
100 rem	Would cause nausea if received in one acute dose and produces a 1 in 100 chance of delayed cancer among those exposed.
300-600 rem	Causes nausea within hours of acute exposure, followed by vomiting, diarrhoea, hair loss and emaciation.
1000 rem	If received instantaneously, this dosage would cause immediate illness and lead to death within a few weeks if no medical treatment was obtained.

Source: Ontario Hydro, *Powerful Facts about Radiation* (public information booklet)

an increase in their occurrence has been projected because of the extra **ultraviolet radiation** (mainly UV-B) being allowed through to the earth's surface as the **ozone layer** thins. **Snow** blindness is a temporary form of radiation blindness caused by the exposure of the eyes to solar radiation reflected off a white snow surface.

RADIATION SCATTERING

The disruption of the smooth flow of **radiation** through the **atmosphere**, usually as a result of **particulate matter** in the **energy** path. Some of the energy will be reflected back (**backscatter**), but through multiple reflection some may be returned to its original path (forwardscatter). **Solar radiation** scattered in this way may eventually reach the earth's surface in the form of **diffuse radiation**.

See also
Atmospheric turbidity.

Further reading
Bohren, C.F. and Huffman, D.R. (1983) *Absorption and Scattering of Light by Small Particles*, New York: Wiley.

RADIATION SICKNESS

Sickness caused by exposure to a sub-lethal dose of **ionizing radiation** (between 100 and 300 rem). Symptoms include the rapid onset of nausea and vomiting followed some days later by diarrhoea, hair loss, sore throat, haemorrhaging and bone marrow damage. Delayed effects such as miscarriages and stillbirths are common. In the longer term, sub-lethal doses of radiation also increase the risk of leukemia and other forms of **cancer**.

Further reading
Turner, J.G. (1995) *Atoms, Radiation and Radiation Protection*, New York: Wiley.

RADIATION SPECTRUM

See **electromagnetic spectrum**.

RADIATIVE FORCING AGENT

Any factor capable of disturbing the **energy** balance of the earth/atmosphere system. Radiative forcing is the change in average net **radiation** – incoming **solar radiation** compared to outgoing **terrestrial radiation** – at the **tropopause**. Changes in natural forcing agents such as solar radiation, planetary **albedo** and atmospheric **aerosol** concentrations continue to disrupt the energy balance, but anthropogenic forcing agents involving **ozone depletion, atmospheric turbidity** and the enhancement of the **greenhouse effect** are also causing significant disruption to the system's energy budget. Forcing may be negative or positive. Aerosols, for example, generally have a negative impact. When **Mount Pinatubo** erupted in 1991, it reduced the global net radiation by 3–4 Wm^{-2}. In contrast, **greenhouse gases** are positive radiative forcing agents, and a forcing of +2.45 Wm^{-2} has been attributed to the increase in greenhouse gas levels since pre-industrial times.

Further reading
Houghton, J.T., Meira Filho, L.G., Bruce, J., Lee, H., Callander, B.A., Haites, E.F., Harris, N. and Maskell, K. (1994) *Climate Change 1994: Radiative Forcing of Climate Change and an Evaluation of the IPCC 1992 Emission Scenarios*, Cambridge: Cambridge University Press.
Shine, K., Derwent, R.G., Wuebbles, D.J. and Morcrette, J.J. (1990) 'Radiative forcing of climate', in J.T. Houghton, G.J. Jenkins and J.J. Ephraums (eds) *Climate Change: The IPCC Scientific Assessment*, Cambridge: Cambridge University Press.

RADIATIVE-CONVECTIVE MODELS

One-dimensional **climate** models incorpor-

ating global scale radiative and convective processes at different levels in the **atmosphere**. The main inputs into these models are incoming **solar radiation** and returning **terrestrial radiation**. They can be used to estimate **temperature** change initiated by changing atmospheric **aerosol** levels, such as those produced by volcanic eruptions. The original **TTAPS scenario** of **nuclear winter** was based on results from a radiative-convective model. One-dimensional models such as this treat the earth as a uniform surface with no geography and no seasons. As a result, they are inadequate to deal with the uneven surface **energy** distribution associated with the differences in heat capacity between land and ocean. At best, they are useful for the preliminary investigation of global scale radiative and convective processes at different levels in the **atmosphere**. However, they cannot deal with seasonal or regional scale features, and require so many assumptions that their ability to provide accurate predictions is limited.

See also
Models.

Further reading
Schneider, S.H. (1987) 'Climatic modeling', *Scientific American* 256: 72–89.

Table R-2 Radiative forcing agents

FORCING AGENTS	RADIATIVE FORCING (WM^{-2})	COMMENTS
Greenhouse gases	+ 0.56 + 0.41	Business-as-usual Major emission controls
Solar variability	+ 0.1 – 0.1	e.g. orbital changes and changes in solar irradiance – sunspot cycles
Large volcanic eruption	– 0.2	e.g. El Chichon, Mount Pinatubo
Anthropogenic sulphur emissions	+ 0.15? – 0.15?	Difficult to estimate – total emissions are declining, regional differences remain
Stratospheric H$_2$O	+ 0.02	

Source: Based on data in Houghton *et al.* (1994)

RADICAL

See **free radical**.

RADIOACTIVE WASTE

See **nuclear waste**.

RADIOACTIVITY

Radiation emitted as the result of the decay of the atomic nuclei of certain **elements**. Nuclei capable of producing radiation in this way are called **radionuclides**. The radiation takes the form of elementary particles – such as alpha and beta particles – or **gamma rays,** and emissions continue until a stable state is attained. Certain elements such as **uranium** (U) and thorium (Th) are naturally unstable and their atomic nuclei disintegrate spontaneously, releasing radioactivity in the process. As the radioactivity is released, a series of new products is formed with the final member in each series being a stable element. With uranium and thorium, that stable element is **lead**. Radioactivity can be induced in many elements by bombarding their nuclei with particles such as **neutrons**.

This is the process that takes place in a **nuclear reactor,** and the induced radioactivity contributes to the problems of **nuclear waste** disposal. Although exposure to radioactivity can have harmful consequences, it is also used in medicine to treat certain **cancers** and as a tracer in physical, biological and chemical **systems**.

See also
Atom, Nuclear fission, Nucleus, Radiation sickness, Radioisotope.

Further reading
Miller, E.W. and Miller, R.M. (1990) *Environmental Hazards: Radioactive Material and Wastes: A Reference Handbook*, Santa Barbara, CA: ABC-Clio.

RADIOCARBON DATING

A method used to date organic materials. It relies on the intake of radioactive **carbon**-14 (^{14}C) by living organisms and the subsequent decay of the carbon after the organism dies. Most carbon exists as one of two stable **isotopes** ^{12}C and ^{13}C, but a small amount is present in the **environment** as radioactive ^{14}C. Radiocarbon is produced in the **atmosphere**

Figure R-2 The radioactive decay sequence of Uranium-238

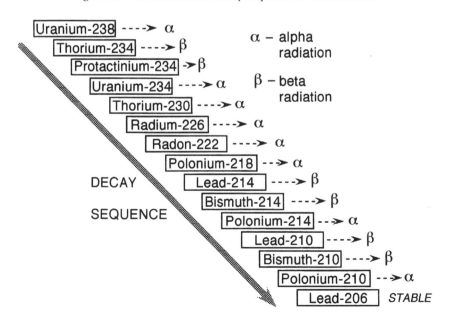

when cosmic rays collide with **nitrogen** (N) **atoms** and transform them into carbon-14 atoms. These atoms mix with the other carbon isotopes and are absorbed as **carbon dioxide** (CO_2) by living plants during **photosynthesis**. Consumption of the plants by animals passes the ^{14}C through the system. As a result, all living organisms contain ^{14}C in their tissues. When the organisms die, ^{14}C is no longer absorbed and the proportion present in the tissue begins to fall as the radiocarbon decays. Radiocarbon has a **half-life** of about 5700 years, and by comparing the proportion of ^{14}C remaining in a **fossil** organic sample with what might be expected in a modern living organism it is possible to estimate the age of the sample. This method is suitable for material up to 70,000 years old. It assumes that the proportion of ^{14}C in the atmosphere has remained constant, but human activities have introduced variations which have reduced the accuracy of the technique. The increased use of **coal** as a **fuel** following the **Industrial Revolution**, for example, increased the output of the other isotopes of carbon and the proportion of radiocarbon decreased. Conversely, the detonation of nuclear devices in the atmosphere increased the amount of ^{14}C produced, at least until 1964 when above-ground testing of nuclear weapons was banned. This combination of **fossil fuel** carbon and bomb carbon resulted in inaccurate dating, particularly of younger samples, but recalibration using dendroclimatological techniques has allowed the inaccuracies to be resolved. Radiocarbon dating has been used extensively to establish the timing and sequence of events during the **Pleistocene** and **Holocene**.

See also
Dendroclimatology.

Further reading
Bowman, S. (1990) *Radiocarbon Dating*, London: British Museum Publications.
Broecker, W.S., Peng, T.-H., Ostlund, G. and Stuiver, M. (1985) 'The distribution of bomb radiocarbon in the ocean', *Journal of Geophysical Research* 90: 6953–70.
Stuiver, M. and Quay, P.D. (1981) 'Atmospheric ^{14}C changes resulting from fossil fuel CO_2 release and cosmic ray flux variability', *Earth Planetary Science Letters* 53: 349–62.

RADIOISOTOPE

An **isotope** of an **element** which emits **radioactivity**. Carbon-14 (^{14}C) is a radioisotope of **carbon** (C), for example.

RADIONUCLIDE

A **nuclide** which disintegrates spontaneously, releasing **radioactivity** in the process. The nuclei of **uranium** (U), **radium** (Ra) and strontium (Sr) **atoms**, for example, are radionuclides.

RADIOSONDE

An instrument package carried up into the **atmosphere** by a balloon. It includes sensors for measuring such **elements** as **temperature**, **atmospheric pressure** and **humidity** and a radio transmitter through which the information is sent back to a ground station. By tracking the radiosonde, either visually or electronically, it is possible to calculate **wind** speed and direction. Radiosonde observations provide the data for creating temperature, pressure and wind profiles of the **troposphere**, which can be used for **weather forecasting**.

RADIUM (Ra)

A rare, radioactive **element**, chemically similar to **barium** (Ba). It occurs as several **isotopes**, the most stable of which has a **half-life** of 1620 years. The main source of radium is pitchblende, an **ore** of **uranium** (U).

See also
Radioactivity, Radon.

RADON (Rn)

A **noble gas** produced by the decay of **radium** (Ra). Radon is radioactive and, as it disintegrates, it produces additional radioactive decay products – the daughters of radon. It is suspected of causing lung **cancer** among **uranium** miners, and studies have suggested it is a common indoor pollutant in some areas. Radon may seep into houses from the decay of radium in the local bedrock or in building materials, creating an environmental hazard for the residents. Exposure limits have been suggested in Britain, Canada, the United States and Sweden for residential property, and these can be met by installing relatively simple ventilation systems that prevent the accumulation of the **gas**.

See also
Daughter product.

Further reading
Nagda, N.L. (1994) *Radon: Prevalence, Measurements, Health Risks and Control*, Philadelphia, PA: ASTM.

RAIN

Precipitation in the form of **liquid water** droplets. Droplets vary in size, but exceed 0.5 mm in diameter. Smaller droplets are considered to be drizzle. Several theories have been developed to explain the formation of raindrops, but the two most commonly accepted involve the presence of **ice crystals** (the Bergeron-Findeisen process) and the coalescence of droplets of different sizes. When ice crystals and supercooled water droplets (temperature < 0°C) are present in a **cloud,** the droplets tend to vaporize and the **vapour** is deposited directly on the crystal. The ice crystals continue to grow as the result of further vapour deposition or through the aggregation of crystals until they are sufficiently heavy to overcome upcurrents in the **air**. They then begin to fall and when they reach warmer air they melt into raindrops. Collision or coalescence theories see larger droplets sweeping up or absorbing smaller droplets as they move through the clouds, until they are large enough to fall out.

See also
Condensation nuclei, Rain making.

Further reading
Lutgens, F.K. and Tarbuck, E.J. (1989) *The Atmosphere: An Introduction to Meteorology*, Englewood Cliffs, NJ: Prentice-Hall.
Mason, B.J. (1975) *Clouds, Rain and Rainmaking*, Cambridge/New York: Cambridge University Press.

RAIN MAKING

The artificial augmentation of **precipitation**, based on the Bergeron-Findeisen theory of the formation of **rain**, in which the deposition of **water vapour** on **ice crystals** in super-cooled **clouds** ultimately produces water droplets large enough to fall from the clouds as rain. Rain making involves seeding clouds with ice crystal nuclei to encourage the growth of ice crystals and initiate the Bergeron process. The most common types of agents for cloud seeding are solid **carbon dioxide** (dry ice) or silver iodide (AgI). The very low **temperature** of the dry ice (−78°C) cools the **air** so rapidly that ice crystals are produced spontaneously, whereas the crystalline structure of silver iodide is similar to that of ice which encourages the initial deposition of vapour from the super-cooled water droplets in the clouds to produce the ice crystals. The seeding of clouds at temperatures between −5°C and −15°C appears to be most effective, with increases in precipitation of about 10–15% being claimed for such conditions. Seeding is most commonly done from aircraft flying through or above cloud, but cloud seeding materials have also been introduced into the clouds in artillery shells fused to burst at a particular height or by **convection** from surface fires or other generators. Rain making has been used in precipitation management schemes in the United States, Australia and Russia, and any future developments are likely to be in these areas also. However, it has received less attention in recent years because of the costs involved, the difficulty of measuring the effectiveness of the process, potential legal problems associated with the timing and distribution of the additional precipitation and the possible effects of silver iodide on the **environment**. Since precipitation augmentation seems to be most successful in clouds that are already likely to produce rain, some thought has been given to seeding clouds in areas of orographic rainfall. The additional precipitation could be stored in **reservoirs** and distributed by pipeline to areas requiring additional water. Such a programme has been considered for the western United States, where rain making would be used to augment the snowpack in the Sierra Nevada and the Rocky Mountains, with the additional water being supplied naturally when the **snow** melted in the spring. There is some concern that if successful, these schemes might cause a deterioration in the **habitats** of existing plants and animals and cause flooding or excess **erosion** in the downstream sections of the rivers flowing out of the mountains. Cloud seeding techniques have also been employed to

dissipate **fog**, suppress hail and in attempts to reduce the power of **hurricanes**.

Further reading

Calder, N. (1974) *The Weather Machine and the Threat of Ice*, London: BBC Publications.
Eagleman, J.R. (1985) *Meteorology: The Atmosphere in Action*, Belmont, CA: Wadsworth.
Mason, B.J. (1975) *Clouds, Rain and Rainmaking*, Cambridge/New York: Cambridge University Press.

RAINFOREST

Broad-leaved, mainly evergreen forests found in the tropics, sub-tropics and some temperate regions where moisture is abundant all year round. Tropical rainforests are best developed 10° north and south of the equator in the Amazon and Congo basins, West Africa and in parts of south-east Asia. Outside of that band, **temperatures** and rainfall are sufficiently high to allow rainforest to flourish in areas such as Central America, Bangladesh, Burma and north-eastern Australia. Tropical rainforests cover some 12 million km² and represent nearly one-third of the world's forests. Temperate rainforest exists in the coastal regions of western Canada, the United States Pacific north-west and New Zealand, where the prevailing westerlies provide enough moisture to maintain the rainforest **ecosystem**. Rainforests are arguably the most persistent, most stable and most complex of the earth's ecosystems. Tropical rainforests alone contain as many as 30 million **species** of plants and animals.

Because the rainforests are capable of producing large quantities of timber, they represent a large capital asset, particularly for developing countries in South America, Africa and Asia. This poses a major dilemma for these areas. In the short term at least, the developing nations would benefit economically from the exploitation of the rainforest, but only at some environmental cost. Changes in microclimatic temperature and moisture regimes inevitably follow the removal of a forest cover. Higher soil temperatures and more active **leaching**

combine to remove **nutrients** from the upper levels of the forest **soil** and increased **runoff** leads to flooding and **soil erosion**. The net result is that the natural regeneration of the forest cover is very difficult, causing long-term disruption of the ecosystem, threatening the survival of certain plants and animals, and causing a decline in **biodiversity**. On the human side, the indigenous inhabitants of the rainforest have been displaced from their traditional hunting and cultivating territories through destruction of **habitat** or in some cases through forced resettlement. Cases of indigenous peoples being murdered to allow access to an area have also been reported from South America.

Many of the threats to the rainforest **environment** originate in the developed nations. The demand for tropical hardwoods remains high, despite attempts at boycotts by environmental groups. Where the export of hardwoods is declining, it appears to be as a result of the exhaustion of **resources** rather than a reduction in demand. Agricultural development also threatens the forest. Trees are cleared to open up land for the cultivation of a variety of cash crops ranging from cassava and cocoa to pepper and pineapples, which are exported to produce a greater profit than could be made by growing food for domestic use. Land is also cleared to provide pasture for cattle ranching, particularly in Latin America. Much of the beef from this area is exported to the United States to be made into hamburger patties

Figure R-3 The global distribution of rainforest

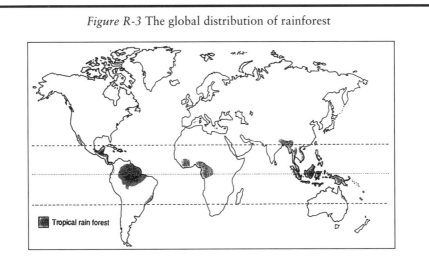

Tropical rain forest

Source: After Park, C.C. (1992) *Tropical Rainforests*, London/New York: Routledge

for the fast food industry. This '**hamburger connection**', as it has been called, is seen by environmental and conservation groups as a particularly invidious misuse of resources. Mineral exploitation, hydroelectric developments and other mega-projects, often using foreign aid or instigated by multinational companies, also contribute to the destruction of the forest.

Although the developed nations are placing the survival of the rainforest in jeopardy through such demands, in the same countries environmental groups are attempting to reverse the situation. Motivated to protect one of the few remaining natural environments and the rights of its inhabitants, they have used a variety of approaches from lobbying to direct action in an attempt to reduce the threat to the forest. Some groups have been accused of using misinformation tactics in their zeal to protect the forest. In the scientific **community**, there is concern over the potential impact of the removal of the rainforest on **global warming**. The clearing of the forest raises **carbon dioxide** (CO_2) levels indirectly through reduced **photosynthesis**, but carbon dioxide is also added directly to the **atmosphere** by burning, by the decay of **biomass** and by the increased oxidation of **carbon** (C) from the newly exposed **soil** of

the forest floor. The increase in atmospheric carbon dioxide resulting from decreased photosynthesis and the clearing of vegetation is equivalent to about 2 billion tonnes per year. An attempt was made to deal with such problems at the **United Nations Conference on Environment and Development** (UNCED) held in Rio de Janeiro in 1992, when a group of nations discussed problems associated with the **sustainable development** of forests. The result was an agreement acknowledging the need to balance the exploitation of forests with their conservation, but not binding on the signatories, mainly as a result of the unwillingness of the developing nations to accept international monitoring and supervision of their forests.

Further reading

Furley, P.A. (ed.) (1993) *The Forest Frontier: Settlement and Change in the Brazilian Roraimia*, London: Routledge.

Mabberley, D.J. (1992) *Tropical Rainforest Ecology* (2nd edition), Glasgow/New York: Blackie/Chapman and Hall.

Moore, P.D., Chaloner, B. and Stott, P. (1996) *Global Environmental Change*, Oxford: Blackwell Science.

Park, C.C. (1992) *Tropical Rainforests*, London/New York: Routledge.

Whitmore, T.C. (1990) *Introduction to Tropical Rain Forests*, Oxford: Oxford University Press.

RAINFOREST ACTION NETWORK (RAN)

An environmental organization working to protect the earth's **rainforests** and the human rights of their inhabitants through education, grassroots organization and non-violent direct action. Launched in 1985, RAN has campaigned to make the problems of the rainforest better known through the organization of action groups and conferences, through the publication of newsletters and reports, and through direct action in the form of consumer boycotts and fund-raising. It led a successful consumer campaign to force the fast food retailer, Burger King, to cancel its contracts for rainforest beef, and has been very active in campaigns to stop the clear cutting of temperate rainforest in British Columbia, Canada. In the human rights field, RAN has supported the indigenous inhabitants of the rainforest in Borneo, Ecuador and Brazil.

RAMSAR CONVENTION

See **wetlands**.

RAYLEIGH SCATTERING

The scattering of **light** by very small particles without loss of **energy**, named after Lord Rayleigh who first developed the theory in 1871. It is responsible for the blue colour of the sky, for example, with white light from the sun being scattered by **gas** molecules in the **atmosphere**. Since small particles are particularly efficient at scattering shorter wavelength **radiation**, the light shining through an atmosphere containing large quantities of small **aerosols** will appear red because the shorter blue wavelengths have been scattered more effectively than the longer red wavelengths.

Further reading
Bohren, C.F. and Huffman, D.R. (1993) *Absorption and Scattering of Light by Small Particles*, New York: Wiley.

RECHARGE RATE

The rate at which **water** withdrawn from the **groundwater** system is replaced by **precipitation**. In many areas, water can be pumped out of the system more rapidly than it can be replaced. As a result, the **water table** declines and extraction becomes more difficult and costly. In parts of the Ogallala **aquifer** in the south-western United States, for example, the water table is being drawn down at a rate of 1 m per year while the effective recharge rate may be as low as 1 to 4 mm. Elsewhere, as in the Sahara region, there is negligible recharge and the groundwater being used is inherited from periods when the region was wetter. The artificial recharge of aquifers has been considered, but in areas where groundwater use is high there is usually no excess water available. In addition, pumping surface water into an aquifer creates the potential for introducing **pollution** into the groundwater system or disrupting the **permeability** of the aquifer by introducing sediments.

Further reading
Walton, W.C. (1991) *Principles of Groundwater Engineering*, Chelsea, MI: Lewis.

RECLAMATION

The treatment of land made derelict by industrial or agricultural activities, with the aim of returning it to productive use. Companies involved in activities such as gravel extraction, **strip-mining** and quarrying, which once created large areas of derelict land, are now required to rehabilitate the areas affected once the activities cease. The term is also applied to the creation of dry land in coastal areas by a combination of dyking and infilling, or to the draining of **wetlands** to allow agricultural development.

Further reading
Agassi, M. (ed.) (1996) *Soil Erosion, Conservation and Rehabilitation*, New York: M. Dekker.
Plotkin, S.E. (1986) 'From surface mine to cropland', *Environment* 28 (1): 1–20, 40–4.

RECURRENCE INTERVAL

The return period of a particular event, represented as a statistical estimation of the frequency with which an event of a specific magnitude is expected to occur. The approach is commonly used in evaluating environmental hazards such as flooding. The recurrence interval for a **flood** of a specific magnitude can be estimated from historic stream flow data, for example, allowing reference to the fifty-year flood, the hundred-year flood or flooding at some other interval. Such information is important for **floodplain** management.

RECYCLING

Since the earth/atmosphere **system** is a closed system in material terms, it includes a number of very efficient natural recycling systems that allow **elements** such as **water**, **carbon** (C), **nitrogen** (N) and **sulphur** (S) to be used many times over. However, in current usage recycling refers to the recovery of **waste** material for reprocessing into new products, or the reuse of discarded products.

Although often seen as a modern phenomenon brought on by the need to conserve **resources**, recycling has a long history, particularly in the **metal** industries. There have been flourishing scrap metal markets for such commodities as **iron** (Fe), steel, **aluminum** (Al) and **copper** (Cu) for many years, and a very high proportion of all the **gold** (Au) ever mined remains in circulation,

Table R-3 Some widely recycled commodities and problems associated with them

COMMODITY	PROBLEMS
Glass – mainly unbroken containers	Clear glass is most valuable. Mixed glass – clear and coloured – not suitable for direct recycling. Ceramics, light bulbs, pyrex glass and mirrors also contaminate the glass.
Paper – clean, dry newspapers easiest to recycle	Rubber bands, plastic bags, dirt, food waste and moulds often contaminate otherwise recyclable paper.
Metal products – cans, caps and lids	Full cans, aerosol spray cans and those containing hazardous waste can cause problems, but metal products are commonly recycled many times.
Plastic – thermoplastics such as PET and HDPE	Coloured pigments, caps that are of non-recyclable material, may contaminate a melt and prevent the proper recycling of the plastic.
Mixed paper – magazines, glossy paper, cereal boxes, shredded office paper, corrugated cardboard	Separation of the different types of paper required. Waxed paper, carbon paper, milk cartons, foil-covered boxes and thermal fax paper make recycling difficult.
Motor oil – filtered and re-refined	Companies participate in recycling programmes – often by law – but disposal of oil by 'do-it-yourself' oil changers is a major source of environmental contamination.
Automotive batteries	Contain lead which can contaminate the environment if disposed of improperly.

Source: Consumer Recycling Guide – 1996: WWW Page of Evergreen Industries

Figure R-4 A recycling depot. The initial collection and sorting of the recyclable material is done by the users of the depot which represents a saving in time and money for the municipal authority.

Photograph: The author

because of its high value and resistance to **corrosion**. Increasingly, **plastics**, glass and paper products are being recycled rather than being incinerated or dumped in **sanitary landfill sites.**

Recycling can take different forms. An object, such as a glass bottle or compressed **gas** cylinder, might be reused several times, for example. Eventually the bottle or the cylinder will no longer be reusable, however, but might then be reprocessed to produce new containers. In many cases the objects cannot be recycled directly. They can often be reprocessed for other uses, however. Glass has been used to produce skid resistant road surfaces, for example, and plastics converted into 'plastic lumber', **garbage** cans and other containers.

While recycling is generally considered to be environmentally appropriate and the technology exists to achieve it, it is less widely developed than it might be.

Although there are savings in extraction and processing costs when materials are recycled, they might be offset by the costs of collecting them and transporting them to an appropriate processing plant. In attempts to encourage recycling, governments at all levels have used a wide range of incentives and disincentives. These usually take the form of subsidies, taxes or direct legislation controlling the disposal of certain products. Subsidies are used to allow recycling where it would not otherwise be economically feasible, but with time is expected to become self-financing. Taxes may be direct, with the revenue being use to subsidize the recycling or disposal process. A number of jurisdictions in North America have taxes in place aimed at dealing with the problem of scrap car tyres which tend to accumulate in massive dumps, trapping water which provides breeding places for mosquitoes and, in case of **fire**, having the potential to cause serious environmental problems through the **smoke**, **gases** and **oil** released. Taxes on the use of landfill sites indirectly encourage recycling by making it a less costly alternative to dumping. Legislation directed at encouraging recycling can also include the banning of certain non-recyclable products, such as 'throw-away' bottles and other containers. Individuals are most often involved in recycling through curbside pick-up – the so-called 'blue-box' programme – or depot recycling where the waste is dropped off at a central location. The main problem with this approach is the sorting of the collected waste. The inclusion of inappropriate materials can make recycling impossible and every year large quantities of plastics collected for recycling are dumped in landfill sites for that reason.

Although the amount of material being collected for recycling is growing rapidly, it is not being used as rapidly. Changing markets alter the demand for the recyclables and as a result they accumulate in warehouses, and the financial and environmental benefits are not realized. Thus, although the potential benefits of recycling are high, and the collection of appropriate materials can be legislated, the process cannot be divorced from the broader considerations of the economic system.

See also
Conservation, Plastics recycling.

Further reading
McKinney, R.W.J. (ed.) (1995) *Technology of Paper Recycling*, London/New York: Blackie/Chapman and Hall.
Powelson, D.R. and Powelson, M.A. (1992) *The Recycler's Manual for Business, Government and the Environmental Community*, New York: Van Nostrand Reinhold.
Selke, S.E. (1990) *Packaging and the Environment: Alternatives, Trends and Solutions*, Lancaster, PA: Technomic.
(See also the WWW Home Pages of such environmental groups as the Environmental Recycling Group and the Recycling Council of Ontario.)

RED DATA BOOKS

A series of books sponsored by the World Conservation Union, a private organization based in Switzerland, that provide a global listing of **endangered species**. Available Red Data Books include issues on birds, mammals of the New World and Australasia, invertebrates and plants.

Further reading
Cox, G.W. (1993) *Conservation Ecology: Biosphere and Biosurvival*, Dubuque, IA: Wm C. Brown.

RED LIST

Based on European Commission directives, this is a British list of twenty-three substances whose discharge into **water** bodies should be minimized because of their hazardous nature. The list includes such substances as **mercury** (Hg) and its **compounds**, DDT, PCBs and a variety of common **pesticides**.

See also
Black List.

RED RAIN

See **precipitation scavenging**.

REDUCTION

The removal of **oxygen** (O) from a **compound** or the addition of **hydrogen** (H) to it. A reducing agent is a chemical which brings about reduction.

REFLECTANCE

A measure of the ability of a surface to reflect **radiation**, defined as the ratio of the amount of radiation reflected to the amount incident on the surface.

See also
Albedo, Reflection.

REFLECTION

The return of **light** waves, or other forms of electromagnetic **radiation**, when they strike a surface.

See also
Albedo, Reflectance.

REFRACTION

The bending of a ray of **light** or other form of **radiation** that occurs when the ray travels from one **transparent** medium into another. Light rays bend when they cross the boundary between **air** and **water**, for example. This occurs because slight differences in the density of the media cause changes in the wavelength of the light. Rainbows result from the refraction of light by raindrops and **desert** mirages are caused by the refraction of light rays when they pass through air of different densities. In coastal areas, water waves are refracted when the wave front is slowed more rapidly in one area than another. This occurs when the waves approach the coast at an angle or when the depth of the offshore water is variable.

REFUSE-DERIVED FUEL INCINERATOR

See **waste-to-energy incinerator**.

REGIONAL ACIDIFICATION INFORMATION AND SIMULATION (RAINS)

A computer **model** originally developed to study **acid rain** in Europe. Designed by the IIASA, it allows users to visualize the future impacts of current actions, or inactions, and to design strategies to deal with them.

RELATIVE DENSITY

Formerly called specific gravity, it is the ratio of the **density** of a substance at a given **temperature** to the density of **water** at 4°C, i.e. its maximum density. If a substance has a relative density of less than 1, it will float on water, if greater than 1 it will sink. Cork, with a relative density of 0.24 will float, **gold** (Au), with a relative density of 19.3 will sink.

RELATIVE HUMIDITY

A measure of the amount of **water vapour** in the **air**, compared to the amount that the air can hold at that **temperature** and **pressure**. It can also be expressed as the ratio of the measured **vapour pressure** to the saturated vapour pressure of the air at the same temperature. Relative humidity is usually expressed as a percentage.

See also
Humidity.

REM

Acronym for Roentgen Equivalent Man – a measure of the relative biological effect of **radiation** on the human body. One rem is the dose of **ionizing radiation** that will have the same biological effect on an individual as one **roentgen** of X-rays. The rem has been replaced by the **sievert** (Sv), with one Sv being equal to 100 rem.

See also
Rad.

REMOTE SENSING

The observation of the surface of the earth from a distance by means of sensors. Aerial photography was the earliest form of remote sensing, but satellite observation is now most common, involving the creation of direct photographic images or the collection of data in digital form. Remote sensing can provide information for large areas very rapidly, but accurate analysis requires adequate **ground control**.

Further reading
Barrett, E.C. and Curtis, L.F. (1992) *Introduction to Environmental Remote Sensing* (3rd edition), London/New York: Chapman and Hall.
Gurney, R.J., Foster, J.L. and Parkinson, C.L. (1993) *Atlas of Satellite Observations Related to Global Change*, Cambridge: Cambridge University Press.

RENEWABLE ENERGY

Energy from natural sources, that can be replaced as it is used or at least within a very limited time frame. Renewable energy is supplied by flowing **water, wind, biomass** and the sun. The sun is the ultimate renewable energy source, and is a very effective nuclear furnace, able to provide the earth/atmosphere **system** with a relatively steady flow of energy as a result of on-going **nuclear fusion**. Solar energy can be used directly through solar panels or photoelectric **cells**, and once in the earth/atmosphere system it drives the **hydrological cycle**, which ensures the renewal of water power, it helps create the pressure differences, which cause the wind to blow and brings about biomass renewal through **photosynthesis**. The energy contained in **fossil fuels** is also stored solar energy, and in theory will be restored after use. However, the replacement processes take so long that once used these fuels cannot be restored within the human time frame, and therefore in human terms they are considered to be non-renewable resources.

Until the **Industrial Revolution**, society depended almost entirely on renewable energy. The more concentrated, more reliable and more efficient energy available from fossil fuels caused the renewable source to decline in importance. Only water, through its ability to provide electrical energy, retained and even increased its importance. Interest in renewable energy increased in the 1960s, when it appeared that **oil** and **gas** sources were facing depletion, and in the 1970s, following the **petroleum** price increases initiated by the **OPEC oil embargo** of 1973. Growing environmentalism also favoured the re-adoption of renewable energy resources, since they were perceived as causing less environmental damage than fossil fuels. They are capable of **sustainable development**, but they are not completely environmentally friendly. Biomass burning can create air **pollution** and in some developing countries such as Nepal and parts of the **Sahel**, where renewable **resources** did not completely succumb to the dominance of fossil fuels, serious problems of **soil erosion** and **desertification** have followed the removal

Figure R-5 (a) The components of a residential active solar heating system

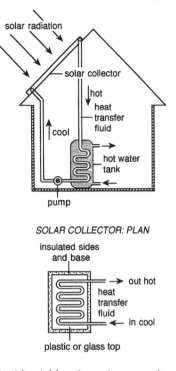

(b) Residential heating using a passive solar heating system

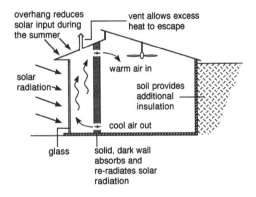

of forest, bush and scrub to meet the fuel needs of rapidly growing **populations**. Although the generation of **hydroelectricity** is less damaging to the **environment** than its production in thermally powered stations, hydroelectric schemes have a significant effect on the hydrological cycle through the creation of large **reservoirs** and changes in stream-flow patterns. The development of wind energy systems is often accompanied by aesthetic and **noise** pollution.

Renewable energy sources cannot supply energy in the quantity and with the efficiency demanded by modern society. Thus, although they have many desirable traits, they are unlikely to replace or even seriously challenge the dominance of non-renewable energy resources.

See also
Non-renewable resources.

Further reading
Boyle, G. (1996) *Renewable Energy: Power for a Sustainable Future*, Oxford: Oxford University Press/Open University.
Darmstadter, J., Landsberg, H.H., Morton, H.C. and Coda, M.J. (1983) *Energy Today and Tomorrow*, Englewood Cliffs, NJ: Prentice-Hall.
Howes, R. and Fainberg, A. (eds) (1991) *The Energy Sourcebook: A Guide to Technology, Resources and Policy*, New York: American Institute of Physics.

RENEWABLE RESOURCE

A **resource** that is replaced at a rate which is faster than, or at least as fast as, it can be used. The **oxygen** (O) in the **air**, the plants and animals in the **environment**, the **water** in the hydrological system and **energy** from the sun are all renewable, for example. Modern pressures on such resources have created situations in which they are unable to renew rapidly enough to meet the demands placed on them. Oxygen is unlikely to be used more rapidly than it can be replaced and the sun will continue to supply energy for millions of years. Such resources are sometimes referred to as perpetual resources. In many places, however, plant and animal resources are being consumed at rates that exceed their natural rates of reproduction. Similarly, in an increasing number of areas around the world, water consumption rates have outstripped the ability of the **hydrological cycle** to maintain the natural supply. Thus the classification of resources as either renewable or non-renewable is less clear-cut than it once was.

Further reading
Pearce, D.W. and Turner, R.K. (1990) *Economics of Natural Resources and the Environment*, Baltimore, MD: Johns Hopkins University Press.

RESERVES

Resources not currently being exploited, but available for use and able to be extracted or harvested using existing technology under the prevailing economic conditions. These are sometimes referred to as proven reserves to distinguish them from potential reserves. The latter can be considered as identifiable resources that could be exploited if and when physical, economic and technological conditions, individually or in combination, became appropriate.

RESERVOIRS – ENVIRONMENTAL EFFECTS

Reservoirs are lakes created artificially to meet specific needs such as the provision of **water** for domestic consumption, for **irrigation**, for industrial use, for **hydroelectricity** production, for **flood** control and for recreation. Reservoirs have many of the attributes of natural lakes, but because of the way in which they are created and used, they also give rise to a number of environmental problems. Reservoirs are created by building **dams** across rivers and allowing the stream flow to pond up to the required level. This almost immediately changes the **hydrology** of the area. Stream flow is altered and the existing balance between deposition and **erosion** is destroyed. Sediment collects in the reservoir, whereas, below the dam, the absence of sediment in the flowing water allows greater erosion. **Nutrients** are deposited along with the sediments, and areas which depended upon the spread of sediments during **floods** to provide nutrients become less fertile. The building of the Aswan Dam and the creation of Lake Nasser on the River Nile, for example, resulted in reduced fertility in the **soil** downstream from the dam for that reason. Upstream from the reservoir, the streams have to adjust to the presence of the lake and the **groundwater** hydrology is also affected. **Habitat** is destroyed by flooding and by water levels that change more rapidly than in the **natural environment**. Animals are forced out or are drowned by the rising water levels behind the dam and in the adjacent streams the existing **flora** and **fauna** may be unable to adjust to the new fluvial regimes. Fish such as salmon, for example, that migrate upstream to spawn will have their migratory patterns disrupted even when fish ladders are included in the project. On the human side, when reservoirs are formed the flooding of agricultural land, towns and villages may make it necessary to relocate people. The building of the Three Gorges Dam on the Yangtse River in China, for example, will flood some 40,000 hectares of land and displace more than 2 million people. Even where the numbers of people are smaller, the disruption of traditional lifestyles can be serious. The construction of reservoirs for the development of the James Bay Project in Canada and hydroelectric megaprojects in the Brazilian Amazon have disrupted the environment to such an extent that the traditional **hunting and gathering** activities of the indigenous people of these regions are no longer viable.

The development of increasingly larger reservoirs has increased their potential to have a global impact. The Altamira Project on the Xingu River in Brazil will involve the flooding of some 18,000 km^2 of rainforest. Past experience with similar projects suggests that such an area is unlikely to be logged before flooding and millions of cubic metres of prime timber will be lost. Flooding the forest will prevent it from **recycling carbon dioxide** (CO_2), causing it to contribute indirectly to **global warming**, and when the drowned timber begins to decay it is likely to produce considerable quantities of **methane** (CH_4) which is also linked to warming. Problems such as these have led environmental groups to oppose schemes which require the creation of large reservoirs. They have met with some success, but in many nations, such as Brazil and China, where a reliable and relatively cheap supply of energy is seen as a prerequisite for future development the planning and construction of such schemes is likely to continue.

The term reservoir is also used to refer to storage facilities for other commodities in the environment. In the **carbon cycle**, for example, **carbon** (C) is stored in a number of reservoirs such as the **oceans, atmosphere** and terrestrial vegetation from which it flows in response to interactions in the **system**.

Further reading
Fearnside, P.M. (1995) 'Hydroelectric dams in the Brazilian Amazon as sources of "greenhouse" gases', *Environmental Conservation* 22: 7–19.
Friends of the Earth (1989) *Damming the Rainforest: Indian Peoples' Summit of Altamira*, London: FOE.
Pearce, F. (1992) *The Dammed: Rivers, Dams and the Coming World Water Crisis*, London: The Bodley Head.

RESOURCE CONSERVATION AND RECOVERY ACT (RCRA)

An act passed by the US Congress in 1976 to deal with the problem of **waste disposal**. The provisions of the act are administered by the **EPA**, which was required to develop criteria for designating **hazardous waste** and establish a nationwide reporting system for companies involved in the production and disposal of such waste. The original law was found to contain many loopholes and in an attempt to close them an amendment was passed in 1984 that extended the waste disposal guidelines to a greater number of companies and broadened its provisions to include not just **landfill** sites, but also underground storage tanks. The 1984 amendments were aimed at encouraging the reuse, recycling and neutralization of toxic wastes, with disposal in **landfill** seen as the last resort. Many environmentalists consider the act's definition of hazardous or toxic wastes to be too narrow and would include **sewage**, agricultural wastes containing **pesticides** and mine or mill **tailings** in the system.

See also
Acid mine drainage.

Further reading
Smith, Z.A. (1995) *The Environmental Policy Paradox*, Englewood Cliffs, NJ: Prentice-Hall.

RESOURCES

In the broadest sense, resources are any objects, materials or commodities that are of use to society. The concept is entirely anthropocentric. Different groups of people value different materials as resources, and as society changes, so does its concept of what constitutes a resource. In the energy field, for example, wood was replaced by **coal**, and coal by **petroleum** as the main resource, but the change has not been universal and some groups still depend on wood as their main source of energy. Perhaps the ultimate example of the way in which materials are valued differently by different groups occurs in **recycling**, where the **waste** products of a **community** become resources for the recycler. Resources can be classified in a number of ways, depending upon the context in which they are being considered. A basic classification would include the following:

Perpetual resources:	Will always exist in relatively constant supply no matter how they are used, e.g. solar energy.
Renewable resources:	Replaced by natural processes once they are used, e.g. forests, animals, water.
Non-renewable resources:	Finite supply and cannot be replaced as rapidly as they are being used, e.g. fossil fuels, mineral ores.
Potential resources:	Will only become resources when the economic, cultural or technological conditions in a society create a demand for them, e.g. waste water, scrap car tyres.

When used in an economic context, **renewable** and **non-renewable resources** are commonly referred to as flow and stock resources respectively. Resources can also be categorized according to their existing availability and their potential future discovery. **Reserves** are known quantities which can be obtained economically at existing prices and with existing technology. Slightly less available are conditional resources which are known to exist, but which are not suitable for exploitation until economic and/or technological conditions change. Unidentified resources are those which have not yet been discovered, but

Figure R-6 A classification of the different forms of resources and their relationships

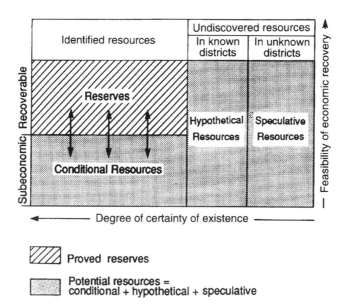

Source: After Park, C.C. (1997) *The Environment: Principles and Applications*, London: Routledge

through past experience or preliminary exploration are expected to become available. The boundaries between the different categories are not static, but change as the socio-economic, cultural and technological nature of society changes.

Further reading
Becht, J.E. and Belzung, L.D. (1975) *World Resource Management*, Englewood Cliffs, NJ: Prentice-Hall.
Cutter, S.L., Renwick, H.L. and Renwick, W.H. (1991) *Exploitation, Conservation, Preservation: A Geographic Perspective on Natural Resource Use*, New York: Wiley.
Rees, J. (1990) *Natural Resources: Allocation, Economics and Policy*, London: Routledge.

RESPIRATION

The process by which **aerobic** organisms take **oxygen** (O) from the **air** and use it to oxidize organic compounds from which they obtain **energy**. By-products of the process are **carbon dioxide** (CO_2) and **water**. Respiration sometimes refers only to the process by which the oxygen is taken into the body initially – through lungs or gills – and carbon dioxide is given out. **Anaerobic** respiration is a form of respiration used by organisms that do not have access to oxygen for producing energy.

RETROFITTING

The adaptation of an existing structure or appliance to meet needs that did not exist when the structure or appliance was first constructed. For example, coal-burning thermal power plants may be retrofitted with **scrubbers** to meet the requirement of acid **gas** emission levels that are now lower than they were when the plants were first built. The addition of extra **insulation** to residential properties to save **energy** and cut costs in the 1970s, as energy prices rose, is another example of retrofitting.

RETURN PERIOD

See **recurrence interval**.

REVERSE OSMOSIS

See **osmosis**.

RICHARDSON, L.F.

A British meteorologist who was a pioneer in the field of numerical atmospheric modelling as a means of providing **weather** forecasts. He published the results of his work in 1922 in his book *Weather Prediction by Numerical Process*. At the time, his methods were not practicable. The complexity of the computations required and the existence of only rudimentary methods of mechanical calculation meant that the predictions could not keep ahead of changing weather conditions. As a result, no real forecast could be made. It took almost forty years and the development of more sophisticated data collection and analysis techniques before the theory could be put into practice.

Further reading
Ashford, O.M. (1981) 'The dream and the fantasy', *Weather* 36: 323–5.
Ashford, O.M. (1992) 'Development of weather forecasting in Britain, 1900–40', *Weather* 47: 394–402.

RIO DECLARATION

A declaration of global principles on the theme of economically and environmentally sound development. Along with **Agenda 21**, it represented the culmination of the activities of the **United Nations Conference on Environment and Development (UNCED)**. The Rio Declaration contains twenty-seven principles which have the collective goal of establishing new levels of global co-operation to achieve **sustainable development** without further jeopardizing the **environment**. Containing reference to such elements as the special needs of developing nations, the eradication of poverty, the particular roles of women, youth and indigenous peoples and the development of public awareness of the problems, as well as promoting the rationalization and expansion of traditional approaches such as environmental legislation, **environmental impact assessment** and economic, scientific and technological policy developments, the Rio Declaration has the potential to have a major impact on global environmental and economic systems. However, there can be no assurance that these efforts will be successful. They were the result of compromise among some 150 nations, and attaining sufficient common ground to make this possible inevitably weakened the language and content of the Declaration, leaving it open to various interpretations and therefore less likely to be effective.

Further reading
Pearson, E.A., Hass, P.M. and Levy, M.A. (1992) 'A summary of the major documents signed at the earth summit and the global forum', *Environment* 34: 12–15 and 34–6.
Pearce, F. (1992) 'Despondency descends on Rio', *New Scientist* 134 (1824): 4.

ROENTGEN

A unit formerly used to measure X- or **gamma radiation**. Named after Wilhelm Konrad Roentgen (1845-1923), the discoverer of **X-rays**, it has been replaced by coulombs per kilogram of dry air in the **SI system**. One roentgen is equivalent to 2.58×10^{-4} C/kg.

See also
Curie, Rad, Rem.

ROSSBY WAVES

Long waves in the circumpolar westerly airflow in the upper **atmosphere**, first described by Carl Rossby in the 1930s. They are brought about by latitudinal variations in the **Coriolis** parameter and their effect on vorticity. **Air** moving polewards takes on an anticyclonic flow which tends to bring it back towards the equator. As the air moves towards the equator it takes on a cyclonic flow which causes it to swing back towards the pole. The net result is the north–south–north oscillation of the air stream as it moves around the earth, and through these oscillations, Rossby waves make a major contribution to meridional **energy** transfer in mid- to high latitudes The flow pattern commonly resolves itself into a range of three to six waves, with a number of preferred locations created by the presence of major mountain ranges such as the Rockies, Andes and the Tibetan Plateau or by the thermal differences caused by differential heating of

land and sea. During the transition between the stable patterns within that range, the paths followed by the waves are quite variable and difficult to forecast. The influence of Rossby waves extends to the lower atmosphere, through their contribution to the development of such features as mid-latitude low **pressure** systems, for example. Wave patterns similar to those of the atmospheric Rossby waves have also been identified in the **oceans**.

See also
Jet streams, Zonal index.

Further reading
Barry, R.G. and Chorley, R.J. (1992) *Atmosphere, Weather and Climate* (6th edition), London: Routledge.
Harman, J.R. (1967) *Tropospheric Waves, Jet Streams and United States Weather Patterns*, Washington, DC: Association of American Geographers.

ROTENONE

See **insecticides.**

RUNOFF

The **water** that flows across the land surface into rivers and lakes when **precipitation** exceeds **evapotranspiration** and the **soil moisture storage** is full. It can take the form of sheet flow, but ultimately forms and follows rills, gullies or channels. Runoff may be increased by human activities such as agriculture and forestry which reduce the infiltration capacity of the **soil,** and increase the potential for **soil erosion** and flooding. In urban areas the presence of large areas of impervious surface material leads to high runoff rates.

See also
Hydrological cycle.

Further reading
Agassi, M. (ed.) (1996) *Soil Erosion, Conservation and Rehabilitation*, New York: M. Dekker.
National Research Council (1993) *Hydrology, Hydraulics and Water Quality*, Washington, DC: National Academy Press.

S

SAARBRUCKEN INTERNATIONAL CONFERENCE (1990)

A conference on the role of **energy** in **climate** and development. It recognized that the problems involving energy and **climate change** were global in nature and therefore needed global solutions. These solutions, however, would have to take into account the different socioeconomic conditions in different countries and be modified accordingly. A major concern involved dealing with the natural aspirations of developing nations. A blanket curb on **carbon dioxide** (CO_2) emissions, for example, would be particularly damaging to these nations, and to arrive at a global average for emission reductions, it might be necessary to introduce above-average restrictions on developed nations, while developing nations would be allowed more leeway, but with progress tied to greater **energy efficiency**. The question of assessment of environmental costs was also discussed. Participants in the conference agreed that the prices of the different forms of energy should reflect the environmental costs which their use incurred, and suggested that taxes should be rearranged to encourage the use of environmentally friendly energy systems.

SAFE DRINKING-WATER ACT (SDWA)

A US federal government act passed in 1974 and modified in 1977, which set national drinking-water standards and established maximum contaminant levels (MCL) for pollutants in water likely to be used for human consumption. The act also introduced regulations which controlled the underground disposal of **wastes** that might contaminate **groundwater** and endanger drinking-water supplies.

See also
Water quality, Water quality standards.

Further reading
Gilbert, C.G. and Calabrese, E.J. (eds) (1992) *Regulating Drinking Water Quality*, Boca Raton, FL: Lewis.
Kaufman, D.G. and Franz, C.M. (1993) *Biosphere 2000: Protecting our Global Environment*, New York: HarperCollins.

SAHEL

A semi-arid to arid area, subject to seasonal and long-term **drought**, in West Africa south of the Sahara Desert. Named from the Arabic

Figure S-1 The location of the Sahel

word for 'border', since it borders the **desert**, the Sahel proper consists of six nations – Senegal, Mauretania Mali, Burkina Faso, Niger and Chad – but the name has come to include adjacent nations that suffer from problems of drought, **famine** and **desertification** that are characteristic of the Sahel. It is an area of **seasonal drought** in which **pastoral farming** depends upon **precipitation** from the **Intertropical Convergence Zone** for the growth of grass and other **forage**. When the rains are late or less than expected, little vegetation grows and there is insufficient food for the animals. During longer periods of drought, such as in the 1960s and 1970s, the animals die and famine follows. Longer periods of drought also encourage desertification through the southward spread of desert-like conditions from the adjacent Sahara Desert.

Further reading
Bryson, R.A. and Murray, T.J. (1977) *Climates of Hunger*, Madison, WI: University of Wisconsin Press.
Glantz, M.H. (ed.) (1976) *The Politics of Natural Disaster: The Case of the Sahel Drought*, New York: Praeger.
Gritzner, J.A. (1988) *The West African Sahel: Human Agency and Environmental Change*, Chicago: University of Chicago; Committee on Geographical Studies.
Hulme, M. (1989) 'Is environmental degradation causing drought in the Sahel? An assessment from recent empirical research', *Geography* 74: 38–46.

SALINITY

A measure of the proportion of dissolved **salts** and other **solids** present in **water**, usually expressed in parts per million (ppm). **Sodium chloride** (NaCl) is the main constituent of saline water, but salts of other **metals** such as **potassium** (K) and **magnesium** (Mg) are also common. Sea water typically contains in excess of 30,000 ppm of dissolved solids, but in some areas, such as the Persian Gulf, for example, the salinity may exceed 40,000 ppm, whereas in the Baltic the salinity is less than 7000 ppm. The highest salinity readings come from the Dead Sea which contains 240,000 ppm of dissolved solids. Underground water pumped up from deep wells is also often salty. To be potable it must contain less than 500 ppm and for **irrigation** less than 1000 ppm of total dissolved solids.

Further reading
Strahler, A.H. and Strahler, A.N. (1992) *Modern Physical Geography* (4th edition), New York: Wiley.

SALINIZATION

The build-up of **salts** in **soil** as a result of the capillary flow of saline **water** towards the surface. Salinization is a common problem in areas where agriculture requires **irrigation**. There, the natural process is exacerbated by the **evaporation** of irrigation water that not only adds salts directly to the soil, but also encourages sub-surface water to be drawn from deeper levels to the surface where it is evaporated. At best, it can lead to a reduction in crop yields; at worst, it makes the land sterile and unsuitable for agriculture. Most of the world's irrigated lands are subject to some degree of salinization, and as much as 30 per cent may have reached the stage of **desertification**. Salinization also occurs where a change in vegetation cover reduces the level of **evapotranspiration** and allows an increase in the **groundwater** recharge rate. The **water table** rises towards the surface, making it easier for capillary action to draw water up through the soil to be evaporated and deposit salt. A combination of good drainage and sufficient water to leach the salts back down through the soil may prevent or even reverse the process, but it is costly and the drainage of salt-laden water may damage the **environment** in adjacent areas. In the United States some 300,000 hectares suffer salinization, reducing crop productivity by as much as 25– 30 per cent. In many developing nations such as Pakistan, Iran, Iraq and Mexico it is a growing concern and in India alone 20 million hectares experience some degree of salinization. Salinization is a problem of long standing in some areas however, having caused problems for the irrigation-based **agrarian civilizations** of southern Mesopotamia some 4000 years ago.

See also
Leaching.

Further reading

Chiras, D. (1994) *Environmental Science: Action for a Sustainable Future* (4th edition), Redwood City, CA: Benjamin/Cumming.
Jacobsen, T. and Adams, R.M. (1971) 'Salt and silt in ancient Mesopotamian agriculture', in T.R. Detwyler (ed.) *Man's Impact on Environment*, New York: McGraw-Hill.

SALT

A **compound** formed when the **hydrogen** (H) in an **acid** is replaced by a **metal**. When an acid reacts with a **base**, for example, a salt is formed and **water** released.

$$NaOH + HCl \rightarrow NaCl + H_2O$$

| sodium hydroxide (base) | hydrochloric acid | sodium chloride (salt) | water |

Salts are named after the acid and metal from which they are formed – sodium chloride (NaCl) from **sodium** (Na) and hydrochloric acid (HCl), for example, and calcium sulphate ($CaSO_4$) from **calcium** (Ca) and sulphuric acid (H_2SO_4). Sodium chloride is known as common salt.

SALT MARSH

A wetland area found in the intertidal zone of low-lying coasts, mainly in temperate regions. Salt marshes develop close to the maximum tidal limits on mud-flats and are flooded only intermittently. They support a range of salt-tolerant or halophytic plants, which contribute to the aggradation of the coast by trapping and colonizing sediment brought into the marsh during tidal flooding.

See also
Wetlands.

SAND

Unconsolidated sediment consisting of mineral granules ranging between about 60 μm and 2 mm in diameter. Particles of **silica** or **quartz** (SiO_2) are common components of sand.

SANITARY LANDFILL

A North American term for the disposal of domestic refuse or **garbage** by controlled tipping or dumping. Sanitary landfill sites are operated in such a way that they reduce or remove the problems associated with uncontrolled garbage disposal in rubbish dumps or waste tips. They are designed to minimize the problems of litter and **odours** that constitute a public nuisance at uncontrolled sites, and to deal with the potential public health threats from rats, birds and flies which are attracted to dumps. In North America, visits by larger animals such as bears are also reduced by controlled dumping. If properly managed, sanitary landfill sites also reduce the indiscriminate dumping of **hazardous wastes**.

The form of individual sanitary landfill sites varies according to such factors as topography, geology, **groundwater hydrology**, land availability and the volume of waste requiring disposal, but most fit into two types – the trench system or the area system. In some sites, the ramp system, which is a combination of the other two, is used. All are operated using similar techniques, however. Incoming garbage is spread and compacted in a selected working area and at the end of the day's operation the waste is covered with a layer of compacted **soil** some 15 cm thick. When the available space at the site has been completely filled, a layer of soil between 50 and 60 cm thick is spread over the surface, vegetation is reintroduced and landscaping is carried out to complete the rehabilitation. The rehabilitated land is most commonly used as open space parkland or for a variety of recreational uses, but in some places agricultural uses such as grazing are allowed.

Despite the many advantages of the sanitary landfill approach to **waste disposal**, it is not without its problems. Burying the

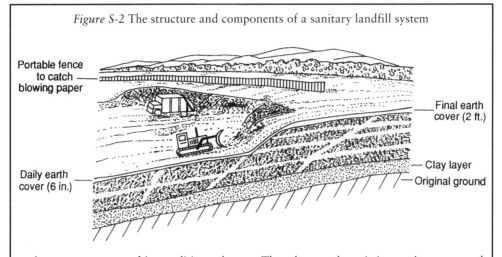

Figure S-2 The structure and components of a sanitary landfill system

Portable fence to catch blowing paper

Final earth cover (2 ft.)

Daily earth cover (6 in.)

Clay layer

Original ground

garbage creates **anaerobic** conditions that lead to the formation of **methane** (CH_4) **gas** during the decomposition process. In the past, seepage of methane has caused explosions in buildings sited on former landfill sites. As a result, building on such locations is seldom permitted, and some form of ventilation system is required at most sites to allow the gas to dissipate. In large landfill sites, the amount of methane produced may be sufficient to make its collection and use as a **fuel** worthwhile. The drainage of **liquids – leachate –** from landfill sites has the potential to contaminate the local surface **water** and **groundwater** supply. This is normally dealt with by appropriate planning that takes into account local drainage patterns, but in some cases it may be necessary to seal the base of the landfill site, using **clay**, for example, to prevent the drainage of leachate into the groundwater system. The leachate is pumped to storage for subsequent safe disposal. Despite their many advantages over traditional waste disposal techniques, sanitary landfill sites are not completely environmentally friendly.

They destroy the existing **environment** and may take agricultural land out of commission for an extended period. They cause **noise**, and increased traffic flow puts pressure on the local road network.

Perceived as reducing adjacent land values, and often identified with old-style open tips, sanitary landfill sites are frequent targets for public opposition, particularly in the planning stages when they are subject to the effects of the **NIMBY** syndrome. As a result of strong political and environmental opposition to the creation of new landfill facilities, many jurisdictions are beginning to reconsider their waste disposal priorities. **Recycling**, for example, can remove large amounts of paper and **plastic** from domestic and industrial garbage, reducing the volume of waste being placed in landfill and allowing the existing sites to remain in operation for a longer period of time.

Further reading
Bagchi, A. (1994) *Design, Construction and Monitoring of a Sanitary Landfill* (2nd edition), New York: Wiley.
Environment Canada (1983) *Stress on Land*, Ottawa: Canadian Government Publishing Centre.

SANTA BARBARA OIL SPILL

A spill which originated in leakage from a drilling operation off the coast of California near Santa Barbara in January 1969. The **oil** continued to leak from the site for a further

eight months and in total introduced some 10,000 tonnes of crude oil into the ocean. It washed ashore to pollute the Santa Barbara beaches, kill some 3500 sea birds and seriously damage other marine life. Although a relatively small spill compared to many **oil** tanker accidents, the Santa Barbara spill was important because it was the first serious **oil spill** associated with offshore **petroleum** development in the United States. The questions raised by the spill led to an examination of the policies and procedures adopted by government and industry for the exploitation of offshore petroleum **resources**, and contributed to improved drilling and clean-up guidelines.

See also
Oil pollution, Oil tanker accidents.

Further reading
Steinhart, C.E. and Steinhart, J.S. (1972) *Blowout: A Case Study of the Santa Barbara Oil Spill*, North Scituate, MA: Duxbury Press.

SATURATED ADIABATIC LAPSE RATE (SALR)

See **adiabatic process**.

SATURATION

A state in which a **solution** contains the maximum amount of **solute** that can be dissolved and remain in solution at a given **temperature**. The term also applies to atmos-pheric conditions. A saturated **air mass** is one which has absorbed as much **water vapour** as it can at a specific temperature. Similarly, saturation in **soils** is reached when all the available pore spaces have been filled with **water**.

See also
Absolute humidity.

SAVANNA

A major **biome** of the semi-arid tropics, in which annual grasses predominate, but with a scattering of trees. The grasses grow rapidly during the wet season and die off in the dry season, whereas the trees have developed **drought**-resistant characteristics which allow them to survive the **seasonal drought**. **Fire** is common in the dry season and that tends to favour the survival of the grass over the trees. Towards their poleward boundaries, the savannas give way to the **desert** biome and equatorwards, they ultimately merge with the tropical forest. The greatest extent of the savanna is immediately south of the Sahara Desert in Africa, but it is also present in parts of South America and Australia. The animal population of the savanna is dominated by large **herbivores** – wildebeest, antelope,

Figure S-3 The global distribution of savanna grasslands

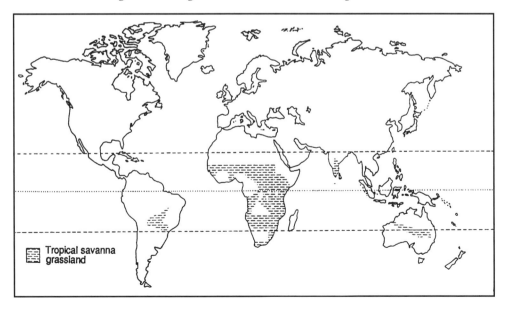

Tropical savanna grassland

kangaroo – and the predators – lions, hyenas, dingoes – that prey on them.

See also
Grasslands.

Further reading
Bourlière, F. (ed.) (1983) *Tropical Savannas*, Amsterdam: Elsevier.
Huntley. B.J. and Walker, B.H. (eds) (1982) *Ecology of Tropical Savannas: Ecological Studies 42*, Berlin: Springer-Verlag.

SCATTERING

The deflection of **radiation** by small particles suspended in its path. Although it applies to all types of radiation it is particularly evident in the case of **visible light**. The blue colour of the sky, for example, is caused by the scattering of white light by the **gas molecules** in the **atmosphere,** and red skies are caused by the scattering of blue light by small **aerosols** suspended in the **air.**

SCHISTOSOMIASIS

A tropical or subtropical intestinal disease of humans caused by parasitic flatworms or flukes. Using snails·as intermediate hosts, the flukes are spread from person to person through polluted **water** and insanitary living habits. Deteriorating health and reduced resistance usually cause those infected to die of secondary diseases rather than the schistosomiasis itself. Schistosomiasis is prevalent in Africa, South America and tropical Asia, where as many as 200 million people are infected.

See also
Parasite.

Further reading
Hickman, C.P., Roberts, L.S. and Larson, A. (1996) *Integrated Principles of Zoology*, Dubuque, IA: Wm C. Brown.
Spencer, H. (1976) *Tropical Pathology*, New York: Springer-Verlag.

SCIENTIFIC COMMITTEE ON PROBLEMS OF THE ENVIRONMENT (SCOPE)

A committee created in 1969 by **ICSU** to conduct interdisciplinary analyses of environmental problems, particularly those that are global in scale. SCOPE does not support research directly, but works to ensure that the knowledge supplied by current environmental research is appropriately evaluated. It has investigated and published reports on a wide range of issues including **biogeochemical cycles,** the environmental consequences of nuclear war and **climate change.**

Further reading
Bolin, B., Doos, B.R., Jager, J. and Warrick, R.A. (eds) (1986) *The Greenhouse Effect, Climate Change and Ecosystems*, SCOPE 29, New York: Wiley.
SCOPE-ENUWAR (1987) 'Environmental consequences of nuclear war: an update', *Environment* 29: 4–5 and 45.

SCREE

Talus. Loose, unconsolidated angular rock fragments, commonly produced when mechanical **weathering** such as frost shattering causes rock to break off a steep slope or cliff and accumulate at its base.

See also
Mass movement.

SCRUBBERS

Structures used to reduce acid **gas** emissions from industrial plants, with the ultimate aim of preventing **acid rain.** Wet or dry techniques can be used, but all involve bringing the gases in contact with alkaline or basic substances which neutralize their acidity. In dry scrubbers, for example, acidity is reduced by passing the gases through crushed **limestone,** whereas in wet scrubbers the neutralization takes place when the gases come in contact with lime-rich **solutions.** Scrubbers have an advantage over other techniques of acid reduction in that they can be employed in existing plants by **retrofitting.** Efficient wet scrubbers can remove up to 95 per cent of the acidity from emissions (see Figure F-4).

Further reading
Park, C.C. (1987) *Acid Rain: Rhetoric and Reality*, London: Methuen.

Ridley, M. (1993) 'Cleaning up with cheap technology', *New Scientist* 137 (1857): 26–7.

SEA ICE MODELS

Models which attempt to simulate the role of sea ice in global **climates**. The simplest models represent sea ice as a motionless uniform layer, but more sophisticated models incorporate **snow** on the **ice**, fractional ice coverage, multilayer ice, the effects of **salinity** and sea ice dynamics. They may be incorporated in **ocean models** or coupled directly to **general circulation models**.

Further reading
IPCC (1996) *Climate Change 1995: The Science of Climate Change*, Cambridge: Cambridge University Press.

SEA LEVEL

The level that a calm sea, unaffected by **tides** or waves would assume. Mean sea level (MSL) at any coastal location is taken as the average of high and low tide levels. MSL is used as a datum from which altitudes on land and beneath the ocean are measured.

Further reading
Woodroffe, C. (1994) 'Sea level', *Progress in Physical Geography* 18: 436–51.

SEA-LEVEL CHANGE

Coastlines around the world provide abundant evidence that sea level has not always occupied its present position. Raised beaches several metres above the current sea level and kilometres from the present coastline indicate higher sea levels in the past, while river valleys that continue out on to the continental shelf beneath the ocean provide evidence that past sea levels were also lower. Rising and falling sea levels are brought about globally by eustatic change, and locally or regionally by isostatic and tectonic change. Any sea-level change is typically the result of a combination of these factors. Tectonic changes such as earth movements or volcanic activity, for example, may change the shape of the ocean basins and contribute to eustatic

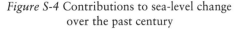

Figure S-4 Contributions to sea-level change over the past century

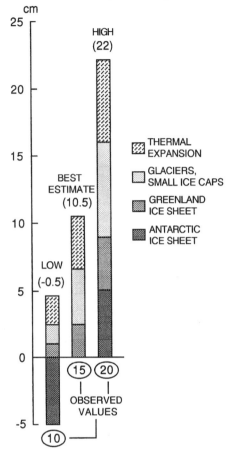

Source: Derived from data in IPCC (1990) *Climate Change: The IPCC Scientific Assessment*, Cambridge: Cambridge University Press

change. Towards the end of the last **ice age** some 10,000 years ago, melting **ice** returned **water** to the **oceans** causing a eustatic rise in sea level. At the same time, the removal of the ice from the land allowed isostatic uplift which altered the relative positions of land and sea. Thus, the net sea-level change was brought about by a combination of eustatic and isostatic factors. Currently there is considerable interest in the potential impact of **global warming** on sea level. Change would result from the thermal expansion of the ocean waters and the melting of **glaciers** and ice sheets as global temperatures rose. The potential for change has been examined by the **IPCC** and it has reported that there is evidence that sea level may already have risen by 10–25 cm in the past century and that the trend is likely to continue. According to its projections, sea level may rise by 38–55 cm by the year 2100, with most of the increase attributable to the melting of **snow** and ice. Should such an increase come to pass, it would create major environmental, political and economic problems for all coastal nations.

See also
Eustasy, Flood, Isostasy.

Further reading
Goudie, A.S. (1992) *Environmental Change*, Oxford: Oxford University Press.
IPCC (1996) *Climate Change 1995: The Science of Climate Change*, Cambridge: Cambridge University Press.
Sahagian, D.L., Schwartz, F.W. and Jacobs, D.K. (1994) 'Direct anthropogenic contributions to sea level rise in the twentieth century', *Nature* 367: 54–7.
Tooley, M.J. and Shennan, I. (eds) (1987) *Sea Level Changes*, Oxford/New York: Blackwell.

SEA-SURFACE TEMPERATURES (SSTS)

A source of information on potential change in the earth/atmosphere **system**. Anomalous warming or cooling of the sea surface, for example, may be followed some time later by changing **pressure, wind** and **precipitation** patterns. Once the initial change has been identified, such time-lags allow subsequent events to be predicted. Sea-surface **temperature** anomalies in the Atlantic Ocean during the spring, for example, have allowed the forecasting of summer rainfall patterns in the **Sahel**. SSTs also play a role in the development of **tropical cyclones**, which require **water** temperatures to be at least 27°C before they form. There is some concern that, with future **global warming**, this threshold will be exceeded more frequently, thus increasing the number of storms.

See also
El Niño.

Further reading
Folland, C.K., Palmer. T.N. and Parker, D.E.

(1986) 'Sahel rainfall and worldwide sea temperatures 1901–85; observational, modelling and simulation studies', *Nature* 320: 602–7.
Owen, J.A. and Ward, M.N. (1989) 'Forecasting Sahel rainfall', *Weather* 44: 57–64.

SEASONAL DROUGHT

One of the forms of **drought** identified by **C.W. Thornthwaite**. Areas experiencing seasonal drought have distinct dry and wet seasons during the year, usually associated with changing **air mass** distribution. Seasonal drought is common in the tropical and subtropical **grasslands** of Africa, India and Australia where dry conditions are brought on by the arrival of **continental tropical air** as the **Intertropical convergence zone (ITCZ)** advances equatorwards. The rainy season arrives with the **maritime tropical air mass** as the **ITCZ** returns. If the maritime air fails to arrive, seasonal drought may extend into the normal wet season, disrupting human activities and causing potential hardship to the inhabitants of these areas.

See also
Sahel, Savanna.

Further reading
Thornthwaite, C.W. (1947) 'Climate and moisture conservation', *Annals of the Association of American Geographers* 37: 87–100.

SECOND WORLD CLIMATE CONFERENCE

A conference sponsored by the **WMO, UNEP** and **UNESCO**, held in Geneva, Switzerland in 1990. Deliberations at the scientific and technical sessions produced the consensus that, since **climate** issues extended beyond the physical sciences into social and economic systems, increased attention should be given to improving knowledge of climate and human interactions. In addition, although they allowed that many uncertainties remained, the participants at the conference suggested that regional and national programmes be initiated to reduce sources and increase sinks of **greenhouse gases** to slow predicted **global warming**. The Ministerial Declaration issued at the end of the conference reaffirmed these suggestions and urged the developed nations to analyse the available options with a view to developing programmes, strategies and targets for the reduction of greenhouse gas emissions, for presentation at the **UN Conference on Environment and Development** (UNCED) at Rio in 1992.

See also
First World Climate Conference.

SECONDARY AEROSOLS

Aerosols formed as a result of chemical and physical processes in the **atmosphere**. They include aggregates of gaseous **molecules**, water droplets and chemical products such as **sulphates** and **hydrocarbons**. They are concentrated in the size range of 0.01–1 μm and may make up as much as 64 per cent of total global **aerosols**. Eight per cent are of anthropogenic origin, from **combustion** systems, vehicle emissions and industrial processes, while the remaining 56 per cent are from natural sources such as **volcanoes**, the **oceans** and a wide range of organic processes.

See also
Primary aerosols.

Further reading
Fennelly, P.F. (1981) 'The origin and influence of airborne particles', in B.J. Skinner (ed.) *Climates, Past and Present*, Los Altos, CA: Kauffmann.

SECURE LANDFILL

A development of the **sanitary landfill** concept for the safe disposal of **hazardous waste**. To prevent the **waste** from escaping into the **environment** through leakage, secure landfill sites are usually sealed with a heavy **plastic** liner and the products stored in containers. This segregation is necessary to prevent the mixing of wastes should individual containers leak. When full, the site is sealed with a plastic and **clay** cap to prevent the **percolation** of **precipitation**. The contents of secure landfill sites must be recorded and the sites monitored for **leachate** and **gas** production for an extended period of time to ensure that the integrity of the system is maintained.

SEDIMENT YIELD

The total amount of particulate matter leaving a drainage basin in the form of bed load or suspended load. It reflects the combined effects of a number of physical conditions and activities in the drainage

Figure S-5 The requirements of a secure landfill site

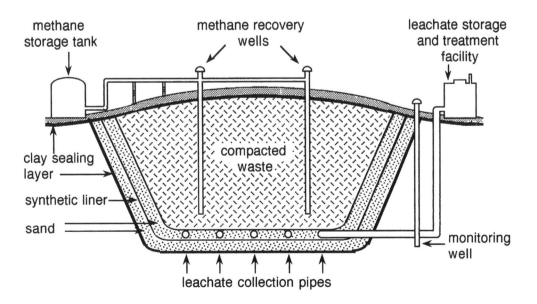

basin. Such factors as **climate**, topography, rock type, vegetation cover and land use control **erosion**, and although not all of the eroded material is carried out of the basin, any changes in these factors are usually followed by changes in sediment yield. In areas where vegetation is sparse and **precipitation** variable, sediment yields will increase in wetter years and decline in drier years. Human activities such as logging and **arable agriculture**, which encourage increased **soil erosion** also contribute to higher sediment yields.

Further reading
Ritter, D.F., Kochel, C.R. and Miller, J.R. (1995) *Process Geomorphology* (3rd edition), Dubuque, IA: Wm C. Brown.

SEDIMENTARY ROCK

Rock produced by the consolidation of sediments deposited by **wind, water** or **ice**. Sedimentary rocks may be created from fragments or particles eroded from older rocks (for example, sandstone and shale), from organic material (for example, **coal**) or from chemical precipitates (for example, gypsum and certain **limestones**). Typically, sedimentary rocks are deposited in distinct layers or strata, and many contain **fossils**.

Further reading
Carozzi, A.V. (1993) *Sedimentary Petrology*, Englewood Cliffs, NJ: Prentice-Hall.

SEDIMENTATION

The settling out of **solid** particles from the **liquid** in which they are suspended, usually under the effects of **gravity**. Sediments carried into a lake by a river, for example, gradually fall to the lake bottom, when the low **energy environment** of the lake no longer allows them to remain in suspension. A similar process takes place in **sewage** settling ponds. In some chemical and industrial processes, sedimentation is encouraged by using a centrifuge, in which **centrifugal force** causes suspended particles to separate from a suspension.

SELECTIVE CATALYTIC REDUCTION

An efficient but costly process developed to reduce the emission of **oxides of nitrogen** (NO_x) from power plants. With the help of a **catalyst**, the oxides of nitrogen are broken down into harmless **nitrogen** (N) and **oxygen** (O). The process is efficient – removing up to 80 per cent of the oxides of nitrogen released – but it is costly to install and maintain.

Further reading
Ellis, E.C., Erbes, R.E. and Grott, J.K. (1990) 'Abatement of atmospheric emissions in North America: progress to date and promise for the future', in S.E. Lindberg, A.L. Page and S.A. Norton (eds) *Acidic Precipitation, Volume 3, Sources, Deposition and Canopy Interactions*, New York: Springer-Verlag.

SEMI-PERMEABLE MEMBRANE

See **osmosis**.

SENSIBLE HEAT

Heat which can be felt or sensed. The sensible heat of a substance or object represents the heat content of the substance per unit mass. The transfer of sensible heat in **air masses** makes an important contribution to the redistribution of heat from the tropics to the poles.

See also
Latent heat.

SEPARATE SEWER SYSTEM

A **waste water** disposal system in which domestic and industrial **sewage** is kept separate from urban **runoff**, usually by installing two sets of drainage pipes. Since the volume of runoff can increase rapidly, for example, during storms, separation ensures that the capacity of the **sewage treatment** system is not exceeded. In the combined or only partially separate systems, common in the older sections of many cities, major rain storms produce such a volume of runoff that the systems cannot cope, and areas are flooded with a mixture of sewage and rainwater. At such times, to minimize the flooding and to prevent damage to the treatment plant, it may also be necessary to release untreated sewage into rivers or lakes.

Further reading
Tchobanoglous, G. and Burton, F.L. (1991) *Wastewater Engineering, Treatment, Disposal and Reuse* (3rd edition), New York: McGraw-Hill.

SEPTIC SYSTEM

A **sewage** disposal system consisting of a collection tank and a drain field. Septic systems are common in rural areas that do not have access to municipal sewage systems. Sewage and **waste water** are discharged into an underground concrete or fibreglass tank in which **sedimentation** and bacterial **digestion** of the waste takes place. The liquid waste then drains into a perforated pipe system that gradually releases it into crushed rock and gravel, where it is filtered before being absorbed by the **soil**. A well-maintained septic system is an effective way of dealing with limited amounts of sewage. If not properly maintained, however, it can lead to

contamination of soil and the local **groundwater** system.

SERE

See **succession**.

SEWAGE

Liquid or semi-solid **waste** from domestic or industrial sources. It is predominantly organic, including human waste and food processing residue, for example, but it also contains **detergents** and other cleansers, plus a variety of industrial chemicals, sometimes including **heavy metals**. Because of the toxicity of many industrial wastes, their release into municipal sewage systems is now very much restricted. Sewage is a ubiquitous pollutant, which continues to be released untreated into many of the world's waterways, disrupting the **aquatic environment** through its impact on the **biochemical oxygen demand**, causing health problems and constraining human use of rivers and lakes. Even in developed nations, where environmental regulations are

Figure S-6 The components of a modern septic system

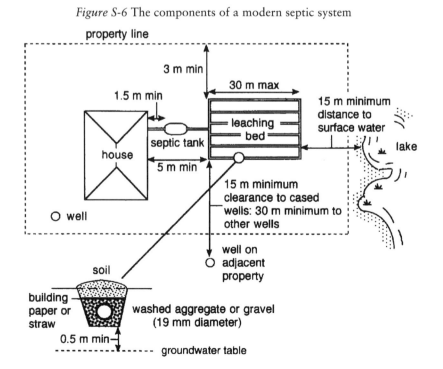

in place, millions of litres of raw or partially treated sewage are released into waterways every year.

See also
Oxygen sag curve, Sewage treatment.

Further reading
Harrison, R.M. and Hester, R.E. (1995) *Waste Treatment and Disposal*, Cambridge: Royal Society of Chemistry.
Miller, G.T. (1994) *Living in the Environment*, Belmont, CA: Wadsworth.

SEWAGE TREATMENT

In the past, **sewage** was released into waterways without the benefit of any treatment. While quantities remained small, the **environment** was able to purify the water through **oxidation** and the activities of a variety of micro-organisms which broke down the pollutants into less noxious by-products. As **populations** grew and lifestyles and technology changed, the volume of sewage increased and the natural approach was no longer feasible. The direct treatment of sewage became necessary, although it was not always available where and when most needed. Modern sewage treatment systems can process waste water so well that the effluent coming out of the

treatment plant can be released into rivers or lakes and have no chemical or biological impact on the **aquatic environment**.

Sewage treatment is normally classified as primary, secondary or tertiary. Primary treatment involves little more than the removal of **solids**. Sticks, rags and metal objects are screened out of the effluent, grease and **oil** are skimmed off the top and suspended sediments are allowed to settle. Undissolved organic material collects in settling ponds as sludge. The effluent remaining after this purely mechanical treatment is commonly released directly into rivers, lakes or the sea, although it is sometimes chlorinated prior to release.

Figure S-7 The elements of a primary, secondary and tertiary sewage treatment system

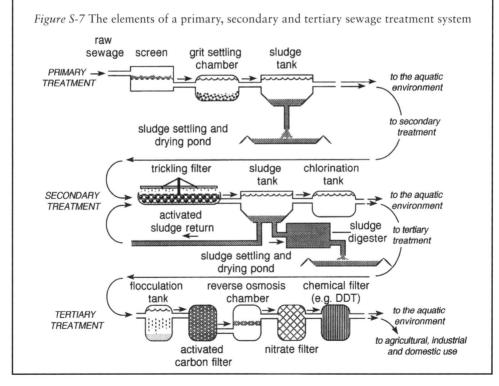

Primary treatment does not remove **nutrients**, dissolved organic material, **bacteria** or potentially toxic chemicals. They are released to become a burden on the environment. In plants providing secondary treatment, the primary effluent is subjected to biological purification. Two methods are used – trickling filters or **activated sewage sludge**. In the former, the primary effluent is allowed to filter slowly through a thick bed of rocks, during which bacteria consume the dissolved organic matter. In plants using the activated sludge method, high **oxygen** (O) levels maintained in the effluent by aeration encourage the rapid digestion of the sewage by bacteria. In both cases, subsequent **sedimentation** allows more of the remaining suspended sediments to settle out. Following secondary treatment, the effluent may have regained 85 per cent of its original quality, and direct discharge into the environment would normally cause few problems. In some cases, the secondary effluent may still retain suspended solids, or include dissolved chemicals of various forms. Tertiary treatment using **activated carbon** filters, reverse **osmosis** systems and chemical **coagulation** techniques can deal with most of the remaining contaminants and provide **water** that is up to 99 per cent pure.

Further reading
Horan, N.J. (1990) *Biological Wastewater Treatment Systems: Theory and Operation*, New York: Wiley.
Reed, S.C., Crites, R.W. and Middlebrooks, E.J. (1995) *Natural Systems for Waste Management and Treatment* (2nd edition), New York: McGraw-Hill.

SHIFTING CULTIVATION

A form of cultivation common in primitive agricultural societies in the tropics and subtropics, also referred to as 'bush fallow' or 'slash-and-burn agriculture'. Forest and **grassland** is cleared, usually by **fire**, which provides ash to fertilize the land to be cultivated. After several years of cultivation, the declining fertility of the **soil** leads to a significant reduction in crop yields and the **community** moves to a new location to start again. The abandoned land is allowed to revert to nature, returning ultimately to its original state. The environmental impact of shifting cultivation is limited if **populations** remain small and the area over which they can migrate is large. As populations increase or movement is restricted, once abandoned land cannot recover before it is needed again, soil fertility continues to decline and the land loses its ability to support the population.

Further reading
Goudie, A. (1984) *The Human Impact*, Oxford: Blackwell.
Park, C.C. (1992) *Tropical Rainforests*, London: Routledge.

SHIFTING SANDS

A popular image of **desertification** in which **desert** sand dunes migrate into an area, covering arable land and pasture and sometimes settlements, thus creating a desert. Although this does happen, it is much less common than once thought. In the 1960s and 1970s, for example, it was claimed that the Sahara Desert was advancing southwards at a rate of as much as 48 km per year, along a 300 km front. Such figures are no longer generally accepted. See Figure S-8 overleaf.

Further reading
Nelson, R. (1990) *Dryland management: the 'desertification' problem*, World Bank Technical Paper No. 16, Washington, DC: World Bank.
Pearce, F. (1992) 'Miracle of the shifting sands', *New Scientist* 134 (1817): 4.

SHORT-WAVE RADIATION

Radiation from the high **energy** end of the electromagnetic **spectrum** with wavelengths less than 5 μm. The term is commonly applied to **solar** **radiation** which consists of **ultraviolet** and **visible light** rays, both of which have shorter wavelengths than the **terrestrial radiation** emitted by the earth.

Figure S-8 Migrating sand dunes overlooking a desert highway

Photograph: Courtesy of Susan and Glenn Burton

SI UNITS

Table S-1 Basic and derived SI units

BASIC UNITS

metre	–	length
kilogram	–	mass
second	–	time
ampère	–	electric current
kelvin	–	temperature
mole	–	amount of substance
candela	–	luminous intensiity

DERIVED UNITS (examples)

newton	–	force
joule	–	work or energy
watt	–	power
volt	–	electric potential
becquerel	–	radioactivity
sievert	–	dose equivalent (ionizing radiation)

Système International d'Unités. An internationally accepted system of measurement units based on the m.k.s. (metres.kilo-grams.seconds) system. It is now used for all scientific purposes having replaced the c.g.s. (centimetres.grams.seconds) and f.p.s. (feet. pounds. seconds) systems.

SICK BUILDING SYNDROME

A condition brought on by poor indoor **air quality. Dust, gases,** paint and **solvent** fumes, **metal** particles and a variety of other contaminants have always been present in industrial buildings, and occupational health standards have been introduced to deal with them. More recently, concern over indoor air quality has shifted to office and residential buildings where contaminants can include **volatile** chemicals from photocopiers, paint, stain, glue and cleansing agents, **odours** from cooking and cleaning, **dust** particles, **ozone** (O_3) from laser printers and copiers and possibly **bacteria** and **viruses**. In the past, such pollutants escaped to the outside through open windows, poorly fitting doors and cracks in the fabric of the buildings. In their escape, they were accompanied by large amounts of heat **energy**. With the energy crisis in the mid-1970s, energy and cost savings were accomplished by sealing cracks and weatherstripping doors and windows, which not only kept the heat in, but also prevented the contaminants from escaping. The net result was a deterioration in air quality and the creation of the sick building syndrome. Symptoms included headaches, watery eyes, sore throats, asthma attacks and general malaise. Because contaminant levels in such a situation are frequently too low to be measured accurately, and the symptoms often non-specific, confrontation between employers and employees is not uncommon. Improved ventilation, regular maintenance of air conditioning systems and better control over the release of contaminants can reduce the incidence of sick building syndrome.

Further reading
Baarschers, W.H. (1996) *Eco-facts and Eco-fiction*, London/New York: Routledge.
Baker, H. (1997) 'Chemical warfare at work', *New Scientist* 154 (2087): 30–5.

SIERRA CLUB

A non-profit environmental organization founded in California in 1892. Its 182 charter members, led by its president **John Muir**, were committed to preserving the North American **wilderness**, which even at that time was coming under increasing threat from development. One of the club's first activities was to participate in the preservation of the Yosemite area of the Californian Sierra Nevada, and since then it has been involved in the founding, preservation and expansion of parks and wilderness reserves throughout the western United States, from Arizona to Alaska. It led the fight for the **Wilderness Act**, and through direct political action – Sierra Club members stood for and were elected to seats in Congress – law suits and public education also contributed to the passing of such environmental legislation as the **Clean Air Act** and **Clean Water Act** and the **Superfund**. More recently, it has raised concerns about the US contribution to **global warming**. Most of the activities of the Sierra Club are directed at environmental problems in the US, but it also has an active Canadian chapter, and has worked through organizations such as the **World Bank** to confront problems in the Brazilian **rainforest**.

See also
Conservation.

SIEVERT

The **SI unit** for the dose equivalent of **ionizing radiation**. One sievert represents a dosage of 1 **joule** per kilogram of tissue, absorbed from the ionizing radiation. The sievert has replaced the **rem** as the unit of dose equivalent.

$$1 \text{ rem} = 10^{-2} \text{ sievert (Sv)}$$

SIGNAL-TO-NOISE RATIO

A measure of the efficiency of communication, in which the signal can be considered as the message being transmitted and the noise as any disturbance that obscures that message. It can be applied directly to electronic communication, but it is also applicable to environmental phenomena. In the study of **global warming**, for example, the long-term temperature trend (signal) may be obscured by short-term temperature variability (noise).

SILENT SPRING

A best-selling book by Rachel Carson, an American writer and naturalist, who was among the first to draw attention to the impact of chemicals on the **environment**. Its title, *Silent Spring*, refers to the silence that would fall over the land as birds succumbed to the chemical poisons released by the growing and often indiscriminate use of **pesticides**, **herbicides** and **fertilizers**. DDT, which up to that time had been viewed as almost a miracle pesticide, was identified as one of the main culprits. When it was published in 1962, the book was denounced by the chemical industry as alarmist, and many biologists treated it with some scepticism, but its concerns proved to be justified. It gave the environmental movement a major boost, and inspired an increasing amount of research over the next two decades into the problem of environmental **pollution** by chemicals.

Further reading
Carson, R. (1962) *Silent Spring*, New York: Houghton Mifflin.

Cox, G.W. (1993) *Conservation Ecology: Biosphere and Biosurvival,* Dubuque, IA: Wm C. Brown.

SILICA

Silicon dioxide (SiO_2), a hard, white or colourless **compound**, common in the rocks of the earth's **crust**.

See also
Quartz.

SILICOSIS

See **pneumonoconiosis.**

SILT

Fine, unconsolidated sediments consisting of particles with diameters in the range 2.0 μm to 60.0 μm, and thus intermediate in size between **sand** and **clay**.

SILVER (Ag)

A white, soft, precious metal. Silver occurs free in nature and as the ores argentite, acanthite and horn silver. Being extremely malleable, ductile and a good conductor of electricity, it is widely used in coinage, jewellery and electronic equipment. In the form of its **compounds** silver bromide (AgBr) and chloride (AgCl), it is used in photography, and silver iodide (AgI) is used to seed **clouds** during **rainmaking**.

Further reading
Bakewell, P. (ed.) (1996) *Mines of Silver and Gold in the Americas*, Brookfield, VT: Variorum.

SIMULATION

The representation of complex phenomena using physical or mathematical **models**. By using appropriate physical and time-scales, simulation can be used to test hypotheses in the laboratory which could not be tested in the field because of time or cost constraints. Computer-based simulation models are currently the most effective means of studying **climate change**, for example. The accuracy of the simulation approach is limited by the inability of models to replicate the complexity of the **environment** exactly.

See also
General circulation models.

SINK

Natural reservoir or store for materials circulating through the earth/atmosphere **system**. The **oceans** are a major natural sink for many substances from **heavy metals** to **carbon** (C). If the storage capacity of a sink is altered, the impacts may be felt throughout the system. For example, growing forests act as a sink for carbon. If they are cut down and not replaced, additional carbon, in the form of **carbon dioxide** (CO_2) remains in circulation to contribute to **global warming**.

Further reading
Moore, B. and Bolin, B. (1986) 'The oceans, carbon dioxide and global climate change', *Oceanus* 29: 9–15.
Van Kooten, G.C., Arthur, L. and Wilson, W.R. (1992) 'Potential to sequester carbon in Canadian forests: some economic considerations', *Canadian Public Policy* 18: 127–38.

SKIN CANCER

A disease indicated by the alteration of skin **cells** and associated with damage to the genetic make-up of the cells. The damage may be caused by a variety of **carcinogens**, but current concern has focused on exposure to **ultraviolet radiation** as the primary cause. Levels of skin **cancer** have been rising since the late 1970s, apparently in parallel with the thinning of the **ozone layer**, and the consequent increase in the amount of ultraviolet radiation reaching the earth's surface (see Figure M-3). Some researchers also consider societal factors, which promote greater exposure of the skin to the sun, to make a major contribution to the problem. Many skin cancers respond to treatment, although those in the **melanoma** group are often fatal.

See also
Ozone depletion.

Further reading
Concar, D. (1992) 'The resistible rise of skin cancer', *New Scientist* 134 (1821): 23–8.
Mackie, R.M. (1993) 'Ultraviolet radiation and the skin', *Radiological Protection Bulletin* 143: 5–9.
Mackie, R., Hunter, J.A.A., Aitchison, T.C., Hole, D., McLaren, K., Rankin, R., Blessing, K., Evans, A.T., Hutcheon, A.W., Jones, D.H., Soutar, D.S., Watson, A.C.H., Cornbleet, M.A. and Smith, J.F. (1992) 'Cutaneous malignant melanoma, Scotland 1979–89', *The Lancet* 339: 971–5.

SLAB MODELS

Interactive ocean-atmosphere circulation models in which the **ocean** is represented by only the uppermost layer or 'slab' of **water**. This is necessary to accommodate the different response times of **atmosphere** and ocean. The ocean deeps may take centuries to react to change, whereas the atmosphere may respond in days. A model incorporating the ocean deeps would take a long time to reach equilibrium, and would therefore be costly to run. Using only the upper layer of the ocean, where response times are closer to those in the atmosphere, is a compromise to save time and money.

See also
Atmosphere climate models, Coupled ocean.

SMELTING

The processing of metallic **ore** to produce the free **metal**, commonly accomplished by heating the ore sufficiently so that the metal becomes molten and can be separated from the unwanted materials which are also present in the ore. During smelting, pollutants such as **sulphur dioxide** (SO_2) and **particulate matter** may be given off and large amounts of **waste** may be left at the end of the process. As the better quality ores are used up, and more ore has to be processed to produce the same amount of metal, the volume of waste increases.

Further reading
Craddock, P.T. (1995) *Early Metal Mining and Production*, Edinburgh: Edinburgh University Press.
Chatterjee, A. (1994) *Beyond the Blast Furnace*, Boca Raton, FL: CRC Press.

SMITH, R.A.

An English chemist credited with first recognizing the link between **air pollution** and the acidity of **precipitation**, by observing **rain** falling on industrial Manchester in the mid-nineteenth century. In 1872, he wrote a book on the subject in which he made the first reference in the literature to '**acid rain**'. He went on to become Britain's first Alkali Inspector responsible for the control of air pollution from industrial sources, but his ideas on acid rain were largely ignored until the mid-twentieth century.

Further reading
Park, C.C. (1987) *Acid Rain: Rhetoric and Reality*, London: Methuen.
Smith, R.A. (1872) *Air and Rain: The Beginnings of a Chemical Climatology*, London: Longmans Green.

SMOG

A combination of **smoke** and **fog** which creates **air pollution**. Originally applied to pollution containing acid smoke from burning **coal**, as in the **London Smog** of 1952, but also used to describe the photochemical pollution of cities such as Los Angeles, where the **hydrocarbons** and **oxides of nitrogen** (NO_x) released into the **atmosphere** from automobile exhausts undergo continuing chemical change made possible by the energy available from abundant **solar radiation**.

See also
Photochemical smog.

Further reading
Brimblecombe, P. (1987) *The Big Smoke: A History of Air Pollution in London since Medieval Times*, London/New York: Methuen.
Patterson, D.J. and Heinen, N.A. (1972) *Emissions from Combustion Engines and their Control*, Ann Arbor, MI: Ann Arbor Science Publishers.

SMOKE

A suspension of fine solid particles (usually < 1 μm in diameter) in the **air**. It is usually produced by the incomplete **combustion** of a **fuel**. The smoke from **coal**, for example, consists of fine particles of **carbon** (C), whereas that from **oil** might include unburned **hydrocarbons**. Smoke can incorporate a variety of solid particles including **silica** (SiO_2), **aluminum** (Al) and **lead** (Pb), acids such as **sulphuric acid** (H_2SO_4) and organic compounds. The presence of smoke in the **atmosphere** can have a significant effect on the earth's **energy budget**, disrupting the flow of both incoming and outgoing **radiation**. Following the Gulf War, smoke from oil fires in Kuwait at one point reduced the incoming **short wave radiation flux** to zero, and led to daytime **temperature** reductions of as much as 5.5°C. The disruption of radiation by smoke also had an important role in the **nuclear winter** hypothesis. The presence of smoke in the atmosphere can lead to health problems when the carbon particles and the other pollutants associated with them – for example, acid droplets – are inhaled and drawn into the lungs.

See also
Atmospheric turbidity, Kuwait oil fires, London Smog, Soot.

Further reading
Johnson, D.W., Kilsby, C.G., McKenna, D.S., Saunders, R.W., Jenkins, G.J., Smith, F.B. and Foot, J.S. (1991) 'Airborne observations of the physical and chemical characteristics of the Kuwait oil smoke plume', *Nature* 353: 617–21.
Shaw, W.S. (1992) 'Smoke at Bahrain during the Kuwaiti oil fires', *Weather* 47: 220–6.

SNOW

Solid **precipitation** in the form of single **ice** crystals or groups of ice crystals which have agglomerated into snowflakes. Snowflakes tend to become larger when the **temperature** is closer to freezing point and when the **air** contains more moisture. When the air is cold and dry, ice crystals are the norm. Snowfall is common in higher latitudes and higher altitudes where temperatures are below freezing for at least part of the year. Snow is difficult to measure accurately since it is prone to blowing and drifting. However, the greatest snow depths and snowfall intensities are recorded in coastal mountainous areas in mid-latitudes which lie across the path of the prevailing onshore wind. Annual snowfalls of

between 750 and 1000 cm are not uncommon at stations in the mountains of western Canada and the United States, Norway, western Japan, the South Island of New Zealand and the southern Andes in South America. A record 2850 cm of snow fell at Mount Rainier, Washington in the United States in the winter of 1971–1972 and at Takada, Japan in 1927, 240 cm of snow fell in three days.

Further reading
Gray, D.M. and Male, D.H. (1981) *Handbook of Snow: Principles, Processes, Management and Use*, Toronto/New York: Pergamon Press.

SODIUM (Na)

A soft, silvery **metal** that belongs to the alkali metal group, along with caesium (Cs), francium (Fr) , lithium (Li), **potassium** (K) and rubidium (Rb). Being highly reactive, sodium is not found free in nature, but it occurs in a wide variety of **compounds**. It reacts so vigorously with **water** that the **hydrogen** (H) released in the process may ignite and cause an explosion. In its elemental form the uses of sodium are limited, but it is used in sodium vapour lamps and as a **coolant** in fast breeder **nuclear reactors**. Its many compounds have a wide range of uses, however, with **sodium chloride** (NaCl) (common salt), **sodium hydroxide** (NaOH) (caustic soda) and sodium bicarbonate ($NaHCO_3$) used extensively in industry and in the home.

See also
Caustic chemicals, Periodic table.

SODIUM CHLORIDE (NaCl)

See **salt**.

SODIUM HYDROXIDE

See **caustic chemicals**.

SOFT WATER

Water that contains only small amounts of dissolved **salts**, particularly **calcium** (Ca) salts. Soft water contains less than 50 ppm of **calcium carbonate** ($CaCO_3$) or its equivalent in other salts. Water containing more than that is considered to be hard. Soaps and **detergents** lather better in soft water, and in certain industries – for example, dyeing, brewing and distilling – soft water is necessary for the production of a quality product. For many residential and industrial uses, where soft water is not available naturally, **hard water** has to be softened, usually through an **ion exchange** process, before it can be used.

See also
Base exchange, Water quality.

SOIL

A mixture of weathered rock particles and organic material on the surface of the land in which plants grow. The composition of soil varies with time, place and use. The rock particles differ in size and chemical composition depending upon such factors as the original bedrock source and the nature and extent of **weathering** that has taken place. Organic matter may be **humus** or other dead and decaying plant or animal remains, but it also includes macro-organisms such as earthworms and micro-organisms such as **bacteria**. In addition, **air** and **water** are usually present within the pore spaces between the various particles. The soil is a dynamic entity, that changes as a result of inputs such as **rain**, organic deposition from plants and animals and **nutrients** released by weathering, and outputs which include moisture **evapotranspiration** and drainage, nutrient uptake by plants and nutrient loss from **leaching**. The balance among these various elements determines the fertility of the soil, which is reflected in its ability to support vegetation. Because weathering, moisture supply and organic activities are to a large extent controlled by **climate**, in the **natural environment** soil dynamics depend upon the prevailing climatic conditions. Where the natural environment has been altered by human activities, however, the addition of nutrients in the form of **fertilizers**, the supply of extra water through **irrigation**

and the removal of organic matter and nutrients through cropping disrupt the natural soil processes with potentially detrimental effects on soil fertility.

Further reading
Coleman, D.C. and Crossley, D.A. (eds) (1996) *Fundamentals of Soil Ecology*, London: Academic Press.
Ollier, C. and Pain, C. (1995) *Regolith, Soils and Landforms*, London: Wiley.
Ross, S. (1989) *Soil Processes: A Systematic Approach*, London: Routledge.
Singer, M.J. and Munns, D.N. (1991) *Soils: An Introduction*, New York/Toronto: Macmillan/Collier Macmillan Canada.

SOIL CLASSIFICATION

The grouping of **soils** according to the characteristics of mature soils. A simple classification is to divide soils according to their relationship to the environmental situation in which they occur. Zonal soils, for example, have characteristics that reflect regional climatic conditions; intrazonal soils are not typical of the climate zones in which they occur, because of the presence of overriding local factors such as geology or drainage; azonal soils are poorly developed and not yet in balance with their **environment**, although it is expected that they will eventually mature into one of the other two types.

The first of the modern soil classifications was developed in the United States in the 1930s, based on earlier Russian models, and divided zonal soils into pedalfers – soils in which **iron** (Fe) and **aluminum** (Al) accumulate – and pedocals – soils in which **calcium** (Ca) accumulates. Pedalfers occur mainly in humid regions, whereas the pedocals are soils of sub-humid, semi-arid and arid climates. Although the system has been superseded, several of the great soil groups identified in that Russian-American classification are well entrenched in the geographical and environmental literature. They include the iron-rich **laterites** of the tropics, the grey-brown leached **podzols** of humid temperate latitudes, and the chernozems or black earths of the sub-humid **grasslands**. In 1975, the US Department of Agriculture produced a new,

Figure S-9 The soil orders of the Seventh
Approximation

ENTISOLS	— young soils with no horizons
INCEPTISOLS	— soils with moderately developed horizons
HISTOSOLS	— soils high in organic matter, common in waterlogged areas
OXISOLS	— highly weathered tropical and sub-tropical soils such as laterites
ULTISOLS	— deeply weathered soils with a clay horizon and low in bases
VERTISOLS	— clay-rich soils which swell in wet conditions and crack in dry causing disruption to the horizons
ALFISOLS	— gray-brown forest soils with a clay-enriched horizon and high in bases
SPODOSOLS	— acid soils such as podsols with a leached upper horizon and an iron-rich depositional horizon
MOLLISOLS	— grassland soils such as black-earths or chernozems rich in organic matter and high in bases
ARIDISOLS	— soils of arid regions, often with accumulations of salt gypsum and carbonates

more complex classification, initially referred
to as the Seventh Approximation for Soil
Classification, because of the number of
revisions necessary, but now more commonly
designated as the Comprehensive Soil Classification System (CSCS). The CSCS is based
on ten soil orders which are further divided
into sub-orders and great groups to provide a
world scale classification, plus additional
subgroups, families and series, which apply
only in the United States. The names of the
orders are intended to provide an indication
of the nature of the soil. Aridisols are dry or
desert soils, for example, and Oxisols contain
high levels of **oxides**, mainly iron and
aluminum. Sub-order names reflect characteristics of the soil or its **environment**. For
example, soils in the sub-order Humox are
oxide-rich soils with **humus** in the upper
horizon. The CSCS has been widely accepted
as a world scale classification, but, to be
useful at the regional level, it requires some

modification. The Canadian soil classification system has incorporated some elements
of the US system, but being concerned with
soils in higher latitudes it has no need to
incorporate the tropical and subtropical elements of the CSCS or any other classification.
Instead, it pays greater attention to cold
region soils such as the cryosols of the far
north which are underlain by **permafrost**.

Further reading

Boul, S.W., Hole, F.D. and McCracken, R.J. (1980)
Soil Genesis and Classification (2nd edition),
Ames, IA: Iowa State University Press.
Briggs, D. and Smithson, P. (1997) *Fundamentals
of Physical Geography* (2nd edition), London:
Routledge.
Paton, T.R., Humphries, G.S. and Mitchell, P.B.
(1995) *Soils: A New Global View*, New Haven:
Yale University Press.
Roth, H. (1991) *Fundamentals of Soil Science*,
London: Wiley.
Strahler, A.H. and Strahler, A.N. (1992) *Modern
Physical Geography* (4th edition), New York:
Wiley.

SOIL CONSERVATION

The preservation of the quality, quantity and
productivity of the **soil** in an area, using techniques that slow the rate of **soil erosion** and
maintain fertility. Conservation commonly
involves the retention of a vegetation cover,
the maintenance of a good **soil structure** and
a reduction in **water** and **wind** speeds. Vegetation cover can be retained by employing
forest practices that do not include **clear
cutting**, by preventing overgrazing or by
adopting agricultural techniques that do not
allow the soil to remain exposed for long
periods of time. In some areas susceptible to
erosion, for example, **pastoral agriculture**
might be more suitable than arable agriculture. A good **soil structure** requires that
the organic content and **nutrient** levels in the
soil are maintained. Both help to bind
individual particles together into larger aggregates that are less easily eroded. Organic
components and **nutrients** also determine the
fertility of the soil, and normally a fertile soil
is less likely to suffer from soil erosion.
Maintaining soil fertility is thus good soil
conservation practice. Contour ploughing –
ploughing across the slope rather than up and

down – of relatively shallow slopes and terracing of steeper slopes reduces the volume and speed of downslope water flow, and therefore reduces erosion. **Windbreaks** which slow the **wind** and cultivation practices which restrict the area of bare soil – for example, strip cultivation – help to reduce soil erosion by wind. The adoption of such conservation practices helped the **Great Plains** of North America to recover from the ravages of soil erosion in the 1930s, but in many areas, particularly on land marginal for agriculture, technological and socioeconomic factors allow poor soil conservation practices, such as overgrazing and excessive cropping, to continue, and soil erosion remains a major global problem.

See also
Desertification.

Further reading
Ellis, S. and Mellor, A. (1995) *Soils and Environment*, London: Routledge.
Morgan, R.P.C. (1995) *Soil Erosion and Conservation*, Harlow: Longman.
Pierzynski, G.M., Sims, J.T. and Vance, G.F. (1994) *Soils and Environmental Quality*, Boca Raton, FL: Lewis.

SOIL EROSION

The removal of topsoil by **water**, **wind** and **gravity**. Soil erosion is a natural part of landscape formation and change, but in its modern usage it usually refers to accelerated erosion in which human activities have caused the topsoil to be eroded at a rate greater than it can be formed. Natural soil erosion is greatest where unconsolidated sediments are directly exposed to the **elements**. Winds easily erode the exposed **sands** of the **desert**, for example, and in sparsely vegetated areas in the semi-arid regions of the world, bare soil is easily eroded by infrequent but often intense **precipitation**. **Erosion** rates are also high in areas with well-marked seasonal wet and dry periods such as those that experience **monsoon** and Mediterranean climates.

The human contribution to soil erosion comes about as a result of activities that lead to the removal of vegetation and the direct exposure of the soil to the elements. **Clear cutting** of forests, for example, particularly on steep slopes, increases the volume and rate of **runoff** and encourages increased erosion. Most arable agriculture involves activities that leave the soil exposed for extended periods of time and therefore vulnerable to erosion. In addition, ploughing, harrowing and rolling contribute to the breakup of soil aggregates, producing smaller particles that are more easily eroded. Compaction by machinery during cultivation and harvesting damages the **soil structure** and reduces the amount of pore space in the soil, which in turn reduces infiltration capacity and increases runoff. Any reduction in soil fertility also encourages soil erosion, through the loss of **humus** and **nutrients** which help to bind the soil particles together. In **pastoral agriculture**, overgrazing may cause sufficient damage to the vegetation cover to initiate erosion. Urbanization and industrial activities such as mining also encourage soil erosion. Whatever the origin, on moderate slopes erosion by water often takes the form of sheet wash in which the fine soil particles are removed by a relatively shallow flow which covers the whole slope, normally because the infiltration rate cannot cope with the intensity of the precipitation. On steeper slopes, or those where the vegetation cover is not completely removed, gully erosion is more common. Gully erosion is also encouraged in row crops, such as corn and potatoes, where the bare soil between the rows provides a natural pathway for the water. Wind erosion mainly involves the finer soil particles, which are small enough to be carried in suspension or drifted along the ground.

Soil erosion causes the productivity of the affected area to be impaired, but its environmental impact can extend into adjacent areas. Soil removed from one area by wind or water can be deposited in

Figure S-10 The distribution of areas subject to soil erosion and their annual sediment loss

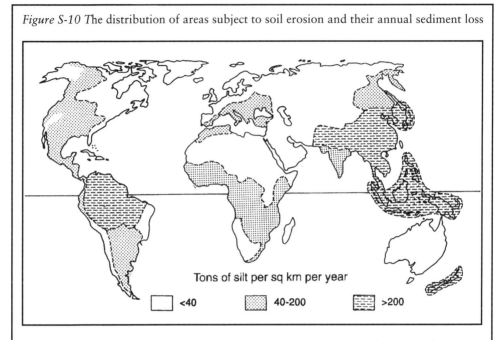

Tons of silt per sq km per year

☐ <40   ▨ 40-200   ▨ >200

Source: After Enger, E.D. and Smith, B.F. (1995) *Envionmental Science: A Study in Interrelationships* (5th edition) Dubuque, IA: Wm C. Brown

sufficient quantities in adjacent areas to cover crops or disrupt existing soil processes. Sediments carried into streams and lakes can cause shoaling, disrupting transportation or damaging fish **habitat**.

See also
Soil conservation.

Further reading
Blaikie, P. (1985) *The Political Economy of Soil*

Erosion in Developing Countries, Harlow: Longman.
Lal, R. (1990) *Soil Erosion in the Tropics: Principles and Management*, New York: McGraw-Hill.
Morgan, R.P.C. (1995) *Soil Erosion and Conservation*, Harlow: Longman.
Pimental, D., Harvey, C., Resosudarmo, P., Sinclair, K., Kurz, D., McNair, M., Crist, S., Shpritz, L., Fitton, L., Saffouri, R. and Blair, R. (1995) 'Environmental and economic costs of soil erosion and conservation benefits', *Science* 267: 1117–23.

SOIL HORIZONS

The series of distinctive layers that make up a **soil profile** (see Figure P-13). They are differentiated according to the physical, chemical and biological properties which give each horizon a characteristic colour, texture or structure. Soil horizons are commonly designated A, B or C. The A horizon is the uppermost, containing **humus** and other organic matter and usually showing evidence of **leaching** or **eluviation**. Beneath it, the B horizon is a zone of deposition for the minerals, such as iron and aluminum oxides, and **clay** particles carried down from the A horizon. The C horizon consists of the parent material for the soil, usually in the form of weathered rock particles. Many classifications include an O horizon – the layer of fresh organic material on the soil surface, and a D or R horizon – the unaltered bedrock lying beneath the C horizon. The different horizons vary in thickness from soil to soil, and the full sequence may not always be present. For some purposes, the individual horizons are subdivided according to minor internal differences.

Further reading
Roth, H. (1991) *Fundamentals of Soil Science*, London: Wiley.
Rowell, D.L. (1994) *Soil Science; Methods and Applications*, Harlow: Longman.
Steilla, D. (1976) *The Geography of Soils*, Englewood Cliffs, NJ: Prentice-Hall.

SOIL MOISTURE DEFICIT (SMD)

When growing plants cannot meet their immediate moisture needs from incoming **precipitation**, they use the moisture reserves that are available in the **soil**. This creates a soil moisture deficit when measured against the **field capacity** of the soil. SMD can be estimated using a **lysimeter** or by comparing **evapotranspiration** and precipitation measurements. Increased accuracy is possible if the nature and state of crop development – which influences moisture requirements – and the storage capacity of the specific soil can be calculated. Thus the SMD across a region will not be a constant value, but will vary according to crop and soil conditions, down to the scale of individual fields. SMD values are important from an agricultural viewpoint, since they give an indication of the amount of **irrigation water** required to bring the soil back to field capacity, which is theoretically the best condition for plant growth.

See also
Soil moisture storage.

Further reading
Kramer, P.J. and Boyer, J.S. (1995) *Water Relations of Plants and Soils*, San Diego, CA: Academic Press.

SOIL MOISTURE STORAGE

Water held in the pore spaces of the **soil**. The amount available varies from soil to soil, but it is available to plants which absorb it through their root **systems** and pass it through to the **atmosphere** during **transpiration**. In arid areas, it offsets the **moisture deficit** caused when **evapotranspiration** exceeds **precipitation**, and thus delays the onset of **drought**. The soil moisture storage is recharged when precipitation is again in excess.

See also
Field capacity, Soil moisture deficit.

Further reading
Iwata, S., Tabuchi, T. and Warkentin, B.P. (1995) *Soil Water Interactions: Mechanisms and Applications* (2nd edition), New York: M. Dekker.

SOIL PROFILE

The vertical sequence of **soil horizons** from the surface down to bedrock.

SOIL STRUCTURE

The form of the aggregates produced when individual soil particles clump together. Aggregates may be crumbs, blocks or plates, for example, depending upon the original **soil texture** and the level of organic matter and **nutrients** in the soil. Organic matter – such as **humus** – and **nutrients** help to bind the individual particles together. The soil becomes susceptible to **erosion** if they are lost.

Further reading
Brady, N.C. (1990) *The Nature and Properties of Soils*, New York/London: Macmillan/Collier Macmillan.

SOIL TEXTURE

A measure of the proportions of **sand**, **silt** and **clay** in a soil. The texture of a soil has a

Figure S-11 A soil texture diagram

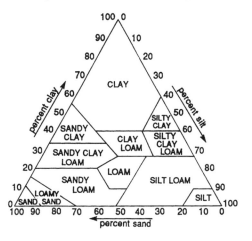

strong influence on its structure and affects such elements as its ability to absorb **precipitation**, its moisture storage capacity and its ease of cultivation.

Further reading
Brady, N.C. (1990) *The Nature and Properties of Soils*, New York/London: Macmillan/Collier Macmillan

SOLAR CONSTANT

The **solar radiation** flux at the outer edge of the **atmosphere** on a surface perpendicular to the incident **radiation**. It is difficult to measure, but its value is usually considered to be close to 1370 Wm^{-2}. Although referred to as a constant, it does appear to have a natural variability, but only within the range of a few watts over time-scales of several decades. Any larger variations have the potential to contribute to **climate change**.

SOLAR ENERGY

See **solar radiation**.

SOLAR FLARES

Short, but intense outbursts of **radiation** that occur in magnetically active regions of the sun. This causes an increase in the **solar wind**, and the emission of **shortwave radiation** such as **X-rays** and **ultraviolet radiation** increases rapidly, to as much as 100 times the norm. Longer, radio frequency wave emissions also increase. Solar flares, and the magnetic storms with which they are associated, increase auroral activity and create major disruptions in telecommunication and navigation systems. Their occurrence is cyclical, with periods of increased frequency similar to that of **sunspots**, and periods of relative quiet in between.

See also
Magnetosphere, Quiet sun.

Further reading
McIntosh, P.S. and Dryer, M. (eds) (1972) *Solar Activity: Observations and Predictions*, Cambridge, MA: MIT Press.

SOLAR RADIATION

Radiant energy given off by the sun. Since the sun is a very hot body, radiating at a **temperature** of about 5700° K, the bulk of the **radiation** is high energy at **ultraviolet** and **visible light** wavelengths (see figure R-1). Although only a small proportion is intercepted by the earth, it provides the energy that drives the earth/atmosphere **system**. About 50 per cent of the solar energy arriving at the outer edge of the **atmosphere** passes through to the earth's surface, and the remainder is absorbed – for example, by the **ozone layer** – or reflected back into space by atmospheric **aerosols**.

See also
Electromagnetic spectrum.

SOLAR WIND

Streams of electrically charged particles emitted by the sun. The volume of particles, or the strength of the wind is increased by **solar flares** and during **sunspot** activity. Interaction between the solar wind and the earth's **magnetosphere** causes the aurora or 'Northern Lights' to form.

See also
Quiet sun.

Further reading
Brandt, J.C. (1970) *Introduction to the Solar Wind*, San Francisco, CA: W.H. Freeman.

SOLID

Along with **liquid** and **gas**, one of the three states of matter. Strong cohesion between the **molecules** that make up a solid allow it to retain its shape. Anything which alters that cohesion – for example, heat or **pressure** – will allow the solid to deform.

SOLID WASTE DISPOSAL

See **waste disposal**.

SOLID WASTES

See **waste**.

SOLSTICE

The time of greatest declination of the sun or the time at which the overhead sun is furthest from the equator. This happens twice a year. At noon on 21 June – the northern hemisphere summer solstice – the sun is directly overhead at the Tropic of Cancer (23.5°N), whereas at noon on 21 December – the winter solstice – it is overhead at the Tropic of Capricorn (23.5°S). At the summer solstice, areas north of 66.5°N experience 24 hours of daylight, while areas south of 66.5°S experience 24 hours of darkness. The situation is reversed at the winter solstice.

SOLUBILITY

A measure of the extent to which a substance (solute) will dissolve in a **solvent** to produce a **solution**. Solubility is usually **temperature** dependent. When **solids** are dissolved in **liquids**, solubility increases as temperature rises, for example, whereas the solubility of **gases** decreases as temperature rises.

SOLUTION

A homogeneous mixture formed when substances in different states (i.e. **solid, liquid** or **gas**) are combined together, and the mixture takes on the state of one of the components. When a solid is dissolved in a liquid, for example, the solution is a liquid. Gases can also be incorporated in liquid solutions.

See also
Solubility, Solvent.

SOLUTION WEATHERING

A form of **weathering** in which minerals are dissolved by acidified **water**. Solution weathering is common in **limestone** areas where water containing dissolved **carbon dioxide** (CO_2) causes the **corrosion** of the rock.

See also
Subsidence – landscape.

Further reading
Sweeting, M.M. (1972) *Karst Landforms*, London: Macmillan.

SOLVENT

A **liquid** capable of dissolving other substances and incorporating them into a **solution**. In a brine solution, for example, **water** is the solvent in which **salt** has been dissolved. Water is the most common natural solvent in the **environment**, but in its pure state it is incapable of dissolving some substances and more complex solvents are required. **Hydrocarbons** provide a wide range of solvents, including acetone, **ethanol**, **methanol** and **trichloromethane**, which are used in the food, **plastics** and pharmaceutical industries. The escape of solvents into enclosed spaces can cause **sick building syndrome**, and improper disposal can lead to water **pollution**. To protect the **environment** and reduce costs many industrial solvents are now recycled.

Further reading
Stoye, D. (ed.) (1993) *Paints, Coatings and Solvents*, Weinheim, NY: VCH.

SONIC BOOM

The **noise** heard when an aircraft or missile travelling at greater than the speed of **sound** (1220 kph at sea-level; 1060 kph in the **stratosphere**) passes an observer. Any object travelling through the **atmosphere** creates **pressure** waves ahead of it. Supersonic aircraft fly fast enough that they overtake the pressure waves and cause a shock wave to develop. The shock wave spreads out as a cone with the aircraft at its apex, and the sonic boom is caused when the cone, spreading out behind the aircraft, reaches the ground. As the line of intersection of the cone with the ground is pulled along by the moving aircraft the sonic boom follows, creating a 'boom carpet'. Thus the sonic boom is not a single event, occurring when the aircraft breaks the so-called sound barrier. It is a continuing event. The pressure change associated with the boom can rattle or even break windows, and the sudden noise is disconcerting to those exposed to it. The sonic boom created by **SSTs** such as **Concorde** places constraints on the flight paths and routes followed by such aircraft.

Figure S-12 The creation of a sonic boom by a supersonic aircraft

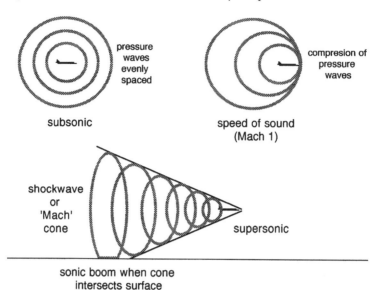

Further reading
Calvert, B. (1981) *Flying Concorde*, London: Fontana.

SOOT

Finely divided particles of **carbon** formed during **combustion**, particularly of **fossil fuels**. Individual soot particles readily combine with each other into clusters or strings, and are effective at absorbing **radiation** across the entire spectrum. Thus soot has a high specific absorption coefficient, and the injection of large amounts of soot into the **atmosphere** has major implications for the earth's **energy budget**.

See also
Aerosols, Nuclear winter.

Further reading
Appleby, L. and Harrison, R.M. (1989) 'Environmental effects of nuclear war', *Chemistry in Britain* 25: 1223–8.

SOUND

The sensation experienced when the ear intercepts vibrations with frequencies between 20 and 20,000 hertz, transmitted through the **air** from a vibrating source. The vibrations cause longitudinal **pressure** waves to form and spread out in all directions from the source. Because of such factors as the composition and structure of the source and the resulting frequency of the vibrations emanating from it, each source will produce a particular sound which with experience is recognizable to the listener. The human ear cannot detect vibrations with a frequency greater than 20,000 hertz, but these higher frequencies can be detected by some animals, such as dogs. Under certain circumstances, sound can become an environmental issue with social and medical implications.

See also
Noise, Ultrasonic waves.

Further reading
Foreman, J.E.K. (1990) *Sound Analysis and Noise Control*, New York: Van Nostrand Reinhold.

SOUTHERN OSCILLATION

The periodic reversal of **pressure** patterns and **wind** directions in the **atmosphere** above the equatorial Pacific Ocean. An indication of the Southern Oscillation is obtained by comparing barometric pressure differences between Tahiti in the eastern Pacific and Darwin in northern Australia. Pressure at these two stations is negatively correlated, high pressure over Tahiti normally being accompanied by low pressure over Darwin, for example. In contrast, low pressure at Tahiti is matched by high pressure at Darwin. These pressure differences induce a strong latitudinal circulation in the equatorial atmosphere – the **Walker Circulation** (see figure W-1). Periodically the regional pressure patterns reverse, and it is this reversal that is referred to as the Southern Oscillation. It has a periodicity of one to five years, and is ultimately responsible for the development of **El Niño**.

See also
ENSO.

Further reading
Lockwood, J.G. (1984) 'The Southern Oscillation and El Niño', *Progress in Physical Geography*, 8: 102–10.
Rasmusson, E.M. and Hall, J.M. (1983) 'El Niño', *Weatherwise* 36: 166–75.

SPACESHIP EARTH

A concept popular in the 1960s and 1970s, that likened the earth/atmosphere system to the **environment** of a spaceship in which **resources** are finite and **waste disposal** is constrained by the space available. For the inhabitants of the spaceship to survive and function, management decisions have to take these factors into account. The idea appears to have been first visualized by the US politician Adlai Stevenson, but was developed by an economist, Kenneth Boulding, to counteract the common view that economic **systems** were open systems unrestrained by environmental limits. He drew attention to the fact that, like a spaceship, the earth/atmosphere system was a closed system in material terms, and if economic decisions did not allow for that, disaster would follow. The concept of environmental responsibility in economic decisions is now widely accepted, and is well established in **sustainable development** policies.

Further reading
Boulding, K.E. (1966) 'The economics of the coming spaceship Earth', in H. Jarrett (ed.)

Environmental Quality in a Growing Economy, Baltimore, MD: Johns Hopkins University Press.
Odum, E.P. (1993) *Ecology and Our Endangered Life-support Systems* (2nd edition), Sunderland, MA: Sinauer.
Park, C.C. (1997) *Environment: Principles and Applications*, London/New York: Routledge.

SPATIAL RESOLUTION

An indication of the detail available from **weather** and **climate models**, determined by the horizontal and vertical distribution of the grid points for which data are available, and at which the appropriate equations are solved. In a model with coarse spatial resolution the grid points are well spaced, and as a result may miss the development of small-scale phenomena. Fine resolution models have many more grid points, and are there–fore more accurate. However, the additional accuracy is costly, and in most models the spatial resolution is the result of compromise between accuracy and cost.

See also
Grid-point models.

SPECIES

A subdivision of a **genus** incorporating a group of organisms with common characteristics, which separate them from other groups. Members of the same species can breed, but breeding is not possible between species.

SPECIFIC GRAVITY

See **relative density.**

SPECIFIC HEAT

The amount of heat required to raise the **temperature** of unit **mass** of a substance through one degree. It can be expressed in several ways – **joules** per kilogram per °K; **calories** per gram per °C; **British thermal units** per pound per °F.

SPECTRAL MODELS

Atmospheric circulation models, used in

weather **forecasting**, that focus on the representation of atmospheric disturbances or waves by a finite number of mathematical functions. The progressive solution of a series of equations allows the development of the atmospheric disturbances to be predicted, normally over a period of between five and ten days.

Further reading

Barry, R.G. and Chorley, R.J. (1992) *Atmosphere, Weather and Climate* (6th edition), London/New York: Routledge.

SPHAGNUM

A **genus** of moss which is a common component of the plant **community** in temperate **peat** bogs. Being acid tolerant, it colonizes the margins of acid lakes.

See also

Wetlands.

SPORE

A microscopic, thick-walled reproductive structure, comparable to a seed, produced by some plants (for example, ferns), **fungi**, **bacteria** and **protozoa**. Spores are generally produced in large numbers, are resistant to adverse environmental conditions and, being microscopic, can be carried great distances by the **wind**, all of which helps to ensure the survival of the **species** that produce them.

Further reading

Ingold, C.T. (1971) *Fungal Spores: Their Liberation and Dispersal*, Oxford: Clarendon Press.

SPRING FLUSH

The rapid **runoff** of **water** from melting **snow** and **ice**, common in mid- to high latitudes at the end of winter. In areas subject to **acid rain**, the winter's accumulation of acidity is flushed

Figure S-13 The rapid decrease in pH levels associated with rapid snow melt in spring

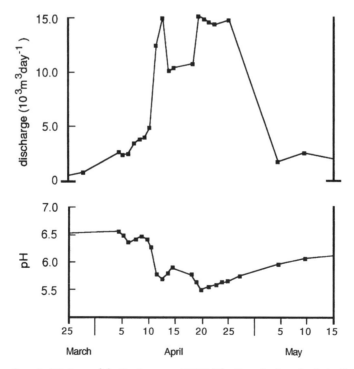

Source: After Ontario Ministry of the Envionment (1980) *The Case Against the Rain*, Toronto: Ministry of the Environment

into rivers and lakes in a matter of days, rapidly reducing the **pH** of these waterbodies and having serious consequences for aquatic organisms. If the increased acidity coincides with the beginnings of the annual fish hatch the newly hatched fry cannot survive the shock, and fish populations in acidic lakes often have reduced or missing age sets which reflect this high mortality.

Further reading
Baker, J.P. and Schofield, C.L. (1985) 'Acidification impacts on fish populations: a review', in D.D. Adams and W.P. Page (eds) *Acid Deposition: Environmental, Economic and Political Issues*, New York: Plenum Press.
Jeffries, D.S. (1990) 'Snowpack storage of pollutants, release during melting, and impact on receiving waters', in S.A. Norton, S.E. Lindberg and A.L. Page (eds) *Acidic Precipitation, Volume 4, Soils, Aquatic Processes and Lake Acidification*, New York: Springer-Verlag.

STANDARD TEMPERATURE AND PRESSURE (STP)

The standard references used to compare the volume of **gases** – a **temperature** of 273.15K (0°C) and a **pressure** of 101,325 **pascals** (760 mm of **mercury**).

STARCH

A white, insoluble **carbohydrate**, consisting of chains of **glucose** units. Starches stored in potatoes, and cereals such as rice, wheat and corn provide some 70 per cent of the world's food supply.

STATE OF THE ENVIRONMENT REPORTING (SOE)

A means of providing current and accessible information on environmental quality and natural **resources**, SOE reports are produced regularly by government agencies or departments in such countries as Australia, Canada and the Netherlands. They provide information on the current state of environmental issues, and through regular publication they allow the monitoring of environmental change with time.

Further reading
Lloyd, R. (1996) 'State of the Environment Reporting in Australia: a review', *Australian Journal of Environmental Management* 3: 151–62.

STATEMENT OF FOREST PRINCIPLES

A product of the 1992 **Earth Summit,** aimed at the **sustainable development** of the world's forests. It is a general statement, not legally binding on the signatories, that acknowledges the need to balance exploitation and **conservation** of forests, but makes no provision for international monitoring or supervision.

Further reading
Pearce, F. (1992) 'Last chance to save the planet', *New Scientist* 134 (1757): 24–8.

STEADY STATE SYSTEM

A **system** in which inputs and outputs are equal and constant and in which the various **elements** are in equilibrium. Any change alters the relationships among the components of the system, creating imbalance, and setting in train a series of responses, or **feedbacks**, which attempt to restore balance.

See also
Dynamic equilibrium, Ecological balance.

STEAM ENGINE

A machine by which the chemical **energy** in a variety of **fuels** – wood, **coal, natural gas,** oil, **uranium** – can be converted into kinetic or mechanical **energy**. Heat released during the **combustion** of the fuels is used to boil **water** and produce steam that under **pressure** can drive a piston or turbine. The development of the reciprocating (piston-driven) steam engine during the early nineteenth century provided the **power** that drove the **Industrial Revolution.** Steam engines were used to pump water, power textile mills and drive hammers, cutters and presses in the **metal** industries. Made mobile in railway locomotives and steamships, they made a major contribution to international trade and allowed the European nations to spread their influence worldwide. The reciprocating steam engine has been replaced by the steam turbine, which, operating at higher **temperature** and pressure, is more efficient. Steam turbines are used extensively to generate electricity in **fossil fuel** and **nuclear power** stations. Steam engines have always been major contributors to **pollution. Particulate matter** and **gases** released during the combustion process pollute the **atmosphere** and hot water produced when the used steam is condensed can cause thermal pollution to waterways. Much of the **acid rain** produced in North America and Europe originated in coal-burning **thermal electric power stations.** Clean air legislation and other environmental regulations have dealt with such problems in most of the developed world, but in developing nations, such as India and China, which continue to depend on coal-fired steam engines, air **pollution** remains a serious problem.

Further reading
Jones, H. (1973) *Steam Engines: An International History*, London: Benn.
Reynolds, W.C. (1974) *Energy: From Nature to Man*, New York: McGraw-Hill.

STEPPE

A semi-arid area, characterized by short-grass vegetation, considered transitional between **desert** and sub-humid climates. In areas closer to the latter, steppe may include woody shrubs. Because of their **aridity** and proximity to deserts, steppe areas are susceptible to **desertification.**

See also
Grasslands.

STOCHASTIC PROCESSES

Processes which include inputs that are in part random in character. They include an element of chance.

STRATOPAUSE

The boundary between the **stratosphere** and **mesosphere**, located at about 50 km above

the earth's surface where the **temperature** is close to freezing point.

See also
Atmospheric layers.

STRATOSPHERE

That part of the **atmosphere** lying above the **tropopause**. It is characterised by an **isothermal layer** (**temperatures** remain constant) up to about 20 km above the earth's surface, beyond which the temperature rises again from about −50 °C to reach close to 0°C at the **stratopause**. This is the result of the presence of **ozone** which absorbs incoming **ultraviolet radiation**, causing the temperature to rise.

See also
Atmospheric layers.

STRATOSPHERIC OZONE

See **ozone.**

STRIP-CROPPING

The practice of cultivating land in long, narrow strips in which, for example, rows of grain may alternate with leafy crops or land lying fallow. Variations in cultivation techniques and timing of different crops ensure that the land always retains some vegetation cover, and is therefore protected from **soil erosion**. Strip-cropping is widely used in the arable farming areas of the **Great Plains**, where the strips may be oriented at right angles to the direction of the prevailing **wind**.

See also
Soil conservation.

STRIP-MINING

Open-cast or **open-pit mining**. The recovery of **coal** or mineral **ore** by stripping the overburden from the surface to expose the coal or ore body, which can then be removed using conventional excavation techniques. Because it is generally cheaper than underground mining, it can be used to develop poorer quality deposits. In the case of **metal** ores, the excavation is usually followed by on-site concentration or **beneficiation** of the ore. As the overburden is removed, it is dumped in long linear mounds that create a large-scale ridge and furrow landscape. In the past, when the mineral deposit was worked out, the site was abandoned and the ridges remained. The resulting landscape was unpleasant to look at, but it also created environmental problems. **Oxidation** of exposed minerals such as pyrites (iron sulphide) created acids which inhibited plant growth. The unconsolidated material, unprotected by a vegetation cover, was easily eroded by **wind** and **water**. Rainwater flowing off the ridges became increasingly acidic and ultimately raised the acidity of the adjacent waterbodies. The **runoff** also carried large amounts of sediment that was transported to streams and rivers where it disrupted the aquatic habitat. Thousands of square kilometres in the Appalachian region of the eastern United States suffered in this way. Currently, in all developed nations, legislation requires that all completed strip-mining operations be followed by the restoration of the landscape. This includes the return of the topsoil and the planting of vegetation. Recovery of the vegetation may take between five and ten years and during that time the site must be monitored and maintained. The serious problems associated with strip-mining activities in the past are much less common, although lax enforcement of the legislation has prevented them from being eliminated completely.

Further reading
Griggs, G.B. and Gilchrist, J.A. (1983) *Geologic Hazards, Resources and Environmental Planning,* Belmont, CA: Wadsworth.
Toole, K.R. (1976) *The Rape of the Great Plains: Northwest America, Cattle and Coal,* Boston, MA: Little, Brown & Co.

STUDY OF CRITICAL ENVIRONMENTAL PROBLEMS (SCEP)

A report produced in 1970 covering a range of atmospheric, terrestrial and aquatic environmental issues, most of which had their immediate origins in the **pollution** problems that had attracted growing concern in the 1960s. SCEP was the first major study to draw attention to the global extent of human-induced environmental issues.

Further reading
SCEP (1970) *Man's Impact on the Global Environment: Study of Critical Environmental Problems,* Cambridge, MA: MIT Press.

STUDY OF MAN'S IMPACT ON CLIMATE (SMIC)

A 1971 report that grew out of issues raised originally in the **SCEP**. It focused on inadvertent **climate** modification, at both regional and global scales, and was widely recognized as an authoritative assessment of all aspects of human-induced **climate change**.

Further reading
SMIC (1971) *Inadvertent Climate Modification: Report of the Study of Man's Impact on Climate,* Cambridge, MA: MIT Press.

SUBLIMATION

The conversion of a **solid** into a **vapour** with no intermediate **liquid** stage. Under conditions of low **relative humidity, snow** can be evaporated directly into **water vapour** without entering the liquid **water** phase. Sublimation is also used to describe the direct deposition of water vapour on to **ice**.

SUBSIDENCE – ATMOSPHERIC

The sinking of **air** in the **atmosphere**. Subsidence may be associated with the cooling of air close to the surface – as in cold **anti-cyclones**, or with the larger scale circulation – for example, in the descending arm of a **convection cell**. Subsidence encourages the retention of pollutants close to the earth's surface and contributes to **drought** by preventing the uplift necessary for **precipitation**.

SUBSIDENCE – LANDSCAPE

The creation of hollows in the landscape following the removal of sub-surface material. Subsidence occurs naturally in **limestone** areas, where **solution weathering** of the underground rock creates a space into which the surface material sinks. It is also common in **coal**-mining areas, where underground mining is practised. The removal of the coal provides the necessary space. Subsidence may take place slowly, being gradually revealed by the ponding of **water** in the hollows it creates or by cracking in the walls of buildings, but it can also take place catastrophically, when it can cause major damage to structures and loss of life.

SUBSISTENCE FARMING

The production of sufficient food and other necessities to meet the requirements of a farm unit, leaving no surplus for sale and little for storage. As a result, subsistence farmers are ill-prepared for crop failure.

See also
Cash cropping.

SUCCESSION

The gradual and sequential change in the structure and content of an **ecosystem** at a particular site. It progresses through distinct stages until the so-called **climax community** is attained, with the complete sequence from the initiation of the community to the climax referred to as a **sere**. The climax community represents the ecosystem that is best suited to the environmental conditions at the site. Succession is best illustrated by changes in vegetation, but it also applies to other elements such as **soils** and animals, which are intimately linked with vegetation. Primary succession begins when a community is established on a previously unvegetated site – for example, a **lava** flow or mud-flats exposed by falling water levels. Secondary succession occurs on a site which has been previously vegetated, but where the natural succession has been disrupted. Natural fires can initiate secondary succession, but human interference with ecosystems is increasingly responsible. Secondary succession will begin on abandoned agricultural land or land cleared by forestry activities, for example.

Further reading
Burrows, C.J. (1990) *Processes of Vegetation Change*, London: Unwin Hyman.
Del Moral, R. and Wood, D.M. (1993) 'Early primary succession on the volcano Mount St Helens', *Journal of Vegetation Science* 4(2): 223–4.
Glenn-Lewin, D.C., Peet, R.K. and Veblen, T.T. (eds) (1992) *Plant Succession: Theory and Prediction*, London: Chapman and Hall.
Shugart, H.H. (1984) *A Theory of Forest Dynamics: The Ecological Implications of Forest Succession Models*, New York: Springer-Verlag.

SUGARS

Relatively simple **carbohydrates**, characterized by their sweetness and **solubility**. They are classified into monosaccharides, containing five or six **atoms** of **carbon** (C) and

Figure S-14 The stages of primary plant succession in a mid-latitude temperate zone in which the climax community is deciduous forest

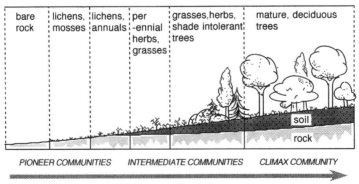

disaccharides containing twelve atoms of carbon. **Glucose** ($C_6H_{12}O_6$), for example, is a monosaccharide and sucrose ($C_{12}H_{22}O_{11}$) is a disaccharide. Sucrose is common household sugar.

SULPHATE PARTICLE

A negatively charged **ion** containing one **atom** of **sulphur** (S) and four of **oxygen** (O). Sulphate particles are released naturally into the **environment**, for example, during volcanic eruptions, but modern industrial activities, particularly those which involve the combustion of **fossil fuels**, are also major producers. Sulphates make up the largest group of **secondary aerosols** in the **atmosphere**, and contribute to a number of environmental issues. For example, they are abundant in **Arctic Haze**; they combine with **water** to contribute to atmospheric acidity; they are very effective at scattering **solar radiation**, and they have been identified as contributing to the thinning of the **ozone layer** over the Antarctic.

See also
Atmospheric turbidity, Heterogeneous chemical reactions, Mount Pinatubo.

Further reading
Barrie, L.A. (1986) 'Arctic air pollution; an overview of current knowledge', *Atmospheric Environment*, 20: 643–63.

Brasseur, G. and Granier, C. (1992) 'Mount Pinatubo aerosols, chlorofluorocarbons and ozone depletion', *Science* 257: 1239–42.
Charlson, R.J. and Wigley, T.M.L. (1994) 'Sulfate aerosol and climatic change', *Scientific American* 270: 48–57.

SULPHUR (S)

A yellow, non-metallic **element** occurring naturally in its elemental form in many volcanic regions and in combination with many **metals** in the earth's **crust**. All **fossil fuels** contain sulphur, which is released in the form of **sulphur dioxide** (SO_2) during **combustion**. Used in the manufacture of sulphuric acid (H_2SO_4) and other chemicals, such as **fertilizers** and **fungicides**, it also has a long history of use in medicine. Sulphur is present in all living matter, usually as a constituent of certain **proteins**.

Further reading
Bates, T.S., Lamb, B.K., Guenther, A., Dignon, J. and Stoiber, R.E. (1992) 'Sulfur emissions to the atmosphere from natural resources', *Journal of Atmospheric Chemistry* 14 (1–4): 315–37.
Meyer, B. (1977) *Sulfur, Energy and Environment*, Amsterdam/New York: Elsevier.

SULPHUR DIOXIDE (SO₂)

An acid gas in which each **molecule** contains one **atom** of **sulphur** (S) and two of **oxygen**

Figure S-15 Sulphur dioxide emissions in selected countries

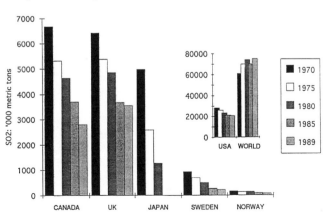

Source: Based on data in World Resources Institute (1992) *World Resources 1992–93: A Guide to the Global Environment*, New York: Oxford University Press

(O). It is a product of the **combustion** of materials containing sulphur. The burning of sulphur-rich **coal**, for example, releases large amounts of sulphur dioxide into the **atmosphere** where it combines with **water** to produce sulphuric acid (H_2SO_4), the main component of **acid precipitation**.

See also
Scrubbers.

SUNSPOTS

Dark spots, associated with strong electro–magnetic activity, that appear on the surface of the sun. The **temperature** within sunspots is as much as 1000 K lower than the gases that surround them. As a result they radiate less **energy** and therefore appear dark. The size and duration of sunspots is quite variable, but their numbers follow a cyclical pattern that includes peaks approximately every eleven years. These sunspot cycles have been linked to events – such as **drought** and glaciation – that occur with some regularity in the earth/atmosphere **system**. Low sunspot numbers, for example, indicating a reduction in solar activity and therefore less **radiation** reaching the earth, have been linked to the advance of glaciers at various times in the past. Some drought years in North America have also been associated with low sunspot numbers.

See also
Maunder Minimum, Quiet sun, Solar flares.

Further reading
Lockwood, J.G. (1979) *Causes of Climate*, London: Edward Arnold.

Figure S-16 Two hundred years of sunspot frequencies

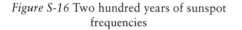

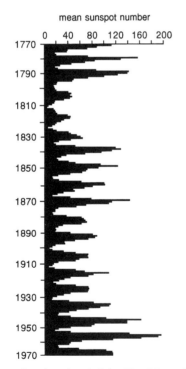

Source: Based on data in Schneider, S.H. and Mass, C. (1975) 'Volcanic dust, sunspots and temperature trends', *Science* 190: 741–6

SUPERFUND

A multibillion dollar fund established in the United States in 1980 under the Comprehensive Environmental Response, Compensation and Liability Act to clean up abandoned **hazardous waste** dump sites that threaten the **environment** and public health. The fund is financed by federal and state governments with contributions in the form of taxes from the **petroleum** and chemical industries. Under the act, fines to pay for the clean-up can be levied on the companies who established the dumps, but the owners cannot always be identified and lengthy litigation can stall the recovery of the costs. Estimates suggest that a total clean-up of all hazardous sites would take between thirty and fifty years and cost in the region of 1.5 trillion dollars, almost 100 times the original size of the fund. The fund is administered by the **EPA**.

Further reading
Smith, Z.A. (1995) *The Environmental Policy Paradox*, Englewood Cliffs, NJ: Prentice-Hall.

SUPERNOVA

A star which over a few days becomes exceptionally bright – more than a million times brighter than normal – before declining again. The levels of **cosmic radiation** released during this process are exceptionally high, and, if the supernova occurs close enough to the earth, may contribute to **ozone depletion** through the creation of **oxides of nitrogen** (NO_x) when the cosmic rays strike the **atmosphere.**

SUPERSATURATION

A condition in which a **solution** contains more solute than is required to saturate it. The term is applied to an **atmosphere** containing more **water vapour** than is necessary to produce **saturation** – in effect its **relative humidity** exceeds 100 per cent. If supersaturated air near the ground encounters a cold surface, **condensation** will take place, but in the free atmosphere well above the ground, the supersaturated state can exist for some time. Condensation will only take place when **condensation nuclei** are introduced into the supersaturated **air mass.**

SUPERSONIC TRANSPORTS (SSTs)

Commercial aircraft that routinely fly faster than the speed of **sound**, and at higher altitudes than subsonic airliners. Flying high in the **stratosphere**, they inject ozone destroying pollutants such as **oxides of niotrogen** (NOx) and **odd hydrogens** (HOx) directly into the **ozone layer**. Only two types of SST have flown since their development in the 1970s – the Russian **Tu-144** and the Anglo-French **Concorde** – and only the Concorde continues in service. High engine noise levels during take-off require that SSTs adopt **noise abatement** procedures at most airports, and the **sonic boom** they generate restricts their use on certain routes. When passing over land, for example, they are usually required to remain at subsonic speeds.

SUSTAINABLE DEVELOPMENT

Development judged to be both economically and environmentally sound, so that the needs of the world's current **population** can be met without jeopardizing those of future generations. In its current form, the concept grew out of the work of the **World Commission on Environment and Development**, which met in the mid-1980s, and became a central issue at the **UNCED** held in Rio in 1992. Sustainability is an integral part of all natural systems, however, maintained by the controlled flow of matter or **energy** through the **systems**, and disruption of these flows or damage to the components of a system can interfere with sustainability. Population pressure or human activities which result in the excess use of available **resources** threaten sustainability. Non-renewable resources, such as most minerals, cannot normally support sustainable development in the long term, although prudent management combined with **recycling** can extend their life-spans. Renewable resources, on the other hand,

should be able to provide a sustainable yield if appropriately managed. However, declining plant and animal stocks, threats to **biodiversity**, **soil** depletion and increasingly widespread **ecosystem** deterioration provide evidence that modern harvesting and resource extraction techniques far surpass the ability of renewable resources to maintain sustainability.

Malthus recognized the dire consequences associated with loss of sustainability, as did the early conservationists such as **John Muir**, although they did not use that term specifically. In modern times, the concept expanded beyond its original links with natural systems, and began more and more to include socioeconomic and developmental elements. The reports of the **Club of Rome**, for example, incorporated the sustainability theme through their consideration of population growth, resource use, food supply and pollution.

Current sustainability issues cover a very broad spectrum, with the concept of sustainable yield applied to agriculture, energy, fishing, forestry and resource development in such a way that the economic factors are integrated with environmental concerns as far as possible. The modern approach to sustainable development is increasingly multifaceted, with a need to integrate a wide range of elements. It has become people centred, for example, with local strategies being developed to meet local problems, while at the same time it is apparent that it must have the scope to foster sustainable patterns of international trade and finance. At the international level, the developed nations must be prepared to provide help to the **Third World** if sustainable development is to be successful. Sustainable development was central to the UNCED's **Agenda 21**, and the signatories to that agreement are working to implement its recommendations. Work is ongoing on several elements, including biodiversity, energy, **deforestation** and land use, under the auspices of the Sustainable Development Commission and such agencies as the UN Food and Agricultural Organization (FAO), **UNEP**, **UNDP** and the **World Bank**. Particular attention is being given to sustainable agriculture and rural development in the Third World, as a means of increasing agricultural production and food supply without threatening the natural resource base of the developing nations. On paper, the sustainability goals of Agenda 21 can be achieved by following a few basic principles such as those listed by Park (1997):

- respect and care for the community of life
- improve the quality of human life
- conserve the earth's vitality and diversity
- minimize the depletion of non-renewable resources
- keep within the earth's carrying capacity
- change personal attitudes and practices
- enable communities to care for their own environments
- provide a national framework for integrating development and conservation
- create a global alliance.

In practice, these involve a wide range of complex, interrelated technical, social and economic issues, which include several areas of potential conflict and tension – for example, the traditional and often very different approaches to resource exploitation and **conservation** will have to be reconciled. Thus, desirable and necessary as it may be, sustainable development will require a long-term commitment, and most observers agree that it is unlikely to be achieved in the near future.

Further reading

Brown, L.R. (ed.) (1997) *State of the World 1997*, New York: Norton.
Elliott, J. (1993) *An Introduction to Sustainable Development*, London: Routledge.
Goodland, R.J.A. (ed.) (1992) *Environmentally Sustainable Economic Development: Building on Brundtland*, Paris: UNESCO.
Meadows, D.H., Meadows, D.L. and Randers, J. (1992) *Beyond the Limits: Confronting Global Collapse, Envisioning a Sustainable Future*, Post Mills, VT: Chelsea Green.
Milbrath, L.W. (1989) *Envisioning a Sustainable Society*, Albany, NY: State University of New York Press.
Park, C.C. (1997) *Environment: Principles and Applications*, London/New York: Routledge.
Plant, J. and Plant, C. (eds) (1990) *Turtle Talk: Fifteen Voices for a Sustainable Future*, Santa Cruz, CA: New Society.

SUSTAINABLE DEVELOPMENT COMMISSION

An institution established as a result of the **Earth Summit**, aimed at monitoring and promoting the approach towards **sustainable development** identified at the Summit.

SYMBIOSIS

A close and permanent relationship between organisms of different **species**. Symbiosis takes a number of forms. In mutualism, both organisms benefit from the relationship; in **commensalism**, one benefits and the other is unharmed; in parasitism, one of the organisms benefits at the expense of the other.

See also
Parasite.

SYNERGISM

A condition in which two substances work together in such a way that their combined effect is greater than the sum of the two separate effects. It applies in air **pollution** episodes, for example, when **smoke** particles and sulphuric acid droplets have a synergistic effect on the respiratory systems of those exposed to **smog**. Synergism can also increase the potency of drugs used for pharmaceutical purposes.

SYNFUELS

Synthetic fuels produced by the conversion of a wide range of organic materials into a **gaseous** or **liquid** form suitable for use as a **fuel**. The raw material may be a conventional fuel such as **coal**, growing plant materials, human and animal **sewage** or **garbage**. Town gas or coal gas produced by the **destructive distillation** of coal was one of the first synfuels produced, and was used in many areas until the advent of more energy efficient **natural gas**. Liquid synfuels have also been produced from coal. Coal liquefication plants were operated in Germany during the Second World War, but cheap postwar **oil** prices made further development uneconomic. The only major development has been in South Africa. **Alcohols** such as **methanol** and **ethanol** produced by the conversion of **carbohydrates** in plants have also been used as fuels, and the production of **methane** (CH_4) – biogas – by the **anaerobic** decomposition of sewage provides an important substitute for conventional fuels in some parts of India. The **pyrolysis** of domestic refuse or garbage can produce small amounts of liquid and gaseous synfuels, in the process helping to reduce the volume of material to be disposed of in **landfill** sites. Synfuels are usually costly to produce in quantity and create the same environmental problems as conventional fuels. Thus, their overall impact on energy supply remains limited, although they may be locally or regionally important.

Further reading
Probstein, R.F, and Hicks, R.E. (1982) *Synthetic Fuels*, New York: McGraw-Hill.
Supp, E. (1990) *How to Produce Methanol from Coal*, Berlin/New York: Springer-Verlag.
World Bank (1980) *Alcohol Production from Biomass in the Developing Countries*, Washington, DC: World Bank.

SYSTEM

An assemblage of interrelated objects organized as an integrated whole. In environmental studies, systems are usually classified as open or closed. Open systems are those which allow an exchange or flow of **energy** and **mass** across the system boundaries in the form of inputs and outputs. Closed systems involve the input and output of energy, but not mass. The earth/atmosphere system, for example, is a closed system, in which there is no material flow across the system boundaries. The closed nature of the system is important, since it means that the amount of matter in the system is fixed, and therefore existing material **resources** are finite. Systems commonly include a series of sub-systems, ranging in scale from microscopic to continental, which may act individually, but within and linked to the main system. Drainage basins are often studied using a systems approach. Individual basins have energy and mass inputs and outputs, but they belong to the larger **hydrological system** driven ultimately by the energy flow in the whole earth/atmosphere system. When outputs balance inputs in a system, it is said to be in a **steady state**. Balance is seldom complete, however, and in most environmental systems some form of **dynamic equilibrium** is the best that can be achieved. Human interference has progressively disrupted even that degree of balance by interfering with the flow of material or energy in systems. In the past, the disruption was mainly at the sub-system level – for example, the addition of pollutants to drainage basins, or the removal of vegetation from **ecosystems** – but the human impact is now being felt at the earth/atmosphere system level. **Global warming**, for example, is in large part a reflection of society's ability to disrupt the flow of energy through the system. The general concept of the system is relatively simple, but its application to specific situations can be difficult. In the complex relationships that exist between the different **elements** of the **environment**, for example, delineating sub-systems is particularly difficult because it is not always possible to identify clear boundaries between them. System theory has been useful in the study of environmental issues, however, through its reinforcement of the concept that the various elements in the environment are interrelated, and, as a result, any changes

introduced do not remain isolated, but ultimately impact on the larger system.

See also
Entropy.

Further reading
Huggett, R.J. (1980) *Systems Analysis in Geography*, Oxford: Clarendon Press.
Huggett, R.J. (1993) *Modelling the Human Impact on Nature: Systems Analysis of Environmental Problems*, Oxford: Oxford University Press.
Huggett, R.J. (1995) *Geoecology: An Evolutionary Approach*, London: Routledge.
White, I.D., Mottershead, D.N. and Harrison, S.J. (1992) *Environmental Systems: An Introductory Text*, London: Chapman and Hall.

Figure S-17 (a) Schematic diagram of the earth/atmosphere system

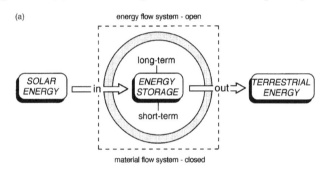

(b) Schematic diagram of a sub-system of anthropogenic origin – a thermal electric power plant

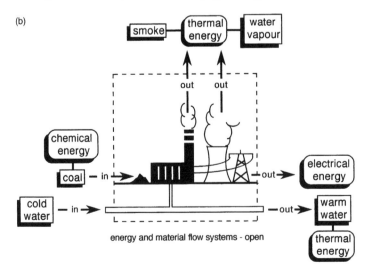

Source: Kemp, D.D. (1994) *Global Environmental Issues; A Climatological Approach*, London/New York: Routledge

T

TAIGA

The Eurasian equivalent of the North American boreal forest (see Figure B-5), stretching from Scandinavia to far eastern Siberia. Lying between **tundra** to the north and temperate **deciduous** forest or **grassland** to the south, the taiga **biome** includes mainly coniferous trees such as spruce, fir and larch, growing in large stands of few **species**. The **climate** of the taiga is harsh, with long cold winters and short summers. **Precipitation**, much of it in the form of **snow**, ranges from 250 to 1000 mm per year, and in combination with low to moderate **temperatures** this produces humid conditions. The landscape therefore includes many lakes, ponds and **peat** bogs. Large mammals such as deer, moose and wolves, which have adapted to the cold winters, inhabit the taiga all the year round, but many of the region's birds are migratory, nesting in the area before returning south for the winter. Characteristic of the region is the large number of biting insects such as mosquitoes and blackfly that hatch in their millions in the short summer season. Human habitation of the taiga often relies upon forest products, and the area is dotted with towns dependent upon lumbering or the **pulp and paper industry**. Threats to the taiga **environment** include clear-cutting of the forest and air and water **pollution** caused by the mills.

Further reading
Larsen, J.A. (1980) *The Boreal Ecosystem,* New York: Academic Press. beyond its local or regional

Figure T-1 The global distribution of taiga

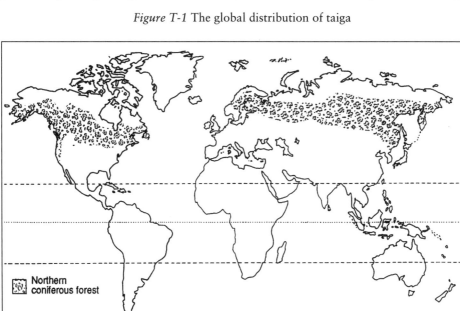

Northern
coniferous forest

TAILINGS

Waste material remaining after the mining and on-site processing of mineral resources. Tailings may be inert and pose few problems to the environment, other than those associated with aesthetics. However, even seemingly innocuous rock waste may contain minerals that, when exposed to surface conditions, undergo physical and chemical changes which allow harmful products such as acids and toxic metals to enter the environment. Many processing techniques cause the production of very fine tailings suspended in water, that may include remnants of the chemicals used in the process. To prevent such materials from reaching the environment, they are stored in tailing ponds until the sediments settle out and the chemicals are neutralized, after which they can be disposed of in an acceptable manner. Such ponds can present a significant environmental hazard if they are breached or even leak consistently.

See also
Acid mine drainage, Ore, Uranium mill tailings.

Further reading
Ritcey, G.M. (1989) *Tailings Management: Problems and Solutions in the Mining Industry*, Amsterdam/New York: Elsevier.

TALL STACKS POLICY

An approach to the problem of local air pollution, which involved the building of tall smokestacks to allow the release of pollutants outside the local atmospheric boundary layer. By the mid-1970s stacks ranging in height from 150 m to 300 m were common on smelters and thermal electric generating stations in Europe and North America. The International Nickle Company (INCO) built the tallest smokestack (400 m) to dispose of exhaust gases from its nickel smelting complex in Sudbury, Ontario. While all of this reduced local pollution, it introduced pollutants into the larger scale circulation and contributed to the long-range transportation of air pollution (LRTAP). A major result of the release of acid gases from tall stacks was the promotion of acid precipitation

Figure T-2 The 400 m high 'superstack' at Sudbury, Ontario, Canada

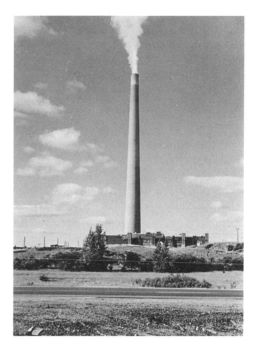

Photograph: The author

beyond its local or regional boundaries to boundaries to become a continental scale problem.

Further reading
Howard, R. and Perley, M. (1991) *Poisoned Skies*, Toronto: Stoddart.
Park, C.C. (1987) *Acid Rain: Rhetoric and Reality*, London: Methuen.

TALUS

See scree.

TANNINS

A group of complex organic compounds derived from organic acids such as tannic acid. Tannins are used in the tanning of leather and in the dyeing industry.

TAR SANDS

Tar sands or oil sands are deposits of sand impregnated with bitumen, a very viscous

Figure T-3 Experimental technique for the *in situ* extraction of oil from tar sands

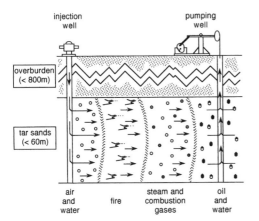

hydrocarbon. Major tar sand deposits are located in Venezuela, the United States and Russia, but by far the largest deposits are those in Alberta, Canada, with estimated reserves equivalent to 800–900 billion barrels of crude oil. Perhaps only 10 per cent of that total is recoverable with existing technology, however. Tar sands consist of 84 per cent sand and 12 per cent bitumen by weight with the remainder mainly water. Approximately 2.5 tonnes of sand are required to produce a barrel of oil. The sands are extracted by strip-mining and treated with hot water, steam and chemicals to separate the bitumen from the sand. Further refining of the bitumen to remove impurities such as sulphur (S) and the fractionation of the hydrocarbons into a mixture of naphtha and gas oils produces a high-quality synthetic crude suitable for use in the production of gasoline and aviation fuel, or as a feedstock for the petrochemicals industry. The processing of one barrel of bitumen produces 15.5 m^3 of gas, 15 kg of coke, 5.9 kg of sulphur and 0.8 barrels of synthetic crude oil. The first oil from the Alberta tar sands was produced in 1967, but the industry faces both economic and environmental problems. The operations are extremely capital, labour and equipment intensive and, although production can be profitable, the level of profitability is very sensitive to the fluctuation of world crude oil prices. Environmental issues arise during the extraction process, which faces the same

problems as all large-scale strip-mining operations, and during refining, which badly pollutes the large volumes of water required for the process. Consideration has been given to the *in situ* extraction of the oil through the liquefication of the bitumen underground and its subsequent pumping using conventional methods, but it is not yet economically and technically feasible. The potential supplies of oil from tar sands are very large, and plans are being made to increase Albertan oil sands petroleum production to 1.2 million barrels per day by the year 2020. However, much will depend upon world oil prices and production, and the tar sands are unlikely to be utilized to their fullest extent as long as crude oil from conventional sources remains readily available.

Further reading
Fitzgerald, J.J. (1978) *Black Gold with Grit*, Sidney, BC: Grays Publishing.
Schumacher, M.M. (ed.) (1982) *Heavy Oil and Tar Sands Recovery and Upgrading: International Technology*, Park Ridge, NJ: Noyes Data Corp.
Smith, J.W. (1980) 'Synfuels: oil shale and tar sands', in L.C. Ruedisili and M.W. Firebaugh (eds) *Perspectives on Energy*, New York: Oxford University Press.

TAXONOMY

The classification of organisms into hierarchical groups. In addition to the description and cataloguing of individual plants and animals, modern taxonomy includes consideration of the causes and effects of the variations among organisms.

TELECONNECTION

The linking of environmental events in time and place. The concept is based on observations that the various elements in the earth/atmosphere system are sufficiently interconnected that changes in one will automatically set in motion changes in others. The changes often involve a time-lag and include locations that may be well separated from each other. For example, an El Niño in the eastern Pacific late in one year may be linked to the failure of the Indian monsoon in the following year. The time-lag between events may also make it possible to predict the consequences of a specific event, and this aspect of teleconnection is being closely examined for its potential in **drought prediction**.

Further reading

Glantz, M.H., Katz, R.W. and Nicholl, N (eds) (1991) *Teleconnections Linking Worldwide Climate Anomalies: Scientific Basis and Societal Impact*, Cambridge: Cambridge University Press.
Mooley, D.A. and Parthasarathy, B. (1983) 'Indian summer monsoon and El Niño', *Pure and Applied Geophysics* 121: 339–52.

TEMPERATURE

In popular terms, temperature is the measure of the warmth or coldness of an object. In scientific terms, it is a measure of the molecular kinetic **energy** of matter and represents the speed at which the **molecules** in the matter move or vibrate. Temperature can be measured using several scales, depending upon the purpose of the measurement and the conditions under which it is being taken. **Celsius**, **Fahrenheit** and **Kelvin** are the most common scales used in modern times.

TEMPERATURE INVERSION

The reversal of the normal **temperature** decline with altitude in the **troposphere**. In an inversion, the temperature rises with altitude because of the presence of a layer of warm **air** above the cooler surface air. The warm layer acts to dampen convective activity and

mixing. **Smoke** and other emissions are therefore trapped beneath the inversion and, as a result, the surface layer may become highly polluted. Inversions are commonly caused by strong **radiation** cooling of the ground at night, particularly in valleys or basins where the cool air drains on to the lower ground and reinforces the radiation cooling. Temperature inversions are also caused by the adiabatic warming of descending air in **anticyclones**.

See also

Adiabatic process, Convection.

TENNESSEE VALLEY AUTHORITY (TVA)

A United States government agency created in 1933 to control flooding, improve navigation, rehabilitate land and foster regional development in the Tennessee River valley through the production of **hydroelectricity**. It is an integrated system which includes parts of seven states through which the Tennessee River flows. The system includes fifty **dams**, most of which produce electricity, and together have a capacity of more than 30 million Kw (30 Gw). The authority also produces electricity from **coal**-fired stations and until 1985 operated two nuclear power plants. Opposition from environmental groups helped to halt the Authority's nuclear programme, and it has also come under attack for environmental disruption caused

Figure T-4 Lapse rates associated with a temperature inversion

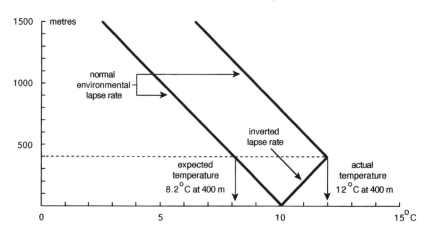

by **strip-mining** and burning coal. As a product of the integration of the system, the TVA has helped to improve agricultural and forestry practices, control flooding and reduce **soil erosion**. Some of the **reservoir** areas have been developed as recreational facilities and the provision of electricity has contributed to industrial development in the valley.

See also
Nuclear energy.

Further reading
Owen, M. (1983) *The Tennessee Valley Authority*, New York: Praeger.
Schaffer, D. (1989) 'Managing water in the Tennessee Valley in the post-war period', *Environmental Review* 13(2): 1–16.

TERATOGEN

An agent or substance that produces birth defects by causing malformation of the tissues in a developing foetus. It may be a relatively common substance like **alcohol**, for example. Foetal alcohol syndrome is caused by the consumption of alcohol by pregnant women. The children affected may display a wide range of defects from brain damage to cleft palate. More complex chemicals such as **dioxin** have also been identified as teratogenic. Children born to mothers exposed to **Agent Orange** are suspected of having suffered **chromosome** malformation while in the womb, which produced a variety of defects.

Further reading
Smith, D.W. and Jones, K.L. (1982) *Recognizable Patterns of Human Malformation: Genetic, Embryologic and Clinical Aspects*, Philadelphia, PA: Saunders.

TERPENE

One of a group of fragrant **hydrocarbons** found in the essential oils of some plants, particularly conifers. Released from these plants, terpenes may be responsible for the haze common over such areas as the Blue Ridge Mountains of Virginia in the United States.

Further reading
Newman, A.A. (1972) *Chemistry of Terpenes and Terpenoids*, London: Academic Press.

TERRESTRIAL ENVIRONMENT

That part of the **environment** that includes the components of the land surface – such as rock and **soil** – and the plants and animals that live on it. The quality of the terrestrial environment is being threatened by such factors as **pollution** from **solid** and **liquid waste**, mining and agricultural activities, **soil erosion, overpopulation, desertification** and **deforestation**. Attempts at maintaining quality have involved the application of **environmental impact assessment** and the introduction of appropriate **land use planning**.

See also
Aquatic environment, Atmospheric environment.

Further reading
Aber, J.D. and Melillo, J.M. (1991) *Terrestrial Ecosystems,* Philadelphia, PA: Saunders.

TERRESTRIAL RADIATION

Radiation emitted from the earth's surface. Since the earth is a relatively low **temperature** body, terrestrial radiation is low **energy, long-wave radiation** from the **infrared** sector of the spectrum. In theory, the terrestrial radiation emitted by the earth balances the amount of **solar radiation** entering. However, that is a long-term balance, based on average input and output of radiation. Various factors intervene to disrupt the balance. For example, terrestrial radiation is trapped by **greenhouse gases** in the **troposphere** and is thus retained in the system. In so doing, it performs the very important function of bringing about the warming of the earth/atmosphere **system**.

See also
Electromagnetic spectrum, Greenhouse effect, Heat budget.

Further reading
Barry, R.G. and Chorley, R.J. (1992) *Atmosphere, Weather and Climate* (6th edition), London/New York: Routledge.

TETRACHLOROMETHANE

See **carbon tetrachloride**.

THERM

A unit of heat equivalent to 100,000 Btus. Once commonly used in Britain as a measure of **gas** or steam consumption, but now being replaced by the megajoule.

See also
British Thermal Unit, Joule.

THERMAL CONDUCTIVITY

The rate of heat transfer through a body by **conduction**.

THERMAL EFFICIENCY

See **Thornthwaite, C.W.**

THERMAL ELECTRIC POWER STATION

An electricity generating station in which the electricity is produced by burning **coal, oil** or natural **gas**, with coal being the most common **fuel** used. The thermal **energy** released when the coal is burned is used to heat **water** and produce steam, which is directed under **pressure** through turbines. These in turn power generators to produce the electricity. Because of the many energy conversions involved – chemical to thermal to kinetic to electrical – thermal power stations are not particularly efficient. Even large, modern, well-maintained plants have **energy efficiency** ratings of less than 40 per cent. Energy is lost through **friction** in the turbines and generators and through flue gases, but the bulk of it is lost in the conversion of water to steam and back to water. Once the steam has passed through the turbines it is condensed back into water, which loses its thermal energy to the **environment** as it cools. The initial high **temperatures** are brought down in on-site cooling ponds or by cooling towers, but even with that the water is commonly warmer than the waterbody into which it is released and therefore creates **thermal pollution**. Thermal power plants

Figure T-5 A coal-fired thermal electric power station. The waterside location, the fuel storage space requirements and the release of pollutants set such power stations in conflict with the environment, in this case a lakeshore marsh ecosystem

Photograph: The author

have also been linked to air **pollution** through the release of **particulate matter** and **acid gases** from their smokestacks, to increase **atmospheric turbidity** and cause **acid rain**. More efficient **combustion** and the fitting of **scrubbers** to the plants have reduced these problems. The production of **carbon dioxide** (CO_2) as a by-product of the combustion process in thermal electric power stations contributes to the enhancement of the **greenhouse effect** and therefore **global warming**. As yet there is no widely acceptable solution in sight for this carbon dioxide emission problem. **Fossil fuel** combustion currently accounts for some 60 per cent of the total world electricity production, with coal-burning power plants alone producing 40 per cent, and these figures are likely to increase as the demand for electricity grows in areas such as China and India, when there are large coal **reserves**. There is concern among environmentalists that the growing demand will be met from thermal plants, most of which will not be designed to meet the increasingly strict environmental standards imposed on plants in the developed nations. That being so, thermal electric power plants will make an increasing contribution to the problems of air pollution, acid rain and global warming.

See also
Hydroelectricity, Nuclear energy, Tall stacks policy.

Further reading
Flavin, C. and Lenssen, N. (1994) 'Reshaping the power industry', in L.R. Brown (ed.) *State of the World – 1994*, New York/London: Norton.
Shannon, R.H. (1982) *Handbook of Coal-based Electric Power Generation*, Park Ridge, NJ: Noyes Data Corp.

THERMAL ENERGY

See **energy**.

THERMAL LOW

Low **pressure** system produced by the heating of the earth's surface and the **air** immediately above it. Thermal lows or **depressions** are best developed under clear skies and strong

insolation which causes the surface air to become more buoyant and exert less pressure. They encourage cyclonic inflow, but only close to the surface, since they tend to be shallow features at most only a few kilometres deep. Thermal lows range in scale from local or regional features to the large-scale low pressure systems that form over Asia and Australia during the strong solar heating of the summer months.

See also
Atmospheric pressure, Cyclone.

Further reading
Hidore, J.J. and Oliver, J.E. (1993) *Climatology: An Atmospheric Science*, New York: Macmillan.

THERMAL POLLUTION

In theory a **temperature** increase in any part of the **environment** brought about by human activities. Thus, the **urban heat island** phenomenon might be considered an example of thermal pollution. However, it is applied mainly to waterbodies. The major source of thermal pollution in rivers, streams and lakes is the release of cooling **water** from **thermal electric power stations** and other industrial enterprises. Cutting down trees that provide shade along water-courses can also lead indirectly to thermal pollution. In the absence of shade, direct **solar radiation** can cause the temperature of the river or stream to rise. The main impact of thermal pollution is through a reduction in the availability of dissolved **oxygen** (O) as temperatures increase, which creates problems for fish and other aquatic organisms. The problem is usually greatest at times of low flow in waters already polluted by organic substances. Rising temperatures can also encourage bacterial growth, creating potential health problems where the water is used for drinking or for recreational activities such as swimming.

See also
Biochemical oxygen demand, Oxygen sag curve.

Further reading
Langford, T.E.L. (1990) *Ecological Effects of Thermal Discharges*, London: Elsevier.

THERMISTOR

A semiconductor in which there is a strong negative correlation between electrical resistance and **temperature**. Thus, it can be used as a sensor for measuring temperature. The sensitivity, rapid response time and small size of thermistors have made them ideal for many microclimatological studies and in other areas such as medicine and electronics, where these characteristics are important.

See also
Thermocouple, Thermometer.

THERMOCLINE

The layer of **water** in the **oceans** and in lakes that separates the warmer surface layer (**epilimnion** in lakes) from the deeper colder layer (**hypolimnion** in lakes). The thermocline is usually absent from the oceans polewards of 60° latitude. It ranges in thickness from a few metres in lakes to more than 100 metres in the tropical oceans and is characterized by a **temperature** gradient that exceeds that of the layers above and below. The upper surface of the thermocline marks the limit of penetration of **solar energy** either directly or through mixing, and since that varies with the seasons, the depth of the thermocline also varies, particularly in mid- and high latitude lakes. During the summer in these areas, warming of the water may push the thermocline down and in shallow lakes it may disappear completely. Conversely, in the winter, the lake may become uniformly cold and no thermocline will be evident. The presence of a thermocline tends to inhibit the interchange of water, **gases** and **nutrients** between the warm and cold layers. It limits the downward mixing of **carbon dioxide** (CO_2), for example, which may explain in part the ability of higher latitude oceans to absorb more of that gas than those in lower latitudes. See Figure E-8.

Further reading
Strahler, A.H. and Strahler, A.N. (1992) *Modern Physical Geography* (4th edition), New York: Wiley.

THERMOCOUPLE

An instrument for measuring **temperature**. It consists of wires of two different **metals** joined to form a continuous circuit. When one of the junctions is maintained at a constant temperature (reference junction) and the temperature of the other (measuring junction) is allowed to rise, an electromotive force (emf) or current is created in the circuit. By measuring that emf, the temperature of the measuring junction can be calculated. In the past, the temperature of the reference junction was maintained by placing it in a freezing solution, and a galvanometer placed in the circuit provided a measure of the current from which the temperature was calculated. In modern thermocouple systems, electronic circuitry maintains the temperature of the reference junction and allows the temperature of the other junction to be displayed on a screen or dial directly in °C or °F rather than in millivolts. It shares the characteristics of sensitivity, rapid response and small size with the **thermistor**.

See also
Thermometer.

THERMOMETER

An instrument for measuring **temperature**. Thermometers vary in size, shape and construction, depending upon such factors as the range of temperature involved, the accuracy required and the conditions under which the instrument will be exposed. The most common type is the **mercury**-in-glass thermometer. The mercury (Hg) is enclosed in a bulb attached to a capillary tube, along which it expands or contracts in response to changing temperature. Where temperatures are likely to fall below −39°C, at which liquid mercury solidifies, **alcohol**-in-glass thermometers are used. Thermometers can be constructed to maintain the maximum or minimum temperature recorded in the time between readings, or to provide a continuous record of temperature conditions over an extended period. Bimetallic strip thermometers, in which the differences in the coefficients of expansion and contraction of the two metals cause the strip to flex and indicate temperature fluctuations, are used in many industries as thermostats to control temperature increases or decreases in an industrial plant or during manufacturing processes. Specialized thermometers such as **thermistors** and **thermocouples** have been developed to allow temperatures to be measured when the more common types are not suitable.

Further reading
Linacre, E. (1992) *Climate Data and Resources: A Reference and Guide*, London: Routledge.

THERMONUCLEAR DEVICE

A powerful bomb in which the explosive force is created by the fusion of the nuclei of **hydrogen atoms** – hence the name 'hydrogen bomb'. The explosion is initiated by the detonation of a fission bomb embedded in a hydrogen-rich **compound** such as lithium deuteride. This creates the high **temperatures** necessary for the fusion process to begin. The power of such a bomb is equivalent to megatons (millions of tons) of **TNT**. The detonation of hundreds of these devices over a short period was considered capable of bringing on **nuclear winter**.

See also
Atomic bomb, Nuclear fission, Nuclear fusion.

Further reading
Morland, H. (1981) *The Secret that Exploded*, New York: Random House.
Pittock, A., Ackerman, T., Crutzen, P., MacCracken, M., Shapiro, C. and Turco, R. (1986) *Environmental Consequences of Nuclear War: Vol. 1, Physical and Atmospheric Effects*, New York: Wiley.

THERMONUCLEAR REACTION

See **nuclear fusion**.

THERMOPLASTICS

See **plastics**.

THERMOSETTING PLASTICS

See **plastics**.

THERMOSPHERE

See **atmospheric layers**.

THIRD WORLD

A term commonly applied to the developing and non-aligned nations of Africa, Asia and Latin America, to distinguish them from the industrial nations of the 'first world' with their developed, capitalist economies, and the communist nations of the 'second world' with their centrally planned economies. There may also be an argument for the presence of a 'fourth world' made up of the oil-rich nations which are not developed in the conventional sense, but do not face the economic problems of the Third World. Third World nations face a number of dilemmas associated with the conflict between their need to develop economically and the environmental consequences of that development. The destruction of the tropical **rainforest** in Latin America and south-east Asia to produce revenue or to allow mining and agricultural activity is indicative of that conflict. Environmental regulations are often less stringent, or less stringently applied, in the Third World, allowing **pollution** from mining and manufacturing plants to become a serious problem in some areas. This encourages companies from the developed nations to follow double standards in which environmental controls for the emission of pollutants or the disposal of **waste** in their plants in Third World countries are more lax than those required in their home plants. Conflicts between developed and developing nations also occur when global environmental issues are considered. The reduction in **rainforest** exploitation, for example, or the proposed imposition of a **carbon tax**, seen by the developed nations as necessary to control **global warming**, are regarded by Third World nations as deterrents to their economic development. Through events such as the **UNCED** and the activities that stemmed from it, Third World nations are being encouraged to work their way out of their problems through **sustainable development**, using appropriate technology scaled to local needs and local **resources**. Many development studies now refer to countries with Third World attributes as the 'South', as opposed to the 'North' with its developed and industrialized nations.

Further reading
Chandra, R. (1992) *Industrialization and Development in the Third World*, London: Routledge.
Gupta, A. (1988) *Ecology and Development in the Third World*, London: Routledge.
Sachs, I. (1976) *The Discovery of the Third World*, Cambridge, MA: MIT Press.
Weatherby, J. (1997) *The Other World: Issues and Politics of the Developing World* (3rd edition), New York: Longman.

30 PER CENT CLUB

A group of thirty-five nations – mainly from the European Community, but including Canada and the United States – which agreed in 1979 to reduce transboundary emissions of **sulphur dioxide** (SO$_2$) in an attempt to deal with the growing problem of **acid rain**. The agreement was not legally binding, however, and it was necessary to prepare an additional protocol in 1985 by which the signatories were required to reduce transboundary emissions of sulphur dioxide by 30 per cent (of their 1980 level) by 1993. Many subsequently improved on the 30 per cent requirement, but fourteen of the original thirty-five signatories refused to sign, among them the United States and Britain. Both subsequently became embroiled with neighbouring states which had signed the protocol.

Further reading
Park, C.C. (1987) *Acid Rain: Rhetoric and Reality*, London: Methuen.

THOREAU, H.D. (1817–1862)

A protégé of **R.W. Emerson**, Henry David Thoreau rejected materialism and sought to improve the quality and meaning of life by the contemplation and study of nature. Perhaps best known for his account of the time he spent living naturally at Walden Pond near Concord, Massachusetts, Thoreau also kept a journal in which for twenty-four years he recorded his philosophical and scientific observations. An ecologist in all but name, his observations made him aware of the concept of forest **succession**, and as early as 1859 he advocated the creation of **wilderness** parks for the preservation of nature.

Further reading
Thoreau, H.D. (1854) *Walden*, Boston, MA: Ticknor and Fields.
Torrey, B. and Allen, F.H. (eds) (1906) *The Journal of Henry D. Thoreau*, Vols 1–14, Boston, MA: Houghton Mifflin.
Cox, G.W. (1993) *Conservation Ecology: Biosphere and Biosurvival*, Dubuque, IA: Wm C. Brown.

THORNTHWAITE, C.W. (1899–1963)

An American applied climatologist who pioneered environmental **water** balance studies, and introduced the concept of **potential evapotranspiration** (PE). His book-keeping approach to water balance, which included regular observation and comparison of inputs, outputs and storage of moisture, allowed the estimation of **moisture deficits** and their relief by controlled **irrigation**. The same approach allowed him to classify **drought**. Although perhaps best known for his pioneering water budget studies, Thornthwaite also worked on the role of the **heat budget** in plant development. He found that by using climate **normals**, he was able to calculate the **energy** available to crops during the growing season. Knowing the energy required by specific crops, it was possible to calculate the time required by the crop to reach maturity. From this he was able to develop a technique for planning the planting and harvesting of crops which improved efficiency and allowed the crops to be harvested at the peak of their quality. Thornthwaite produced two systems of **climate** classification. In 1931, he developed the concept of thermal efficiency, based on mean monthly **temperatures**, which he combined with a **precipitation** effectiveness element to distinguish his climate regions. By

1948, PE, itself based ultimately on temperature, was being used as a measure of thermal efficiency in his second classification.

See also
Contingent drought, Invisible drought, Permanent drought, Seasonal drought.

Further reading
Thornthwaite, C.W. (1931) 'The climate of North America according to a new classification', *Geographical Review* 21: 633–55.
Thornthwaite, C.W. (1947) 'Climate and moisture conservation', *Annals of the Association of American Geographers* 37(2): 87–100.
Thornthwaite, C.W. (1948) 'An approach towards a rational classification of climate', *Geographical Review* 38: 55-94.
Thornthwaite, C.W. and Mather, J.R. (1954) 'Climate in relation to crops', *Meteorological Monographs* 2(8): 1–10.

THREE MILE ISLAND

The site of a nuclear power plant on the Susquehanna River near Harrisburg, Pennsylvania, at which a combination of mechanical failure and human error caused a major accident in March 1979. As a result of a loss of **coolant** and the subsequent rise in **temperature**, almost half of the reactor core melted, leading to the build-up of high levels of **radioactivity** in the containment building. The latter functioned as designed, and only small amounts of radioactivity escaped into the **environment**. Emergency measures brought the system under control again after about sixteen hours. Because of the threat of a complete **meltdown**, however, more than 100,000 people fled the immediate area or were evacuated. Some were exposed to higher than normal levels of **radiation** as the result of the release of radioactive materials into the **atmosphere** and into the river, but authorities claimed that exposure was sufficiently low that they were unlikely to suffer any serious effects. Other studies have disputed that claim, and have provided evidence of an increase in the number of fatal cases of **cancer** and leukemia. Three Mile Island was the most serious nuclear accident in the United States up to that time. The power company running the plant suffered financially as a result of the loss of revenue from the damaged reactor and the major costs associated with the clean-up. The accident also brought about a new focus on nuclear safety, but confidence in the nuclear industry declined significantly, and there has been little additional development of the nuclear energy programme in the United States since then.

See also
Chernobyl, China Syndrome, Nuclear reactor, Radiation sickness.

Further reading
Gofman, J.W. and Tamplin, A.R. (1979) *Poisoned Power: The Case Against Nuclear Power Plants Before and After Three Mile Island*, Emmaus, PA: Rodale Press.
Megaw, J. (1987) *How Safe? Three Mile Island, Chernobyl and Beyond*, Toronto: Stoddart.
President's Commission on the Accident at Three Mile Island (1979) *Report of the President's Commission on the Accident at Three Mile Island*, Washington, DC: Government Printing Office.

TIDAL POWER

Power that can be generated in coastal locations from the twice-daily ebb and flow of the **tides**. Generation is most effective where the tidal range is 5 to 10 m or more, which restricts the number of available sites. Currently tidal power plants are operating in France, Canada, China and Russia, all involving the damming of tidal inlets to create a head of **water**. As the tide rises water is allowed to flow through gates in the **dam** to fill the basin behind it. At high tide the gates are closed and as the tide falls the water in the basin is retained behind the dam. Once a sufficient head of water is built up, the water behind the dam is released and the **potential energy** it possesses is converted into **kinetic energy** which drives generators to produce electricity. The world's first, and still largest, commercial tidal power plant was built on the estuary of La Rance in Brittany, France, where the tidal range is 13 m. It has an installed capacity of 240 MW, and includes reversible generators which allow electricity to be produced as the incoming tide flows into the basin. The Bay of Fundy in eastern Canada, where the tidal range of 17 m is the largest in the world, has considerable potential as a tidal power site. Several schemes have been proposed with a potential installed capacity of 5 GW, but as yet only a demonstration project is operating. An estimated twenty-five to thirty sites are available for development in locations such as Patagonia in South America, the English Channel, the Murmansk coast in northern Russia and the Sea of Okhotsk in the north Pacific. Tidal power has the advantage that it is renewable, non-polluting and has low operating costs. Disadvantages include the limited number of suitable sites, high construction costs, disruption of shipping and the ecological disturbance caused by changes in the normal tidal flow. Supply problems also occur because the output of electricity varies with the tidal flow, and the periods of maximum electricity production do not always coincide with the periods of maximum demand. The latter problem can be overcome by including a **pumped storage system** in the operation, but that increases the overall cost of the scheme.

See also
Renewable energy.

Further reading
Aubrecht, G. (1989) *Energy*, Columbus, OH: Merrill.
Baker, A.C. (1991) *Tidal Power*, London: Peregrinus.
Boyle, G. (ed.) (1996) *Renewable Energy: Power for a Sustainable Future*, Oxford: Oxford University Press/Open University.

TIDES

Twice-daily rise and fall of sea level caused by the effect of the gravitational pull of the moon, and to a lesser extent the sun, on the

Figure T-6 The method by which electricity is generated using tidal power

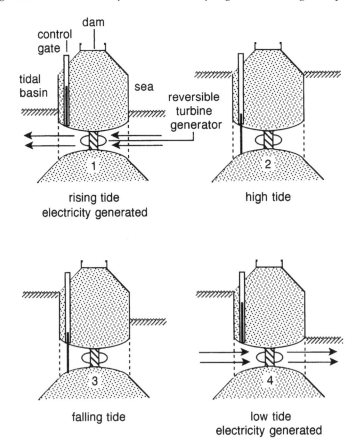

oceans. The difference between high tide and low tide is the tidal range. Tidal ranges are higher than normal during the so-called 'spring tides' because the forces exerted by the sun and moon complement each other. At the 'neap tides', the gravitational pull of the sun tends to offset that of the moon and the tidal range is lower. Differences in the timing of the rotation of the earth and the orbit of the moon cause high and low water to occur fifty minutes later on each successive day, and create a tidal cycle of approximately twenty-seven days.

See also
Gravity, Tidal power.

Further reading
Russell, R.C.H. and Macmillan, D.H. (1970) *Waves and Tides*, Westport, CN: Greenwood Press.

TITRATION

The addition of measured amounts of one **solution** to a specific amount of a second solution until the chemical reaction between them is complete. The completion of the reaction may be indicated by a change in the colour of the second solution or by the chemical **precipitation** of an insoluble substance. The level of acidity in a solution can be obtained through titration with an alkaline solution, and the process has been used to study the acidity of **precipitation**.

See also
Acid rain.

TNT

Trinitrotoluene, a yellow, crystalline **solid** widely used as a high explosive.

TOPSOIL

The uppermost layer of the **soil** containing the bulk of its organic material, **nutrients** and living organisms. It is the most fertile part of the soil, and its loss through **erosion** creates problems for agriculture and may lead to **desertification**.

TORNADO

An intense rotating storm usually no more than 100 to 500 m in diameter, accompanied by **winds** that commonly exceed 200 kph.

Figure T-7 The distribution of tornadoes in the United States in 1995

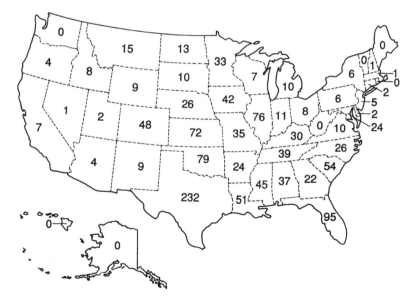

Source: Coutesy of the NSSL/NOAA Storm Prediction Center Web Page

Tornadoes originate in areas where cold and warm, moist **air masses** collide, creating steep **temperature** gradients, and causing thunderstorms. They are also associated with **hurricanes**. The most obvious characteristic of a tornado is the funnel **cloud** that descends to the ground from the base of a severe thunderstorm to produce some of the most violent **weather** in the **natural environment**. Tornadoes are a major **natural hazard** in the central plains of the United States, from Texas north to Nebraska, with Texas experiencing in excess of a hundred in some years. Elsewhere they are less frequent, but tornadoes do occur in all other states, and north into the Canadian plains they are less common but no less deadly. Even in normal years, tornadoes cause several dozen deaths and millions of dollars' worth of damage, and in some years these amounts can be exceeded in a single tornado outbreak.

Further reading
Robinson. A. (1993) *Earthshock: Climate, Complexity and the Forces of Nature*, London: Thames and Hudson.

TOXIC WASTE

See **waste classification**.

TRACE ELEMENTS

Elements that are essential for the proper well-being of an organism, but are needed only in very small quantities. They are often constituents of **vitamins** and **enzymes**. Plants need traces of such elements as **copper** (Cu), zinc (Zn) and manganese (Mn). Cattle and sheep fall ill if they become deficient in cobalt (Co).

Further reading
Kabata-Pendias, A. and Pendias, H. (1984) *Trace Elements in Soils and Plants*, Boca Raton, FL: CRC Press.

'TRAGEDY OF THE COMMONS'

The title of an essay by Garrett Hardin, published in 1968. It considered how the historical concept of the commons (property available to all) was no longer tenable in modern society, mainly as a result of rapid **population** growth. Threats to the commons were seen to be direct – for example, the impact on national parks of a rapidly increasing number of users – or indirect – the addition of **waste** to **air** and **water**. Hardin considered two possible choices to deal with the problem, both involving the infringement of personal freedoms. Freedom of access to the commons would have to be restricted, but that was unlikely to be enough without society relinquishing its 'freedom to breed'. By reducing the rate of population growth, the latter would reduce the threat to other more important freedoms. Hardin's essay attracted widespread attention and spurred debate on the broad issues of population and environmental ethics.

See also
Population – environmental impacts.

Further reading
Hardin, G. (1968) 'The tragedy of the commons', *Science* 162: 1243–8.
Hardin, G. and Baden, J. (eds) (1977) *Managing the Commons*, San Francisco: Freeman.

TRANS-ALASKA PIPELINE

A 1200 km long pipeline built between Prudhoe Bay on the North Slope of Alaska and Valdez on the south shore to bring Alaskan **oil** to market in the United States. From Valdez the oil is transported by tanker to refineries in the Pacific North-west and California. Construction began in 1974 and was completed in 1977 at a cost of some $7.1 million. Although both ends of the pipeline are at sea level, it rises some 1600 m above Prudhoe Bay to pass through the Brooks Range and crosses some thirty-four major rivers before proceeding through the Alaska Range at 1200 m and descending to sea level again at Valdez. During construction and in its operation, the pipeline faced major physical and environmental constraints. To reduce damage to the fragile ecological balance of the **tundra** and the underlying **permafrost**, for example, most of the construction was confined to the winter months. The presence of permafrost also necessitated construction techniques not normally required in temperate

Figure T-8: The route followed by the Trans-Alaska pipeline

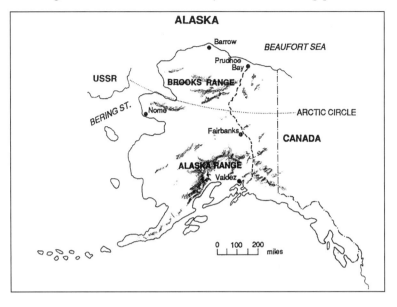

regions. The buried pipe had to be insulated to prevent the heat from the flowing oil melting the surrounding **ice**, and in places, where the permafrost was particularly unstable, the pipe required additional refrigeration or had to be raised above ground on trestles. Near its southern end the pipeline crosses several **earthquake**-prone areas. There the pipe had to be placed on a series of elevated moveable saddles which would allow it to flex without breaking in the event of an earthquake. During the planning stage, environmentalists expressed major concerns about the impact of the construction and operation of the pipeline on the wildlife **population** of the area. The pipeline crosses a number of caribou migration routes, and it was felt that the activity and noise associated with the operation would cause the animals to change their migration patterns, affecting not only the viability of the herds, but also the livelihood of the indigenous peoples of the region who depended upon them as a food source. Similarly, the upheaval caused at pipeline river crossings was seen as likely to damage fish **habitat** such as spawning beds. A major concern was the possible rupture of the line, and the escape of thousands of barrels of oil into and on to the tundra. The inclusion of

valves to stop the flow of oil following a break in the line plus regular inspections have been able to cope and as yet there have been no serious spills associated with the operation on land. Although it is impossible to undertake a project of the magnitude of the Trans-Alaska Pipeline without altering the **environment** to some extent, after nearly twenty years of operation it has been remarkably problem free.

See also
Exxon Valdez, Oil pollution.

Further reading
Coates, P.A. (1991) *The Trans-Alaska Pipeline Controversy: Technology, Conservation and the Frontier*, Bethlehem, PA/London/Cranberry, NJ: Lehigh University Press/Associated University Presses.
Roscow, J.P. (1977) *800 miles to Valdez: The Building of the Alaska Pipeline*, Englewood Cliffs, NJ: Prentice-Hall.

TRANSIENT MODELS

General circulation models that attempt to provide information at intermediate stages during the model run, unlike **equilibrium models** which provide only one final result. Most current transient models incorporate a coupled ocean/atmosphere system with full ocean dynamics, but retain a coarse resolution because of inadequate computer power and high running costs. Since the conditions associated with climate change may well have important environmental impacts long before equilibrium is reached, there has been a move towards the development of transient models. In the second assessment of the IPCC in 1995, for example, ten transient experiments were run compared to only one in the 1990 assessment.

Further reading
IPCC (1990) *Climate Change: The IPCC Scientific Assessment*, Cambridge: Cambridge University Press.
IPCC (1996) *Climate Change 1995: The Science of Climate Change*, Cambridge: Cambridge University Press.
Russell, G.L., Miller, J.R. and Rind, D. (1995) 'A coupled atmosphere-ocean model for transient climate change experiments', *Atmosphere-Ocean* 33: 683–730.

TRANSLUCENT

Permitting the passage of **light**, but not without some scattering or **diffusion**. As a result, objects viewed through a translucent medium such as frosted glass cannot be seen clearly.

See also
Scattering of light, Transparent.

TRANSMITTANCE

A measure of the reduction in the intensity of electromagnetic **radiation** as it passes through a medium. It is the ratio of the intensity of the radiation after it has passed through a unit distance of the medium to the original intensity.

See also
Opacity.

TRANSPARENT

Permitting the passage of **light** without disruption. As a result, objects viewed through a transparent medium such as clear glass are clearly visible.

See also
Translucent.

TRANSPIRATION

The loss of **water** from vegetation to the **atmosphere** by its **evaporation** through leaf pores or stomata in individual plants. The replacement of transpired water by way of the root system helps to carry **nutrient** solutions through the plant tissues.

See also
Evapotranspiration, Hydrological cycle.

TREE DIEBACK

The gradual wasting of a tree from the outermost leaves and twigs inwards. Leaves turn prematurely yellow, dry out and fall well before autumn. The affected branches may fail to leaf out in the following spring. Over several seasons, the tree weakens, becomes vulnerable to insect attack, disease and **weather**, and eventually dies. Dieback has been linked to the effects of **acid precipitation**, but there is no conclusive proof that this is the only factor involved.

See also
Waldsterben.

Further reading
Blank, L.W., Roberts, T.M. and Skeffington, R.A. (1988) 'New perspectives on forest decline', *Nature* 336: 27–30.
Ulrich, B. (1983) 'A concept of forest ecosystem stability and of acid deposition as driving force for destabilization', in B. Ulrich and J. Pankrath (eds) *Effects of Accumulation of Air Pollutants in Forest Ecosystems*, Dordrecht: Reidel.

TREE RINGS

See **Dendrochronology, Dendroclimatology.**

TRIATOMIC OXYGEN

The **gas ozone** (O_3), in which each **molecule** consists of three **atoms** of **oxygen** (O).

TRICHLOROMETHANE

See **chloroform.**

TRITIUM

A radioactive **isotope** of **hydrogen** (H). With a **mass number** of 3 it is three times the mass of ordinary hydrogen. Tritium is not abundant in nature but can be produced artificially in **nuclear reactors**.

See also
Deuterium, Radioactivity.

TROPHIC CHAIN

See **food chain**.

TROPHIC LEVELS

Energy levels within a **trophic chain** or **food chain**. The energy assimilated at each trophic level declines from the primary producers at the base of the chain through the **herbivores** to the carnivores. Some of the primary energy consumed is used to allow the consumers to function, for example, and is lost to the **environment** in the form of heat. A general rule of thumb is that 10 per cent of the energy available at any one level is transferred to the next level up the chain, but the actual value may range from 5 to 20 per cent. The relationship between trophic levels is usually represented in the form of a pyramid, with a broad base provided by the primary producer level and a few top carnivores forming the narrow apex.

See also
Ecosystem, Food chain, Food web.

Further reading
Enger, E. and Smith, B. F. (1995) *Environmental Science: A Study of Interrelationships*, Dubuque, IA: W.C. Brown.
Kormondy, E.J. (1984) *Concepts of Ecology* (3rd edition), Englewood Cliffs, NJ: Prentice-Hall.

TROPICAL CYCLONE

A generic name for a type of violent tropical storm, known as a **hurricane** in the Atlantic and Caribbean, **typhoon** in the western Pacific and **cyclone** in the Indian Ocean and Bay of Bengal. In all of these, the **atmospheric pressure** is low (typically 92–95 kp) and the circulation pattern is cyclonic with the **air** flowing in towards the centre of the storm. Thick banks of **clouds** form a circular pattern around a central area - the eye of the storm – in which skies are clear, **winds** are light and there is no **precipitation**. Tropical cyclones may be 500–600 km in diameter and reach 10–15,000 m into the atmosphere. They are accompanied by extreme winds, routinely exceeding 160 km/hour and sometimes reaching as much as 300 km/hour, plus heavy **rain** in amounts ranging from 30–50 cm over the duration of the storm. Tropical cyclones form between 10–15° north and south of the equator, where the conditions that support their development – **sea surface temperatures** above 27°C, high **humidity** and unstable **lapse rates** – are most common. They are absent from the immediate vicinity of the equator, probably because there the **Coriolis effect** is too small to initiate the necessary circulation. Once formed, tropical cyclones tend to travel from east to west, before tracking north (in the northern hemisphere) or south (in the southern hemisphere) as they mature. Every year tropical cyclones cause extensive damage to the **environment**. Coastlines are eroded and trees are blown down by the strong winds, and the heavy precipitation combined with storm surges from the sea causes serious flooding. Damage to property can cause losses amounting to billions of dollars and major loss of life is common particularly in low lying, heavily populated areas such as Bangladesh. Nothing can be done to prevent such storms, but improved forecasting, for example, using satellite imagery, and better emergency planning have at least reduced the loss of life from such storms. There is some concern that with the projected **global warming**, storminess in the environment will increase. There is some evidence that the frequency and intensity of tropical storms has increased in the first half of the 1990s, but records are as yet insufficient to prove or disprove that.

Further reading
Eagleman, J.R. (1985) *Meteorology: The Atmosphere in Action*, Belmont, CA: Wadsworth.
Pielke, R.A. (1990) *The Hurricane*, London: Routledge.

TROPICAL OCEAN AND GLOBAL ATMOSPHERE PROJECT (TOGA)

A project of the **World Climate Research Program**, TOGA was set up in 1992 to examine the relationship between the behaviour of the tropical **oceans** and the global **atmosphere** with a view to determining the predictibility of the system. Basic data have been collected using satellites, instrumented aircraft, survey ships and oceanographic buoys, and these are now being used in a **coupled ocean/atmosphere** response experiment (TOGA-COARE) from which a predictive **model** can be developed.

TROPICAL RAINFOREST

See **equatorial rainforest**.

TROPOPAUSE

The upper boundary of the **troposphere**. It varies in height from about 8 km at the poles to 16 km at the equator.

See also
Atmospheric layers.

TROPOSPHERE

The lowest layer of the **atmosphere**, in which **temperatures** decrease with altitude at a rate of about 6.5°C per kilometre, to reach between –50°C and –60°C at the **tropopause**. The troposphere contains as much as 75 per cent of the gaseous **mass** of the atmosphere and is the zone in which most **weather** systems develop. It is also the part of the atmosphere which suffers most human intervention, which ensures that many current global environmental problems have their origin in the troposphere.

See also
Atmospheric layers.

TSUNAMI

Popularly referred to as tidal waves, tsunamis are sea waves propagated by submarine **earthquake** activity or volcanic eruptions. On the open ocean, tsunamis travel at several hundred kilometres per hour, but with relatively low wave heights (c. 1 m). When they reach shallower coastal waters, however, the individual waves close up, heighten and become steeper and their **energy** is concentrated to produce extremely destructive waves several metres high. Offshore earthquakes in Japan and Alaska have generated damaging tsunamis in the past, as have major volcanic eruptions such as that of **Krakatoa**. Pacific Ocean coasts are particularly prone to tsunamis, and since potential loss of life and property damage is high there is now a system of tide gauges and seismographs to give advanced warning of their occurrence.

Further reading
Bernard, E.N. (1991) *Tsunami Hazard: A Practical Guide to Tsunami Hazard Reduction*, Dordrecht: Kluwer.
Murty, T.S. (1977) *Seismic Sea Waves: Tsunamis*, Ottawa: Department of Fisheries and the Environment.

TTAPS SCENARIO

The scenario developed to explain the onset of **nuclear winter**. TTAPS is an acronym based on the initials of the scientists who developed the original hypothesis – Turco, Toon, Ackerman, Pollack and Sagan.

Further reading
Turco, R.P., Toon, O.B., Ackerman, T.P., Pollack, J.B. and Sagan, C. (1983) 'Nuclear winter: global consequences of multiple nuclear explosions', *Science* 247: 166–76.

TU-144

A **supersonic transport** developed by Tupolev in the Soviet Union in the 1970s. Similar to the Anglo-French **Concorde**, but less successful as a commercial venture, it aroused the same environmental concerns. The type is no longer in operation.

TUNDRA

A major **biome** located polewards of the **taiga** or boreal forest. Tundra-like **ecosystems** are

Figure T-9: The global distribution of tundra

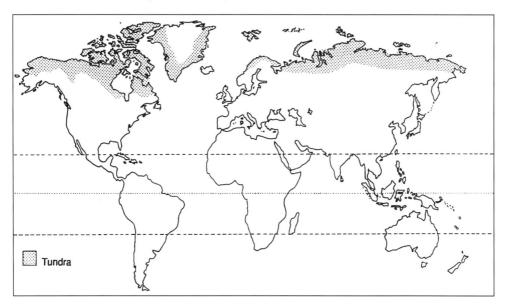

Tundra

also found above the treeline in mountainous areas – alpine tundra. The vegetation in the tundra consists of **lichens**, mosses, grasses and short herbaceous perennials such as heaths and various berry bushes. Mosses, sedges and dwarf willows are common where the tundra is wet or swampy. The tundra is underlain by **permafrost** which restricts **soil** development and hampers drainage. Combined with the low summer **temperatures**, limited **precipitation** – c. 25 mm per year – and strong **winds**, the poor soil and drainage effectively prevent the growth of trees, except in a few sheltered locations where favourable microclimatic conditions may allow dwarf versions of birch and willow to survive. During the short summers, the upper few centimetres of the tundra thaw, but since the **water** is unable to drain it remains on the surface, producing waterlogged soils and shallow ponds. The tundra supports few permanent animal residents. Only musk oxen, caribou, Arctic hare, Arctic foxes, wolves and various small rodents are hardy enough to survive the winter. The bird population is mainly migratory, with a large influx of ducks and geese taking place every summer when melting creates a large number of ponds and abundant insects are available for food. Because of the short growing

season, damage to the tundra ecosystem is slow to repair, and exploration for minerals and **petroleum**, the development of mines or the construction of pipelines must be undertaken with great care.

Further reading
Bliss, L.C., Heal, D.W. and Moore, J.T. (eds) (1981) *Tundra Ecosystems: A Comparative Analysis*, Cambridge/New York: Cambridge University Press.
Smith, R.L. (1980) *Ecology and Field Biology* (3rd edition), New York: Harper & Row.

TURBIDITY

See **atmospheric turbidity**.

TURBULENT FLOW

Irregular, unco-ordinated motion in a fluid. During turbulent flow the direction and velocity of the particles in a fluid vary continuously. Turbulent flow is common in the natural **environment**, where it contributes to mixing in flowing **water** and in the **atmosphere**.

See also
Eddy diffusion, Laminar flow.

TYNDALL, J. (1820-1893)

British scientist who described how the
scattering of light by particles in its path
causes a visible light beam to form – the
Tyndall effect. He was also one of the first to
investigate the link between increasing
atmospheric **carbon dioxide** (CO_2) and **global
warming**.

See also
Arrhenius, S.

TYPHOON

See **tropical cyclone**.

U

ULTRASONIC WAVES

Waves that are similar in form to **sound** waves, but have frequencies above the audible range. Ultrasonic scanning is widely used in the medical field, where other methods of diagnosis might cause damage to the object being examined. Ultrasound has replaced exploratory surgery, for example, for the examination of soft tissue or internal organs and is much safer than **X-rays** for the observation of foetuses in the womb. Ultrasonic waves are used in industry to clean objects too delicate for normal cleansing and ultrasonic scanners are used to examine the integrity of **metal** objects. Pipeline welds are routinely tested using ultrasound, for example.

Further reading
Ensminger, D. (1988) *Ultrasonics: Fundamentals, Technology and Applications*, New York: M. Dekker.

ULTRAVIOLET RADIATION

High **energy**, **short-wave radiation** lying between **visible light** and **X-rays** in the **electromagnetic spectrum**. It is usually divided into ultraviolet-A (UV-A) with wavelengths of 320–400 nanometres (nm), ultraviolet-B (UV-B) with wavelengths of 280–320 nm and ultraviolet-C (UV-C) with wavelengths of 200–280 nm. Ultraviolet rays are an important component of **solar radiation**. At normal levels it is an important germicide and is essential for the synthesis of Vitamin D in humans. At elevated levels, it causes sunburn and skin **cancer**, and can produce changes in the genetic make-up of organisms. It also has a role in the formation of **photochemical smog**. Most of the UV radiation which reaches the earth from the sun is absorbed by the **ozone layer** in the **stratosphere**. Thinning of the ozone layer, however, has increased the proportion of ultraviolet radiation – particularly UV-B – reaching the earth's surface, giving rise to fears of an increasing incidence of skin cancer and other radiation-related problems.

See also
Antarctic ozone hole, Melanoma, Ozone depletion.

Further reading
Cutnell, J.D. and Johnson, K.W. (1995) *Physics* (3rd edition), New York: Wiley.
Gribbin, J. (1993) *The Hole in the Sky* (revised edition), New York: Bantam.

UNITED NATIONS CONFERENCE ON DESERTIFICATION (UNCOD)

A conference held in Nairobi, Kenya in 1977 that established the modern approach to the problem of **desertification**. The role of human activities in creating the land degradation which led to desertification was considered to be firmly established and the contribution of **climate change** through **drought** was seen as secondary. Over a period of fifteen years following the conference, some $6 million were spent through the **United Nations Environment Program** (UNEP) in an attempt to alleviate the problem, but few of the countermeasures were effective. Current thinking on desertification suggests that the data on which UNCOD's findings were based were flawed. In concentrating on the human element, they failed to give sufficient weight to natural factors such as rainfall variability and the extent of annual fluctuations in vegetation boundaries. The net result was that the extent of irreversible desertification was overestimated. Failure to appreciate the various potential causes of desertification may also have limited the response to the problem. Different causes would normally elicit different responses, and UNEP's application of the societal response to all areas without distinguishing the cause may in part explain the lack of success in dealing with the problem.

See also
Desertification Convention.

Further reading
Hulme, M. and Kelly, M. (1993) 'Exploring the links between desertification and climate change', *Environment* 35: 4–11 and 39–45.
Nelson, R. (1990) *Dryland Management: The Desertification Problem, World Bank Technical Paper No. 16*, Washington, DC: World Bank.

UNITED NATIONS CONFERENCE ON ENVIRONMENT AND DEVELOPMENT (UNCED)

An international conference – the **Earth Summit** – held in Rio de Janeiro, Brazil in 1992, with the theme of **sustainable development**, based on economically and environmentally sound principles. To uphold that theme, UNCED established a Sustainable Development Commission and produced two main documents – the **Rio Declaration** and **Agenda 21** – plus a variety of treaties and conventions on topics ranging from **climate change** to **biodiversity**, reached as the result of compromise among the 150 participating nations. Many of the agreements currently lack the legal and financial commitment required for successful implementation, but the widespread participation in the Conference by politicians and scientists and its reporting by journalists added momentum to the growing concern for global environmental issues without which future progress will not be possible. UNCED was essentially a conference dominated by heads of state and government ministers, but to diversify the discussion and to illustrate the depth of concern for the **environment,** a parallel conference of non-government

organizations (NGOs) – the **Global Forum** – was held at the same time.

Further reading
Parson, E.A., Hass, P.M. and Levy, M.A. (1992) 'A summary of the major documents signed at the Earth Summit and the Global Forum', *Environment* 34: 12–15 and 34–6.
Hass, P.M., Levy, M.A. and Parson, E.A. (1992) 'The Earth Summit – How should we judge UNCED's success?', *Environment* 34: 7–11 and 26–33.

UNITED NATIONS CONFERENCE ON THE HUMAN ENVIRONMENT (UNCHE)

Held in Stockholm, Sweden in 1972, UNCHE ushered in the modern era in environmental studies. Participants in the Conference recognized that steps had to be taken to deal with growing threats to the **environment,** and formalized that recognition by signing a Declaration of the Human Environment. UNCHE also established the **United Nations Environment Program** (UNEP) and initiated a number of programmes aimed at gathering the raw data required to understand the nature of the environmental issues facing the world at that time. Projects were developed to examine **air** and **water pollution, climate,** drought forecasting, **desertification** and water **resources.** UNCHE was the first of a series of UN initiatives, including the **World Commission on Environment and Development,** which led ultimately to the **Earth Summit** in Rio in 1992. A public Environment Forum and a People's Forum, organized by concerned environmentalists, were held at the same time as UNCHE, but separate from it.

Further reading
Friends of the Earth (1972) *The Stockholm Conference – Only One Earth: An Introduction to the Politics of Survival*, London: Earth Island.
Rowland, W. (1973) *The Plot to Save the World: The Life and Times of the Stockholm Conference on the Human Environment*, Toronto: Clarke, Irwin.

UNITED NATIONS CONVENTION TO COMBAT DESERTIFICATION (UNCCD)

The **United Nations Conference on Environment and Development** (UNCED) held at Rio in 1992 included discussion on a proposal aimed at addressing the problems of those areas suffering from **desertification**. Subsequent negotiation led to a UN Convention to Combat Desertification being signed by 110 nations in 1994. The Convention recognizes the importance of both local and international activities in the fight against desertification and emphasizes the role of education, training and participation at all levels in that fight. Developed nations are expected to mobilize funding for action programmes which will involve co-ordination of their activities with those of the recipients. In the longer term, the movement to combat desertification is seen to be part of the broader objective of attaining **sustainable development** espoused at Rio.

See also
UN Conference on Desertification.

Further reading
Pearce, F. (1992) 'Miracle of the Shifting Sands', *New Scientist* 136 (1851): 38–42.
Williams, M., McCarthy, M. and Pickup, G. (1995) 'Desertification, drought and landcare: Australia's role in an international convention to combat desertification', *Australian Geographer* 26 (1): 23–32.

UNITED NATIONS DEVELOPMENT PROGRAM (UNDP)

A UN agency established in 1951 to provide development aid for **Third World** nations. Initially, its approach was to provide traditional economic and technical aid to advance the economies of the developing countries and in so doing help to eradicate poverty. In recent years it has added an environmental element to its aid, regularly funding projects aimed at dealing with environmental problems. It also encourages funding of projects that have the potential to advance **sustainable development**. In conjunction with other agencies such as the **World Bank** and the **UN Environment Program** (UNEP), UNDP also provides funds to offset the extra costs often faced by **Third World** nations when they embark on schemes for the protection and rehabilitation of the **environment**.

Further reading
Starke, L. (1990) *Signs of Hope: Working Towards Our Common Future*, Oxford/New York: Oxford University Press.

UNITED NATIONS EDUCATIONAL, SCIENTIFIC AND CULTURAL ORGANIZATION (UNESCO)

A UN agency established in 1945 to promote collaboration among member states in the areas of education, scientific research, the arts and other aspects of culture. It is affiliated with the **UN Environment Program** (UNEP) and contributes to environmental studies through its promotion of such activities as the **Man and the Biosphere Program** (MAB) which was launched in 1971 to study human impact on the **biosphere**, the International Geological Correlation Program (IGCP) formed in 1972 to encourage international collaboration for the solution of geological problems, the **International Hydrological Program** (IHP) formed in 1975 to improve the management of **water resources** and the Intergovernmental Oceanographic Commission established in the early 1960s to co-ordinate global ocean science programmes.

Further reading
Hajnal, P. (1983) *Guide to UNESCO*, London/New York: Oceana Publications.

UNITED NATIONS ENVIRONMENT PROGRAM (UNEP)

Formed at the **UN Conference on the Human Environment** (UNCHE) in 1972, UNEP co-ordinates international measures for monitoring and protecting the **environment**. It is responsible for the **Earthwatch** programme, which collects environmental data and monitors trends, and supports Earthscan, an environmental news and information agency. Its mandate is broad, producing affiliations with the **World Health Organization** (WHO), the Food and Agriculture Organization (FAO), the World Conservation Union (IUCN) and the **World Climate Program**

(WCP). Together with the **World Meteorological Organization** (WMO), UNEP sponsored the **Intergovernmental Panel on Climate Change** (IPCC), whose 1995 report on **climate change** was arguably the most comprehensive account of the impact of society on past, present and future climates. It was also involved in the development and implementation of the **Montreal Protocol**. The UNEP is based in Nairobi, Kenya with numerous regional offices in other UN member nations.

(UNITED STATES) DEPARTMENT OF ENERGY (DOE)

Created in 1977, through the amalgamation of a number of other agencies involved with **energy** matters, the Department of Energy is responsible for a wide range of activities including the regulation of energy prices, the enforcement of **conservation** measures and the control of licences and permits. In addition to its involvement with conventional **fuels**, the DOE is responsible for the US **nuclear energy** programme. Many of these activities involve an environmental element, and the department's interest in that area is reflected in its research projects, which include the investigation of solar, **geothermal** and other **renewable energy** sources, clean **coal** technology, **nuclear waste** disposal and **energy conservation**. It also supports research on technologies and strategies to mitigate increase in **carbon dioxide** (CO_2) and other energy-related **greenhouse gases**.

(UNITED STATES) DEPARTMENT OF THE INTERIOR

The Department of the Interior comprises the **US National Park Service**, the US Fish and Wildlife Service, the US Geological Survey, the Bureau of Land Management, the Minerals Management Service, the Office of Surface Mining, the Bureau of Mines and the Bureau of Reclamation. As such it has a wide-ranging responsibility for environmental issues in the United States. For most of its existence, the Department of the Interior has been broadly conservationist in its approach

to the **environment**, except in the 1980s, when the Reagan and Bush administrations installed Secretaries of the Interior who advocated greater development of **resources** in areas controlled by the Department.

See also
Conservation.

Further reading
Smith, Z.A. (1995) *The Environmental Policy Paradox*, Englewood Cliffs, NJ: Prentice-Hall.

(UNITED STATES FOREST SERVICE) MULTIPLE USE SUSTAINED YIELD ACT (MUSYA)

Passed in 1960, MUSYA was the first of a series of legislative attempts to deal with a growing conflict in the use of US national forests. In the act, the traditional use of forests – lumbering – was seen as only one of several activities that had to be considered in balanced forest management. The other activities recognized as important were watershed management, wildlife **habitat** preservation and recreation, all of which had to be taken into account when forest management plans were being developed. Timber harvesting was still considered important, but it had to be conducted in such a way as to allow a sustained yield. The Forest and Rangeland Renewable Resource Planning Act (FRRRPA) of 1974 and the National Forest Management Act (NFMA) of 1976 clarified and developed the elements set out in MUSYA. Both of these acts provided more practical guidance on how the provisions of MUSYA could be met. In NFMA, for example, the extent of clear cutting, harvest rates and the preservation of stream corridors were considered and **wilderness** reviews established, all with the basic purpose of reconciling the concerns and needs of environmentalists, timber companies and recreationalists.

See also
Conservation, Sustainable development.

Further reading
Cutter, S.L., Renwick, H.L. and Renwick, W.H. (1991) *Exploitation, Conservation, Preservation:*

A Geographic Perspective on Natural Resource Use, New York: Wiley.

UNITED STATES GLOBAL CLIMATE PROTECTION ACT (1987)

An act aimed at supporting research into the nature, timing and likely impact of **global warming**. Despite its title, it contained no practical proposals for **climate** protection, and was quite typical of the approach to global warming at that time, which advocated 'business-as-usual' until additional research had been done.

See also
'Business-as-usual' scenario, 'Wait-and-see' scenario.

UNITED STATES NATIONAL PARK SERVICE

The National Park Service was established in 1916 to administer previously existing national parks such as Yellowstone and Yosemite and to promote and regulate the use of new parks. Central to its mandate was the conservation of the natural beauty, history and **flora** and **fauna** of the parks and monuments under its control, in such a way that they might be enjoyed by the existing **population** and also by future generations. When populations were small and access more difficult, this potential conflict between preservation and use caused few problems, but it now creates major headaches for the Service in many areas.

Further reading
Cox, G.W. (1993) *Conservation Ecology: Biosphere and Biosurvival,* Dubuque, IA: Wm C. Brown.

UNITED STATES WILDERNESS ACT (1964)

An act that came into being in large part because of pressure from environmental groups in the 1960s to preserve **wilderness**. As defined by the act, wilderness areas had to be large enough (> 5000 acres) to make preservation practicable, have no noticeable human impact, possess outstanding oppor-

tunities for solitude and primitive recreation and include physical features of scientific, scenic, educational or cultural value. Most of the areas initially designated as wilderness were in the western states, with the largest in Alaska, but in 1974 an additional act allowed for the creation of wilderness areas in the eastern states where the human impact on the **environment** is generally greater and the potential wilderness areas therefore smaller. The total officially designated wilderness area in the United States now exceeds 90 million acres (36 million hectares), and environmental groups continue to push for an increase in that total.

See also
Muir, J., Sierra Club.

Further reading
Frampton, G.T. (1988) 'Wilderness Act, 25 years', *Wilderness* 52 (183): 2.

UPPER WESTERLIES

Westerly **winds** that blow in the upper **atmosphere** close to the **tropopause** in mid- to high latitudes. It is in these upper westerlies that **Rossby waves** form and the changing paths that they follow are responsible for the **zonal index**. Faster flowing sections in the upper westerlies make up the **jet streams**.

URANIUM (U)

A naturally occurring radioactive **metal**, which is the principal **element** used in the production of **nuclear energy**. Its principal **ore** is pitchblende, in which the uranium is present in the form of uranium oxide (U_3O_8). Natural uranium consists of two main isotopes ^{238}U (99.28 per cent) and ^{235}U (0.7 per cent) with very small amounts of other isotopes such as ^{234}U. Only ^{235}U is capable of sustaining a nuclear **chain reaction**, and is therefore important as a **fuel** for **nuclear reactors** and nuclear weapons. ^{238}U can be treated or enriched so that it becomes fissionable, after which it can be used as a nuclear fuel. The Canadian **CANDU nuclear reactor** has been designed so that it can use natural uranium without treatment or enrichment.

Figure U-1 Radioactive waste production during the mining, refining and utilization of uranium

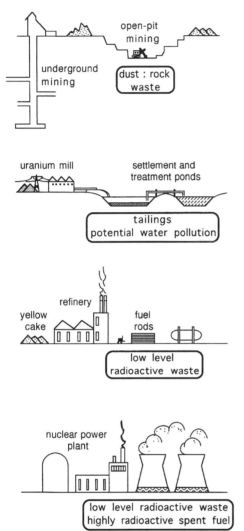

Figure U-1 Radioactive waste production during the mining, refining and utilization of uranium

One kilogram of uranium contains the thermal **energy** equivalent of about 12,000 barrels of crude **oil**.

See also
Nuclear fission, Radioactivity.

Further reading
De Vivo, B. (1984) *Uranium Geochemistry, Minerology, Geology, Exploration and Resources*, London: Institution of Mining and Metallurgy.

URANIUM MILL TAILINGS

Radioactive **waste** material created during the processing and production of enriched **uranium** (U) from **uranium ore**. For every kilogram of reactor **fuel** produced, between 500 and 900 kilograms of waste are produced. In the past this was commonly piled up on the mill property where it could be picked up by **water** or **wind** and carried away to contaminate the surrounding area. In some places the waste was used in road construction or as landfill in residential and recreational areas. Although the levels of **radioactivity** in the tailings are low, such uses have now been discontinued, since there is some concern that exposure to even low levels of radioactivity can have cumulative effects leading to illness.

Further reading
Ritcey, G.M. (1989) *Tailings Management: Problems and Solutions in the Mining Industry*, Amsterdam/New York: Elsevier.

URBAN HEAT ISLAND

The name given to the situation in which at certain times the **temperatures** within the urban built-up area are higher than those in the surrounding rural areas. Average values of the temperature difference range between 1–3°C, with maximum recorded values in excess of 10°C. The heat island is a night-time phenomenon, best developed when **wind** speeds are low and skies clear, and most common in the summer. Its development comes about as a result of changes produced when the built **environment** is created. The replacement of vegetation and **soil** by brick, concrete and asphalt changes the **heat budget** of the area. The new materials have a greater heat capacity than the natural ones, and overnight they continue to release heat into the urban environment for some time after the natural surfaces. In addition, **evaporation** and **transpiration** rates in the built-up area are lower than in the surrounding rural area, because of the lack of vegetation and the rapid surface **runoff**, which leaves little surface **water** in the city. As a result, the **energy** normally required for these processes

Figure U-2 The morphology of an urban heat island

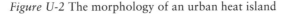
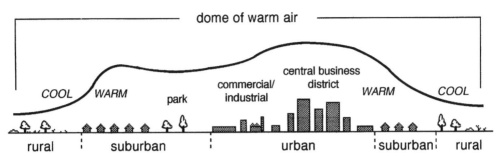

is available for raising the temperature of the urban **air**. Urban areas also release heat from the **combustion** of **fossil fuels** and the ventilation of buildings which contributes to the heat island.

Further reading
Landsberg, H. (1981) *The Urban Climate*, New York: Academic Press.
Oke, T.R. (1987) *Boundary Layer Climates* (2nd edition), London: Methuen.

URBAN RENEWAL

The renewal or redevelopment of rundown urban areas. It applies particularly to the central business districts of larger cities, where buildings have been abandoned or allowed to deteriorate. Renewal involves the renovation of existing buildings or their replacement by completely new structures plus reconsideration of such urban elements as the road network or open spaces such as parks or plazas. If successful, the net effect of urban renewal is an overall improvement in the urban **environment**, an increase in residential property in the downtown area, the revitalization of the retail and service sectors and an expansion of the city's tax base. The urban renewal approach can also be applied to industrial areas, to replace the facilities used by heavy industries with new units which support modern 'high-tech' companies involved in the electronics industry or other light manufacturing enterprises.

Further reading
Gibson, M.S. and Lanfstaff, M.J. (1982) *An Introduction to Urban Renewal*, London: Hutchinson.

V

VACUUM

A space that is completely empty of **molecules** or **atoms**. Such a situation is impossible to achieve, and in most cases the term indicates a space containing **air** or a **gas** at very low **pressure**. Vacuum pumps and **distillation** units that take advantage of that low pressure are used in industry and vacuum filters are used to dewater sludge during **sewage treatment**.

VALENCE

Valency. A measure of the combining capacity of an **atom**, usually defined as the number of **hydrogen** (H) atoms that an atom will combine with or replace. The valence of **oxygen** (O) as indicated when it combines with hydrogen, for example, in **water** (H_2O) is two, and for **carbon** (C) as indicated when it combines with hydrogen in **methane** (CH_4) is four.

Further reading
Winter, M.J. (1994) *Chemical Bonding*, New York: Oxford University Press.

VAN ALLEN RADIATION BELTS

Two belts of charged particles trapped by the earth's magnetic field at heights of 3000 and 16,000 km above the earth's surface. Particles in the inner belt are probably created when **cosmic radiation** strikes the **atmosphere** whereas those in the outer layer originate from the sun. The belts were discovered in 1958 by James Van Allen using data provided by some of the first satellites and space probes sent through the upper atmosphere

See also
Magnetosphere.

Further reading
Roederer, J.G. (1970) *Dynamics of Geo–magnetically Trapped Radiation*, Berlin/New York: Springer-Verlag.

VAPOUR

A substance in a gaseous state.

See also
Gas, Gas laws.

VAPOUR PRESSURE

The **pressure** exerted by **vapour**. In the **atmosphere**, for example, **atmospheric pressure** includes the pressure exerted by **water vapour**. Close to the surface, it contributes a partial pressure of between 0.5 and 3.0 kilopascals (kpa) to the total atmospheric pressure. For any given **temperature**, there is a limit to the amount of water vapour the atmosphere can hold, which creates an upper limit for vapour pressure. This is the saturated vapour pressure. Above this level, the introduction of additional vapour will produce **condensation**. Saturated vapour pressure is linked to temperature, however, and any increase in temperature will allow it to increase. Conversely, any decrease in temperature will reduce the saturated vapour pressure and cause condensation to occur.

See also
Humidity.

Further reading
Peixoto, J.P. and Oort, A.H. (1992) *Physics of Climate*, New York: American Institute of Physics.

VARVES

Thin layers of laminated sediments deposited in waterbodies in glacial **environments**. Each varve consists of two layers, which together represent deposition over one year. The lower and slightly thicker layer consists of relatively coarse-grained **sand** or **silt**, produced by greater meltwater activity during the warm season. It grades into a thinner layer of finer silt and **clay**, which represents the settling out of sediments in suspension during the cold season, when the streams which would normally carry sediment into the waterbody are frozen. Since varves are formed annually, they can be used for relative dating in much the same way as **tree rings**. The technique was developed by a Swedish geomorphologist, Gerard De Geer, who in the early part of the twentieth century established a chronology of glacial retreat in Sweden based on the counting of varves. Varve-like laminated sediments with an annual periodicity are also deposited in other environments. They are usually referred to as rhythmites.

Figure V-1 The correlation of varve sequences

VARVE SEQUENCES
(from cores or exposed sections)

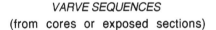

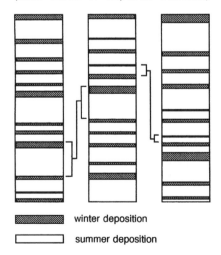

winter deposition

summer deposition

Further reading
Flint, R.F. (1971) *Glacial and Quaternary Geology*, New York: Wiley.
Rutter, N.W. (ed.) (1985) *Dating Methods of*

Pleistocene Deposits and their Problems, St John's, Nfd: Geological Society of Canada.
Thornes, J.B. and Brunsden, D. (1977) *Geomorphology and Time*, London: Methuen.

VARVE DATING

See **varves**.

VELOCITY

A combination of speed and direction. The velocity of **wind**, for example, is the wind speed and the direction from which it flows. In popular usage, velocity is usually taken to refer to speed alone.

VIENNA CONVENTION FOR THE PROTECTION OF THE OZONE LAYER

The product of a meeting held in Vienna, Austria in 1985. Signed by twenty nations, it promised international co-operation in research, monitoring and the exchange of information on the problem of **ozone depletion**. Discussion arising out of the convention led ultimately to the signing of the **Montreal Protocol** on Substances which Deplete the Ozone Layer in 1987.

VILLACH CONFERENCE

The first of the major, modern environmental conferences to deal with the rising levels of **greenhouse gases** in the **atmosphere** and their impact on **climate**, held at Villach, Austria in 1985. Organized by the **WMO**, **UNEP** and **ICSU**, it examined existing data on the role of **carbon dioxide** (CO_2) and other greenhouse gases in climate variation, recognized that understanding of the problem was incomplete and suggested ways in which scientists, policy makers and governments should proceed in dealing with the issue. An **Advisory Group on Greenhouse Gases** (AGGG) was established to ensure that the recommendations of the Conference were followed up, and sponsored a technical workshop at Villach in 1987 to encourage the development of policies for responding to climate change. The Villach Conference, the AGGG and the Villach Technical Workshop laid the groundwork for the increasing number of studies of **global warming** in succeeding years which led directly to the production of the first **IPCC** climate change assessment in 1990.

VILLACH TECHNICAL WORKSHOP (1987)

See **Villach Conference**.

VIRUSES

Sub-microscopic, non-cellular particles which cause disease. They are generally less than 200 nanometers in diameter and are only visible using an electron microscope. Viruses consist of a core of **nucleic acid** covered by a coating of **proteins**. They are intracellular **parasites** that cannot reproduce outside living **cells**, and each type of virus requires its own host. The tobacco mosaic virus infects only plants cells, for example, and the rabies virus only mammals. Some viruses are particularly specialized and will only infect specific cells, such as liver cells, spinal nerve cells or blood cells. They are spread by insects, direct contact, droplet infection and the exchange of body fluids, and cause a great variety of diseases, including mosaic diseases of cultivated plants, **myxomatosis** and foot-and-mouth disease among animals, the common cold, measles, and Acquired Immune Deficiency Syndrome (AIDS) among humans. Viral diseases can be controlled in society by preventing transmission or by using vaccines and a range of new antiviral drugs. Many viruses mutate quite rapidly and regularly, however, making it difficult to combat them. The eradication of the smallpox virus was a spectacular success for modern disease control, but the AIDS virus continues to evade control and even common viruses such as those that produce colds and flu remain difficult to deal with.

See also
Bacteria.

Further reading
Matthews, R.E.F. (1992) *Fundamentals of Plant Virology*, San Diego, CA: Academic Press.
Murphy, W.B. (1981) *Coping with the Common Cold*, Alexandria, VA: Time-Life Books.

VISCOSITY

A measure of the resistance of a fluid to internal flow. The viscosity of a fluid reflects the strength of the forces that hold the **molecules** together and give it substance. The greater the strength of the intermolecular forces, the greater will be the viscosity. Viscosity varies from fluid to fluid – compare **water** with treacle, for example – and also varies with **temperature**, being less when temperatures are higher. Since it determines the flow rate of fluids, viscosity has practical environmental implications. The viscosity of **oil** in oil spills or the environmental conditions under which the spill occurs – arctic or tropical – may influence its initial impact or place constraints on the methods available for clean-up.

See also
Oil pollution.

Further reading
White, F.M. (1991) *Viscous Fluid Flow* (2nd edition), New York: McGraw-Hill.

VISIBLE LIGHT

Radiation from that part of the spectrum, with wavelengths between 0.4 μm and 0.7 μm, to which the human eye is sensitive. Visible light varies in colour, the shorter wavelengths being blue and the longer wavelengths red, indicative of its position between **ultraviolet** and **infrared radiation** in the **electromagnetic spectrum**.

VITAMINS

Organic substances which are a necessary part of the human diet. Thirteen essential vitamins contribute to a variety of functions including the formation of red blood cells, the **metabolism** of **carbohydrates** and **amino acids**, the fixing of **calcium** (Ca) and **phosphorus** (P) for bone development and the maintenance of **cell** membranes. All vitamins are available from plant or animal tissue – for example, vitamin B_1 from beans, peas and **yeast**, vitamin B_{12} from liver, vitamin C from citrus fruits, vitamin E from cereals and green vegetables – and in addition vitamin D is manufactured in body surface tissues following the absorption of **ultraviolet radiation** by the skin. Although vitamins are required in only small quantities for the normal health and development of the body, deficiencies can lead to a variety of diseases. Insufficient vitamin C, for example, leads to scurvy, vitamin B_{12} deficiency can cause pernicious anaemia and inadequate amounts of vitamin D, particularly in children, cause rickets and other bone diseases. Some of the health problems associated with **malnutrition** in **Third World** nations are compounded by vitamin deficiencies. Vitamins are readily available in the form of pills or tablets and vitamin therapy has become an integral part of a healthy lifestyle for many individuals in developed nations.

Further reading

Carpenter, K.J. (1986) *The History of Scurvy and Vitamin C*, Cambridge/New York: Cambridge University Press.

VOLATILE

Changing easily into a **vapour**. Many **organic compounds** are volatile and contribute to **air pollution** either directly or indirectly through combination with other substances after vaporization. Volatile organic compounds (VOCs) include acetone, ethylene, **benzene** and propylene. When used as **solvents** in cleaning fluids, paints and varnishes and in industrial processes, VOCs can be an important source of indoor pollution, possibly contributing to the so-called **sick building syndrome**.

VOLCANIC EXPLOSIVITY INDEX (VEI)

An index for comparing individual volcanic eruptions. It is based on volcanological criteria such as the intensity, dispersive power and destructive potential of the eruption, as well as the volume of material ejected.

See also

Dust veil index, Glaciological volcanic index.

Further reading

Chester, D.K. (1988) 'Volcanoes and climate: recent volcanological perspectives', *Progress in Physical Geography* 12: 1–35.
Newhall, G.C. and Self, S. (1982) 'The volcanic explosivity index (VEI): an estimate of the explosive magnitude for historical vulcanism', *Journal of Geophysical Research* 87: 1231–8.

VOLCANO

A vent or fissure in the earth's **crust** through which **magma**, **gases** and **solids** such as volcanic ash are ejected during a volcanic eruption. The proportions of these various products and the rate at which they are ejected is quite variable. The volcanoes of the Hawaiian Islands, for example, produce quiet eruptions characterized by large amounts of mobile **lava** and some gas, whereas others such as Mount Vesuvius erupt explosively, resulting in lava, **dust**, ash and cinders being ejected to great heights in the **atmosphere**. Volcanoes are associated with zones of weakness in the earth's crust, such as those along tectonic plate boundaries (see Figure P-10). They contribute directly to landscape formation through the materials they bring to the surface, creating the typically conical shapes associated with deposition around volcanic vents or the more subdued landscapes of fissure eruptions and the resultant extensive lava flows. Volcanoes can also disrupt the **energy** flow in the earth/atmosphere **system**, through their contribution to **atmospheric turbidity**, and thus bring about **climate change**. In human terms, volcanoes have been the cause of major disasters with considerable loss of life, but at the same time volcanic **soils** are often very fertile and can support a variety of arable agricultural activities.

See also

Dust veil index, Igneous rocks. Krakatoa, Mount Agung, Mount Pinatubo, Mount St Helens, Mount Tambora, Plate tectonics.

Further reading

Chester, D.K. (1993) *Volcanoes and Society*, London: Edward Arnold.
Decker, R. and Decker, B. (1995) *Volcanoes*,

Figure V-2 The nature of volcanic eruptions and the volcanoes they produce

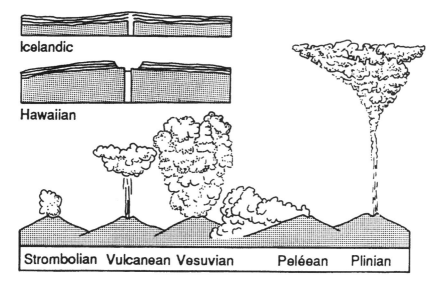

Source: After Goudie, A. (1989) *The Nature of the Environment* (2nd edition), Oxford: Blackwell

Oxford: W.H. Freeman.

Lamb, H.H. (1970) 'Volcanic dust in the atmosphere; with a chronology and assessment of its meteorological significance', *Philosophical Transactions of the Royal Society A* 266: 435–533.

Ollier, C.D. (1988) *Volcanoes*, Oxford: Blackwell.

'WAIT-AND-SEE' SCENARIO

The maintenance of the status quo until the nature and extent of an environmental change can be verified. It has been applied particularly to **global warming**, in which causes, timing and potential impact still include a considerable degree of uncertainty. Given the uncertainty involved, it can be argued that a premature response, in both environmental and socioeconomic terms, might do more harm than good. However, responding only when all uncertainties have been resolved may allow detrimental impacts to become well established, making mitigation more difficult and costly. The 'wait-and-see' scenario has much in common with the 'business-as-usual' scenario in which problems are not considered significant enough to warrant any change in existing activities.

Further reading
IPPC (1990) *Climate Change: The IPCC Scientific Assessment*, Cambridge: Cambridge University Press.
Waterstone, M. (1993) 'Adrift in a sea of platitudes. Why we will not resolve the greenhouse issue', *Environmental Management* 17: 141–52.

WALDSTERBEN

The destruction of the forests. A term coined in Germany to describe the damage caused to forests by **acid rain**.

See also
Tree dieback.

Further reading
Ulrich, B. (1990) 'Waldsterben: forest decline in West Germany', *Environmental Science and Technology* 24 (5): 436–41.

WALKER CIRCULATION

A strong latitudinal or zonal circulation in the equatorial **atmosphere** which contrasts with the normal meridional circulation. It is particularly well marked in the Pacific Ocean where it was first recognized by Sir Gilbert Walker in the 1920s as he sought to develop methods for forecasting rainfall in the Indian **monsoon**. The fluctuations in **pressure** that drive the Walker Circulation are referred to as the **Southern Oscillation**.

Figure W-1 The changing nature of the Walker Circulation and its influence on El Niño events

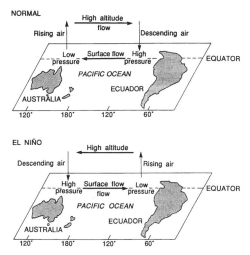

Source: After Goudie, A. (1989) *The Nature of the Environment* (2nd edition) Oxford: Blackwell

See also
El Niño, ENSO, La Niña.

Further reading
Goudie, A. (1989) *The Nature of the Environment* (2nd edition), Oxford: Blackwell.
Lockwood, J.G. (1979) *Causes of Climate*, London: Edward Arnold.

WASTE

Any material, **solid, liquid** or **gas**, that is no longer required by the organism or **system** that has been using it or producing it. Waste is an integral part of the **environment**, and the environment has developed a series of very efficient **waste disposal** systems which involve the **recycling** of the products. Organic waste, such as that produced by animals, is reduced by insects and **bacteria** into its constituent chemicals which are reabsorbed into the environment. The leaves discarded by **deciduous** trees in the autumn are treated in a similar fashion, and the process of **photosynthesis** prevents a build-up of the **carbon dioxide** (CO_2) given off as a waste gas by animals. Problems arise when waste is produced in such quantity that the normal disposal systems cannot cope or when the waste takes such a form that existing systems can dispose of it only slowly or in some cases not at all. **Population** growth, new lifestyles and a rapidly changing technology have contributed to an increase in the generation of waste and created serious waste disposal problems.

See also
Recycling, Sewage, Waste classification.

Further reading
Jones, B.F. and Tinzmann, M. (1990) *Too Much Trash?*, Columbus, OH: Zaner-Bloser.
Packard, V. (1968) *The Waste Makers*, Harmondsworth, Middlesex: Penguin Books.

WASTE CLASSIFICATION

Since almost any substance can become **waste**, there are an infinite number of ways of classifying it. There are some common groupings, however. Wastes can be classified according to their origin (for example, clinical waste, domestic refuse, agricultural waste, industrial waste, **nuclear waste**), form (**solid, liquid, gas**), or properties (**inert**, toxic, carcinogenic). Most wastes will fit into a number of such groupings.

Government organizations also develop classifications for special purposes such as waste management, **pollution** control, safety or taxation. One class of waste that receives much attention is that of hazardous waste, defined as waste particularly harmful to the **environment** or to society. Hazardous wastes may be dangerous because they are toxic, biologically active, flammable, corrosive, radioactive or a combination of these factors. The extent of the hazard posed by the waste will depend on the amount involved, its durability – for example, short-term or long-term toxicity – and particularly on the methods used to store or dispose of it. Most problems caused by hazardous waste can be traced to ignorance of or disregard for these factors.

See also
Carcinogen, Domestic waste, Garbage, Radioactivity.

Further reading
Berkhout, F. (1991) *Radioactive Waste: Politics and Technology*, London: Routledge.
Nemerow, N.L. and Dasgupta, A. (1991) *Industrial and Hazardous Waste*, New York: Van Nostrand Reinhold.

WASTE DISPOSAL

The storage or destruction of **waste** materials in such a way that the impact on the **environment** and on society is minimal. Dumping and **incineration** are well-established methods of waste disposal that have been modified to meet modern standards. In the past, domestic **sewage** was dumped directly into lakes, rivers and the sea and natural processes were allowed to integrate it back into the environment. Today, the volumes of sewage produced are so great that natural disposal is no longer an option, and it has to be treated mechanically and chemically before it is released. Similarly, the uncontrolled dumping of solid waste in gravel pits, on waste land and at sea has been replaced by disposal in **sanitary landfill** sites. Modern incinerators can attain a higher **combustion** efficiency and can be fitted with **scrubbers** or filters to eliminate hazardous emissions and reduce the **air pollution** that was once characteristic of waste incinerators. Waste disposal and **energy**

Figure W-2 Approaches to waste minimization

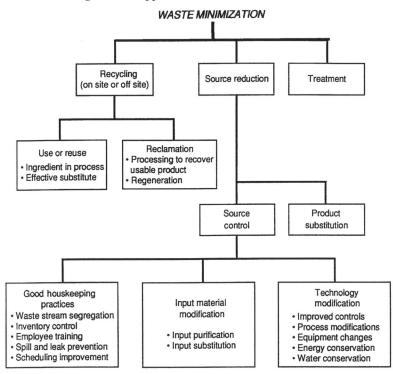

production are combined in **waste-to-energy incinerators**. Certain hazardous wastes require special disposal techniques. Hospital waste and **PCBs**, for example, are burned in specially developed incinerators, while **nuclear waste** requires its own unique systems to prevent the escape of **radioactivity**. Waste disposal is frequently combined with **recycling**. Glass, **metal** and **plastic** can often be reused, and are removed from the waste before final disposal, which eases pressure on sanitary landfill sites and reduces the amount of incinerator ash. Another approach designed to reduce the need for waste disposal is waste minimization in which an attempt is made to reduce the amount of waste produced in the first place. It may involve recycling or waste-to-energy incineration, but minimization can also be achieved by re-engineering the process that generates the waste or redesigning the product.

Further reading
Denison, R.A. and Ruston, J. (1990) *Recycling and*
Incineration, Washington, DC: Island Press.
Harrison, R.M. and Hester, R.E. (1995) *Waste Treatment and Disposal*, Cambridge: Royal Society of Chemistry.
Kharbanda, O.P. and Stallworthy, E.A. (1990) *Waste Management: Toward a Sustainable Society*, New York: Auburn House.
Kupchella, C.E. and Hyland, M.C. (1993) *Environmental Science: Living within the System of Nature* (3rd edition), Englewood Cliffs, NJ: Prentice-Hall.

WASTE-TO-ENERGY INCINERATOR

An **incinerator** that uses **waste** products as **fuel**, to provide **energy** for space or **water** heating. Various types of refuse are used, from simple paper products to **plastic** and scrap car tyres. In many cases they are used as fuel supplements, since on their own they have an **energy** content that may be only 30 to 50 per cent that of solid fuels. Advantages include the reduction of waste products deposited in **landfill sites** and a reduced demand for conventional fuels. However, these incinerators face handling and sorting problems, and in some cases contamination of the waste can cause hazardous emissions. The incineration of waste **oil** or **solvents** contaminated with **PCBs** can lead to the emission of **dioxins**, for example.

Further reading
Domino. F.A. (ed) (1979) *Energy from Solid Waste*, Park Ridge, NJ: Noyes Data Corp.

WASTE MINIMIZATION

See **waste disposal**.

WATER (H₂O)

Pure water is a colourless, odourless **liquid** that is a **compound** of **hydrogen** and **oxygen** (H_2O). Natural water in the **environment** is never pure, but contains a variety of dissolved substances. Sea water, for example, is a solution of **sodium chloride** (NaCl – common salt) and other **salts**; rainwater can be acidic because of the **carbon dioxide** (CO_2) that it contains and the water in rivers may include minerals dissolved from the rocks over and through which it has flowed. Water can exist as a **solid** (**ice**), **liquid** (water) or **gas** (**water vapour**) and changes readily from one to the other, either releasing or taking up **energy** as it does so. This property of water allows it to contribute significantly to the earth's **energy budget**.

Water is the largest constituent of all living organisms – human bodies comprise about 65 per cent water. It helps **cells** to maintain their form, and the chemical processes that are involved in **metabolism** take place in a watery **solution**. The **digestion** of food in mammals, the transportation of bodily **wastes** and the maintenance of a stable body **temperature** through perspiration and **evaporation** all require water. Plants require water to carry **nutrients** from the root zone into the body of the plant, to allow **photosynthesis** to take place and to support **transpiration**. Without a regular supply of water, organisms are unable to survive, as is evident during prolonged **drought**, when plants and animals become dehydrated and die. Water is not evenly distributed across the earth's surface. In some places there is too little, in others too much, and human beings spend much time, money and energy redistributing it. The major demand is for fresh water, but the proportion of fresh water on and in the earth's surface is severely limited.

Some 97 per cent of the world's water is in the **oceans**, while a further 2 per cent is in the form of ice and **snow**, which leaves only 1 per cent available as fresh water for plants and animals. Survival on such a small amount is made possible by the natural **recycling** of the water in the **hydrological cycle**, which not only replaces the water once it has been used, but also cleans it.

The demand for water is growing so rapidly that in some areas the hydrological cycle cannot replace it fast enough to meet the needs of domestic, industrial, agri-

Figure W-3 The physical distribution of the world's water

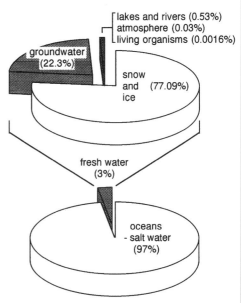

Source: Based on data in Peixoto, J.P. and Ali Kettani, M. (1976) 'The control of the water cycle', in F. Press and R. Seiver (eds) *Planet Earth*, San Francisco: W.H. Freeman

cultural, transportation and recreational consumers. Nor is it able to deal with the **pollution** created by them. Local schemes to offset the limited availability and poor quality of water are not uncommon, but the dimensions of the problem are now so great that continental scale water diversions are being considered. Disputes over water can become so serious that they lead to conflict or aggravate existing conflicts between states.

See also
Acid precipitation, Evapotranspiration, North

American Water and Power Alliance, Water quality.

Further reading
Agnew, C. and Anderson, E. (1992) *Water Resources in the Arid Realm,* London: Routledge.
Gleick, P.H. (1994) 'Water, War and Peace in the Middle East', *Environment* 36 (3): 6–15 and 35–42.
Postel, S. (1995) 'Facing a future of water scarcity', in J.L. Allen (ed.) *Environment 95/96,* Guilford, CN: Dushkin Publishing.
Speidel, D.H. (1988) *Perspectives on Water: Uses and Abuses,* New York: Oxford University Press.

WATER BALANCE

A book-keeping approach to the moisture budget, involving the comparison of moisture input (**precipitation**) and output (**evapotranspiration** and **runoff**) to provide a value for the net surplus or deficit of water at a specific location. Although some of the **elements** involved are difficult to measure or estimate accurately, water balance data are useful in water resource planning.

See also
Lysimeter, Moisture deficit, Moisture index, Thornthwaite, C.W.

WATER CYCLE

See **hydrological cycle.**

WATER POLLUTION

See **environmental pollution.**

WATER QUALITY

The definition of water quality will vary to some extent depending upon the proposed use. **Water** intended for **irrigation** or for certain industrial purposes will not have to meet the same quality standards as water intended for drinking, for example. However, there are certain factors that determine water quality whatever the use. They can be classified as physical properties, chemical properties and biological properties.

Water quality is under threat everywhere. Even in remote areas the chemical properties of water are being changed by **acid precipitation. Pollution** by domestic **sewage** is ubiquitous and although the release of industrial effluents into water bodies is subject to increasing control, it remains a serious problem. Water quality can also be reduced catastrophically as a result of major chemical spills, **oil spills** or the accidental release of untreated sewage.

Further reading
Gray, N.F. (1994) *Drinking Water Quality,* London: Wiley.
McDonald, A. and Kay, D. (1988) *Water Resources: Issues and Strategies,* Harlow: Longman.
Maybeck, M., Chapman, D.V. and Helmer, R. (eds) (1990) *Global Freshwater Quality: A First Assessment,* Cambridge, MA: WHO/UNEP/ Blackwell.

WATER QUALITY STANDARDS

Acceptable standards for **water quality** have been developed at both the national and international level. The **World Health Organization** (WHO) has published guidelines for drinking-water quality, for example, and the European Commission has produced a directive which applies to the quality of water intended for human consumption. Elsewhere, all the developed nations have established standards for water quality, and some progress has been made among the developing nations. Published standards typically include **maximum allowable concentrations** (MACs)

Table W-1: Properties that determine water quality

PHYSICAL PROPERTIES	CHEMICAL PROPERTIES	BIOLOGICAL PROPERTIES
colour: not necessarily harmful; aesthetic concern; streams flowing through peat often brown	*acidity and alkalinity:* fresh water tends to be neutral (pH 6–7); deviation from that may indicate contamination	*micro-organism content:* presence of algae and bacteria can make water unusable; *E. coli* bacterium indicates faecal contamination
turbidity: cloudiness caused by silt or bacteria; may require treatment such as filtration before use	*hardness:* caused by the presence of dissolved calcium and magnesium compounds; prevents soap from lathering and causes scale buildup in boilers	
taste and odour: presence of dissolved solids of biological, mineral or human origin; very small amounts of some chemicals make water unpalatable	*dissolved oxygen:* important for biological and chemical processes; levels indicated by chemical and biochemical oxygen demand	
temperature: influences the dissolved oxygen content		

Table W-2 Drinking water quality objectives for selected substances – Ontario, Canada

SUBSTANCE	LIMIT
Nitrate	10 ppm (MAC)
Sodium	20 ppm (AO)
Chloride	250 ppm (AO)
Hardness	80–200 ppm (OG)
Iron	0.3 ppm (AO)
Manganese	0.05 ppm (AO)
Coliforms	
Total coliforms	1 CFU/100 ml
Fecal coliorms	0 CFU/100 ml
Petroleum hydrocarbons	
Benzene	0.005 ppm (MAC)
Toluene	0.024 ppm (AO)
Ethylbenzene	0.0024 ppm (AO)
m, p-Xylene	0.30 ppm (AO)
o-Xylene	0.30 ppm (AO)

ppm = parts per million   CFU = colony forming units
MAC = maximum acceptable concentration
AO = aesthetic objective

of specific toxic **elements** – for example, **heavy metals** – and guide levels (GL) for those considered less harmful – for example, acidity or **calcium** (Ca). There is no guarantee, however, that established standards will be attained or enforced. In China, for example, standards for drinking-water are in place, but tap water is seldom safe for human consumption. Water quality standards can only be made to work if they are established in conjunction with pollution control programmes. In Britain, the UK **Environmental Protection Act (1990)** and the Inspectorate of Pollution work to deal with such problems and in the United States the **Safe Drinking-Water Act** (1974), the Water Quality Act (1987) and the **Environmental Protection Agency** (EPA) serve similar functions. Although water quality standards are considered mainly in human terms, they are also applied to agriculture, fisheries and certain industries.

Further reading
Edmonds, R.L. (1994) *Patterns of China's Lost Harmony,* London: Routledge
Keller, A.Z. and Wilson, H.C. (1992) *Hazards to*

Drinking Water Supplies, London/New York: Springer-Verlag.
Wheeler, D., Richardson, M.L. and Bridges, J. (eds) (1989) *Watershed 89. The Future for Water Quality in Europe, Vol II*, Oxford: Pergamon Press.

WATERSHED

See **catchment**.

WATER TABLE

The upper level of the saturated or **ground-water** zone in the rocks beneath the earth's surface. In general, the shape of the water table follows that of the surface, but in places it may reach the surface, creating ponds or natural springs. The depth of the water table in any one area varies with such factors as input from **precipitation**, loss through sub-surface flow and pumping of **groundwater** from wells.

WATER VAPOUR

Water in its gaseous state, produced from liquid water by **evaporation** or by **respiration** from animals and **transpiration** from plants. Its presence in the **atmosphere** contributes to **humidity** and through subsequent **condensation** to **precipitation**. Water vapour is also a **greenhouse gas**.

WATT

The **SI** unit of power, equivalent to 1 **joule** per second.

WEATHER

The current or short-term state of the **atmosphere** expressed in terms of such variables as **temperature, precipitation**, airflow and cloudiness.

WEATHER FORECASTING MODELS

Models which use fundamental equations representing atmospheric processes to predict short-term changes in meteorological elements. Currently this is the most common method of weather forecasting ranging from periods of several hours to five or six days ahead. The quality of the results is constrained by the compromise that has to be struck between the accuracy required and the cost of running the programme.

See also
Richardson, L.F.

WEATHER MODIFICATION

Any change in weather conditions caused by human activity, whether by design or accident. **Rain making** by seeding **clouds** is one of the most common attempts at bringing about intentional change in the weather, while the development of **smog** as a result of **air pollution** is an example of unintentional weather modification. The building of cities and the flooding of **reservoirs** can also cause local weather modification.

See also
Urban heat island.

Further reading
Dennis, A.S. (1980) *Weather Modification by Cloud Seeding*, New York: Academic Press.
Oke, T. R. (1987) *Boundary Layer Climates* (2nd edition), London: Methuen.

WEATHERING

The physical (mechanical), chemical and biological breakdown of rocks at the earth's surface, brought about by exposure to **air, water**, temperature change and organic activity. It is essential for the formation of **soil**. The weathering process does not involve transportation of the rock particles except under the effects of **gravity** – for example, on a slope – but by destroying the integrity of the rock surface, weathering provides material that can be readily removed by agents of **erosion**, such as water, **wind** and ice. Physical and chemical weathering work together, but at different rates governed by such factors as climatic conditions and the nature of the rock surface. Weathering rates are greatest in hot, moist conditions and least in cold, dry conditions.

See also
Hydrolysis.

Further reading
Colman, S.M. and Dethier, D.P. (eds) (1986) *Rates of Chemical Weathering of Rocks and Minerals*, Orlando, FL: Academic Press.
Kittrick, J.A. (1986) *Soil Mineral Weathering*, New York: Van Nostrand Reinhold.

WET DEPOSITION

The most common form of **acid precipitation**, in which the acids are present in **solution** in the **atmosphere** and reach the earth's surface in **rain, snow**, hail and fog.

See also
Dry deposition.

WETLANDS

Swamps, marshes, **fens**, tidal marshes, **peatlands** and other **ecosystems** which are dominated by **water**. The presence of water may be permanent, temporary or seasonal and it may be fresh or salt, but the plant and animal organisms in wetlands have adapted to that situation to create unique **communities** that reflect the conditions at a specific site. Wetlands provide habitat for fish and wildlife, act as staging areas for migrating wildfowl, filter sediments and control flooding in stream systems and protect the shore from **erosion** in coastal areas. In human terms they have often been considered of little value, providing breeding grounds for insects such as mosquitoes, restricting overland and water transportation, reducing the amount of land available for **agriculture** and limiting the growth of settlements. As a result, wetlands have been extensively drained and filled or reclaimed for human use. In the United States alone, some 4 million hectares of interior wetland were lost between 1955 and 1975. Major wetlands such as the Florida Everglades in the United States, the Camargue in France and the Okavango Delta in Botswana continue to survive, but they are under threat, and in 1971 a List of Wetlands of International Importance was proposed at a conference held under the auspices of **IUCN** and **UNESCO** in Ramsar, Iran. By 1989, under the terms of the Ramsar Convention, more than 400 wetland areas covering more than 193 million hectares had been designated as worthy of protection. Wetlands which are not considered internationally important are being preserved or rehabilitated by **conservation** authorities or environmental groups.

Further reading
Williams, M. (ed.) (1990) *Wetlands: A Threatened Landscape*, Oxford: Blackwell.
Cox, G.W. (1993) *Conservation Ecology: Biosphere and Biosurvival*, Dubuque, IA: Wm C. Brown.

WILDERNESS

An area still in its natural state, that has not been significantly disturbed by humans. Very few such areas exist, since modern technology allows human activities to take place anywhere on the earth's surface. However, some areas – for example, the high Arctic or mountainous areas such as the Himalayas – are sufficiently remote or difficult of access that humans are only temporary visitors, often for recreational purposes. Once designated as wilderness areas they can be protected, and any change away from their natural state prevented.

Further reading
Hendee, J.C., Stankey, G.H. and Lewis, R.C. (1990) *Wilderness Management*, Golden, CO: North American Press.
Oelschlager, M. (ed.) (1992) *The Wilderness Condition: Essays on Environment and Civilization*, Washington, DC: Island Press.

WILDERNESS ACT (1964)

See **United States Wilderness Act (1964)**.

WILDERNESS SOCIETY

Founded in the United States in 1935, the Wilderness Society promotes the **conservation** of **wilderness** areas. It has been particularly successful in lobbying the government to designate areas suitable for preservation. The society's mandate also includes support of other conservation and environmental issues.

See also
Leopold, A., Thoreau, H.D., United States Wilderness Act (1964).

WILDLIFE AND COUNTRYSIDE ACT (1985)

An act designed to improve the protection of wildlife and the **conservation** of nature in Britain. It provides information on protected **species**, for example, and includes consideration of existing national parks and countryside parks. More controversially, it provides a financial compensation system for farmers and landowners who take care not to damage valuable sites/habitats known as sites of special scientific interest (SSSI) on their land.

WIND

Air in motion over the earth's surface. Its main component is horizontal, but vertical airflow can also be involved. Wind is produced as a result of **pressure** differences in the **atmosphere**, and its strength is determined by the magnitude of the difference and the distance over which the difference occurs – the pressure gradient. Wind direction (the direction from which the wind blows) is initially determined by these pressure differences, with the air flowing from high pressure to low pressure. Once the air begins to move, however, it comes under the influence of the **Coriolis effect**, which causes it to deviate from the pressure gradient path. When the pressure gradient force and the Coriolis force balance each other, the air will tend to flow parallel to the isobars. This is referred to as the geostrophic wind. Below about 500 m, **friction** between the moving air and the earth's surface comes into play. Friction slows the airflow and tends to counteract the Coriolis effect. As a result, near the surface the winds do not blow parallel to the isobars, but at an angle of between 10° and 30° to them, the impact being greater over the land than over the **oceans**. Friction also contributes to **turbulent flow** in the lower layers of the atmosphere. On a global scale, the circulation of the atmosphere, at the surface and in the upper atmosphere, reflects the interplay among pressure differences, the Coriolis effect and friction. Superimposed on the global winds are local and regional phenomena, ranging from land and sea breezes or mountain and valley winds, which are normally harmless, to **tornadoes** and **hurricanes**, where windspeeds regularly exceed 150–200 kph and create extremely hazardous conditions. Wind has an impact on environmental conditions at all scales. The heat and moisture transported by global winds play an important role in the earth's **energy budget**, and influence the distribution of major **ecosystems**. Winds in the upper atmosphere – for example, the **jet streams** – also help to redistribute **energy** and they transport **aerosols** such as those emitted during major volcanic eruptions. The upper westerlies in the northern hemisphere have also been implicated in the transportation of **acid rain**. In urban areas, calm conditions encourage the build-up of air **pollution**, whereas turbulence aids mixing and helps to dilute pollutants. The important role of wind in the human experience is reflected in the way in which specific winds have been named. The Chinook and Santa Ana of North America, the Sirocco, Bora and Mistral of the Mediterranean or the **Harmattan** of West Africa, for example, have had sufficient impact on human activities in the areas in which they blow to merit special identification.

Further reading
Ahrens, C.D. (1994) *Meteorology Today* (5th edition), Minneapolis/St Paul: West Publishing.

WIND ENERGY

Windmills have long been used to convert the **kinetic energy** available in the **wind** into mechanical **power**, that power then being used to grind grain or pump **water**. More recently, the energy in moving **air** has been converted into electricity, by using the wind to turn electrical generators. Wind energy has a number of advantages over conventional forms of energy. It is pollution-free and renewable, for example, and once the costs of building and maintaining a windmill have been covered, the wind energy is available at no direct cost. Problems occur because winds blow only intermittently and may not be able to supply the demand for power at any given time. Modern wind systems can overcome this problem to some extent by using storage batteries or by using electricity to compress air for use when the wind is not blowing. Most windmills are used by individuals to pump water or to provide enough electricity for lighting or powering small appliances. Such generators typically have a capacity of 10–12 kilowatts. Much larger wind turbines are rated at more than 100 kilowatts, and some have been designed with capacities of several megawatts. Set up in perpetually windy areas, such as mountain passes, ridges or the sea coast, groups of such generators – wind farms – can make a significant contribution to the production of electricity. Where they are already in place, they have produced complaints about **noise** pollution and the aesthetic deterioration of the landscape. By the middle of the twenty-first century, wind could supply some 10 per cent of the world's electricity. Currently, although the generation of electricity using wind energy is important locally, for example, in Britain, Denmark, Germany and California, its global impact is likely to remain small scale in the foreseeable future, at least until the direct and indirect costs of **fossil fuel** generation are judged to be too high.

Further reading
Johnson, G.L. (1985) *Wind Energy Systems*, Englewood Cliffs, NJ: Prentice-Hall.
Park, J. (1981) *The Wind Power Book*, Palo Alto, CA: Cheshire Books.
Spera, D.A. (ed.) (1994) *Wind Turbine Technology*, New York: ASME Press.

WINDBREAK

A row of trees or shrubs planted at right angles to the prevailing **wind**. The consequent reduction in windspeed helps to protect sensitive plants, reduces the rate of **evapotranspiration** and helps to prevent **soil erosion**.

WORLD BANK

See **International Bank for Reconstruction and Development**.

WORLD CLIMATE APPLICATIONS PROGRAM (WCAP)

See **World Climate Applications and Services Program (WCASP)**.

WORLD CLIMATE APPLICATIONS AND SERVICES PROGRAM (WCASP)

A component of the **WCP**, developed in 1991 by the expansion of the **WCAP**, and designed to assist in the collection and analysis of **climate** data that might be applied to such socioeconomic sectors as agriculture, forestry, fisheries, **water resources**, **energy** and health.

WORLD CLIMATE PROGRAM (WCP)

Established in 1979 under the auspices of the **WMO, UNEP, IOC** and **ICSU**, the WCP has terms of reference which include improving knowledge and understanding of global **climate** processes and facilitating the application of such information to human activities.

See also
World Climate Research Program, World Climate Applications and Services Program.

WORLD CLIMATE RESEARCH PROGRAM (WCRP)

One of the components of the **WCP**, the WCRP is concerned mainly with research into the dynamic and physical aspects of the earth/atmosphere **system**. It complements the activities of the **IGBP** which concentrates on the biological and chemical aspects of the system. The main aims of the WCRP are to determine the extent to which **climate** can be predicted and the degree to which human activities influence climate.

See also
TOGA.

WORLD COMMISSION ON ENVIRONMENT AND DEVELOPMENT

A commission set up in 1983 by the UN General Assembly to consider issues involving the relationship between **environment** and development. It was chaired by Gro Harlem Brundtland, the Norwegian Prime Minister, and as a result became known popularly as the Brundtland Commission. Economy and environment were firmly combined through its promotion of the concept of **sustainable development**. Part of the Commission's mandate was to explore new methods of international co-operation that would foster understanding of the concept and allow it to be developed further.

To that end, it proposed a major international conference to deal with the issues involved, which led directly to the UNCED in Rio de Janeiro in 1992. The commission's final report, *Our Common Future*, was made public in April 1987 and presented to the General Assembly of the United Nations later that year.

Further reading
Starke, L. (1990) *Signs of Hope: Working Towards our Common Future*, Oxford/New York: Oxford University Press.

WORLD HEALTH ORGANIZATION (WHO)

A UN agency created in 1948 to deal with global health issues and to achieve as high a

Table W-3 Summary of the main recommendations of the World Commission on Environment and Development (1987)

Revive growth:	Stimulate growth to combat poverty, particularly in developing nations. Industrialized nations must contribute.
Change the quality of growth:	Growth to be sustainable and related to social goals such as better income distribution, improved health, preservation of cultural heritage.
Conserve and enhance the resource base:	Conserve environmental resources – clean air, water, forests and soils. Improve the efficiency of resource use and shift to non-polluting products and techniques.
Ensure a sustainable level of population:	Population policies to be formulated and integrated with economic and social development programmes.
Reorient technology and manage risks:	Capacity for technological innovation to be enhanced in developing countries. Environmental factors to receive more attention in technological development. Promotion of public participation in decision-making involving environment and development issues.
Integrate environment and economics in decision-making:	Responsibility for impacts of policy decisions to be enforced to preserve environmental resource capital and promote sustainability.
Reform international economic relations:	Basic improvements in market access, technology transfer and international finance to allow developing nations to diversify economic and trade bases.
Strengthen international co-operation:	Higher priorities to be assigned to co-operation on environmental issues and resource management in international development.

Source: Based on information in Starke (1990)

level of physical, mental and social well-being as possible for the peoples of the world. It is involved in a variety of environmental studies, including the impact of **climate change** and **ozone depletion** on health, in conjunction with other agencies such as the WMO and UNEP.

WORLD METEOROLOGICAL ORGANIZATION (WMO)

A specialized agency of the UN based in Geneva, Switzerland, the WMO was created in 1951 to co-ordinate worldwide weather data collection and analysis. It facilitates co-operation and exchange of information among national meteorological organizations and among hydrological and geophysical agencies which provide observations related to **meteorology**. The WMO promotes standardization in the observation and publication of meteorological data and encourages research and training in meteorology and related fields. Through its sponsorship of such programmes as the **WCRP** and **IPCC**, the WMO has also become involved in climate change research.

WORLD RESOURCES INSTITUTE

An organization funded by the UN, some national governments and private organizations to study the relationship between economic development strategies and environmental issues.

WORLD WEATHER WATCH (WWW)

A system set up by the WMO for the global collection, analysis and distribution of information on the **weather** and other environmental **elements**. It consists of an observing unit, which collects ground-based and satellite observations, a data-processing unit, which manages the database and undertakes analysis of the data, and a telecommunications unit, which provides for the rapid exchange of observational data as well as analyses and forecasts produced by the data-processing unit. Together, these provide the information required for general and specialized weather forecasts and contribute to the longer range requirements of other WMO activities, such as the **World Climate Program** (WCP). Research on global environmental issues such as those involving **pollution** or food and **water** supply also depends upon data supplied by World Weather Watch.

WORLDWATCH INSTITUTE

A United States-based organization which monitors the impact of economic development on the **environment** and the world's progress towards **sustainable development**. It is funded mainly by private trusts and foundations. The Worldwatch Institute publishes the results of its research in a bimonthly journal and, since 1984, in an annual report entitled *State of the World*.

Further reading
Brown, L.R. (1997) *State of the World 1997*, New York: W.W. Norton

WORLDWIDE FUND FOR NATURE (WWF)

A private international organization founded in the 1960s as the World Wildlife Fund, it has been primarily concerned with the survival of **endangered species** – both plant and animal – but has now expanded its interests to include all aspects of **conservation**. The WWF currently has projects ongoing in over 130 countries ranging from Hong Kong to Northern Europe. It merged with the Conservation Foundation in 1990, and now claims some five million supporters worldwide, working to stop the accelerating degradation of the **natural environment**.

X, Y, Z

XENON

See **inert** (gases).

XEROPHYTIC VEGETATION

Plants adapted for life in arid conditions. Adaptations may include long or enlarged roots and fleshy leaves and stems which allow as much moisture as possible to be obtained and stored. Other adaptations are designed to reduce moisture loss by **transpiration**. These include leaves with thick skins and sunken pores, leaves replaced by thorns and the development of thick bark or outer skin on the trunks and stems of trees and other plants.

Further reading
Hocking, D. (ed.) (1993) *Trees for Drylands*, New York: International Science Publishers.

Figure X-1 Cactus and sage brush in Arizona desert

Photograph: Courtesy of Susan and Glenn Burton

XEROSPHERE

The part of the earth's surface covered by hot or cold **deserts**. It is characterized by arid conditions, and supports **flora** and **fauna** that have adapted to the dryness.

See also
Xerophytic vegetation.

Further reading
Furley, P.A. and Newey, W.W. (1983) *Geography of the Biosphere*, London: Butterworth.

X-RAYS

High **energy** electromagnetic **radiation** with a wavelength between 0.006 and 5 nanometres. They were discovered by Wilhelm **Roentgen** in 1895. X-rays are produced naturally when high energy particles such as **electrons** collide with other particles or **atoms**. They can be produced artificially by accelerating electrons from a heated **cathode** through a vacuum tube to strike a metallic **anode**. When the **electrons** strike the **atoms** in the **anode**, X-rays are emitted. In general, X-rays can penetrate **solids**, but that ability varies with the nature of the material. Matter that is less dense, for example, is more **transparent** to the rays. As a result, it is possible to prepare X-ray photography of human bones, because they are more dense than the surrounding flesh. Similarly this ability to penetrate and reveal the internal structure of materials has been used in industry to locate structural defects that are invisible to the naked eye. Radiography is widely used as a diagnostic tool in medicine and dentistry, and is applied therapeutically in the treatment of certain types of **cancer**. However, X-rays are a form of **ionizing** radiation and overexposure can cause tissue damage.

See also
Electromagnetic spectrum.

Further reading
Cutnell, J.D. and Johnson, K.W. (1995) *Physics*, New York: Wiley.

YEAST

Single-celled **fungi** that occupy a wide range of **habitats**. They are common in **soil**, for example, where they contribute to the decomposition of dead plant materials. Some yeasts cause skin disease and irritation of the mucous membrane in humans. Under **anaerobic** conditions, certain yeasts encourage **fermentation**, and commercial strains of *Saccharomyces cervisiae* – brewers' or bakers' yeast – have long been used to ferment the **sugars** in rice, wheat, barley and maize to produce **alcohol** (**ethanol**). In baking, sugars are decomposed under **aerobic** conditions to produce **water** and release **carbon dioxide** (CO_2), the process being used to cause bread dough to rise.

Further reading
Harrison, J.S. and Rose, A.H. (eds) (1987) *The Yeasts* (2nd edition), London: Academic Press.

YELLOW-CAKE

Concentrated **uranium** oxide (U_3O_8). The first stage in the refinement of **uranium ore**. Yellow-cake is usually produced at the mine site before being shipped to the nuclear fuel manufacturer.

See also
Beneficiation, Nuclear reactor.

ZERO POPULATION GROWTH (ZPG)

A condition in which the crude birth rate equals the death rate and as a result **population** becomes stable. It was a popular concept in the 1960s and early 1970s and led to the creation of a group in the United States called Zero Population Growth. Its main aim was to draw attention to problems of **resource** depletion, human suffering and environmental degradation associated with rapid population growth, and it advocated planned progress towards stability. Technological and lifestyle factors complicate the relationships between population growth and resource use or environmental impact, however, and zero population growth on its own cannot solve the problems. Among the rapidly growing nations, both India and China have instituted family planning programmes to reduce population growth. There has been some success in both countries, but they remain far from zero population growth. Several nations in Europe, however, – for example, Denmark, France, Germany – have attained unplanned zero population growth.

See also
Carrying capacity, Malthus.

Further reading
Ehrlich, P.R. and Ehrlich, A.H. (1990) *The Population Explosion*, New York: Simon & Schuster.
Mesarovich, M. and Pestel, E. (1975) *Mankind at the Turning Point*, London: Hutchinson.

ZINC (Zn)

A hard, bluish-white **metal**. The common **ores** of zinc are zincite, calamine and zinc blende.

Zinc is used in **alloys** such as brass, and as a coating on **iron** (Fe) to produce corrosion-resistant galvanized iron. Zinc is also an important micronutrient or **trace element**.

ZONAL CIRCULATION

The movement of **air** in the **atmosphere** in an east–west or west–east direction, i.e. parallel to the lines of latitude.

See also
Meridional circulation, Zonal index.

ZONAL INDEX

A measure of the intensity of the mid-latitude atmospheric circulation pattern. During a period of high zonal index, the airflow is west to east or latitudinal. A low zonal index

Figure Z-1 The index cycle associated with the meandering of the mid-latitude westerlies in the northern hemisphere

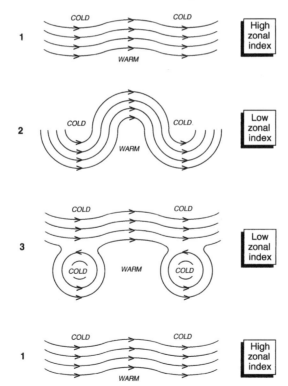

Source: Kemp, D.D. (1994) *Global Environmental Issues: A Climatological Approach*, London/New York: Routledge

involves a south–north–south or meridional component as a result of waves forming in the airflow. Over a period of between three and eight weeks, the index cycles from low to high and back to low, and the net impact of this changing circulation pattern is a significant latitudinal **energy** transfer.

Further reading
Barry, R.G. and Chorley, R.J. (1992) *Atmosphere, Weather and Climate* (6th edition) London: Routledge.

ZOOPLANKTON

Microscopic animal **plankton**, which live in the upper layers of fresh and salt-water environments. Zooplankton includes microscopic crustaceans, shrimp-like forms such as krill and the eggs and larvae of a variety of shellfish. Some graze upon **phytoplankton** and occupy the first consuming stage in the aquatic **food chain**. These primary consumers are in turn consumed by other zooplankton through a series of three or four **trophic levels**, before fish and aquatic mammals enter the chain. Some zooplankton, such as krill which occur in heavy concentrations, are an important element in the diet of certain whales, and some thought has been given to harvesting them for human use also. However, the economic feasibility of such an operation, and concern for the potential damage to aquatic food chains has as yet prevented it.

Further reading
Parsons, T.R. (1980) 'Zooplankton production', in R.S.K. Barnes and K.H. Mann (eds) *Fundamentals of Aquatic System*s, Oxford: Blackwell Scientific.

NAME INDEX

SUBJECT INDEX

Main entries in **bold**.